D1072500

The Concise Encyclopedia

of the

Ethics

of

New Technologies

The Concise Encyclopedia

of the

Ethics

of

New Technologies

Edited by

Ruth Chadwick

ACADEMIC PRESS

A Harcourt Science and Technology Company

San Diego   San Francisco   New York   Boston   London   Sydney   Tokyo

Academic Press
A Harcourt Science and Technology Company
525 B Street, Suite 1900, San Diego, California 92101-4495, USA
http://www.academicpress.com

Academic Press
Harcourt Place, 32 Jamestown Road, London NW1 7BY, UK
http://www.academicpress.com

Library of Congress Catalog Card Number: 00-107670

International Standard Book Number: 0-12-166355-8

PRINTED IN THE UNITED STATES OF AMERICA
00 01 02 03 04 05 MM 9 8 7 6 5 4 3 2 1

Contents

Birth-Control Technology
JAMES W. KNIGHT

Brain Death
C. A. DEFANTI

Cloning
THOMAS H. MURRAY AND GREGORY KAEBNICK

Computer and Information Ethics
SIMON ROGERSON

Consequentialism and Deontology
MATTHEW W. HALLGARTH

Embryology, Ethics of
SØREN HOLM

Environmental Impact Assessment
MARJA JÄRVELÄ

Feminist Ethics
ROSEMARIE TONG

Geneticization
ROGEER HOEDEMAEKERS

Genetic Research
R. O. MASON

Genetics and Behavior
GARLAND E. ALLEN

Genetic Screening
RUTH CHADWICK

Genetic Technology, Legal Regulation of
TONY McGLEENAN

Nuclear Power
KRISTIN SCHRADER-FRECHETTE

Nuclear Testing
STEVEN LEE

Organ Transplants and Xenotransplantation
RUTH CHADWICK and UDO SCHÜKLENK

Playing God
WILLIAM GREY

Precautionary Principle
JENNETH PARKER

Reproductive Technologies
LUCY FRITH

Science and Engineering Ethics, Overview
R. E. SPIER

Slippery Slope Arguments
WIBREN VAN DER BURG

Contributors

GARLAND E. ALLEN
GENETICS AND BEHAVIOR
   Washington University
   St. Louis, Missouri

RICHARD E. ASHCROFT
HEALTH TECHNOLOGY ASSESSMENT
HUMAN RESEARCH SUBJECTS, SELECTION OF
   Imperial College School of Medicine
   London, United Kingdom

ROBERT H. BLANK
FETAL RESEARCH
   University of Canterbury
   Christchurch, New Zealand

RUTH CHADWICK
GENETIC SCREENING
ORGAN TRANSPLANTATION AND
XENOTRANSPLANTATION
   Lancaster University
   Lancaster, United Kingdom

ANGUS J. CLARKE
GENETIC COUNSELING
   University of Wales
   Cardiff, Wales

C. A. DEFANTI
BRAIN DEATH
   Ospedali Riuniti di Bergamo
   Bergamo, Italy

STRACHAN DONNELLEY
HUMAN NATURE, VIEW OF
   The Hastings Center
   Garrison, New York

LUCY FRITH
REPRODUCTIVE TECHNOLOGIES
   The University of Liverpool
   Liverpool, United Kingdom

RAANAN GILLON
BIOETHICS, OVERVIEW
   Imperial College
   London, United Kingdom

WILLIAM GREY
PLAYING GOD
   University of Queensland
   Brisbane, Queensland, Australia

HETA GYLLING
GENETIC ENGINEERING
   University of Helsinki
   Helsinki, Finland

MATTHEW W. HALLGARTH
CONSEQUENTIALISM AND DEONTOLOGY
   United States Air Force Academy
   Colorado Springs, Colorado

MATTI HÄYRY
GENETIC ENGINEERING
   University of Helsinki
   Helsinki, Finland

ADAM M. HEDGECOE
GENE THERAPY
GENOME ANALYSIS
   University College London
   London, United Kingdom

ROGEER HOEDEMAEKERS
GENETICIZATION
   University of Nijmegen
   Nijmegen, The Netherlands

SØREN HOLM
EMBRYOLOGY, ETHICS OF
   University of Copenhagen
   Copenhagen, Denmark

MARJA JÄRVELÄ
ENVIRONMENTAL IMPACT ASSESSMENT
   The University of Jyväskylä
   Jyväskylä, Finland

NICHOLAS JOLL
INTRINSIC AND INSTRUMENTAL
VALUE
   University of Essex
   Essex, United Kingdom

GREGORY KAEBNICK
CLONING
   The Hastings Center
   Garrison, New York
JAMES W. KNIGHT
BIRTH-CONTROL TECHNOLOGY
   Virginia Polytechnic Institute and State University
   Blacksburg, Virginia
LORETTA M. KOPELMAN
MEDICAL FUTILITY
   East Carolina University School of Medicine
   Greenville, North Carolina
STEVEN LEE
NUCLEAR TESTING
   Hobart and William Smith Colleges
   Geneva, New York
R. O. MASON
GENETIC RESEARCH
   Southern Methodist University
   Dallas, Texas
TONY McGLEENAN
GENETIC TECHNOLOGY, LEGAL REGULATION OF
   Queen's University
   Belfast, Northern Ireland
BEN MEPHAM
NOVEL FOODS
   The Food Ethics Council
   Southwell Notts, United Kingdom
THOMAS H. MURRAY
CLONING
   The Hastings Center
   Garrison, New York
JENNETH PARKER
PRECAUTIONARY PRINCIPLE
   South Bank University
   London, United Kingdom
MAURIZIO MORI POLITEIRA
LIFE, CONCEPT OF

   Center for Research in Politics and Ethics
   Milan, Italy
MICHAEL REISS
BIOTECHNOLOGY
   Cambridge University
   Cambridge, United Kingdom
SIMON ROGERSON
COMPUTER AND INFORMATION ETHICS
   De Montfort University
   Leicester, United Kingdom
KRISTIN SCHRADER-FRECHETTE
HAZARDOUS AND TOXIC SUBSTANCES
NUCLEAR POWER
   University of South Florida
   Tampa, Florida
UDO SCHUKLENK
ORGAN TRANSPLANTS AND XENOTRANSPLANTATION
   University of the Witwatersrand
   Johannesburg, South Africa
R. E. SPIER
SCIENCE AND ENGINEERING ETHICS, OVERVIEW
   University of Surrey
   Surrey, United Kingdom
G. E. TOMLINSON
GENETIC RESEARCH
   University of Texas Southwestern Medical Center
   Dallas, Texas
ROSEMARIE TONG
FEMINIST ETHICS
   The University of North Carolina at Charlotte
   Charlotte, North Carolina
WIBREN VAN DER BURG
SLIPPERY SLOPE ARGUMENTS
   Tilburg University
   Tilburg, The Netherlands
ROBERT WACHBROIT
HEALTH AND DISEASE, CONCEPTS OF
   University of Maryland
   College Park, Maryland

Guide to the Encyclopedia

In order that you, the reader, will derive maximum benefit from your use of the *Concise Encyclopedia of the Ethics of New Technologies*, we have provided this Guide. It explains how the Encyclopedia is organized and how the information within it can be located.

ARTICLE FORMAT

The articles in this encyclopedia are arranged in a single alphabetical list by title. Each new article begins at the top of a right-hand page, so that it may be quickly located. The author's name and affiliation are displayed at the beginning of the article. The article is organized according to a standard format, as follows:

• Title and Author
• Outline
• Glossary
• Defining Statement
• Main Body of the Article
• Bibliography

OUTLINE

Each article in the Encyclopedia begins with an Outline section that indicates the general content of the article. This outline serves two functions. First, it provides a brief preview of the article, so that the reader can get a sense of what is contained there without having to leaf through the pages. Second, it serves to highlight important subtopics that are discussed within the article.

The Outline section is intended as an overview and thus it lists only the major headings of the article. In addition, second-level and third-level headings will be found within the article.

GLOSSARY

The Glossary section contains terms that are important to an understanding of the article and that may be unfamiliar to the reader. Each term is defined in the context of the particular article in which it is used. Thus the same term may appear as a glossary entry in two or more articles, with the details of the definition varying slightly from one article to another.

The following example are glossary entries that appear with the article "Genetic Research."

genome The set of genes characteristic of each species, revealed by the base set of chromosomes which is species specific. In humans, the genome consists of 23 pairs of chromosomes.

mutation Local change in the genetic information carried by the DNA or a change in DNA sequence. This change may be responsible for causing a disease.

DEFINING STATEMENT

The text of each article in the Encyclopedia begins with a single introductory paragraph that defines the topic under discussion and summarizes the content of the article. For example, the article "Novel Foods" begins with the following statement:

NOVEL FOODS is a term used in the European Union (EU) which refers principally, but not exclusively, to foods produced by modern biotechnology, such as those employing genetically modified (GM) microbes, plants, and animals. This article discusses ethical and legal issues associated with their (prospective) use, in both the EU and the United States of America (USA), and describes the relevant government regulatory and advisory committees.

BIBLIOGRAPHY

The Bibliography appears as the last element in an article. It lists recent secondary sources to aid the reader in locating more detailed or technical information. Review articles and research papers that are important to an understanding of the topic are also listed.

The bibliographies in this Encyclopedia are for the benefit of the reader, to provide references for further reading or research on the given topic. Thus they are not intended to represent a complete listing of all the materials consulted by the author in preparing the article, as would be the case, for example, with a journal article.

For example, the article "Organ Transplants and Xenotransplantation" lists as references (among others) the works *Animal Tissue Into Humans*, *Organ Transplants and Ethics*, and *Animal to Human Transplants: The Ethics of Xenotransplantation*.

COMPANION WORKS

This encyclopedia is part of a continuing program of scholarly reference works published by Academic Press. This program encompasses many different areas of science, ranging from life science (e.g., *Encyclopedia of Biodiversity, Encyclopedia of Microbiology*) to biomedical topics (*Encyclopedia of Reproduction, Encyclopedia of Stress*), to physical science (*Encyclopedia of the Solar System, Encyclopedia of Volcanoes*), with special emphasis on social and political issues (*Encyclopedia of Applied Ethics, Encyclopedia of Creativity, Encyclopedia of Nationalism, Encyclopedia of Violence, Peace, and Conflict, Encyclopedia of Mental Health*).

For more information on Academic Press reference publishing, please see the Website at:

www.academicpress.com/reference/

Preface

Developments in technology inevitably give rise to ethical questions—this is nothing new. They bring with them advantages and disadvantages, and hence affect the interests of individuals and groups. The last decades of the 20th century, however, witnessed some of the most remarkable events in the history of new technological developments in, for example, information technology and genetics. They also witnessed a proliferation of committees and commissions to respond to these developments. I will give just two examples of such bodies. In the United States the National Bioethics Advisory Commission has addressed issues such as cloning and stem cell research, and in Europe, the European Group on Ethics in Science and New Technologies has issued Opinions on such topics as ethical issues of healthcare in the information society. A series of international summits of national bioethics commissions has also been established. Ethics as a field of academic study, therefore, feeds into a considerable amount of activity in ethics, perhaps particularly bioethics, in its public policy aspect.

This volume on the ethical assessment of new technologies is of necessity selective. An attempt has been made to include not only relevant topics that have had a relatively longer history, such as nuclear power and testing, but also some of the more recent issues under ethical debate, such as gene therapy and cloning, along with articles concerning frameworks for assessment, such as the discussions of environmental impact assessment, consequentialism and deontology, and slippery slope arguments. It has not been possible to include every technological issue currently confronting society, or every ethical theory. What I want to suggest in this preface, however, is that the ethical assessment of new technologies is made problematic by the fact that the new technologies themselves present challenges for ethics.

These challenges, I suggest, fall into two categories.

First, there is the rapid development not only in technology but also in the opportunities and potential for use of that technology. The speed of change requires a similarly swift response on the part of society in terms of ethics, policy, and legislation. The difficulty of this first challenge is complicated by the second: the fact that the development of technologies poses challenges for ethical frameworks themselves. These technologies push existing ethical frameworks to the limits. It is therefore not simply a matter of "applying" ready-made theories to the possible implementation of technological advance—developments in the life sciences and other technologies can lead us to rethink theories and even concepts. Ethics is of course always engaged in critical self-reflection on its own methodologies, but currently there is a particlular focus on the multidisciplinary aspect of this field of study. The interplay between philosophy and law and the relationship between philosophical principles and empirical data are topics of ongoing debate.

Furthermore, from these challenges emerge several issues that must be confronted. The first issue concerns the contribution of ethics to discussions of options for action at different levels—in relations between health care professionals and clients or patients, in national policymaking, and at an international level. Questions to be considered include the following: in what circumstances should population screening be introduced and how should the results be managed in terms of disclosure? Should we use genetic technology to try to delay the aging process? Should we use genetically modified animals as sources of organs or should we instead try to come to terms with the finitude of human life?

Second, the social and political implications of technological developments need to be considered. These include arguments concerning the potential for discrimination and how different political philosophies and legal systems might approach these questions. This might have application in relation to different health technologies, such as worries about the use of genetic information by insurers and employers in ways that introduce new forms of discrimination, fears of eugenic policies, and issues of justice in access to the benefits of technology.

Third, there are philosophical and conceptual questions about the potential impact of technologies on how we understand human beings and human life. These can include implications for our understanding of health, illness, and disease. Given the advent of predictive testing (including presymptomatic and predisposition testing), what does a positive result suggest about one's status as "healthy" or not? Again, while there is a view that population genetic screening should be introduced only for diseases that can be classified as serious, the meanings of both "serious" and "disease" are still contested. So articles that deal with views of concepts of health and disease, as well as with views of human nature, have been included in this volume. Some people have taken the view that the completion of the Human Genome Project will finally settle aspects of the long-standing nature–nurture debate, especially as more information is forthcoming about the role of genetics in behavioral traits. This aspect of ethical assessment could be taken further. The issue of xenotransplantation, for example, which is frequently discussed in terms of safety (clearly an important issue), also raises questions about our relationship with other species and our collective identity as a species. The article "Organ Transplants and Xenotransplantation" in this volume does discuss some of the ethical issues concerning our use of animals; further reading on animal rights is included in the *Encyclopedia of Applied Ethics*.

Finally, I want to return to the implications of new technologies for ethics itself. When new developments occur, they not uncommonly give rise to anxiety about possible undesirable consequences. In part, this may arise from previous bad experiences. The public response to genetically modified food in the United Kingdom, for example, has been attributed to some extent to experience with previous food scares such as BSE ("mad cow" disease). However, anxiety may arise precisely because there is no experience on which to draw. In other words, what is feared is the unknown. This factor has arguably also contributed to the response to genetically modified food: the long-term effects on both human health and the natural environment are not foreseeable. The response to certain developments, that human beings should not "play God," may be at least in part an expression of this fear of the unknown. In fact, the "playing God" objection to new developments has been voiced so frequently that an article analyzing it has been included in the volume. It is not always clear what the objection amounts to or how seriously it should

be taken, but insofar as it points to undesirable consequences of the exercise of new powers, it may suggest caution. The advocacy of caution also finds expression in the "precautionary principle," which has been used by a number of policymaking bodies and is the topic of an article in this volume. Slippery slope arguments, also discussed here, typically point to something unpleasant and undesired at the bottom of the slope.

The tendency for new technological developments to give rise to anxiety about their potential consequences, making certain responses and the voicing of certain arguments more likely, however, is only one of the implications for ethics. At a deeper level, they raise questions about the way we perceive the world—they have the potential to challenge our very concepts. A prominent example of this occurred in the context of reproductive technology. Concepts of "mother," "father," and "family" are no longer as clear-cut as they once were, given the use of techniques such as donor insemination, egg donation, and surrogacy. Cloning has complicated the picture still further. The technique of somatic cell nuclear transfer forces us to ask not only whether it is advisable to proceed with the use of this technique in its various possible applications, including research, therapy, and reproduction, but also whether an embryo produced by this technique *is* an embryo as previously understood. In other words, the boundaries of our concepts sometimes have to be rethought.

This rethinking does not apply solely to biological categories; it also applies to our ethical concepts. Autonomy, for example, is a concept that remains the focus of considerable discussion. There are different aspects to the debate, including the issue of cultural specificity, the criticism of individualism in much of the discussion of autonomy, and the attempt to produce a relational interpretation, as in some feminist work in bioethics. The aspect that I would argue deserves particular attention in the current context, however, is the way in which the emphasis on autonomy and its relation to notions of being "informed" in order to make "autonomous" decisions are challenged by developments that are producing so much "information" that it may become too onerous for individuals to deal with. This applies, for example, to the forecasts of the amount of genetic information about ourselves that will be available; we have seen arguments for rights not to know increasingly voiced, with in turn a questioning of the extent to which autonomy is facilitated or hampered by increasing amounts of information.

We may, likewise, be unwise to assume that fairly long-standing doctrines such as informed consent can, without some rethinking, do the work required of ethical frameworks as we find ourselves in new contexts. In part, reconsideration of some ethical issues has been brought about by globalization and by the conduct of international research. Technological developments, however, have produced a situation in which increasingly large amounts of information about individuals, and, indeed, populations, are held in tissue and data banks of various kinds and in which increasingly complex mechanisms for coding and anonymizing information are being devised. Cultural differences in attitudes about genes and about genetics and its implications produce sensitivities concerning these developments that may not be easily susceptible to treatment by application of a principle developed in another context.

Elsewhere I have argued for taking seriously the notion of "value impact assessment" in applied ethics. I use the term to refer to the conscious attempt to examine the interplay between technology and values. Value impact assessment would be a multidisciplinary activity on three levels. The first level would involve empirical research on social values and attitudes concerning new technologies and their applications. Research has been undertaken, for example, on whether genetic test results affect reproductive decisions or attitudes toward disability, as well as on the comparative perceived acceptability of genetic technology in the context of food as opposed to medical applications. The second level concerns how technological developments test the adequacy of ethical theories and principles in ways such as those outlined above. As I have tried to suggest, we cannot take some of these developments as just another case to which we might apply, for example, a consequentialist analysis, without also considering how the boundaries of our concepts are being redrawn. For example, novel foods designed to have health benefits require us to rethink the boundary between food and medicine, as the term "nutraceuticals" suggests. The ways in which concepts of health and disease are subject to revision have been mentioned above. The concept of disability is also in dispute, and principles of justice in access to resources may require us to consider these classification questions first. Moreover, it is not simply a matter of revisiting isolated concepts—ways of thinking may be affected on a wider scale. The phenomenon of "geneticization," to which an article in this volume is devoted, is one example; another is the prevalence of "instrumental reason" discussed in the article on intrinsic versus instrumental value.

It may be objected that it is part of the nature of philosophical activity that debate about concepts such

as these is ongoing in any case. This is not in dispute. What I am suggesting is that in "applying" ethics to technological developments we need to pay conscious attention to the phenomena I am describing as well their interactions, in order to make progress in ethical assessment.

The third level in value impact assessment concerns the relationship between the first and second levels. What is the relationship between empirical findings and ethical theories? What, if anything, follows from findings about acceptability of genetically modified food? Such findings may be important for policymakers; the issue also, however, links into one of the oldest debates in moral philosophy about the relationship between "is" and "ought." Clearly the fact that 90% of people think that x is the case does not show that they are right. It is possible that the 10% who think that x is not the case have better grounds for their view—grounds of which the 90% are unaware or to which they have accorded insufficient weight. Nevertheless, developments in applied ethics need to be sensitive to the ways that new developments are framed or conceptualized, as well as to empirical evidence about their implementation, impact, and effectiveness.

It might be argued that if it is necessary to take into account the possibility that the very frameworks we are applying may be subject to modification by technological advances themselves, then what is to be understood by the notion of "application" in applied ethics? This argument, however, depends on a particular view of what is involved in application—one of having a particular theory to apply to a clearly defined issue. This is one view of applied ethics, to which there are alternatives. The first task may be to get a clearer idea of the ethical issue to which we need to address ourselves—and to do this we need to be aware of different ways of constructing the problem, which demands input from, at least, conceptual analysis, law, and empirical data.

The articles selected for this volume, therefore, are part of the ongoing development of applied ethics, including analyses of topical ethical issues, articles on frameworks and ethical perspectives, and articles on concepts such as geneticization. I am grateful to all the contributors and to the team at Academic Press, particularly Scott Bentley, for their support of this project.

Ruth Chadwick

Bioethics, Overview

RAANAN GILLON

Imperial College, London University

GLOSSARY

autonomy Literally self-rule, the ability to make decisions for oneself on the basis of deliberation. Self-determination is an alternative term. Respect for people's autonomy, to the extent that this is consistent with equal respect for the autonomy of all affected, is a component of many ethical theories in health care bioethics.

beneficence Acting so as to benefit others—a limited but universal moral obligation in many moral theories, and widely regarded as a fundamental moral obligation in health care bioethics. Acting so as to benefit oneself is also, strictly speaking, beneficence, but, given people's natural self-interested tendency to benefit themselves, self-beneficence is of less ethical interest than beneficence for others. However, in some ethical theories promoting self-beneficence/self-interest is seen as the way to maximize overall welfare.

justice The moral obligation of fairness, common to many moral theories, including much of bioethics. Essentially justice is about treating people equally in relation to criteria acknowledged to be morally relevant. However, while that much is commonly agreed upon within most theories of ethics and bioethics, when it comes to specifying what the relevant criteria are (for example, treating people equally in relation to their needs, rights, merits and demerits, ability to benefit, or autonomous desires), there is marked disagreement in philosophy, ethics, religion, and politics.

nonmaleficence Not harming others. A moral obligation in many moral theories including much of bioethics. Needs to be taken into account together with beneficence whenever it is intended to benefit others, but is widely accepted as an independent moral obligation even when no obligation or intention to benefit is acknowledged.

person A moral category into which all readers of this encyclopedia will be agreed, by the norms of probably all moral theories, to fall. Persons, or people, owe other persons or people the highest level of moral respect. There is far less agreement about the attributes needed to be a person. Such disagreement is typified in bioethics by major disputes about whether or not human embryos, fetuses, newborn babies, patients who are permanently unconscious, and even brain dead patients on ventilators are persons. Similar disputes arise about whether any nonhuman animals are persons, and if so which. More theoretical philosophical debate concerns whether or not machines could be developed with the attributes of persons, and about the attributes that life-forms from other planets would need to be persons.

scope of application Even when agreement about moral obligations is achieved, there may remain radical disagreement about their scope or range of application—to whom or to what are the obligations owed? For example, while it may be agreed that there is a universal obligation that we must not unjustly kill each other, the question of what counts as "each other" may be vigorously disputed. While it may be agreed that we have an obligation to respect others' autonomy, there remains disagreement about who counts as autonomous—or sufficiently autonomous to fall within the scope of this obligation. Similarly with distributive justice, even if we accept an obligation to distribute scarce resources justly, who or even what falls within the scope of this obligation? Questions of scope are relevant to many issues in ethics generally and bioethics in particular.

BIOETHICS (as the etymology of its Greek roots implies—*bios* means life and *ethike* ethics) is the study of ethical issues arising in the practice of the biological disciplines. These include medicine; nursing; other health care professions, including veterinary medicine; and medical and other biological or life sciences. Bioethics is "applied ethics" in the sense that it is the study of ethical issues that arise or might be anticipated to arise, in the context of real activities. While medical and other health care ethics are a major component of bioethics, the latter is now widely—though not universally—acknowledged to extend well beyond health care ethics to include not only the ethics of research in the life sciences but also

- Environmental ethics, encompassing such areas as environmental pollution and consideration of the proper relationships between humans, other animals, and the rest of nature.
- Ethical issues of sexuality, reproduction, genetics, and population.
- Various sociopolitical moral issues, including the adverse effects on people's health of unemployment, poverty, unjust discrimination (including sexism and racism), crime, war, and torture.

As well as its breadth of subject matter, bioethics is characterized by the wide variety of people and disciplines actively involved. Apart from the relevant professionals such as doctors, nurses, and life scientists, and their patients and research subjects, academic disciplines involved in bioethics include moral philosophy, moral theology, and law (perhaps the "big three" disciplines in bioethics); economics; psychology; sociology; anthropology; and history. And of course the public in general, both as individuals and in various interest groupings, and their political representatives increas-ingly take a direct interest in bioethical issues, as do the media.

I. HISTORICAL NOTES ON BIOETHICS

The term "bioethics" seems to have been invented—or at least first used in print—in 1970 by an American biologist and cancer researcher, Van Rensselaer Potter of the University of Wisconsin in the USA. However, the word was also used, apparently independently, shortly afterward and in a somewhat different sense, by a Dutch fetal physiologist and obstetrician working in Washington, DC, Andre Hellegers, and others who with him founded the Kennedy Institute of Human Reproduction and Bioethics at Georgetown University in 1971. Van Rensselaer used the term to refer to a "new discipline that combines biological knowledge with a knowledge of human value systems" which would build a bridge between the sciences and the humanities, help humanity to survive, and sustain and improve the civilized world. Hellegers and his group, on the other hand, used the term more narrowly to apply to the ethics of medicine and biomedical research—and indeed, in reporting these two different conceptions, Warren Reich, editor of the massive *Encyclopedia of Bioethics,* tells us that when it was first being planned in 1971 it was to have been called the *Encyclopedia of Medical Ethics.*

This division and debate are instructive in various ways about the discipline of bioethics. First, it recalls that a major component of the field of bioethics is medical and other health care ethics. Second, it shows that there is substantive disagreement about how far into what might be called general applied ethics the discipline of bioethics should extend. Third, it indicates the major developments even within the narrower subject of medical ethics that were occurring in the 1960s, in the decade prior to this debate. Before the 1960s the traditional approach to medical ethics, which was then largely limited to ethical issues arising in clinical medical practice, was for doctors in training to be told or even simply to be expected to pick up from the example of their seniors what the ethical norms of professional conduct were. Doctors were rewarded by professional acceptance if they behaved "appropriately" and punished by sanctions ranging from expressed disapproval via reprimands to, at worst, expulsion from the profession if they transgressed these norms.

In the 1960s, to this continuing norm of "professionalization" began to be added other components. The first was the involvement in the previously largely closed world of medicine of "outsiders" such as philosophers, theologians, lawyers, sociologists, and psychologists looking in on the medical profession and offering their expertise and their views. The second was the concomitant beginnings of acceptance within the medical profession that the insights offered from these varying outside perspectives could be helpful in the development of medicine. The third was an increasing realization that medical ethics needed to extend its sphere of interest beyond the clinical encounter into broader issues of social ethics in such contexts as fair and beneficial distribution of health care facilities within societies—areas of direct and necessary concern in countries such as the United Kingdom with their existing national health services—and of potential concern in countries such as the USA with their comparatively poor provision for those who could not pay for health care.

Thus by the end of the 1960s medical ethics itself was beginning to change away from being almost entirely concerned with ethical rules and codes of conduct governing clinicians to also including ethical aspects of health and illness in society. And it was also beginning to accept, however cautiously and tentatively, that people and disciplines other than doctors and medicine could have instructive and useful things to say about the broad subject area of medical ethics. In other words, traditional medical ethics was tentatively beginning to encompass both aspects of the new bioethics: the philosophically more critical, analytic, and multidisciplinary approach to ethical issues arising within the clinical practice of medicine, and the understanding that new developments within medicine and the life sciences were raising ethical issues for society as a whole. In addition, a third strand of bioethics activity was beginning to be acknowledged by some doctors and other health care professionals, notably, a sense of their obligation to become involved, as health care professionals, in trying to remedy social factors that impinged adversely on people's health—whether through lifestyle factors such as unhealthy diet, tobacco smoking, and lack of exercise; environmental pollution and other environmental hazards; overpopulation; or, even more politically contentiously, unemployment, poverty, crime, and warfare in its various forms.

Underlying all of these various strands of bioethics from its recent origins is a further distinction, sometimes clear and sometimes fuzzy, between bioethics as previously defined as the intellectual activity of study of, reflection on, and inquiry into a range of ethical issues, and bioethics as a reforming activity intended to achieve substantive moral reforms whether at a personal or at a political level (examples might be public exposure of unacceptable health care practices, stronger entrenchment of patients rights, or abolition of nuclear weapons or land mines). While it is probably true that the large majority of people pursuing contemporary bioethics are

at least in part motivated by a desire to change the world for the better, there is also a fairly clear divide between those who would do so by the pursuit and promotion of ideas, arguments, and ways of thinking and those who would add to these intellectual activities exhortation, emotional pressure, and political activity at a variety of levels. It is unclear whether the term, "the bioethics movement," which appears occasionally in the literature is intended to apply to both groups of people or more narrowly to the latter group of reformers.

The explosion of interest in medical ethics and bioethics in the 1970s was most marked in the USA where, as well as the Hastings Center (founded in 1969, originally as the Institute of Society Ethics and the Life Sciences, it started its *Hastings Center Report* in 1971) and the Kennedy Institute, founded at Georgetown University in 1971, much academic activity developed in universities and private institutes.

However, although ahead of the field, the USA was not alone in this development, and critical medical ethics activity was also beginning to be widespread if sporadic in Europe. In 1963 in the United Kingdom the multidisciplinary London Medical Group and its successors, the Society for the Study of Medical Ethics and the Institute of Medical Ethics were founded. Starting with discussion groups and study groups in UK medical schools, the Institute founded its *Journal of Medical Ethics* in 1975 and its *Bulletin of Medical Ethics* (subsequently becoming independent of the IME) in 1985. Academic courses in medical and later health care ethics first started in 1978 and began to flourish in the 1980s. Similar developments in the 1970s occurred in the Netherlands and other Benelux countries and in the Nordic countries. Development of critical medical ethics arose somewhat later in Germany, in ex-Soviet-bloc countries, and in southern Europe. In each of these three latter groups different explanations are offered for the relatively late start of modern bioethics. In Germany the experiences of the Nazi era had created a widespread reluctance to discuss critically and openly (rather than with simple and firm opposition) some of the issues being addressed in the "new" medical ethics—such as experimentation on human subjects; euthanasia; abortion, especially for genetic defects; sterilization, especially without the patient's informed consent; and the "new genetics" with all its echoes and perceived echoes of eugenics. Indeed in Germany a positive hostility was to grow toward the new "bioethics," with one leading Australian bioethicist having invitations to lecture withdrawn or withheld as a result of raucous minority protest and threats.

In the Roman Catholic countries critical medical ethics were also slow to get off the ground. One reason was that Roman Catholic medical ethics were already very well established as an important aspect of medical education in many medical schools that were within the Catholic tradition. There, the medical ethics taught was "largely a branch of traditional Catholic moral theology" (p. 982. C. Blomquist, 1978. In W. T. Reich (Ed.), *The encyclopedia of bioethics* (pp. 982–987). New York: Macmillan), and it took some time for such teaching to adapt to the new mode of philosophically critical ethics. Similar links between medical ethics and the prevailing religious culture existed where medical schools were closely integrated within other religious traditions. Thus doctors who shared the religious traditions often felt no need to accommodate the new critical approach to medical ethics teaching, which might be perceived as threatening, while doctors who did not share the religion had often turned away from medical ethics altogether, perceiving it to be a guise for the imposition of a particular religious stance which they did not share. Some such doctors went even further in their rejection of medical ethics; participating in the pervasive spirit of postwar scientific positivism, they saw medicine increasingly more as science than as art, and they perceived science to be a value-free enterprise. Ethics was thus nothing to do with science, and indeed for some of the more extreme positivists ethics was in any case strictly nonsense.

Finally, in what used to be the "Iron Curtain" countries, study of the new critical medical ethics was also slow to take off, being discouraged primarily by the prevailing state orthodoxy of Marxist–Leninism, in which medical practice was a function of the state in developing and maintaining communism. Underground opposition to Marxist ideology by the many doctors in these countries who continued to adhere to Roman Catholicism or to the Orthodox Christian faiths and their medical ethical norms was also not conducive to the new critical medical ethics.

In other parts of the world, including Africa and Asia, bioethics was also slower to develop, but by the 1990s the new multidisciplinary area of inquiry and study had become a worldwide phenomenon.

II. SUBSTANTIVE ISSUES IN BIOETHICS

As already indicated, the range of substantive issues now considered to be legitimate substrates for bioethics is vast.

A. Issues Stemming from Health Care Relationships

At one end of the scale are moral issues stemming from the relationship between patients and their doctors, nurses, or other health care workers. These include the following:

- Issues of paternalism. Is it morally acceptable for doctors to do things to patients in order to try to benefit them without obtaining the patients' informed consent? Who should decide what is in the patient's best interests if a patient and his or her doctor disagree? So far as respecting patients' decisions is concerned, is there a morally relevant difference when a patient refuses a treatment and when a patient demands a treatment?
- Issues of confidentiality. Is it morally legitimate to reveal information stemming from the consultation without the patient's consent? If so, in which circumstances and why?
- Issues of honesty and deceit. When and why, if at all, might a doctor or nurse properly lie to or otherwise deliberately deceive a patient?
- Issues stemming from patient's impaired or inadequate autonomy. When and why should children at various stages of development make their own health care decisions? When they should not, who should do so on their behalf, using what criteria, and why? How should decisions be made on behalf of adults who are substantially mentally impaired or disordered, either temporarily or permanently, and by whom? Can great distress sufficiently impair a patient's autonomy to justify overriding his or her refusal of treatment? Can the autonomy of "frail elderly" patients be legitimately overridden in their interests? If so, in which circumstances, how, and why?

B. Issues of Life and Death

Is abortion ever justified, and if so in what circumstances and why? How are moral tensions between the interests of a pregnant woman and those of her fetus—or unborn child—to be properly resolved when they arise? Why? Is the moral status of the human embryo, fetus, or newborn baby different from the moral status of more developed human beings? Why? Is it ever morally justified to kill patients? Is it ever morally justified to allow them to die? Is there ever any morally relevant distinction to be made between killing and allowing to die? Why? What is it to die? Is "brain death," with the rest of the body apparently alive as a result of being sustained by a ventilator and other interventions, morally equivalent to death in the usual sense where, as well as brain death, heart action and breathing have also ceased? What are the doctor's moral obligations to patients diagnosed as permanently unconscious but not brain dead, for example, patients in persistent or "permanent" vegetative state? How far are doctors obliged to try to keep patients alive when the probability of recovery is very low? Why? What should count as "recovery"? Why?

C. The Patient's Interests versus the Interests of Others

Should doctors always give moral priority to the best interests of the individual patient with whom they are then concerned, or may the interests of others sometimes take precedence? If so in which circumstances and why? Specific examples of such tensions include emergencies versus routine consultations or operations, and many other situations where outsiders have greater needs than the patient of the moment; medical research, where the interests of future patients may conflict with the best interests of the patient of the moment; health promotion and disease prevention where the needs of those who are not currently ill may conflict with the needs of those who are; and the requirements of medical education, both undergraduate and postgraduate (for example, the need to teach students how to examine patients and how to carry out various procedures, including operations). More obviously, tensions between the interests of the individual patient and others increasingly arise in the context of inadequate availability of resources of meet medical needs. Should doctors participate in rationing inadequate resources to their individual patients? If so why and using which criteria and processes? If not, who should carry out such rationing, why, and using which criteria and processes?

D. Issues of Distributive Justice

In asking questions like this the need to step away from the doctor–patient setting becomes particularly obvious. Distribution of scarce resources is a problem at several levels, only one of which is at the doctor–patient interaction (so-called micro allocation). At the other end of the spectrum governments must decide how much of their available national budgets to allocate to health care rather than other welfare programs, education, defense, or the arts (macro allocation). In between these two ends of the spectrum of allocation are distribution decisions at the organizational level: between different sorts of health care and other health-related activities, including teaching and research; between different hospitals or primary care organizations; and between different sectors and groups within organizations. Here bioethics becomes relevant at societal and organizational levels rather than at the level of the clinical encounter. At all these levels, however, there is a need for basic theoretical tools. In the context of fair distribution of scarce resources, for example, there is a need for an acceptable working theory—or working theories—of justice. How should the relevant agents decide that this way of deciding to distribute scarce resources is fair and that another is unfair?

E. Conceptual Analysis

In development of basic theoretical tools, conceptual analysis of the meaning—or more often meanings—of a particular concept or set of concepts is clearly a fundamental component. As obvious examples, what is meant by the terms disease, health, life, human being, person, death, brain death, and vegetative state? What is the difference, if any, between the meaning of "human being" and that of "human person"? What is meant by needs, rights, duties, and obligations? What is meant by benefit and harm in health care? What is justice in health care? What is autonomy and what conceptual distinctions are needed between it and respect for autonomy? What is meant by "care" in the context of health care? What is meant by "virtue" in the context of virtue theory, or by "nature" and "natural" when the natural is extolled and the unnatural opposed?

F. Ethical Issues in the Practice of Medical Science—The Impetus of Nuremberg

Thus even though bioethics started with critical analysis of ethical issues arising from clinical encounters, the internal intellectual momentum of that analytic endeavor has taken it far beyond its starting points. The same can be said of the critical analysis of moral issues arising from medical science. From at least the 19th century ethical issues of medicine included ethical issues of medical science, fundamentally ethical issues concerning the treatment of human (and to some extent animal) subjects of experimentation. This aspect of medical ethics was given a shocking impetus after the second world war by the revelations at Nuremberg of atrocities by Nazi doctors. This rapidly led, through the newly created World Medical Association, to an international agreement known as the Helsinki Declaration in which were enshrined the principles that informed consent had to be obtained from research subjects and that the interests of the individual patient should never be subordinated to the interests of society. Since then the ever more astounding exploits of science, and recently especially of the biological sciences in the context of organ transplantation and genetic engineering, have also pushed the concerns of bioethics well beyond their starting point within the medical sciences.

G. Bioethics, Science Technology, and Society

Quite apart from harm–benefit analyses, respect for people and their choices, and justice in the context of fair allocation of scarce resources, respect for people's rights, and respect for just laws, some strands of bioethics have become concerned with the natural and un-natural, with the effects of science and technology on the environment and on the biosphere, and with critical evaluation of that version of the scientific ethos whose self-proclaimed reductionism and purported freedom from values is perceived as more of a threat than a benefit to humanity. Such critiques in bioethics include concerns about the "new genetics," organ transplantation, especially the projected use of animal organs, and the ever increasing efforts of "high-tech" medicine and the medical equipment and pharmaceutical industries to develop methods for prolonging the "natural" life span of human beings. Sometimes such critiques are based on excessive cost, sometimes on their "unnaturalness," and sometimes as part of a broader concern about the environmental sustainability of the contemporary growth of scientific and technological interventions and their potential damage to earth's or Gaia's (Lovelock, 1979) environments and integrity.

H. Environmental Ethics

Springing from earlier roots (for example, in the transcendentalism and idealism of Thoreau, Emerson, Aldo Leopold, and John Muir), an extensive contemporary environmental ethics movement and literature has developed (e.g. Attfield 1983, Callicott 1989, 1995, Hargrove 1989, Johnson 1991, Naess 1989, Taylor 1986). Much of this environmental ethics movement considers itself to be part of bioethics, or, as in the case of Deep Ecology (Naess 1989) and other "ecocentric" environmental ethical perspectives, considers bioethics to be part of it. Not content with the limiting of the scope of much of traditional ethical concern to the interests of moral agents, potential moral agents, or human beings (anthropocentrism), environmental ethics seeks to expand the scope of ethical concern. Disagreement arises as to what should be included as having moral status—is it all sentient animals, all living animals, or all living beings, including plants (biocentric environmental ethics), or are inanimate entities also to be included within the scope of ethical concern, for example, the biosphere as a whole, ecosystems, species, land, water, and air (ecocentric environmental ethics)?

Interleaved with varieties of environmental ethics are varieties of feminist environmental ethics, of which one group—ecofeminism—claims that adequate theories for both feminism and environmental ethics need to understand the connections between woman and nature and between the domination of women by men and that of nature by man.

Thus the range of substantive issues encompassed by bioethics is indeed vast, and some have advocated that the subject area and discipline be explicitly subdivided

into relevant subdisciplines. One proposal is for the subdivision of bioethics into theoretical bioethics, concerned with the intellectual foundations of bioethics; clinical bioethics, concerned with ethical issues arising from interactions between patients and those who care for their health; regulatory and policy bioethics, concerned with rules, regulations, and laws in the context of bioethics; and cultural bioethics, which seeks "systematically to relate bioethics to the historical ideological cultural and social context in which it is expressed" (D. Callahan, 1995. In W. T. Reich (Ed.), *The encyclopedia of bioethics* (2nd ed., pp. 247–256). New York: Simon & Schuster–Macmillan).

III. DISCIPLINARY APPROACHES TO BIOETHICS

In what follows, some generalizations about various disciplinary approaches to bioethics are made without the qualifications, often extensive, that they deserve. This is an intrinsic pitfall within the "overview" enterprise, perhaps justified by the attempt to give a broad picture of the woods, even though it may fail to show the fine detail and variety of the trees and shrubs and other plants composing those woods. Overviews offer impressions and invite reflection, along with criticism and analysis, including consultation of the relevant specific entries in this encyclopedia and in the bibliography. The disciplines sketched are clinical, scientific, religious, legal, sociological, and psychological.

A. Clinical Approaches

Characterized by the immediacy of the ethical issues, a personal relationship often akin to friendship or even love (a relationship described by Campbell as, at its best, "moderated love"), clinical approaches to bioethics tend to be highly particular, situational, contextual, and partial in both senses of the term. Positive aspects of clinical approaches to bioethics at their best include the typical ethical commitment of the clinician to the individual patient, a commitment that ideally draws together all involved in the patient's health care; detailed awareness of the patient's individual problems and situation; and an ability and readiness to draw on clinical experience for predictive and management purposes. Clinical approaches to bioethics are perhaps the most ancient, and firmly established, stemming as they do from the existence and nature of clinical practice itself, and from the earliest codified deontological medico-moral obligations. Indeed aspects of the Hippocratic Oath of classical Greece remain integral parts of contemporary international and national codes of medical ethics, and also at the heart of contemporary clinical ethics.

At their worst, clinical approaches to bioethics may lack theoretical underpinnings, both scientific and ethical; they may succumb to the potential injustice inherent in excessive partiality on behalf of the individual patient; excessive paternalism is an ever present moral hazard, with patients being treated like young children and having things done to them without adequate consultation for what their clinicians regard as the patients' own good; and clinical ethics are vulnerable to inconsistency of approach, with action sometimes being too variably determined by the stance, personality, knowledge, skills, and attitudes of the individual clinician. Clinical approaches vary not only between individual clinicians and between clinicians of different cultures, but also between types of clinicians, for example, as between doctors and nurses. Such variations and their attendant conflicts can, when badly managed, cause confusion, distress, and damage to the patient, even when individual clinicians all believe themselves to be acting in the patient's best interests.

B. Scientific Approaches

Typically manifested by medical researchers, scientific approaches to bioethics aim to be as consistent as possible with scientifically established evidence and theory. At their best, when they focus on development of new treatments and diagnostic methods and on rigorous assessment of the efficacy of existing treatments and methods, they benefit patients by protecting them from unproven and potentially dangerous "remedies" and other interventions, and protect future patients by insisting on subjecting potential interventions to scientific assessment (especially, in the context of the development of new medications, by use of what medical scientists in the second half of the 20th century have regarded as the "gold standard" of such assessment, the randomized controlled clinical trial).

At their worst, scientific approaches to bioethics have a number of faults, many stemming from a reductionism whereby people and their activities, thoughts, and feelings are reduced to more or less complex combinations of scientifically analyzable components and processes. When clinicians adopt such an approach (and many contemporary doctors and increasingly nurses are also scientists, engaging in scientific research) patients may be unpleasantly confronted by what in popular parlance is termed a "clinical attitude"—cool, detached, and investigative, treating patients as biophysical problems to be solved rather than as people with problems to be solved. Such approaches have no time for unproven remedies even if the patients believe them to be helpful, and often involve hostile rejection of "alternative" health care approaches such as acupuncture and osteop-

athy, in the absence of scientific evidence of efficacy. In this and other contexts scientific approaches to methods of evaluation that do not involve scientific measurement—including religious, spiritual, and aesthetic evaluation—are, at their worst, highly intolerant and disrespectful. Finally, a not uncommon concomitant of a certain scientific approach to bioethics is the assimilation, and even identification, of ethics with the scientific theory of evolution. Survival of the fittest, and of the "selfish gene," becomes not only the genetic explanation for the development of ethics in humankind, but also its mistakenly reductionist substantive content.

C. Religious Approaches

Though it is even more difficult to generalize about religious approaches to bioethics, some broad positives and negatives may be discerned. At their best, religions offer a firm grounding of firmly established positive general ethical stances in which people are educated to have clear and substantive general and specific ethical obligations whose fulfillment is a religious duty. Bioethical obligations are situated within these general obligations. Moral respect for God's creation—the universe and all therein—is a common religious theme and obligation, and bioethical obligations are encompassed within such respect, sometimes under the specific obligation of stewardship for that creation. Beneficence to others is a common religious obligation, supported by the moral obligation to learn to overcome or temper self-interest by a concern to help others. Just distribution of scarce resources that aims at helping not only kith and kin and co-religionists but also all in greater need than self is another widespread religious concern of obvious relevance in the context of bioethics. Recognition of free will as a characteristic of humankind and a concern to nurture and respect it is tempered by the obligation to love God, help others in need, and treat all people as of equal moral importance—again substantive moral positions of obvious potential relevance to bioethics.

In the analysis of particular bioethical issues, religous thinkers and thinking have often been highly influential not only for their co-religionists but also for those of other religions and none. Examples include religious analyses of ordinary and extraordinary means in the context of prolongation of life; of the significance of intention in the moral analysis of action; and of the potential importance of distinguishing between acts and omissions, especially in contexts where the absence of specific prior moral obligations of beneficence does not negate general moral obligations of nonmaleficence. In addition, religious approaches tend to combine within practical morality the need for general moral principles with the need for specific applications of those general principles (casuistry); the need to combine these with obligations to educate the character to behave well (virtue ethics); and the need to make specific moral judgments only after careful attention to the stories of the people involved and to details of the specific context (narrative ethics). Indeed religious bioethicists, of different faiths, may sigh wearily, even impatiently, as they see many of the wheels of bioethics being laboriously and separately reinvented by contemporary secular thinkers.

On the other hand, at their worst, religious approaches to bioethics are intolerantly, sometimes even fanatically, rigid about received doctrines—of whichever variety they happen to be—and incapable of adjusting to new developments, or to different moral perspectives, albeit conscientiously and thoughtfully held and defended, that are opposed to their own.

D. Legal Approaches

Widely acknowledged (with the exceptions of a few legal positivists) to be based themselves on moral obligations, legal approaches to bioethics tend to reflect the moral norms of the societies concerned. At their best such legal approaches are imbued with a concern for societal benefit and harmony, along with strong commitments to the equality of all under the law. Such legal approaches enshrine the rights of the weak against being exploited and harmed by the powerful, and the rights of the individual against being victimized, whether by other individuals, by groups, or by the state. Indeed contemporary legal contributions to bioethics have been strong in developing rights-based theories of justice as underpinnings for bioethics. At their best legal approaches to bioethics argue carefully the pros and cons of contentious bioethical issues and resolve them in ways that respect to the greatest extent possible the conflicting sincerely held and carefully reasoned moral views represented in the relevant societies.

At their worst legal approaches to bioethics facilitate and enhance state oppression—for example, under German National Socialism in relation to compulsory euthanasia and to human experimentation, and under Soviet Communism in relation to misuse of psychiatry against political dissidents.

E. Sociological Approaches

Seeing themselves as scientists of societies, social scientists (sociologists) tend to try to approach bioethics in descriptive scientific mode, explaining how societal factors result in the substantive bioethical features of different societies and social groupings. At their best

such approaches help to broaden the gaze of bioethics so as to look at and understand not only ethical issues arising from the personal relationships of individuals but also ethical issues stemming from societal features that cause harm and ill health. At their worst they can underestimate the moral importance of individuals as moral agents, and combine a purportedly value-free descriptive approach to social functioning, in which all perspectives on morality are alleged to be of equal value or none, with a simultaneously prescriptive political stance.

F. Psychological Approaches

Seeing themselves, for the most part, as scientists of individual psyches, at their best psychologists in their approaches to bioethics illuminate it by showing how individuals come to develop their personal stances to moral issues. At their best they facilitate self-understanding, as well as understanding of others, by all involved in bioethics, especially perhaps an understanding that much of an individual's personal stance to ethical, including bioethical, issues is a function of emotional or other nonintellectual aspects of his or her psyche stemming from personality and from environmental influences, including those of early childhood. Ethical *reasoning* is recognized to be but one component of a person's ethical stance—a component that if not necessarily the slave of the passions, as Hume put it, is at least heavily influenced by nonreasoning aspects of the mind. At their worst psychological approaches to bioethics can also manifest a relativistic, deterministic approach to morality in which the moral stance of any individual is seen entirely as a function of influences beyond his or her control, an approach that tends to negate any purpose in bioethics (or indeed in any human endeavor).

IV. FOUNDATIONAL ETHICAL ASSUMPTIONS IN BIOETHICS—PHILOSOPHICAL/ETHICAL "SCHOOLS" OF BIOETHICS

As well as being pursued from many different disciplinary perspectives, bioethics is pursued from a variety of foundational theoretical assumptions, especially philosophical/ethical foundational assumptions. Typically articulated by philosophers independently of the various religious foundations for bioethics, such foundational ethical assumptions offer the secular "floating voter" a basis for ethical appraisal in bioethics. They also seek to offer those who are already firmly based upon a particular religious or cultural theoretical foundation a way of communicating about bioethics with those who do not share their religion, either because they have different religious beliefs or because they have none. Such foundations seek to provide a widely acceptable set of ethical theoretical assumptions, a widely acceptable approach to ethical analysis, and at least elements of a widely agreed upon ethical language suitable for the multicultural international contexts in which bioethics is pursued.

Even while reiterating earlier concerns about the dangers of generalization (and careful reading of the relevant sources will show that the generalizations that follow are no more than indications of the emphases of the relevant authors), nonetheless some groupings or "schools" of bioethics may be discerned on the basis of the importance they ascribe to different moral foundations for bioethics. Among the most important of these schools are those emphasizing the foundational importance of respect for people and their autonomy; welfare or other utility maximization; social justice; the "four principles"; and a variety of foundational approaches that either reject moral principles as foundational or find them inadequate—these include casuistry, virtue ethics, narrative ethics, various feminist ethics, and an increasing variety of "geocultural" ethics, of which only three are outlined.

A. The Foundation of Respect for Autonomy

Among early American approaches to bioethics several gave special emphasis to the foundational moral importance of respect for autonomy. Thus Veatch, within his social contract theory for medical ethics, emphasized the priority of respect for the autonomy of moral agents. Similarly (though for very different reasons) Engelhardt, developing a theory for bioethics that would enable those of different moral backgrounds to cooperate in matters of bioethics, emphasized the foundational centrality of the principle of respect for autonomy (which he later renamed the principle of permission). Important positive aspects of such emphasis are its recognition of the moral importance of respect for people, whether patients or not, as ends in themselves, not to be treated by others instrumentally, merely as means to an end. Problems with such foundational emphasis on autonomy include the tendency for this to be interpreted as encouraging atomistic, selfish, individualism, and a lack of concern and care for others.

B. The Foundation of Utilitarian Welfare Maximization

Welfare maximization—the obligation to maximize benefits and minimize harms—is a widespread foundational assumption in bioethics, and in the context of medical ethics reflects the widespread perceived obligation of doctors and other health care workers to produce as much health benefit as they can with as little harm as possible. One of the most influential (and meticulous) utilitarian philosophers in the area of bioethics is R. M. Hare, and underlying utilitarian foundational assumptions are to be found in the work of Singer and Harris. A utilitarian perspective also underlies and is challengingly argued for by Parfit. Positive aspects of utilitarianism for bioethics include its requirement of a universal duty to benefit others and to avoid harming them, and to do as much good and as little harm as possible.

Among problems with utilitarianism as a moral foundation for bioethics is that "common morality" widely perceives it to deal inadequately with several types of moral obligation. These include obligations resulting from special relationships (for example, parents' obligations to their children and doctors' and nurses' obligations to their patients), as a result of which some people should not, it is widely held, be treated merely as of equal importance with all others but should be given special priority by those who have special relationships with them and therefore special obligations to them. Similarly those in great medical need ought, it is widely thought, to be given moral priority over those in less medical need, yet welfare maximization may be ready to ignore such needs if more total benefit is achieved (for others) by doing so. And utilitarianism is widely perceived by its opponents to be too ready to subordinate respect for individuals' autonomy and other individual rights where overall maximal benefit (to others) is achieved by doing so.

C. The Foundation of Social Justice

A third moral foundation offered for bioethics is social justice. As with the other foundational principles, a wide variety of versions of social justice have been proposed, but a particularly important one in bioethics is the ideal social contract theory of J. Rawls (1971. *A theory of justice*. Oxford: Oxford Univ. Press), drawn on for his specifically health-orientated theory of justice by Daniels. For Rawls the theory of justice that rational people would arrive at behind a "veil of ignorance" (i.e., impartially because of not knowing what their own specific social roles or circumstances would be) would be based on two fundamental moral principles. The first would be an obligation to respect everyone's liberty to the maximal extent compatible with equal respect for the liberty of all. The second would be to aim at equality for all, and for deliberately created inequalities to be just only if they were both to the greatest benefit of the least advantaged and attached to offices and positions open to all under conditions of fair equality of opportunity (Rawls, 1971, 60 and 83). Daniels emphasizes and develops the "fair equality of opportunity" component of the Rawlsian account.

Advantages of the Rawlsian approach include its combination of liberty and differential benefit to those most disadvantaged. Problems include straightforward rejection in competing theories of justice of the Rawlsian principles themselves (and their theoretical justification) and/or the "lexical ordering" ascribed to them by Rawls whereby liberty takes priority over egalitarianism. Libertarian theories of justice, for example, tend to reject any obligation to attain equality or to benefit the disadvantaged—that would be good but not obligatory. Marxist socialist theories of justice subordinate liberty to the meeting of need and the attempt to attain equality. Communitarian theories of justice may reject a Rawlsian approach on the grounds that it does not sufficiently specify a positive conception of the good linked to the needs and interests of human communities or the human community. And rights-based theories of justice may oppose a Rawlsian approach on the grounds that it is inadequately grounded in and supportive of human rights, with variations in such opposition depending on which rights are regarded as of particular importance. In brief, foundations for bioethics grounded in a moral concern for justice share a concern for treating people justly, but differ widely over the substantive theory of justice that should be applied.

D. A Quasi-foundational Approach—The "Four Principles"

An attempt to offer not a foundational approach to bioethics but an approach that tries to combine some fundamental or foundational moral principles in a way that is compatible with a variety of mutually incompatible foundational theories is the "four-principles" approach offered and developed since the 1970s by the Americans Beauchamp and Childress, and enthusiastically adopted and promoted in Europe by Gillon. Developing upon an earlier triad of three principles produced as a working framework for the ethics of medical research by a group of American bioethicists in the "Belmont Report" (A. R. Jonsen & A. Jameton, 1995. In W. T. Reich (Ed.), *The encyclopedia of bioethics,* 2nd ed., pp. 1616–1632, New York: Simon & Schuster–Macmillan), themselves drawing on a long tradition of post-Enlightenment moral theory, the four-principles approach (4PA) starts from the claim that acceptance of four *prima facie* moral principles is common to a wide range of theoretical perspectives on bioethics, and also to much of "common morality."

Thus the 4PA is offered as a common working approach to bioethics, compatible with and neutral be-

tween a wide range of competing moral theories. It is also sometimes seen as an approach that lies in between the level of relatively abstract (and usually mutually incompatible) moral theories on the one hand, and highly specific moral situations, cases, problems, and judgments on the other. The principles are respect for autonomy, beneficence, nonmaleficence, and justice. Gillon additionally emphasizes the importance of consideration of the scope of application of each (to whom or what is the *prima facie* duty owned, and why?). While there is little substantive rejection of any one of these *prima facie* principles, opposition to the approach, pejoratively dubbed "principlism" or the "Georgetown mantra," has been considerable (for example, Clouser and Gert 1994, Wulff 1994).

Criticisms of principles as foundational in bioethics emerge from a variety of alternative schools of contemporary bioethics. Some, like those already mentioned, do not reject principles but argue that these need to be grounded in a theory of ethics. Thus utilitarians ground their principles or rules within the over-arching principle of utility, or in a logical analysis of the meaning of moral terms such as "ought" (Hare); Kantians ground their principles, rules, or maxims in a Kantian moral theory; and many religions ground their own bioethical principles within their own religious ethical framework.

E. Casuistry

The newly revived school of casuistry not only points out that reliance on potentially conflicting moral principles often fails to provide a decision procedure for when those principles conflict in particular contexts, but also adds that principles emerge from consideration of cases, not the other way around. Thus it is particular cases, and decisions about particular cases, rather than principles that are foundational for bioethics. Casuistry, of which Jonsen and Toulmin are leading contemporary proponents, is the application of general moral norms to specific cases in particular contexts in the light of comparisons and contrasts with previously determined clear or "paradigm" cases.

F. Virtue Ethics

Another school of bioethics rejects moral principles as foundations for bioethics on the grounds that virtues, not principles, are the proper moral base for bioethics. Virtues, or character dispositions to act or otherwise respond well *as people*, and then as people of a certain sort (for example, doctors, parents, scientists, or accident investigators), are, according to this approach, the fundamental concerns of ethics and therefore of bioethics. This was the approach taken to ethics by Aristotle, and Aristotelian virtue ethics is enjoying a contemporary revival, with the work of MacIntyre being highly influential. One variant of virtue ethics, again importantly influenced by MacIntyre and emphasizing the necessarily socially embedded and committed nature of virtue, is communitarian ethics, a movement that at the end of the second millennium was gaining impetus in the USA, perhaps partly in reaction to that country's prevailing libertarian individualism (for example, Emmanuel 1991).

G. Narrative Ethics

Associated with virtue ethics is another school of bioethics, the school of narrative ethics, that again finds reliance on moral principles inadequate. Fundamental to bioethics in this approach is the narrative or story of particular cases, and the story is both highly specific and highly culture bound, for every culture also has its story. Everyone involved in the story has interests in its outcome and, as Brody puts it,

> The "right course of action" to resolve a problem is not necessarily the action that conforms to an abstract principle; rather, it may be the action which, without violating any moral principles, most successfully navigates all the contextual factors to move the situation in a direction that best serves the major interests of all involved parties. (p. 215. H. Brody, 1994. In R. Gillon and A. Lloyd (Eds.), *Principles of health care ethics* pp. 205– 215. Chichester/New York: Wiley).

H. Feminist Ethics

Many, though not all, strands within contemporary feminist ethics also oppose reliance on moral principles, though given the wide variety of feminist approaches to bioethics, the criticisms vary. Common feminist criticisms are that moral reasoning in terms of principles is excessively abstract; fails to acknowledge the importance of the particular, of the subjective and the emotional, of the moral importance of caring and empathy, and of the responsibilities stemming from relationships; and above all fails to acknowledge and redress the oppression of women, not least in their medical care (e.g. Lebacqz 1995, Sherwin 1992).

I. Geocultural Bioethics

Just as individual religions tend to have their schools of bioethics, so too are various geocultural regions establishing their own schools of bioethics. For example, Gracia refers to a "Latin model" of bioethics appropriate to southern European nations and based more on virtues than on principles; insofar as principles are seen as relevant foundations, they may not be the quartet from Georgetown. A different quartet offered by

Gracia for Latin bioethics comprises the fundamental value of life, therapeutic wholeness, liberty and responsibility, and sociality and social subsidiarity (whereby social problems are always best addressed through the smallest relevant social unit).

Further north in Europe, Wulff, while agreeing that the Georgetown principles individually "cannot be contested," claims that in practice they are used to support typically American cultural approaches to bioethics which "do not accord with the prevailing moral tradition in other parts of the western world, eg the Nordic countries" (p. 277. H. Wulff, 1994. In R. Gillon and A. Lloyd (Eds.), *Principles of health care ethics* (2nd ed., pp. 277– 286). Chichester/New York: Wiley). Instead Wulff argues that the essentially Christian and Kantian Golden Rule is and should be the foundation of Nordic bioethics—and in an earlier work Wulff and his coauthors also emphasized the importance of the Danish philosopher Kierkegaard and more generally of the continental tradition of philosophy with its concerns for phenomenology, hermeneutics, and existentialism.

On a different continent, the East Asian Association for Bioethics was established in 1995 partly because bioethicists from Japan and China felt that Western approaches to bioethics were inappropriate as moral foundations for their own countries, where Buddhist and Confucian ethical norms so firmly underlie everyday morality, even in China where Maoist Marxism has had such a powerful social influence.

V. CONCLUSION

While future developments in bioethics are unpredictable, one prediction can be safely made. Whether or not we go so far as to accept Gracia's assertion to the International Association of Bioethics that "bioethics, I believe, is going to be the civil ethics of all our societies" (D. Garcia, 1993. *Bioethics* **7**(2/3), 97–107), we can confidently predict that it will continue to provide a range of absorbing and important ethical concerns for which an ever expanding audience of interest can equally confidently be anticipated.

Bibliography

Attfield, R. (1983). The ethics of environmental concern. Columbia University Press, New York.

Beauchamp, T., and Childress, J. (1994). Principles of biomedical ethics 4th ed. New York Oxford: Oxford University Press.

Blomquist, C. (1978). Medical ethics history: Western Europe in the twentieth century. In *The Encyclopedia of Bioethics* (W. T. Reich, Ed.), pp. 982–987. New York: Macmillan.

Brody, H. (1994). The four principles and narrative ethics. In *Principles of Health Care Ethics*. (R. Gillon and A. Lloyd, Eds.). Chichester/New York: Wiley.

Callahan, D. (1995). Bioethics. In *The Encyclopedia of Bioethics* (W. T. Reich, Ed.), 2nd ed., pp. 247–256. New York: Simon & Schuster–Macmillan.

Callicott, J. (1989). *In Defense of the Land Ethic: Essays in Environmental Philosophy.* Albany: State Univ. of New York Press.

Callicott, J. (1995). Environmental ethics: Overview. In *The Encyclopedia of Bioethics* (W. T. Reich, Ed.), 2nd ed., pp. 676–687. Simon & Schuster–Macmillan, New York.

Campbell, A. (1984). *"Moderated Love—A Theology of Professional Care."* London: SPCK.

Clouser, K., and Gert, B. (1994). Morality vs. principlism. In *Principles of Health Care Ethics* (R. Gillon and A. Lloyd, Eds.), pp. 251–266. Chichester/New York: Wiley.

Daniels, N. (1985). *Just Health Care.* Cambridge, MA: Cambridge University Press.

Emanuel, E. (1991). *The Ends of Human Life—Medical Ethics in a Liberal Polity."* Cambridge, MA: Harvard University Press.

Engelhardt, H. T. (1986, 1996). The Foundations of bioethics, 1st and 2nd ed. New York/Oxford: Oxford University Press.

Gillon, R. (1981). The function of criticism. *British Medical Journal, 282,* 1633–1639.

Gillon, R. (1986). *Philosophical Medical Ethics.* Chichester/New York: Wiley.

Gillon, R., and Lloyd, A. (Eds.) (1994). *Principles of Health Care Ethics.* Chichester/New York: Wiley.

Gracia, D. (1993). The intellectual basis of bioethics in southern European countries. *Bioethics, 7*(2/3), 97–107.

Hare, R. M. (1981). *Moral Thinking: Its Levels, Method and Point.* Oxford: Clarendon Press.

Hargrove, E. (1989). *Foundations of Environmental Ethics.* Englewood Cliffs, NJ: Prentice Hall.

Harris, J. (1985). *The Value of Life.* London: Routledge and Kegan Paul.

Johnson, L. (1991). A Morally Deep World: An Essay on Moral Significance and Environmental Ethics. Cambridge University Press, Cambridge.

Jonsen, A. R., and Jameton, A. (1995). Medical ethics, history of; the Americans; the United States in the twentieth century. In *The Encyclopedia of Bioethics* (W. T. Reich, Ed.), 2nd ed., pp. 1616–1632. New York: Simon & Schuster–MacMillan.

Jonsen, A. R., and Toulmin, S.E. (1988). *The Abuse of Casuistry: A History of Moral Reasoning.* Berkeley: University of California Press.

Lebacqz, K. (1995). Feminism. In *The Encyclopedia of Bioethics* (W. T. Reich, Ed.), 2nd ed., pp. 808–818. New York: Simon & Schuster–MacMillan.

Lovelock, J. (1979). *Gaia: A New Look at Life on Earth.* Oxford: Oxford University Press.

MacIntyre, A. (1981). After Virtue: A Study in Moral Theory. Notre Dame, IN: University of Notre Dame Press.

MacIntyre, A. (1990) *Three Rival Versions of Moral Inquiry: Encyclopedia, Genealogy, and Tradition.* Notre Dame, IN: University of Notre Dame Press.

Naess, A. (1989). Ecology, Community and Lifestyle: Outline of an Ecosophy. (Transl. D. Rothenberg). Cambridge: Cambridge Univ. Press.

Oakley, J. (1995). Medical ethics, history of: Australia and New Zealand. In The Encyclopedia of Bioethics (W. T. Reich, Ed.), 2nd ed., pp. 1644–1646. New York: Simon & Schuster–Macmillan.

Parfit, D. (1984) *Reasons and Persons.* Oxford: Clarendon Press.

Rawls, J. (1971). "A Theory of Justice." Oxford: Oxford University Press.

Reich, W. T. (Ed.) (1978). *The Encyclopedia of Bioethics.* New York: Macmillan.

Reich. W. T. (1994). The word "bioethics": Its birth and the legacies of those who shaped it. *Kennedy Institute of Ethics Journal, 4*(4), 319–335.

Reich, W. T. (Ed.) (1995). *The Encyclopedia of Bioethics,* 2nd ed. New York: Simon & Schuster–Macmillan.

Reich, W. T. (1995). The word "Bioethics": The struggle over its earliest meanings. *Kennedy Institute of Ethics Journal, 5*(1), 19–34.

Sherwin, S. (1992). *No Longer Patient: Feminist Ethics and Health Care.* Philadelphia: Temple University Press.

Singer, P. (1979). *Practical Ethics.* Cambridge: Cambridge University Press.

Taylor, P. (1986). *Respect for Nature: A Theory of Environmental Ethics.* Princeton, NJ: Princeton University Press.

Veatch, R. (1981). *"A Theory of Medical Ethics.* New York: Basic Books.

Warren, K. (1990). The power and the promise of ecological feminism. *Environmental Ethics, 12*(2), 125–146.

Wulff, H. (1994). Against the four principles: A Nordic view. In *Principles of Health Care Ethics* (R. Gillon and A. Lloyd, Eds.), pp. 277–286. Chichester/New York: Wiley.

Wulff, H., Andur Pedersen, S., and Rosenberg, R. (1986). *Philosophy of Medicine—An Introduction.* Oxford: Blackwell.

Biotechnology

MICHAEL REISS

Homerton College, Cambridge

GLOSSARY

biotechnology The application of biology for human ends. Often divided into "traditional biotechnology"—farming and the long-established use of microorganisms in the production of foods and drinks—and "modern biotechnology"—which utilizes novel disciplines such as tissue culture, embryo transfer, and genetic engineering.

clone A collection of genetically identical cells or multicellular organisms.

DNA The chemical that carries the genetic information contained in an organism's genes.

gene therapy The intentional alteration of human genetic material for medical ends.

genetic engineering The intentional transfer of genetic material from one organism to another, usually of a different species. Synonyms include "genetic manipulation," "genetic modification," and "recombinant DNA technology."

proteins Molecules, such as the hormone insulin, that are composed of one or more chains of subunits known as amino acids; made by all organisms as a result of genes that code for these animo acids.

BIOTECHNOLOGY is the application of biology for human ends. It involves using organisms to provide humans with food, clothes, medicines, and other products. The phrase "traditional biotechnology" refers to activities like the farming of animals and plants, and the use of microorganisms in the manufacture of beer, wine, bread, yogurt, and cheese. By contrast, modern biotechnology has only become possible within the last 20 years or so through advances in novel disciplines such as tissue culture, embryo transfer, and genetic engineering.

Modern biotechnology, though it has grown out of traditional biotechnology, is distinctive in a number of regards. For one thing, its scope seems near endless. It has been claimed that it will revolutionize agriculture, medicine, the food industry, and much else besides. On the other hand, it has been argued that its potential for harm is immense. Then there is the tremendous, and seemingly ever quickening, pace of change. Finally, many aspects of modern biotechnology, such as genetic engineering, arouse deep feelings. Genetic engineering raises issues about the nature of life itself, about what it is to be human, about the future of the human race, and about our rights to knowledge and privacy.

I. THE HISTORY OF BIOTECHNOLOGY

A. Traditional Biotechnology

Traditional biotechnology has a long history. The domestication of animals and plants seems to have happened independently in the Middle East, Asia, and the Americas about 12,000 to 10,000 B.P.

Around 12,000 to 11,000 B.P., the dog was domesticated in Mesopotamia and Canaan. Within a thousand years of this time goats and sheep were domesticated in Iran and Afghanistan, and emmer wheat and barley were being cultivated in Canaan. Around 10,000 to 9000 B.P., potatoes and beans were domesticated in Peru, rice in Indochina, and pumpkins in middle America.

By 8000 B.P. the pig and water buffalo had been domesticated in eastern Asia and China, the chicken in southern Asia, and cattle in southeastern Anatolia (modern day Turkey). At the same time, einkorn wheat was being cultivated in Syria; durum (macaroni) wheat in Anatolia; sugar cane in New Guinea; yams, bananas, and coconuts in Indonesia; flax in southwestern Asia; and maize and peppers in the Tehuacan valley of Mexico. By 8000 B.P. a type of beer was being made with yeast in Egypt. Indeed, by 4000 B.P the Sumerians brewed at least 19 brands of beer—a whole book on the subject survives.

Four processes are involved in the farming of domesticated animals or plants:

- Breeding of animals or sowing of seeds
- Caring for the animals or plants
- Collecting produce (e.g., harvesting, milking, and slaughtering)
- Selecting and keeping back some of the produce for the next generation.

For more than 10,000 years, therefore, farmers have selected animals and plants. Much of this selection will have been conscious, with farmers often choosing, for example, to breed from larger and healthier individuals. Indeed, genetics is probably a much older science than is generally realized. However, much of the selection by farmers will have been unconscious, as farmers unwittingly chose, for example, animals that were tractable or tolerant of overcrowding.

B. The Relationship of Traditional to Modern Biotechnology

The fact that traditional biotechnology has such a long history might lead one to conclude that perhaps too much concern is generated about genetic engineering and other techniques of modern biotechnology. After all, traditional biotechnology often involves the transfer of genes in a way that would not happen in nature. That happens every time a farmer selects a bull to mate with cows, and every time a plant breeder dusts the pollen from one plant onto the female sex organ of another plant. Indeed, such traditional selective breeding has achieved dramatic results, as is witnessed by the many very different breeds of dogs.

Traditional biotechnology has also changed certain plants very greatly. The modern wheat used in bread making is so different from native wheats that scientists are still uncertain as to its precise ancestry. What is clear, though, is that it results from at least two interspecific crosses. In other words, on at least two separate occasions, thousands of years ago, people succeeded in breeding one species of wheat with another species. The net result is that today's bread wheat contains approximately three times the number of genes as wild wheats found in the Middle East.

However, although traditional biotechnology can result in major alterations in the genetic makeup of organisms, it differs from modern biotechnology in at least three important respects.

First, although traditional biotechnology sometimes involves crossing one species with another, these species are always closely related. To the nonexpert, the plant species crossed to make modern bread wheat all look much the same. Indeed, botanists classify them as being very closely related. This is markedly different from genetic engineering where genes can now be moved from one species to another, however unrelated, almost at will.

Secondly, the pace of change in traditional biotechnology is much slower than that in modern biotechnology. We are already at the point where a gene from one organism can permanently be inserted into the genetic material of another organism within a period of weeks. Traditional biotechnology, by comparison, works on a time scale of years.

Thirdly, genetic change as a result of traditional biotechnology happened to only a relatively small number of species, namely, those that provide us with food and drink, such as crop plants, farm animals, and yeasts. Modern biotechnology is far more ambitious. It seeks to change not only the species that provide us with food and drink, but those involved in sewage disposal, pollution control, and drug production. It also seeks to create microorganisms, plants, and animals that can make human products, such as insulin, and even possibly to change the genetic makeup of humans.

II. TECHNIQUES IN MODERN BIOTECHNOLOGY

A. Genetic Engineering

1. The Significance of Genetic Engineering

By far the most significant development in modern biotechnology, from both a scientific and an ethical perspective, is the practice of genetic engineering, which dates from the late 1970s and early 1980s.

Every organism carries inside itself what are known as genes. These genes are codes or instructions: they carry information which is used to tell the organism what chemicals it needs to make in order to survive, grow, and reproduce. Genetic engineering typically involves moving genes from one organism to another. The result of this procedure, if all goes as intended, is that the chemical normally made by the gene in the first organism is now made by the second.

2. Principles of Genetic Engineering

Suppose one wants a species to produce a protein (i.e., a biochemical consisting of a chain of subunits called amino acids) made by another species. For example, one might want a bacterium to produce human insulin so as to be able to collect and then give the insulin to people unable to make it for themselves. The basic procedure, using genetic engineering, involves the following two steps:

1. Identify the gene that makes the protein one is interested in
2. Transfer this gene from the species in which it occurs naturally to the species in which one wants the gene to be.

The first of these steps is more difficult than it may sound. Even a bacterium has hundreds of different genes, while animals and plants have tens of thousands. Nowadays, though, there are a number of ways of identifying the gene that makes the protein in which one is interested.

Two different types of approaches can be used to carry out the second step, namely, transferring this gene from the species in which it occurs naturally to the intended species recipient. One involves the use of a vector organism to carry the gene; the other, called vectorless transmission, is more direct and requires no intermediary organism.

3. Vectorless Transmission

One way of getting DNA into a new organism is simply to fire it in via a gun, i.e., biolistic (particle gun) delivery. The DNA is mixed with tiny metal particles, usually made of tungsten. These are then fired into the organism, or a tissue culture of cells of the organism. The chief advantage of this method is its simplicity, and it is widely used in the genetic engineering of plants. One problem, not surprisingly, is the damage that may be caused as a result of the firing process. A more intractable problem is that only a small proportion of the cells tend to take up the foreign DNA.

A second way of getting DNA into a new organism is by injecting it directly into the nucleus of an embryonic cell. This approach is quite widely used in the genetic engineering of animals. This method ensures that at least some of the cells of the organism take up the foreign DNA.

4. Vectors

A vector carries genetic material from one species (the donor species) to another (the genetically engineered species). Genetic engineering by means of a vector involves three steps:

1. Obtaining the desired piece of genetic material from the donor species
2. Inserting this piece of genetic material into the vector
3. Infecting the species to be genetically engineered with the vector so that the desired piece of genetic material passes from the vector to the genetically engineered species.

An example of genetic engineering by means of a vector is the infection of certain plants by genetically engineered forms of the bacterium *Agrobacterium. Agrobacterium* is a soil bacterium that naturally attacks certain plants, infecting wounds and causing the development of swellings known as tumors. In 1977 it was found that the tumors were due to the bacterium inserting part of its genetic material into the host DNA. This means that if foreign DNA is inserted into the DNA of *Agrobacterium,* the *Agrobacterium* can in turn insert this foreign DNA into the genetic material of any plants it subsequently attacks.

Viruses can also be used as vectors in genetic engineering. For example, retroviruses have been used in genetic engineering research on humans. Retroviruses are good candidates for this approach as they have millions of years of experience at inserting their genetic material into that of a host. A number of diseases are caused by mutations in genes expressed in bone marrow cells—the cells that give rise to our blood cells. Retroviruses have been used in attempts to insert a functional copy of the faulty gene into these bone marrow cells. The aim is to ensure that all the blood cells that descend from these bone marrow cells are healthy.

One problem with this approach, which limits the number of diseases on which it is being trialed, is that retroviruses only infect dividing cells. Many human diseases, for example, those of the nervous system, are not caused by mutations in dividing cells. A second problem is that, as so often is the case in genetic engineering, there is no control presently available as to where the gene is inserted in the human chromosomes. Instead the retrovirus inserts the desired gene more or less randomly. This has two consequences. First, the new gene may not be as effective as when it is located in its normal place. This is because genes often work best only if they are situated close to certain other genes which help turn them on and off. The second, and more dangerous, possible consequence is that the new gene may, by mistake, be inserted into an important gene, for example, tumor-suppressor genes which help prevent cancer. Disruption of the activity of a tumor-suppressor gene by the insertion of a new gene through the activity of a retrovirus has been shown in monkeys to sometimes lead to the development of cancer.

For these reasons, researchers are experimenting with other viruses. For example, adenoviruses are being used in attempts to insert functional copies of the gene which, in its faulty form, causes cystic fibrosis in humans. Adenoviruses, unlike retroviruses, do not integrate their genes into their host's DNA. This has both advantages and disadvantages. An obvious disadvantage follows from the fact that any descendants of the genetically engineered cells do not carry the functional cystic fibrosis gene. This means that once the genetically engineered cells die, the functional cystic fibrosis gene is lost with them. As a result, this approach is only likely to be effective if people with cystic fibrosis are treated with genetically engineered adenoviruses every few months. On the other hand, there is less risk of the virus inserting its genetic material into the host cells in such a way as to disrupt normal functioning or even cause cancer.

B. Tissue Culture

Tissue culture involves the growing, under sterile laboratory conditions, of cells or tissues derived from animal, plant, or other living material. It is a prerequisite for many of the techniques used in modern biotechnology.

Tissue culture based on plant material is nowadays of huge commercial importance, for example, in the horticultural trade. In particular, tissue culture can be used to produce many identical plants in a short period of time. In a standard procedure, a number of distinct small clumps of cells are taken from a plant, transferred to laboratory containers, provided with water, nutrients, and light, and allowed to grow rapidly into functioning plants. The plants so produced are genetically identical and constitute a clone. This allows features of commercial value found in only one or a small number of plants to be quickly present in larger numbers of plants. In animals, cloning can be achieved in a number of ways. The simplest procedure in farm animals is to divide an embryo, while in tissue culture, in half and then return the halves to the mother. The result is identical twins.

C. *In Vitro* Fertilization

In vitro fertilization is most commonly used as one of the treatments for human infertility, though it also sometimes used to produce large numbers of cattle embryos from desirable parents. In humans, *in vitro* fertilization leads to the production of so-called "test tube babies."

Whether in humans or other animals, *in vitro* fertilization requires the collection of suitable eggs and sperm. Egg production may be stimulated by treatment with hormone-based drugs. In mammals, eggs are collected via a surgical procedure shortly before ovulation. In the simplest form of *in vitro* fertilization sperm are added to laboratory dishes containing the eggs in a suitable medium. Fertilization occurs and after an interval of up to several days the fertilized egg(s)—which by now may have divided to form a cluster of cells—are transferred to a uterus, which may, or may not, be the uterus of the female from which the eggs were obtained.

D. Embryo Transfer

Embryo transfer entails the removal of embryos at an early stage of development from a donor female and transfer to a surrogate female. The procedure—one form of "surrogate motherhood"—is still uncommon in humans, but is quite widely used in cattle and some other farm animals.

III. ETHICAL ARGUMENTS FOR AND AGAINST BIOTECHNOLOGY

A. Consequentialist Arguments

1. What Consequences Does Biotechnology Have for Humans?

The actual (past and present) and possible (future) consequences of biotechnology are legion. Think simply of alcohol production, one of the longest established examples of biotechnology. On the one hand, consequences of alcohol consumption include much human suffering, either directly through cirrhosis of the liver, various cancers, and babies born with fetal alcohol syndrome, or indirectly through alcohol-related accidents and violence. On the other hand, alcohol is widely enjoyed by many people, provides large numbers of satisfying jobs, and is even thought now to promote physical health if taken in small quantities.

Utilitarian attempts to calculate the consequences of specific instances of biotechnology can help to clarify ethical decision making. However, such attempts will probably only rarely allow us unambiguously to decide whether or not to proceed with the technology in question. For one thing, it is difficult to accurately and quantitatively predict the consequences of new technologies.

On a broader canvas, it would be virtually impossible for humans to exist without biotechnology. This is not, of course, to imply that biotechnology should not be regulated, only to point out that even before the advent of modern biotechnology we were heavily dependent on agricultural crops, paper, and clothes made from wool and cotton, medicines derived from plants, etc.

2. Is Biotechnology Safe for Humans?

It is, of course, not possible to answer this question with either a "yes" or a "no." One can only proceed on a case-by-case basis guided by knowledge and experience. In addition, it needs to be remembered that:

- There is little, if anything, in life that is 100% safe
- "Safer" is not necessarily to be equated with "better" (it may, for example, be safer for me to read about the ethics of torture than to take practical steps to strive to reduce the extent of torture)
- By following a safer course of action in the short term (e.g., not proceeding with the genetic engineering of crop plants because of a nonzero risk that this is unsafe) one may end up with a less safe end result (e.g., more starvation or famine-induced wars).

3. Does Biotechnology Lead to Animal Suffering?

Suffering involves susceptibility to pain and an awareness of being, having been, or about to be in pain. Pain here is used in its widest sense and includes stress, discomfort, distress, anxiety, and fear. It is difficult to argue against the contention that vertebrates, and probably certain invertebrates such as octopuses, can experience pain. The extent to which animals are aware of their pain is more open to question. There is little doubt that certain of our closest evolutionary relatives, such as chimpanzees and other apes, have the requisite degree of self-consciousness. Although the extent to which other animals suffer is contentious, a growing number of biologists and philosophers accept that, at the very least, most mammals, and probably most vertebrates, can suffer. This conclusion is unlikely to surprise anyone who has ever kept a pet or has worked with animals.

Does animal biotechnology lead to animal suffering? No overall answer can be given. Take, first of all, the case of conventional farm animals such as sheep, cattle, pigs, and chickens. Under the best management regimes such animals enjoy better health than their counterparts would in the wild. It is true that some might describe their lives as being boring while their movements and certain other natural behaviors (e.g., mating) are restricted, but it is doubtful whether this constitutes suffering. However, there are countless examples of farm animals suffering as a direct result of biotechnology and poor husbandry, especially some of those kept under the most intensive regimes. For example, many chicken varieties have been subjected to such extreme artificial selection for accelerated growth that a high proportion of individuals experience bone fractures and other clinical deformities during their brief existence in battery cages.

4. What Are the Ecological Consequences of Biotechnology?

Ecological consequences of biotechnology need to be taken into account both because they often have consequences for humans and because they have consequences for other organisms too. Some countries nowadays have strict regulations about the introduction into the wild of new animal and plant varieties precisely because of the number of ecological disasters that occurred long before the advent of modern biotechnology. Indeed, literally hundreds of species are known to have gone extinct as a result of the introduction of nonindigenous species.

Fears have been expressed that modern biotechnology will contribute to further ecological damage. For example, it is possible that fish genetically engineered to be able to live in cooler waters will increase their geographic range and so displace native species. Considerable uncertainty still exists as to how significant such fears are.

B. Intrinsic Arguments

1. Is Biotechnology Unnatural?

Since at least the time of Hume it has been accepted that attempts straightforwardly to argue from what is the case in nature to what is the right course of action for humans to take are problematic. It is tempting therefore to react dismissively to objections to modern biotechnology on the grounds that it is unnatural. However, there now exist more nuanced attempts to relate to what ought to be. In addition, there is no doubt that on psychological grounds, politicians and regulators, if not moral philosophers, do well to heed such arguments. Opinion polls in a number of different countries consistently show that many people object to modern biotechnology on the grounds that it is "unnatural."

2. Is Biotechnology Blasphemous?

Some people with a religious faith either reject or are hesitant about the rightness of certain developments within modern biotechnology, maintaining that such developments are blasphemous. This is commonly summed up in the phrase, "We should not play at being God." On the other hand, it has been argued that, in some sense, we are cocreators with God. The reasoning goes as follows. Creation is an ongoing process, the universe having been in a continual state of development for some 10 to 15 thousand million years. Within just the last few thousand years, humans have begun consciously to influence the course of that continued creation in ways never before attained by any species. In any useful sense of the term, therefore, we are already cocreators with God.

3. Does Biotechnology Entail Disrespect for Organisms?

It is often argued that animals have rights and that we are not entitled to use them for our ends. A different approach is to hold that, in Kantian terms, it may be acceptable for us to use an animal's ends as our own (e.g., using a sheep to produce wool and lambs), but it is unacceptable for us entirely to ignore an animal's ends and instead use it solely as an instrument by which we attain our ends.

A number of writers have argued that it is wrong for us to violate the genetic integrity or the *telos,* in an Aristotelian sense, of organisms. Others have concentrated on the distinction between instrumental and intrinsic value. Paul Taylor has argued that all living organisms possess inherent worth, and explores the consequences of this for the resolution of conflicts between humans and other species when there is competition for limited resources.

C. Patenting

Is it right to patent genes, parts of organisms, or whole organisms? Obviously questions to do with the ethical implications of patenting are, logically, distinct from questions to do with the ethical implications of biotechnology. However, the practical reality is that the potential for money to be made from modern technology has led to a rush to patent human and other genes.

The fundamental argument in favor of patenting is that it rewards those who have put time, effort, ingenuity, or money into the invention of a new product or process. For a finite length of time (typically 20 years, though the exact length of time varies in different countries), a patent allows the inventor a monopoly right to exploit the patented invention. After this period the patent ceases.

Those opposed to the patenting of genes, parts of organisms, or whole organisms advance a number of arguments:

- It is wrong to patent life and this means that it is wrong to patent either whole organisms or their genes; the very idea is absurd, obscene, or blasphemous; living things are not "products of manufacture," but rather, the genetic resources of the planet are our common heritage
- Patenting reduces the exchange of information among researchers
- patenting encourages researchers to target their efforts where money is to be made, rather than where work is most needed.

Those in favor of the patenting of genes, parts of organisms, or whole organisms advance the following main arguments:

- Patenting is right in that it rewards the investment and ingenuity of those who develop new products
- Without such patenting there will be fewer benefits to health than would otherwise be the case
- In the absence of patenting, firms will resort to greater secrecy to protect their investment
- Patents do not interfere with pure research since experimental use of an invention does not constitute patent infringement.

Many of the ethical issues raised by the patenting of genes or even organisms are common to those already raised by the patenting of any product or process. However, it can be argued that the ability of organisms—unlike, for example, corkscrews—to reproduce and to have genes which mutate renders the notion of the patenting of organisms or their genes especially problematic from both a legal and a philosophical viewpoint.

IV. CASE STUDIES

A. Cheese Making

Cheese has been made by people for at least 5000 years. The fundamental principles have changed little over the millennia. During cheese making a number of substances are added to sour milk. One of them is rennet. Rennet is a crude extract of enzymes, of which much the most important is chymosin, also known as rennin. These enzymes act on a milk protein called casein. Their effect is to cause the milk to form a soft curd, also known as junket. Without rennet, most cheeses cannot be made.

Traditionally rennet has been obtained from the stomachs of young calves (or piglets, kids, lambs, or water buffalo calves). Rennets can be of vegetable origin, but, until recently, by far the most important source was young calves. Calves' stomachs were ground up in salt water—10 of them being required for one gallon of rennet. Calves' stomachs contain a wide range of substances in addition to rennet, so the purity was not very high. However, the gene for calf chymosin has now been inserted, by genetic engineering, into a yeast which produces a ready supply of chymosin in commercial quantities. As a result, the use of rennet obtained directly from animals has greatly decreased. In addition, genetically engineered chymosin is cheaper than traditional rennet and considerably purer.

The original gene used in the genetic engineering of the yeast came from an animal source. However, many vegetarians have endorsed genetically engineered chymosin on the grounds that its use significantly decreases the slaughtering of calves. Cheese produced through the use of genetically engineered chymosin is also approved by Muslims and Jews.

One important factor in many vegetarians' approval of genetically engineered chymosin is the fact that in practice the rennet actually used by cheese manufacturers will not contain the original calf chymosin gene, but copies of it. However, something of a diversity of views exists on this point among vegetarians. Some accept the copying stage as meaning a host containing a gene from an animal source is acceptable while others view its animal origin as meaning that the product is not acceptable.

By 1994, approximately half of the worldwide market for rennet was being supplied by genetically engineered chymosin. It is tempting to see this example of genetic engineering as a way of saving the lives of the millions of 4- to 10-day-old calves that, until recently, were killed for their rennet each year. The reality, though, is that these calves continue to be produced to keep their mothers—dairy cows—producing milk. Female calves generally themselves become dairy calves. Male calves are usually reared for meat, either as veal calves or as beef cattle.

B. Genetically Engineered Tomatoes

Tomatoes are big business, with sales exceeding those of potatoes or lettuce. However, it is generally acknowledged that today's tomatoes all too often have a poor flavor and texture. The main reason is that tomatoes are usually picked before they are ripe. The benefit of this practice is that it allows the tomatoes to be moved from where they are grown to where they are sold before they go soft. Consequently, the tomatoes are less likely to be damaged in transit. The disadvantage, though, stems from the fact that tomato flavor correlates with the amount of time the tomatoes spend on the parent plant. Picking tomatoes when they are still green leads, therefore, to relatively flavorless tomatoes.

A second problem with picking tomatoes before they go red is that they then have to be treated with ethylene before being sold. Ethylene is a natural plant growth substance and is responsible for the ripening of tomatoes *in situ*. It is supplied to tomatoes that are picked when still green as otherwise they fail to ripen. A final problem with harvesting tomatoes when they are still green is that it is all too easy to pick them when they are still very unripe. These immature green tomatoes taste even less good than the ones that result from tomatoes picked when "mature green."

For a variety of reasons, therefore, there are strong incentives for breeding tomatoes which can be picked when red. Such tomatoes should taste better and be firmer. They would not need to be treated with ethylene and might even end up costing less, as the wastage that comes from picking immature green tomatoes would probably be reduced.

From the mid-1980s a race ensued between several companies trying to manufacture and market genetically engineered tomatoes with these characteristics. In 1994 this research reached commercial fruition when Calgene's "Flavr Savr" tomato went on sale in the USA. Other companies still involved in the genetic engineering of tomatoes include Monsanto and Zeneca seeds.

The approach used relies on the fact that tomatoes take much longer to go soft if they do not produce a protein called polygalacturonase (PG). In a natural tomato, PG synthesis only takes place as the tomato ripens from green to red. Its effect is to soften the fruit by breaking down some of the compounds in cell walls between the cells of the fruit.

Preventing PG synthesis does not affect ripening, but it does mean that the fruit remains firm—just what is wanted by manufacturers and consumers. It also means that tomato sauces made from the tomatoes are more viscous and so flow less readily. This, too, is a desirable characteristic—runny tomato ketchups are not popular. Indeed, tomato paste manufacturers sometimes heat tomatoes to inactivate PG. However, this heating costs money and further reduces the flavor.

PG synthesis can now be prevented by so-called "antisense gene technology." In essence an artificial gene, made in the laboratory to be the reverse of the PG gene, is inserted into the tomato's DNA. This artificial gene effectively cancels out the effect of the PG gene and so prevents the cell from manufacturing PG. It is almost as if the PG gene had been excised from the DNA.

For those who object to such a procedure on the grounds that it is unnatural, it needs to be realized that the modern tomato, *Lycopersicum esculentum,* has already had many features bred into it by hybridizations, through traditional techniques of plant breeding, between a number of different *Lycopersicum* species. It might, therefore, be argued that the genetic integrity of the tomato has already, through conventional plant breeding, been somewhat violated.

A slight complication is that when researchers genetically engineer a species, they often add a "marker" gene which makes the organism immune to a particular antibiotic. The reason for this is that it makes it easier in the laboratory to see whether the genetic engineering has worked. In the case of tomatoes, for example, one simply has to see if the young tomato seedling is unaffected by the presence of an antibiotic such as kanamycin, rather than waiting to see if the fruits the adult tomato eventually produces take longer to go rotten.

It has been pointed out, not least by the United Kingdom's Department of Health's Advisory Committee on Novel Foods and Processes, that there is a possibility, albeit a very small one, that when large amounts of foods containing these antibiotic marker genes begin to be consumed, the gene might move to disease-causing microorganisms in the gut and so make them resistant to the antibiotic too.

Most experts suspect that the chances of this happening are not great. Even if it does happen, the consequences are unlikely to be desperately serious as there are many different types of antibiotics and any one marker gene only conveys resistance to one of them. Nevertheless, the existence of a finite risk slowed the regulatory approval of genetically engineered tomatoes and other foods in the United Kingdom. In the long run, a possible solution is for companies to use other, less problematic markers. Technically this is feasible, though less easy.

C. Reproductive Techniques Applied to Farm Animals

Reproductive techniques have been applied to farm animals for over 10,000 years. Farmers have long been accustomed to breeding selectively from only certain individuals—castrating, isolating, or slaughtering the rest. Traditional biotechnology has already led to the point where turkeys are unable to mate. Instead, the males have to be "milked" by hand to obtain their semen and the females artificially inseminated.

To a considerable extent, the history of animal husbandry has been a history of the increasing physical dependence of domestic animals on humans. Modern biotechnology, which already boasts genetic engineering, *in vitro* fertilization, and embryo transfer, is likely to accelerate this trend, but it need not. It is perfectly possible that consumer pressure and/or appropriate regulations could lead to animals being given back some of their natural freedoms by the use of modern biotechnology.

D. Bovine Growth Hormone (BST)

1. Why Bother to Make Genetically Engineered BST?

Bovine growth hormone, also called bovine somatotrophin (BST for short), is a natural hormone produced by cattle. During lactation, BST causes nutrients derived from a cow's food to be diverted to her mammary glands where they are used to make milk. It is this fact that has led to an extraordinary 10-year battle over genetically engineered BST.

In the 1970s it was suggested that injecting a cow with BST might increase her milk yield. The chemical structure of BST was determined in 1973, and by 1982 genetically engineered BST had been made. A huge amount of basic and applied research has been carried out on genetically engineered BST by a number of companies, i.e., Cyanamid, Elanco, Monsanto, and Upjohn.

Injecting dairy cows with genetically engineered BST increases their milk yields by some 20%. Furthermore, it increases milk to feed ratios by some 15%—that is, the amount of milk made by the cow relative to the food she consumes goes up by around 15%. Monsanto argues that genetically engineered BST offers a number of significant advantages to dairy farmers. The main one is that by raising milk yields and increasing feed efficiency, profits are raised. Further, the technology requires no capital investment: the farmer merely injects the cows every 14 days from the 9th week after calving until the end of lactation. Finally, genetically engineered BST is virtually identical to the BST naturally produced by cows, usually differing by just one amino acid. BST is present in the resulting milk only in trace amounts and Monsanto argues that BST has no physiological effects on humans as it differs in structure from human growth hormone. In any case, the minute amounts present in milk are digested and so do not pass into the human bloodstream.

2. Arguments against Using Genetically Engineered BST

First of all, who wants or needs more milk? The number of dairy cows kept in Europe and the USA—the two regions where profitable sales of BST are most likely—fell throughout the 1980s and 1990s. This was because the demand for milk failed to keep up with the dramatic increases that took place in milk yields through selective breeding and the use of feed concentrates and other argicultural practices. It has been argued that the widespread introduction of genetically engineered BST would lead to even more farmers being put out of business.

Secondly, while it is true that genetically engineered BST itself almost certainly poses no health risks to humans, its use is linked to significantly raised levels of insulin-like growth factor-1 in the cow's milk. The consequences of this is still controversial. It has been argued that the presence of these high levels of insulin-like growth factor-1, which is chemically identical in cattle and in humans, may trigger breast cancer in women and growth stimulation of cells in the colon. While few scientists regard this possibility as a likely one, Ben Mepham, of the Centre for Applied Bioethics at Nottingham University, has argued that legalisation of commercial use of BST in the absence of more extensive information on these questions could lead to a deterioration in public health if widespread rejection of milk were, ironically, to result.

Thirdly, does the use of BST injections harm the cow's health? Even the manufacturers of genetically engineered BST accept that its use may increase the incidence of mastitis, cystic ovaries, disorders of the uterus, retained placentas, and other health problems including indigestion, bloat, diarrhea, and lesions of the knees. In addition, its use may result in permanent swellings up to 10 cm in diameter at the injection site. Mastitis, as many women know all too well, is a painful inflammation of the mammary glands. It has the same effects in cows as in humans. Mastitis in cows is commonly treated by giving antibiotics to infected animals. Some concern has been raised at the consequences of this for human health, though these fears may be exaggerated as antibiotics have been used on farm animals for decades. A related point is that BST injections possibly put even more pressure on a cow's health in countries where farmers have little or no access to high-concentrate feeds.

Fourthly, while everybody knows that dairy farming is big business, for many people the thought that cows will be artificially stimulated by biweekly injections of genetically engineered BST for most of their lives is somehow off-putting. True, genetically engineered BST is almost identical in structure to natural BST, but to some people it seems wrong that it should be used to boost a cow's BST levels beyond what is normal. Is its use analogous to the force-feeding of geese to produce pâté de foie gras? For some people milk still retains a special aura of freshness and naturalness, perhaps because we all start our lives, once born, by living off milk. This image is tarred by the use of genetically engineered BST. It may be hard to reconcile a belief, albeit a naive one, that milk is a "natural" product with the recognition that genetic engineering is being used to direct the process.

3. The Current Legal Position

By 1996 genetically engineered BST had been licensed for use in a number of countries including South Africa, India, Mexico, Brazil, the former USSR, and the USA. Endless debates have taken place within the European Union, but in December 1994 agriculture ministers from European Union countries agreed to continue the ban on its use until the year 2000.

One of the most remarkable features of the lifting, in the USA, of the ban on the use of genetically engineered BST was that the Food and Drug Administration produced guidelines stating that any company proclaiming that its milk was produced without the use of genetically engineered BST would have to carry a long statement explaining that there is no advantage to BST-free milk. The first two American dairies that advertised their milk as "hormone-free" were promptly sued by Monsanto.

E. The Development of Animal Models for Human Diseases

Genetically engineered mice and rats are being used in increasing numbers as models for human diseases. The reason is that they are extremely useful animals on which to try experimental procedures. They have a very short generation time; large numbers can be kept easily, cheaply, and conveniently in a laboratory; and a great deal is known about their genetics. The basic procedure goes as follows. First, genetically engineer your mice (or rats) so that they mimic a human disease. Secondly, study these altered animals either to investigate the disease or to see if it can be alleviated. The hope, of course, is that what is learned about the disease from the mice or rats will be applicable to humans.

Examples of human diseases for which mouse models exist include albinism, Alzheimer's disease, atherosclerosis, β-thalassaemia, cancers, cystic fibrosis, high blood pressure, Lesch–Nyhan syndrome, muscular dystrophy, severe combined immunodeficiency, and sickle-cell anaemia. The first of these genetically engineered mice was the so-called Harvard oncomouse developed by Philip Leder and his colleagues at Harvard Medical School and patented in 1988. The Harvard oncomouse contains certain human genes which result in the majority of the individual mice in the strain developing cancers. It has to be admitted, though, that the Harvard oncomouse has not been a tremendous commercial success. Du Pont, the company which funded the research, has invested millions of dollars in the project. However, it has failed to persuade a single pharmaceutical company to sign up for deals involving the mice. Du Pont had hoped that it would be able to charge a royalty on anticancer drugs developed through studies using the mice.

However, there are instances where genetically engineered mice are proving more useful. In 1993 scientists from the Imperial Cancer Research Fund in Oxford and the Wellcome Trust at the University of Cambridge found that mice that had been genetically engineered to show symptoms of cystic fibrosis could themselves have their symptoms alleviated through genetic engineering.

Controversy exists, though, as to precisely how valuable or necessary the use of any animals, let alone genetically engineered ones, in medical research is. Some people, including the majority of medical researchers, maintain that their use is essential; others maintain that improvements in alternative approaches (including cell culture, tissue culture, and computer modeling) mean that animals are no longer needed for such work.

With regard to the effects of such procedures on the animals themselves, in the case of the oncomice, common sense, the scientific community, and the courts have all concluded that these animals suffer. Oncomice develop tumors in a variety of places including mammary tissue, blood, skeletal muscle, the lungs, the neck, and the groin. Tumors can lead to severe weight loss (40% body weight or more) while large tumors may ulcerate.

It is hardly surprising that the development and patenting of oncomice has been attacked by a large number of animal welfare and animal rights movements around the world. In addition, religious organizations are increasingly speaking out against the suffering that humans cause to animals. In some religions, such as Christianity and Islam, the animals' points of view have only really been put forth with any strength in recent decades. A number of other religions, though, have a much longer history of according priority to the nonsuffering of animals. In Jainism, the concern for *ahisma* (noninjury) goes hand-in-hand with an insistence on a vegetarian diet, while lay members are encouraged to engage only in occupations that minimize the loss of life. Within Jainism it is the monastic practice to carry a small broom with which gently to remove any living creature before one sits or lies down. In Buddhism too there has traditionally been a strong emphasis on animal well-being.

F. Human Growth Hormone

1. The Role of Human Growth Hormone

Throughout our lives our bodies produce human growth hormone. This hormone is a protein and is produced in the pituitary gland. From here it passes into the bloodstream and is carried around the body. Its main effects are on bones and muscles—growth hormone stimulates cells to increase in size or divide.

Some children produce too little growth hormone and end up much shorter than average, typically around 4 ft. in height. This condition is sometimes referred to as pituitary dwarfism, though the term is often avoided on the grounds that some people find the word "dwarfism" insulting.

In most cases, people with an abnormally low production of growth hormone end up physically quite healthy, with a body that is normally proportioned, albeit unusually small. Until the 1950s there was not anything that could be done to change their height—eating more, for example, has no effect. Then Dr. Maurice Raben at the Tufts New England Medical Center began painstakingly to extract human growth hormone from the pituitary glands of corpses. The hope was that if this was given to people with abnormally low levels of growth hormone production, they might benefit by growing more.

One of the first people treated was a Canadian called Frank Hooey. By the time he saw Maurice Raben, Frank Hooey was aged 17 and was only 4'3" in height—the height of a typical eight and a half year old. Over the next five years he received thrice-weekly injections of human growth hormone. By the end of the treatment he stood at 5'6", slightly below average, but well within the normal range.

Frank Hooey's treatment was a success story. However, it takes around 650 pituitary glands to produce the 2 to 3 g of human growth hormone needed for the 5-year treatment. Because of this, human growth hormone obtained from pituitaries costs far more than gold, and only a relatively small number of people benefited from the procedure.

The advent of genetic engineering has changed all this. Once the gene responsible for the production of human growth hormone was found and isolated, it was a relatively simple matter to insert it into laboratory bacteria and get them to synthesize the protein. The hormone is then collected, checked for purity, and given to people suffering from a shortage of it.

One might think that this is a perfect example of biotechnology in action. Bacteria are being used to replace a missing human protein. As being 4 ft. in height is manifestly disadvantageous, the procedure seems extremely useful and surely only someone with an extreme aversion to biotechnology could object. However, the truth is more complicated.

2. Who Uses Human Growth Hormone?

There are three main categories of people who use injected human growth hormone: children who would otherwise suffer from pituitary dwarfism and end up only about 4 ft. in height; children who would otherwise end up round about 5 ft. in height; and sportspeople.

The first of these categories, children who would otherwise suffer from pituitary dwarfism and be only about 4 ft. in height—are the ones for whom synthetic human growth hormone was originally intended. For such children there is a lot to be said for it. Growth hormone obtained by the old method of extraction from the pituitary glands of countless corpses was almost unobtainable. Further, when it was available, it was occasionally contaminated by viruses. Over a dozen cases are known where children who had received human growth hormone extracted from pituitary glands went on to develop Creutzfeldt–Jakob disease—a fatal condition caused by a virus that can infect human brain tissue.

Only about 1 in 100,000 people are pituitary dwarfs, but there are millions of children who, though taller than pituitary dwarfs, are shorter than average. What has happened since the advent of genetic engineering is that some parents have started to put their children onto programs of human growth hormone injections, hoping thereby that they will end up taller.

It might be supposed that this is not a very serious problem. After all, many parents pay for their children to receive music lessons or tennis coaching. What does it matter if some parents pay for their children to receive human growth hormone treatment?

One problem is the cost of the procedure. A full course of treatment lasts between 5 and 10 years and costs, at 1990's prices, up to $150,000. A second problem is that no one knows for certain if the treatment works. We know that pituitary dwarfs benefit, because the injected growth hormone replaces the missing growth hormone. But there are lots of reasons why children can be below average height, and some pediatricians doubt that injections of human growth hormone will help in all cases. Actual findings are unclear. There are some data which suggest that injecting short children who are not growth hormone deficient does not increase their eventual height. On the other hand, there are other data, obtained by Genentech, the leading manufacturer of human growth hormone, which suggest that growth hormone injections do increase the height of both boys and girls.

A third problem is that once you start the treatment, you have to continue. Stopping growth hormone treatment in children who are not growth hormone deficient before they have reached their adult height may cause them to grow more slowly than they did before treatment. This is because taking the extra growth hormone causes the body to temporarily stop making its own growth hormone.

A fourth problem is that even if the full treatment does cause children to grow a few inches taller, the effects on a child's self-esteem and mental health are unknown. Maybe these will be bolstered. However, it has been suggested that quite the opposite may be the case. The injections may cause children to see themselves as abnormal, with subsequent loss of self-esteem.

A final problem with injected human growth hormone is that a number of independent studies have suggested a possible causal relationship between its long-term use and leukemia. As a result, both Genentech and Eli Lilly, another company that produces human growth hormone, have changed their labeling to indicate this.

It is worth bearing in mind that half of us are shorter than the average. It is true that tall people benefit in all sorts of ways—other things being equal, tall people are more likely to be favored at job interviews, while the taller candidate has won 80% of USA Presidential election campaigns. Surely, though, society should be challenging this bias in favor of the tall rather than conniving with it by allowing parents to spend many tens of thousands of dollars in an effort to enable their children to grow a little taller. As Abby Lippman, a professor at McGill University and Chair of the Human Genetics Committee of the Council for Responsible Genetics, puts it, "Why not lower the hoops on a basketball court?"

A third category of people who are injecting human growth hormone are certain sportspeople. Some athletes have used the hormone in an attempt to increase strength, in much the same way as steroids are used illegally for the same purpose. The consequences of larger than normal doses of human growth hormone are still incompletely known, but it will not be surprising if it turns out there are harmful medical consequences. At the same time the practice, if it works, is unfair on athletes who do not inject themselves.

Recently, human growth hormone has been given to a number of people with AIDS and to large numbers of people over the age of 50. These is considerable, though as yet largely anecdotal, evidence that human growth hormone can increase muscle strength, reduce fat deposition, and reduce depression. Perhaps time will tell whether we really have found the elixir of life or whether there are significant side effects.

G. Gene Therapy

1. Somatic and Germ-Line Therapy

It is helpful to distinguish between two classes of cells found in our bodies: germ-line cells are the cells found in the ovaries of a female and the testes of a male and give rise, respectively, to eggs and to sperm; somatic cells are all the other cells in the body. The importance of this distinction is that any genetic changes to somatic cells cannot be passed onto future generations. On the other hand, changes to germ-line cells can indeed be passed onto children and to succeeding generations.

2. Somatic Gene Therapy for Human Diseases

The first successful attempts to genetically engineer humans were carried out in 1990. These attempts involved patients with a very rare disorder known as severe combined immune deficiency (SCID). In someone with SCID, the immune system does not work. As a result the person is highly susceptible to infections. Children with SCID are sometimes known as bubble babies because, until recently, almost the only way to allow them to live more than a few years was to isolate them in plastic bubbles. These bubbles protect the children from harmful germs but also, poignantly, cut them off from all social contact. In any event, at best the bubbles prolong life by only a few years.

SCID can have a number of causes. One cause is an inherited deficiency in a single protein, adenosine deaminase (ADA). The first person with SCID to be treated with gene therapy was a 4-year-old girl. She was unable to produce ADA, and in 1990 some of her white blood cells were removed and functioning versions of the ADA gene introduced into them using a virus as a vector. The improvement in her condition was remarkable. Five years later she was living a comparatively normal life, attending a normal school, and so on.

By 1995 over 200 trials for somatic gene therapy had been approved. In addition to trials on people with SCID, somatic gene therapy is being tried for β-thalassemia, cancers, cystic fibrosis, Duchenne muscular dystrophy, familial hypercholesterolemia, and hemophilia.

3. The Scope of Somatic Gene Therapy

Somatic gene therapy has been paraded as a possible cure for a great range of medical problems. It should be realized, however, that some human diseases caused by faulty genes can already be treated quite effectively by conventional means. For example, phenylketonuria is a condition which, if untreated, leads to the person being severely mentally retarded. Since 1954, though,

it has been realized that the condition can be entirely prevented by giving children with the faulty gene that causes phenylketonuria a special diet. This illustrates a most important truth about human development: both genes and the environment play essential parts.

A second reason why we should not see gene therapy as the likely solution to all medical problems is that diseases such as cystic fibrosis, phenylketonuria, and sickle-cell disease are the exception, not the rule. These conditions are caused by inborn errors in single genes. However, less than 2% of our total disease load is due to errors in single genes. Most human diseases have a strong environmental component, so that genetic defects merely predispose the person to develop the condition. In addition, the genetic component is usually the result of many genes.

4. Somatic Gene Therapy for Other Human Traits

What of gene therapy to affect traits such as intelligence, beauty, criminality, and sexual preference? Will this ever be practicable? There are frequent reports in the popular press of "a gene for homosexuality" or "a gene for criminality," and it may be that much human behavior has a genetic component to it. However, attempts to find genes for homosexuality, intelligence, beauty, or criminality are, at best, the first steps to understanding the rich and complex ways in which we behave. At worst, they are misguided attempts to stigmatize certain members of society. We are more, far more, than our genes.

5. The Ethical Significance of Somatic Gene Therapy

What new ethical issues are raised by somatic gene therapy? The short answer, when we are talking about real human diseases, is "possibly none." This, for example, was the conclusion reached in the United Kingdom by the government-appointed Clothier Committee which produced its report on the ethics of gene therapy in January 1992. Because somatic gene therapy typically involves giving a person healthy DNA to override the effects of their own malfunctioning DNA, it is widely held that this is not very different from giving a person a blood transfusion or organ transplant. Of course, some individuals may choose not to have a transfusion or transplant, but very few people suggest forbidding them entirely. It is the case, though, that somatic gene therapy can be less reversible than most conventional treatments.

It is also the case that somatic gene therapy has the potential to reduce the number of ethically problematic decisions. At present the only "solution" offered to a woman who is carrying a fetus identified as having a serious genetic disorder such as muscular dystrophy is the possibility of an abortion. Somatic gene therapy may be able to offer a more positive way forward.

However, somatic gene therapy may, in time, raise new ethical issues. Suppose, despite what we have said about the complexities of human behavior, it does eventually transpire that somatic gene "therapy" could reduce the likelihood of someone being violently aggressive or of being sexually attracted to others of the same sex. What then? One answer is to throw one's hands up in horror and agree that such "treatments" should be outlawed. However, one logical problem with this response is that most countries already spend a lot of time and effort trying to get people who have been convicted of violent crimes to be less likely to commit these again. Such people may attend education programs or receive state-funded psychotherapy, for instance, in attempts to achieve these aims. Similarly, some psychiatrists and counselors are still prepared to work with homosexuals to help them change their sexual orientation.

These two examples (violent behavior and sexual orientation) highlight two related issues. The first has to do with the social construction of disease. A disease is, in a sense, a relationship a person has with society. Is being 4 ft. tall a disease? The answer tells us more about a society than it does about an individual of this height. Some conditions are relatively unproblematic in their definition as a disease. For instance, Lesch–Nyhan disease is characterized by severe mental retardation, uncontrolled movements (spasticity), and self-mutilation. No cure is at present available and the person dies, early in life, after what most people would consider an unpleasant existence. It is the existence of conditions such as this that have even led to claims in the courts of wrongful life or wrongful birth where a sufferer, or someone acting on their behalf, sues either their parent(s) or doctor(s) on the grounds that it would have been better for them never to have been born. However, years of campaigning by activists for people with disabilities have shown us the extent to which many diseases or disabilities are as much a reflection of the society in which the person lives as they are the product of the genes and internal environment of that person.

The second issue is to do with consent. It is one thing for a person convicted of a violent crime to give their informed consent to receive psychotherapy or some other treatment aimed at changing their behavior, though even these treatments are, of course, open to abuse. It would be quite another for a parent to decide, on a fetus' or baby's behalf, to let it receive somatic gene therapy to make it less aggressive.

6. Germ-Line Therapy

At the moment it is generally acknowledged that human germ-line therapy is too risky. Researchers cannot, at present, control precisely where new genes are inserted. This raises the not insignificant danger that the inserted gene might damage an existing gene, which could lead to diseases, including cancers. We can note, in passing, that the existence, despite these problems, of germ-line manipulation in animals (i.e., non-human animals) illustrates the distinction between what is generally deemed acceptable for animals and for humans.

However, although human germ-line therapy may currently be too risky, it is difficult to imagine that this will continue to be the case indefinitely. It seems extremely likely that scientists will develop methods of targeting the insertion of new genes with sufficient precision to avoid the problems that presently attend such procedures. Nor need these new methods require much, possibly any, experimentation on human embryos. A great deal, perhaps all, of the information could be obtained through the genetic engineering of farm animals.

Further, we should realize that although germ-line therapy is sometimes referred to as "irrevocable," it is more likely, if we ever get to the point where its use is routine, that it will normally be reversible. There is no reason to suppose that if something went wrong with the results of germ-line therapy, this wrong would necessarily be visited on a person's descendants in perpetuity. The same techniques that will permit targeted germline therapy should permit its reversal.

One can ask whether germ-line therapy is necessary. It is no easy matter to demonstrate that something is "necessary." Value judgments are involved, so that there may be genuine controversy about whether something is needed. Is nuclear power necessary? Or the motor car? Or tigers? Or confidentiality between doctors and their patients? It is likely that most improvements that might result from germ-line therapy could also be effected by somatic gene therapy or conventional medicine. However, it may prove to be the case that germ-line therapy allows some medical conditions to be treated better.

Assuming, then, that one day germ-line therapy is both relatively safe and allows certain medical conditions to be treated better than by other approaches, would it be right or wrong?

It is sometimes argued that germ-line therapy will decrease the amount of genetic variation among people and that this is not a good thing, since evolution needs genetic variation. There are several things that are dubious about this objection. First, empirically, it is difficult to imagine in the foreseeaable future that germ-line therapy is going to significantly decrease the amount of useful human variation among people. Secondly, it is possible that germ-line therapy may one day lead to even more genetic variation—as some parents opt for certain genes in their children and other parents for other genes. Thirdly, the argument that evolution needs genetic variation is difficult to sustain faced with some-one suffering as a result of a disease that is largely the result of a genetic mutation. The argument relies on possible, very distant advantages for groups of people being sufficient to override the more immediate, clear disadvantages for individuals.

Then some people have expressed the fear that germ-line therapy might be used by dictators to produce only certain types of people. The emotive term "eugenics" is often used in this context. Perhaps the major problem with this objection is that it assumes too much of genetic engineering. It is easy to overstate the extent to which humans are controlled by their genes. Dictators have had, do have, and will have far more effective ways of controlling people.

A more likely problem is that germ-line therapy will be permitted before people have grown sufficiently ac-customed to the idea. The pace of technological change is so fast nowadays that some people end up feeling bewildered by new possibilities. The theologian Ian Bar-bour has argued that it is important that sufficient time is allowed before germ-line therapy on humans is per-mitted, both to ascertain, so far as is possible, that the procedure is safe and so that people may feel comfort-able with the idea.

A frequently expressed worry about germ-line ther-apy is the extent to which future generations will be affected. Again, it is possible that this fear may be an exaggerated one. As we have said, we can overestimate the importance of our genetic makeup. Then there is the point that people already have and will continue to have a tremendous influence over future generations through everything from child-rearing patterns and fam-ily planning to books and pollution. The philosophers Robert Nozick, Jonathan Glover, and John Harris have been quite bullish about germ-line therapy, and Nozick,

back in 1974, introduced the notion of a "genetic super-market" at which parents, rather than the state, could choose the genetic makeup of their children. Harris has even argued that we may one day have a duty to carry out germ-line therapy.

There remains the worry, though, born of long expe-rience of slippery slopes, that the road to hell is paved with good intentions. After all, suppose that germ-line therapy allowed people to choose the color of their offspring's skin. For many people, the idea is frightening and abhorrent. Indeed, it is possible that the selective use of such a practice, particularly if the procedure is expensive so that only some people can afford it, might reinforce racist attitudes in society. Despite the diffi-culties of distinguishing in all cases genetic engineering to correct faults (such as cystic fibrosis, hemophilia, or cancers) from genetic engineering to enhance traits (such as intelligence, creativity, athletic prowess, or mu-sical ability), the best way forward may be to ban germ-line therapy intended only to enhance traits, at least until many years of informed debate have taken place.

Bibliography

Bains, W. (1993). "Biotechnology from A to Z." Oxford Univ. Press, Oxford/New York.

Cole-Turner, R. (1993). "The New Genesis: Theology and the Genetic Revolution." Westminster–Knox, Louisville, KY.

Dyson, A., and Harris, J. (Eds.) (1994). "Ethics and Biotechnology." Routledge, London/New York.

Glover, J. (1984). "What Sort of People Should There Be?" Penguin, Harmondsworth Middlesex.

Krimsky, S. (1991). "Biotechnics and Society: The Rise of Industrial Genetics." Praeger, New York.

Levidow, L. (1995). Agricultural biotechnology as clean surgical strike. *Social Text 44* **13**(3), 161–180.

Ministry of Agriculture, Fisheries and Food (1995). "Report of the Committee to Consider the Ethical Implications of Emerging Technologies in the Breeding of Farm Animals." HMSO, London.

Nelson, J. R. (1994). "On the New Frontiers of Genetics and Reli-gion." Eerdmans, Grand Rapids, MI.

Reiss, M. J., and Straughan, R. (1996). "Improving Nature? The Sci-ence and Ethics of Genetic Engineering." Cambridge Univ. Press, Cambridge/New York.

Wheale, P., and McNally, R. (Eds.) (1990). "The Bio-revolution: Cor-nucopia or Pandora's Box." Pluto, London/Winchester, MA.

Birth-Control Technology

JAMES W. KNIGHT

Virginia Polytechnic Institute and State University

GLOSSARY

abortifacient An agent that produces an abortion.

abortion Expulsion from the uterus of an embryo or fetus prior to the stage of viability.

birth control The act or fact of preventing birth.

conception Fertilization; the point at which the sperm cell and the ovum (egg) unite.

conceptus The products of conception; this includes embryo/fetus and placenta.

contraceptive An agent for preventing conception.

embryo The product of conception until approximately the end of the 8th week of pregnancy in humans.

fetus The product of conception from approximately the end of the 8th week (in humans) until birth.

gametes The sex cells; ovum (egg) in the female and spermatozoa (sperm) in the male.

implantation The embedding of the conceptus in the uterine lining, occurring 6 or 7 days after fertilization.

interceptive A birth control product that arrests conceptus development prior to implantation.

ovulation The release of an ovum (egg) from the ovarian follicle.

pregnancy The condition of a female during the period from conception until birth.

progestagen Any agent capable of producing physiological effects similar to those of progesterone. Progestagens generally inhibit ovulation and have other effects on the female reproductive system to prevent pregnancy.

BIRTH CONTROL TECHNOLOGY represents the endless struggle of the human species to alter the most basic design of nature—reproduction and perpetuation of the species. The constancy of our desire to control our reproductive capacities has, throughout history, also been the focal point for some of our most contentious ethical, moral, legal, and political struggles. Perhaps never in history has the need for advances in birth control technology been greater, nor the debate attendant to reproductive control harsher and more divisive, than it is currently as we approach the 21st century.

At present in the United States, over half of all pregnancies are unplanned, the teenage pregnancy rate is more than double that of European countries, the abortion rate is the highest in the developed world, and new cases of sexually transmissible diseases are increasing at the rate of over 12 million per year. These facts would seem to indicate a crying need for development of new and improved methods of birth control, greater and more widespread availability and application of existing products, and enhanced education, especially among the young, on the use of birth control products. However, numerous factors combine to thwart the ability of

Americans to control their reproductive destinies. These factors, as well as a description of contemporary and potential future approaches to birth control, will be discussed in this entry.

I. INTRODUCTION AND OVERVIEW

Efforts to control human reproductive processes, from prehistory until the present day, are characterized by two constants: (1) a continuous but as yet unrealized quest for the perfect contraceptive, and (2) controversy, with reproductive control being the central focus of several contentious ethical, moral, political, and legal issues.

Dichotomies abound. For example, although reproductive capacities play such a significant role in our lives, most people know surprisingly little about the details of reproductive processes, including a lack of understanding as to how the birth control methods that they employ work to prevent pregnancy. Advances in birth control technology beginning in the early 1960s provided American women the reproductive freedom to choose societal roles other than mother and homemaker and accorded to women the freedom to engage in sexual activity without the previous fear of pregnancy. However, the promise of the birth control revolution is yet to be realized, especially in the United States where it began. Indeed, women in the United States now have fewer birth control options than women in other countries. Furthermore, the options available to American women are more expensive and more difficult to obtain. These limitations are reflected in the fact that voluntary sterilization is now the most popular form of birth control in the United States (Table I).

TABLE I   Birth Control Choices in the
United States

Option	%
Female sterilization	29.5
Oral contraceptives	28.5
Condoms	17.7
Male sterilization	12.6
Diaphragm	2.8
Periodic abstinence	2.7
Intrauterine devices	1.4
Other methods	4.8

Source: Contraceptive use in the United States (1989–1990). National Center for Health Statistics, *16,* 95–1250 (Released February 14, 1995)

Limitations in contraceptive availability to the young and the economically disadvantaged, although rarely addressed in the political debate of the issues, are in large measure responsible for the facts that teenage pregnancy in the United States is more than double that in European countries, that the rate of abortion in the United States is one of the highest in developed countries, and that "welfare mothers" are available as a group that can be assailed for political advantage. Both federal and private sector funding for reproductive research has decreased almost to the point of being nonexistent. While the governments of many countries encourage family planning and many subsidize the use of birth control, even the public discussion of birth control tends to be taboo in the United States. Although American culture exploits sex and sexual imagery in all walks of life, a prudish hesitancy to discuss the realities and consequences of sex persists and is perhaps intensifying as we approach the 21st century. In the United States, numerous factors combine to reduce the birth control options currently available, to discourage the greater application of the options that we do have, and to provide disincentives for the development of new technologies.

II. APPROACHES TO BIRTH CONTROL AND MORAL DISTINCTIONS

The World Health Organization (WHO) describes the ideal contraceptive as one that is highly effective, easy to apply, readily reversible, inexpensive, easily distributed, has no serious side effects or risks, and is acceptable in light of the religious, ethical, and cultural background of the user. To complete this utopic vision, one may wish to add that it would also prevent transmission of sexually transmissible diseases. Needless to say, there is no product that can meet those exhalted standards. The dissatisfying reality is that sexually active couples must make compromises and trade-offs among these criteria. This dissatisfaction leads to the misuse or nonuse of birth control products. This, in turn, contributes to the fact that nearly one-half of the over 6 million pregnancies in the United States each year are unplanned, and that approximately 1.6 million of these pregnancies are terminated by an elective abortion.

There are three general approaches to birth control: (1) preventing the formation or release of gametes; (2) preventing fertilization; or (3) preventing the continued development of the embryo at some point following fertilization. Although many individuals erroneously use *contraception* and *birth control* as interchangeable terms, we should not the distinction between them. While the first two approaches to birth control just described truly are contraceptive in nature, that is, the union of the ovum and the sperm cell is prevented, the third is not. Depending upon the point of embryonic development targeted, the third approach may be characterized as either *interceptive* or *abortifacient*. Although some individuals may consider the distinctions among contraceptives, interceptives, abortifacients, and induced abortion methods to be academic, to others they represent a morally and ethically crucial distinction. These distinctions largely center around the age-old question, "When does life begin?"

Although most people consider pregnancy to begin at fertilization, many medical practitioners consider pregnancy to begin at implantation (approximately 6 or 7 days after fertilization). This view is supported by the fact that prior to implantation, the tissues of the conceptus are not yet differentiated into those that will give rise to the placenta and those that will become the embryo and subsequently the fetus. Thus, according to this view, until this differentiation occurs there is not yet an embryo. It then follows from this reasoning that any birth control product that intercepts and arrests development prior to implantation is an interceptive and not an abortifacient product. This view would then imply that it is semantically impossible to consider a birth control product to be abortifacient or to induce an abortion prior to implantation. In more common usage, the term embryo is used both before and after implantation. To some, yet another distinction may be made between utilizing a potentially abortifacient product prior to having confirmation of pregnancy and the conscious termination of a known pregnancy by an induced abortion.

Two moral issues merit further elaboration. First, individuals who hold the "life begins at fertilization" view oppose destruction at any point following conception of what they understand to be a developing human being. This view generally holds as morally crucial that a unique human life begins at fertilization and hence that any intervention destructive of that life counts as aborting that life and is morally wrong. To those who hold this view, the fact that the precursor cells of an embryo do not differentiate until after implantation is irrelevant.

Second, distinguishing between true contraceptives and products that *may* employ interceptive or abortifacient modes of action is a matter of considerable importance to many women and men. While many people have no moral objection to preventing conception, they do find the termination of a potential life at any point after fertilization to be morally objectionable. For this reason, individuals need to understand not only the primary but also the *potential* secondary and tertiary modes of action of the various birth control products as they

choose a product for usage. For example, oral contraceptives generally do prevent ovulation. However, so-called "breakthrough" ovulation does sometimes occur and fertilization may result. However, the pregnancy commonly will fail due to other alterations induced in the reproductive tract by the pill. Likewise, an intrauterine device (IUD) may disrupt sperm transport and survival, thereby preventing fertilization. However, fertilization does often occur but the conceptus does not survive due to the disruption within the uterus induced by the IUD. In both examples, the woman would have no way of knowing which mode of action (i.e., contraceptive or interceptive/abortifacient) was the effective one. She would only know that her menstrual cycle recurred as expected.

III. CONTEMPORARY METHODS OF BIRTH CONTROL

A. Steroidal Inhibitors of Ovulation

1. Oral Contraceptives (i.e., "The Pill")

The first oral contraceptive (Enovid, G. D. Searle Pharmaceuticals) was approved by the Food and Drug Administration (FDA) in June 1960. Shortly thereafter, several other pharmaceutical companies rushed their products onto the U.S. market. In essence, this event heralded the onset of the sexual revolution. With the advent of the pill, women at long last had a neat, simple, private, tidy, and highly effective way to avoid unwanted pregnancy. No longer did women who chose to be sexually active have to carry awkward and embarrassing (and unreliable) birth control devices, such as diaphragms and tubes of jelly, nor did they have to depend upon the conscientiousness to their male partners to provide protection. The previously unavailable freedom to enjoy a spontaneous sexual encounter without fear of unwanted pregnancy gave women an approximation of procreational sexual equality with men.

Perhaps partly because of the continued reference to oral contraceptives as *the pill,* many people tend to erroneously assume that there is only a single formulation. There are actually dozens of different formulations among the over 40 different brands on the U.S. market at present. These formulations differ in the particular type of synthetic progestagens and estrogens used, relative quantities of these steroid hormones, overall potency, potential side effects, and relative contraceptive effectiveness. Potential complications and adverse side effects of oral contraceptives can be minimized if care is taken to select the specific pill formulation that is most suitable for each particular woman based upon her age, overall medical history, menstrual history, physical condition, behavioral and health habits, and other individuating factors. This is perhaps the most practical argument against the over-the-counter availability of birth control pills, as some individuals and groups have advocated as a means to increase their availability and usage and potentially decrease their price.

There are two basic categories of oral contraceptives, the combination pill (consisting of a combination of a synthetic progestogen and a synthetic estrogen) and the minipill (progestogen only). The active components of the pill are simply synthetic steroid hormones that are similar to the endogenous hormones progesterone and estrogen produced by a woman's ovaries during her menstrual cycle and which regulate the function of her reproductive tract.

The traditional combined formulation is a constant hormonal dose taken daily for 21 days and followed by a week during which no pills are taken. Many brands include seven inert pills containing iron and vitamins that are taken during the final week of a pill cycle. The decrease in systemic hormone levels that begins after the completion of each 21-day regimen induces the changes leading to menses. These changes are analogous to the decrease in hormones that normally occur at the end of an unregulated menstrual cycle. A new cycle of pills will begin 5 days after the onset of menses.

In simple terms, the hormones in the ingested pill suppress the development of follicles on the ovaries, prevent the maturation of the ovum within the follicle, and inhibit the normal sequence of changes essential for ovulation. Because hormone levels provided by the pill are out of synchrony with normal menstrual cycle events, assorted other changes are also induced throughout the reproductive tract. For example, the characteristics of the mucus produced by the cervix is altered, making it more difficult for sperm to penetrate it. The biochemical secretions within the fallopian tube (the site of fertilization) are altered, and this alteration may result in either the failure of fertilization or the inability for proper development of the early conceptus before it moves into the uterus. Changes induced within the uterus may prevent implantation even in the (rare) instances when the other modes of action fail to be effective.

Biphasic and triphasic pills are a refinement of the combined constant formulation pills. As their names imply, rather than all 21 pills containing a constant hormone dose, biphasic pills employ two dosage levels and triphasic pills three dosage levels due the 21-day regimen. A higher dose of the hormones is contained in the pills taken during the time at which ovulation would normally occur, and a lower dose or doses at other days. One advantage of the phasic pills is the lower total amount of hormones consumed during each cycle. How-

ever, because all pills in a packet are not the same, a woman employing the phasic pills must carefully follow the prescribed sequence.

The minipill is a low-dose progestogen-only pill taken daily without a break. With the lower dosage of the minipill, ovulation is more likely to occur than in women taking the combined formulations. The minipill is especially useful for women who should not be using estrogen, such as women with a history of cardiovascular disease, hypertension, and diabetes. Since the combined pills may have an inhibitory effect on milk production, the minipill is also useful for women who are lactating.

The most significant advancement in oral contraceptives since the initial formulations of the early 1960s has been the enormous reduction in the hormone dosage. Due to the relatively short development and testing period prior to introduction of the pill, the initial pills contained extremely high levels of synthetic hormones in order to ensure their contraceptive effectiveness. Potential long-term effects of these high doses were not known. By the 1970s, evidence linking the pill with an increased statistical risk of cardiovascular disease, stroke, and assorted other health problems in women who began taking the extremely high-dose pills in the early 1960s began to accumulate. Soon after the first oral contraceptives were marketed, researchers began testing formulation with ever-decreasing levels of hormones to determine minimal effective dosages. The present-day pills contain only a small fraction of the hormone dose of those early formulations. Therefore, the health risks of the contemporary pill are far less than those from the early pill.

In addition to its 97–99% effectiveness for pregnancy prevention, the pill has many other benefits, including a more regular, less painful menses entailing a lower blood loss, a decreased likelihood of ovarian cysts and pelvic inflammatory disease, and a decreased risk for ovarian and uterine cancer. Potential negative side effects include a greater statistical risk for blood clots, heart attack, and stroke (especially among women who smoke cigarettes). Although there is not consensus among researchers, some reports indicate that taking oral contraceptives increase the risk for breast cancer, especially among women who begin to use the pill before age 18 and take it for more than 10 years. More minor side effects may include weight gain, mood changes, mild headaches, nausea, temporary irregular menstrual bleeding, and breast tenderness. Issues of potential side effects are clouded by the fact that there are so many different types of oral contraceptives and that there are complex interactions with individuating factors such as family medical history, age, smoking, and consumption of alcohol. Therefore, any generalization of what "the pill causes" must be viewed with skepticism.

Potential negative side effects of the pill are apparently exaggerated in the minds of many women. Perhaps this is attributable to the extensive media coverage given to the negative effects of the early formulations of the pill. A 1995 Gallup poll commissioned by the American College of Obstetricians and Gynecologists found that 75% of women surveyed believed that the pill caused "serious health problems." One-third of the respondents believed that the pill caused cancer and nearly as many felt that it caused heart attacks and strokes. No doubt these erroneous beliefs contribute to many women not using the pill.

2. Subdermal Implants (Norplant)

Although research on the concept of using a silastic implant to continuously deliver a progestagen began in 1967, these implants were not available in the United States until 1990. This was almost a decade after they were available in other countries. The concept of this product, marketed under the brand name of "Norplant" by Wyeth-Ayerst Laboratories, is a simple one. By placing six matchstick-sized silastic capsules filled with the progestagen levonorgestrel in a fan-shaped formation under the skin (generally between a woman's elbow and armpit), the hormone is released in a slow and constant pace, and the daily fluctuations in blood levels that occur with daily ingestion of a pill are avoided. Following placement, the implant provides highly effective birth control protection for approximately 5 years.

The Norplant system overcomes the reason for the greatest number of failures with the pill: the failure to remember to take it daily. Additionally, due to the constant rate of slow release, the total amount of progestagen released over any given time period is less than that from taking a daily pill. Hence, long-term side effects due to exposure to the synthetic hormones should be reduced.

In essence, Norplants merely represent an alternative means to oral consumption for delivering a synthetic hormone that controls the function of the reproductive tract for the prevention of pregnancy. However, because of the lower amount of the progestagen in the woman's circulation at any given time, ovulation is not always suppressed. However, the other alterations of the reproductive system (as described earlier for the pill) prevent pregnancy in a highly efficient manner. For women to whom the distinction between true contraception and birth control *potentially* due to interceptive effects is important, it should be noted that the likelihood of fertilization (but not implantation) is greater with the Norplant system compared to combination oral contra-

TABLE II Lowest Expected (Theoretical) and Typical Failure Rates of Birth Control Options[a]

Method	Lowest expected	Typical
Total abstinence	0	?
Male sterilization	0.1	0.15
Female sterilization	0.2	0.4
Depo-Provera	0.3	0.3
Norplant	0.1	0.3
Oral contraceptives		
Combined pills	0.1	3.0
Minipills	0.5	3.0
Intrauterine device		
Progestasert	1.5	2.0
Copper T 380A	0.6	0.8
Condom (male)	2.0	12.0
Condom (female)	5.0	26.0
Diaphragm	6.0	18.0
Cervical cap	6.0	18.0
Spermicides (alone)	3.0	24.0
Periodic abstinence	1.0–9.0	20.0
Withdrawal	4.0	19.0
"Morning after" pill	—	25.0
Chance	85	85

[a] Expressed as % women experiencing an accidental pregnancy in the first year of continuous use.

ceptives. Since potential user error does not figure into their application, the actual and theoretical effectiveness of Norplants (over 99%) are identical, a major advantage over the pill (Table II).

The initial cheers from American women for the long-awaited arrival of a potentially revolutionary improvement in birth control that greeted Norplant in 1990 has faded to jeers from many of the more than 1 million women who have tried it, and a firestorm of controversy on several fronts now cloud its once rosy future. The first controversial issue was that of cost. Although the actual cost of the capsules containing the steroid is minimal, the company prices the Norplant kits at $345 per patient (A. G. Rosenfield, 1994. *J. Reprod. Med.* **39,** 337–342). And, although the implantation of the capsules is a simple procedure completed in 10 min or so, the average cost to the patient is $600. One reason often cited for this inflated price is the common one for all new pharmaceutical products—the necessity of recovering research and development cost. However, that claim is questionable. Trials on Norplants were conducted in the 1970s and the product was available in at least 16 other countries for up to nearly a decade and in use by over 500,000 women before it was marketed in the United States. Norplant is an excellent example of the primary financial disincentive that continues to make pharmaceutical companies hesitant to introduce new birth control products in the United States. Simply stated, if a company can sell a tested, accepted, and highly profitable product to a consumer on a monthly basis (birth control pills cost approximately 15 cents per cycle but are marketed at approximately $20 per cycle), why sell her a product that she only has to purchase once every 5 years unless it can be sold to her at a somewhat comparable profit? This is especially so when a company must be concerned with potential lawsuits from product liability claims in our highly litigious society. And this fear has been realized.

Although the potential side effects of Norplant are similar to those of the pill (e.g., weight gains, mood swings, headaches, and irregular menstrual bleeding), many women report them to be of a more severe degree. Menstrual disturbances, ranging from no menstrual period at all to continuous spotting, are common to all progestagen-only systems (including the minipill) and an unavoidable consequence of the constant delivery of progestagen without estrogen. The considerable advantage of having no estrogen in the delivery system is a reduced risk of the more serious side effects of circulatory disorders, blood clots, strokes, and perhaps breast cancer. Some Norplant users have reported infections and inflammations at the implant site. Surgical side effects unique to Norplant have caused the greatest dissatisfaction. Since the capsules are not biodegradable, they must be surgically removed. This removal has proved problematic with reports that one removal in five takes an hour or more and that one in four is painful (Nov. 27, 1995. *Newsweek*, p. 52).

Although Wyeth-Ayerst has always listed the potential negative side effects in its literature for Norplant, critics have claimed that these side effects have been downplayed in its marketing of the product. Perhaps there are many individuals who naively assume that all pharmaceutical products approved for marketing are free of any adverse side effects and who are unwilling to make the necessary trade-offs and accept the reality that no birth control product is ideal. However, the surgical complications of Norplant removal were apparently unexpected, even by doctors who implanted the devices. According to a report in *Newsweek* magazine (Nov. 27, 1995. p. 52), in the first 3 years that Norplant was on the market in the United States, there were fewer than 20 out of 800,000 users who filed legal complaints; now over 50,000 women are suing the company. This has led to a precipitous decline in Norplant sales over the past year. The FDA, the WHO, and numerous medical authorities emphatically conclude that Nor-

plant is a safe and highly effective contraceptive and that at least 90% of the women who use it are satisfied. Certainly the fear of lawsuits has been a major reason why pharmaceutical companies have been hesitant to develop and market new birth control options in the United States. The legal assault on Norplant confirms that these fears were justified and clearly serves as a disincentive for future innovations.

The most prickly controversy associated with the Norplant system of birth control is the numerous instances of forcing or attempting to force women to use it as a means of mandating control of the procreational liberty of women as judicial punishment or as a condition to receive government benefits including welfare and public education. Since economically disadvantaged women of color have been the most frequent targets of these mandates, highly inflammatory issues of racism and discrimination are at the center of this ethical debate.

3. Injectables

In June 1993, the FDA approved the use of the injectable progestagen Depo-Provera, another "new" product to the United States market with a long history of investigation as a contraceptive dating back to 1963. Although not officially approved for contraceptive purposes until 1993, this progestagen (medroxyprogesterone acetate, MPA) had already been marketed by the Upjohn Company for 20 years as a palliative treatment for endometrial (uterine lining) cancer. Prior to U.S. approval, Depo-Provera was already being utilized in approximately 90 countries for contraception. No doubt some physicians in the United States had unofficially utilized it for that purpose.

Depo-Provera is administered as an intramuscular injection and provides highly effective (over 99%) birth control protection for a 3-month period. Each of four annual injections costs approximately $30, making it similar in annual cost to the Norplant system and less than the pill. Like Norplant, its actual effectiveness is higher than that of the pill because the user need not do anything during the 3-month period of protection (Table II). Depo-Provera is not readily reversible. Should a woman desire to become pregnant after the injection is administered, she can only wait for it to "wear off" before fertility is restored. The 3-month period of fertility suppression is a conservative estimate, and if a woman desires to become pregnant within a year or two, Depo-Provera would not be a good contraceptive choice.

The mechanisms of action of Depo-Provera and its potential side effects (positive and negative) are similar to those of the pill and Norplant. A study by officials of the WHO suggests that users of Depo-Provera may have a higher than normal statistical risk of breast cancer. However, women who use Depo-Provera have a significantly lower risk of uterine cancer and "come out ahead" when all cancer risks are tallied.

There are other synthetic progestagens (norethisterone enanthate, NET-EN, being the most widely studied) and combined progestagen and estrogen formulations that have been studied since the early 1960s to the present day and that are utilized in other countries as injectables, providing varying durations of protection. None of these have any significant overall advantages compared with Depo-Provera.

B. Inhibitors of Fertilization

1. Physical Barriers

a. Male Condoms

Condoms are truly an ancient birth control device, having been utilized in one form or another for centuries for the dual purposes for which they remain effective— preventing the spread of sexually transmissible diseases and for pregnancy prevention. Gabriello Fallopius (1523–1562), an Italian anatomist perhaps best known for his description of the fallopian tubes (oviducts of the female) that bear his name, is credited with the first published description of the condom. Use of his "linen sheath" has been credited with helping to stem the spread of syphilis in Europe in the 16th century. Development of the process for vulcanization of rubber in the mid-1880s that permitted mass production of condoms led to a tremendous surge in their popularity and usage.

Condoms are the third most widely used form of birth control in the United States, ranking behind only female sterilization and the pill (Table I). Because they are widely available and easy to obtain, relatively low in cost, have no side effects (other than for a very few individuals who are allergic to latex), and have no equal for preventing the spread of microbes that are responsible for sexually transmissible diseases (STDs), condoms play an especially important role as the primary birth control choice among sexually active young people. Condoms represent one of the few contraceptives readily available to minors. And condoms are the most logical birth control option for individuals who are not "pre-prepared" for sexual activity with a continuous protection method (such as taking the pill). Even if other methods of birth control are used, it is certainly advisable to also use a condom in any sexual encounter other than a monogamous relationship between two individuals known to be disease free. One criticism aimed at making hormonal methods of birth control

available to young women with multiple sex partners is that they are less likely to insist that their partners use a condom, either because they feel safe in regard to their primary concern, preventing pregnancy, or because they erroneously believe that the contraceptive will also provide protection against disease transmission. Unfortunately, surveys of people with multiple sex partners— the population most susceptible for acquiring STDs— indicate that less than 10% of them use condoms during every sex act.

The chilling increase in both heterosexual and homosexual male transmission of AIDS (Acquired Immunodeficiency Disease) during the past 15 years has played a major role in spurring the increase in sales of condoms (which now totals over 450 million a year in the United States alone; May, 1995, *Consumer Rep.*) and in hastening development of more brands and options. If properly used, latex condoms reduce to near zero the risk of transmission of STDs, including the human immunodeficiency virus (HIV) that causes AIDS. One bit of good news relative to public education is a survey by the Centers for Disease Control and Prevention which found that three of every four Americans do know that latex condoms, if used consistently and correctly, will prervent the transmission of HIV and other STDs. The obvious problem, however, lies in the "if" clause.

Somewhat ironically, the oldest of our contraceptives is perhaps the one that has undergone the most dramatic improvement and received the greatest amount of research and development attention over the past two decades. Well over 100 brands of condoms are available in the United States in a dazzling array of options (nonlubricated, lubricated, spermicidal, snug, extra-large, contoured, sensitive, ribbed, studded, clear, colored, scented, etc.).

The gap between theoretical and actual effectiveness is greater for the condom than for any other birth control method (Table II). Theoretical effectiveness (effectiveness of the method itself *if* properly used and apart from user error) for the condom is quite high, 97–99%. However, actual effectiveness (effectiveness based upon subtracting from 100 the percent of women who use a given method but nonetheless become pregnant in a year's time) is only approximately 88%. Reasons for the disparity in the two figures include the facts that couples do not use condoms every time they have intercourse and that they do not follow the instructions for proper use of the condom. Breakage of the condom occurs far less frequently than does user error.

The more contentious issues associated with condoms deal with their advertising and the distribution of them, especially to minors. Although condoms are now advertised on cable television, condom ads continue to be banned on network television. This is, of course, despite the ceaseless use of blatant sexuality in advertising and the implicit and often explicit sexual encounters protrayed in television programs. While all of the potential sociological implications will not be explored here, it is certainly not surprising that condom usage is so low among at risk individuals when they are continuously exposed to the portrayal of sex on television (and in movies) as being without consequences and without ever seeing responsible sexual interaction and an intelligent discussion of protection preceding a sexual encounter modeled for them. This is compounded by the Victorian attitude toward sex education in many school systems and the reticence (or inability) of many parents to openly and honestly discuss sexual matters with their children.

b. Female Condom

The female condom is a soft, loose-fitting sheath that lines the vagina. It is made of polyurethane (not latex like the male condom) and has a semistiff plastic ring at each end. The inner ring is used to insert the device inside the vagina and serves as an internal anchor. The outer ring covers the labia area and the base of the penis during intercourse. Following a year of selected test marketing to determine acceptance, the female condom was released for widespread United States marketing by Wisconsin Pharmaceuticals in August 1994 under the brand name Reality. In Europe and Canada, the product was already available under the name Femidom.

Other than the fact that it accords women an over-the-counter product of which they have control, the female condom has few advantages and many disadvantages compared with the male condom. One advantage is that the polyurethane membrane is 40% stronger than latex; resistant to oils, allowing the use of an oil-soluble (rather than water-soluble) lubricant that may be preferred by some couples; and transfers heat and gives a more natural feel than does latex. It is also an option for those allergic to latex. However, it costs considerably more than does the male condom for a one time usage (approximately $2.50 each), requires extensive practice to properly insert, may slip around and need to be repositioned several times during intercourse, and to many is aesthetically unappealing. More importantly, although the contraceptive effectiveness under normal use was expected to be around 87% (similar to the actual effectiveness of the male condom), initial results indicate a pregnancy rate among users of 26% (74% effectiveness). This is no doubt at least partly due to the lack of familiarity with using this rather cumbersome device. Another very important point is that the female condom is not considered to provide protection to either party against STDs, including HIV transmission.

c. Diaphragm

The vaginal diaphragm, a soft, bowl-shaped rubber dome with a flexible metal rim, is an ancient female barrier contraceptive that has remained largely unchanged since the late 19th century. The diaphragm, which must be custom fitted to the particular measurements of each individual woman, is positioned in the posterior vagina, covering the opening to the cervix and serving as a physical barrier to sperm entry. Diaphragms should always be used with spermicidal gel or cream. The spermicide (generally nonoxynol-9) kills or immobilizes a substantial number of the 200 million or so sperm present in the typical ejaculate. It should be realized that regardless of the method in which spermicides are used, they only function to reduce sperm numbers and never totally eliminate all of the sperm. Actual effectiveness, even when used with spermicide, is low (80–83%). The diaphragm is, simply stated, a cumbersome, messy, inconvenient, and outdated device. Its many disadvantages are well known, especially to women who have used it.

d. Cervical Cap

The cervical cap is perhaps the perfect symbol for the state and rate of birth control progress in the United States. The cap, a small, thimble-shaped device with a small amount of spermicidal gel inside the cap, is placed directly over the cervix. It was first approved by the FDA in mid-1988. However, the cervical cap was developed in 1838, 150 years earlier, by the German gynecologist Friedrich Wilde, and has long been available throughout most of the world. It too must be custom fitted to each woman and its effectiveness (or lack thereof) is similar to the diaphragm. It does have the advantage that it can be left in place without the need for application of additional spermicide from one menstruation to just before the start of the next.

e. Contraceptive Sponge

The Today contraceptive vaginal sponge was withdrawn from the United States market by its manufacturer, Whitehall-Robins Healthcare, in early 1995. It was a polyurethane foam device saturated with spermicide and sold over-the-counter. Although it absorbed some of the ejaculate and served as a partial barrier to sperm transport, its primary mode of action was the release of spermicide (nonoxynol-9). Its reported actual effectiveness when used alone was low (75–83%), and it had numerous drawbacks, including vaginal irritation, discomfort to both partners, and causing an increased risk for toxic shock syndrome. The reason provided by the company at the time of withdrawal was that complying with more stringent FDA manufacturing and testing requirements would require modification of the manufacturing process and that these changes would take several years and result in pricing the product beyond the reach of many users.

2. Chemical Barriers: Spermicides

When used alone, the spermicidal foams, creams, gels, suppositories, and tablets currently available are only 70–76% effective at pregnancy prevention. They do add an element of effectiveness when used in combination with condoms and diaphragms. Spermicides are also useful in reducing the likelihood of transmission of agents responsible for STDs, including HIV. Agents that inhibit sperm enzymes, thereby altering sperm motility, and that interfere with other metabolic functions of sperm are under investigation but are not likely to be available in the near future.

C. Interceptive and Abortifacient Methods

1. Intrauterine Devices

The intrauterine device (IUD) has an ancient, storied, and stormy history. Although inserting some sort of object into the female reproductive tract, including the uterus, dates to antiquity, the "modern" IUD as a birth control device traces to the early 1900s. Numerous materials, designs, sizes, and shapes have been utilized over the years. The primary mode of action for all IUDs is the induction of a foreign-body reaction within the uterus that disrupts numerous reproductive events, including sperm transport, sperm capacitation, ovum transport, fertilization, conceptus movement into the uterus, and implantation. Many types also incorporate metallic ions (especially copper) that enhance their disruptive effect on sperm on progestagens that are released to elicit both localized and systemic effects. Because of the large number of events that may be disrupted, IUDs are extremely effective (98–99+% effectiveness). As described earlier, one would never know which of the potential actions of the IUD led to preventing pregnancy during any particular menstrual cycle. Research studies indicate that fertilization does occur in a significant number of cycles but that pregnancy fails prior to implantation. Women to whom that moral distinction matters should be aware of that fact before considering the use of an IUD. At present, only about 2% of women in the United States who employ birth control use the IUD. Worldwide it is one of the most commonly used options.

Despite a history of most brands of IUDs providing an effective and relatively safe birth control option, IUDs were withdrawn from the U.S. market in February 1986, and only recently have they reappeared. Their removal was largely due to the highly justified outrage and lawsuits directed against one of the many types of IUDs, the ill-designed Dalkon Shield manufactured by A. H. Robins Company. The Dalkon Shield was a crab-shaped device with sharp "legs" perfect for harboring bacteria picked up from the vagina during insertion and then injecting the bacteria into the uterin lining. This led to a painful infection that resulted in a large number of cases of pelvic inflammatory disease (PID), infertility, and at least 14 deaths. In addition, the tail of the Dalkon Shield (the string that normally trails from the IUD through the cervix and into the vagina), acted rather like a wick to continually pull bacteria from the vagina into the uterus, perpetuating the infection. The Dalkon shield was voluntarily withdrawn from the market in 1975. The predictable consequence of the lawsuits appropriately lodged against the Dalkon Shield and A. H. Robins Company (that led to their filing for Chapter 11 bankruptcy in August 1985) was a surge in lawsuits against *all* manufacturers of IUDs. Many users and potential users of IUDs failed to draw a distinction between the Dalkon shield and the many other safer varieties of IUDs. This led to a decline in sales of all IUDs and an increase in lawsuits against other manufacturers. The predictable business decision of IUD manufacturers was to remove their products from the U.S. market.

Currently there are two types of IUDs available in the United States, one that releases copper (Copper T 380A) and one that releases a progestagen (Progestasert). Intrauterine release of a progestagen has the advantage of reducing menstrual blood loss, hence overcoming one of the major negative side effects of a greater menstrual flow typical of earlier IUDs.

Certainly the IUD is not an appropriate choice for all women; however, it may be the best choice for some women. Many women experience intense cramping and spontaneously expel their IUD. It is generally most appropriate for older women who have previously had a child (the interior of their uterus is usually larger) and who are in a monogamous relationship (and hence at a greatly reduced risk for PID).

2. Postcoital Contraception (the "Morning After" Pill)

Semantical inconsistencies abound in most discussions of this method. First, as discussed previously, use of the terms "postcoital contraception" and "emergency contraceptive" are oxymoronic when discussing an interceptive method of birth control. Second, the "morning after" pill may be used up to 72 hr after unprotected intercourse (in fact, the window of effectiveness is probably even greater than 72 hr). Third, the treatment regimen typically entails four pills, not a single pill. Fourth, *all* "regular" birth control pills may be used for "emergency" purposes if administered in sufficiently high doses.

The more typical "morning after" pill is a regimen of four high-dose combined progestagen–estrogen oral contraceptives (most commonly Ovral) taken in pairs 12 hr apart beginning within 72 hr after unprotected intercourse. It is estimated to be about 75% effective in preventing pregnancy. This high dose of hormones disrupts a number of reproductive processes that are likely to interfere with either fertilization or conceptus development. Given the timing of application, the woman does not know if conception did or did not occur. This is its primary distinguishing trait from use of a product, such as RU-486, that aborts a known pregnancy.

As implied in the monikers describing this method, it is not considered a method to be used on a routine basis. In addition to being used in cases of rape, it may be appropriate as a "backup" system for cases such as condom breakage, diaphragm slippage, or simply unprotected intercourse. It should be noted that uses of high doses of the pill for this purpose have been recognized as effective for at least 25 years.

University health centers have been primary and more highly publicized dispensers of this method. Especially since student fees paid by all students support these centers and many students object to what they term a form of abortion, the availability of this service has been a recent controversial issues on many campuses.

In a somewhat unusual proactive move, an FDA advisory panel announced in June 1996 that it considered oral contraceptives to be safe and effective when used as "morning after" pills. They largely based their conclusion on the fact that for a decade women in Europe have been using the pill for that purpose. Although the FDA was clearly letting drug companies know that it would welcome an application to market birth control pills for "emergency" purposes in the United States, as of September 1996 no companies have applied to do so. Once again, fears of liability suits, negative publicity, and/or product boycotts by antiabortion groups make pharmaceutical companies reluctant to "take the bait" dangled before them by the FDA.

3. RU-486 (Mifepristone)

Not since the controversy surrounding the initial introduction of the pill has there been a birth control option that has caused the degree of excitement, optimism, outrage, condemnation, and protest as that created by RU-486. Elements of that outrage and controversy may largely be attributed to the success of those who oppose the availability of RU-486 in the United States in having it associated in the minds of many simply as "the abortion pill." Indeed, one rarely sees or hears the product discussed in the media without the moniker, *the abortion pill,* immediately written or spoken following *RU-486.* Although it can be used to induce a postimplantation abortion, RU-486 has many other potential applications that generally get ignored and obscured by the heated debate that follows the mention of this one, and no doubt which many believe to be its only, potential application.

RU-486 (the RU refers to its manufacturer, the French company Roussel-Uclaf, and the number 486 is simply an in-house designation) has the unique ability to bind to cellular receptors for the hormone progesterone at a five times greater affinity than does progesterone itself. Progesterone is essential for the maintenance of pregnancy, and, simply stated, progesterone cannot induce the changes within the uterus that are necessary for pregnancy to continue if RU-486 is occupying its receptors. Hence, the uterine lining is sloughed off, just as during a menstrual period. Since the developing conceptus requires attachment to the uterine lining for continued development, it too would be expelled. In other words, RU-486 has induced an abortion. RU-486 is generally combined with a prostaglandin, a natural product the makes the uterus contract and aids in the sloughing of the uterine contents (prostaglandins are what cause menstrual cramps). Although it can be successfully used later in pregnancy, RU-486 is generally used for induction of abortion within the first 2 months of pregnancy. RU-486 offers numerous advantages as an alternative to traditional aspiration abortions. One practical advantage is that abortion access would not be limited to the relatively few specialized abortion clinics that are currently easily targeted for protests and violence by those opposed to abortion. It is not the purpose of this chapter to address the issue of abortion, and abortion is not and should not be regarded as a method of birth control. In the best of circumstances, abortion is a physically and psychologically traumatic experience for most women. However, by all objective measures, the effectiveness (96–98%) of RU-486 as a product for elective induction of abortion and the relative absence of short-term and long-term health complications make it a highly attractive alternative to traditional methods of abortion.

However, since RU-486 binds to progesterone receptors whenever and wherever it finds them, it could also be used as an interceptive agent before pregnancy has ever progressed to the point of being established or confirmed. Its use as a "once a month" product could induce a menstrual period at its expected time—which may or may not entail an abortifacient action. Scenarios can be developed in which it could be used as a true contraceptive. Perhaps even more important are the numerous other potential and less controversial applications of RU-486 that are vitally important to women's health. For example, it may be a valuable treatment for breast cancer. As many as one-third of the cases of breast cancers contain an abundance of progesterone receptors and these tumors require progesterone to survive. Endometrial cancers and the painful excessive proliferation of endometrial cells known as endometriosis (the third leading cause of infertility in the United States) can be successfully treated with RU-486. RU-486 can cause dilation of the cervix, allowing vaginal delivery in women who now must have a cesarean delivery. Cushing's disease, a disorder of the adrenal glands affecting both women and men that often necessitates removal of the glands, also responds to RU-486. The bottom line is that the health of American women has been imperiled because RU-486 has been banned due to *one* of its many potential applications. If it were available, RU-486 could have saved the lives of or at least benefited the health of thousands of American women. The fact that it remains unavailable is largely, if not solely, due to the hesitancy of politicians to handle the hot-potato issue of abortion.

RU-486 was first made available in France in 1988. Since that time, it has become widely available in numerous countries throughout the world. In France and the United Kingdom, over 200,000 women have had medical abortions with RU-486. Just as politicians wish to avoid the issue of abortion, so too do pharmaceutical companies. The American patent rights to RU-486 have been given to the nonprofit Population Council. No pharmaceutical company was willing to take on the controversy and liability that would greet any future attempts to market RU-486 in the United States. Clinical trials on Ru-486, renamed mifepristone, were completed in September 1995 by the Population Council. More than 2100 women who were less than 9 weeks pregnant participated in trials at 17 clinics throughout the United States. The Population Council reported "overwhelming satisfaction" among participants (Spring, 1996. *Planned Parenthood Today*).

A recommendation for approval of RU-486 for use in the United States was filed with the FDA by the Population Council in the summer of 1996. A panel of scientific advisors to the FDA voted unanimously to recommend approval (with conditions) of RU-486 on July 19, 1996. However, a final decision by the FDA on the marketing of RU-486 in the United States is pending as we go to press.

4. Methotrexate

Methotrexate is a product that is already on the market and approved by the FDA for a number of purposes, including the treatment of tumors and cancers. It was first reported in late 1993 that methotrexate could also be used to induce abortion. Since that time, numerous researchers have experimentally investigated the drug and confirmed its effectiveness as an alternative to RU-486 for induction of a nonsurgical abortion. Many scientists believe that it has outstanding potential as an abortifacient agent. Since methotrexate is already approved by the FDA for other purposes, only an application to the FDA for "supplemental indication" is necessary for consideration of its application for induction of abortion. In general, this is easier to get than is new-drug approval.

However, politics, publicity, and profit again loom as storm clouds on the horizon. To date, no pharmaceutical company has requested supplemental indication approval for methotrexate. Given the likely negative publicity potentially leading to protests and boycotts organized by antiabortion groups and the unresolved potential for future lawsuits, it is unlikely that a major company will seek such approval. In addition, because methotrexate is extremely inexpensive (approximately $4 a dose compared with about $200 for RU-486), it is unlikely to be a big profit maker.

5. Menstrual Regulation (Menstrual Induction, Menstrual Planning, Preemptive Abortion, Endometrial Aspiration, Atraumatic Termination of Pregnancy)

Without venturing too far into the semantical minefield implicit in the multitude of terms applied to this technique, in the minds of many it to some extent fills the gap between the foresight of birth control and the hindsight of abortion chosen following confirmation of pregnancy. The process is a simple, safe, highly effective outpatient procedure involving vacuum aspiration of the uterine lining within a few days to 2 weeks after the failure to commence menstruation. This procedure may or may not be an abortion since the key point that distinguishes it from a "regular" abortion is that the woman does not know for sure if she is pregnant. While pregnancy may have been the reason for her menstrual period not recurring, there are numerous other potential reasons.

Since this procedure is generally performed at a time when it is possible to discern whether or not she is pregnant, one would think that most women would want to know and avoid the procedure if they were not pregnant. However, some women choose to have the procedure performed as soon as possible and without pregnancy confirmation in order to avoid the ethical dilemma that they would see arising from aborting a known pregnancy. This approach is troubling to many, not only because it is a "head in the sand" way of dealing with an issue of considerable moral import, but also because it is important for a woman's psychological well-being that she be clear that she accepts abortion as morally permissible before employing this procedure or any of the numerous birth control methods that may have abortifacient actions.

D. Surgical Sterilization

No doubt in large measure due to the frustration arising from the deficiencies in the methods of birth control previously discussed, surgical sterilization is now the most common form of birth control in the United States (Table I). According to the National Center for Health Statistics, in 1990 (the latest comprehensive figures available) 29.5% of all American women employing birth control had tubal ligations. Add in the 12.6% of men who had vasectomies and the total percentage of women relying on some type of sterilization is over 42%. Of women in their early thirties, 47% rely on either male or female sterilization. For women in their late thirties, the figure increases to 65%. And for women over age 40, 73% rely on sterilization. Sterilization rates of women do not differ significantly by either race or level of family income. However, Caucasian men are 14 times more likely to have had a vasectomy than are African-American men. American women elect tubal ligation at a higher rate than women in any other country, with only women in China, a country that imposes a limit on the number of children that a woman may have, being a close second. The rate of tubal ligations for married women in other Westernized countries, for example, 23% in Great Britain and 15% in Holland, is far lower than that in the United States. Worldwide, 16% of all married women are protected by tubal ligation. With the repeal of state laws over the last couple of decades that previously restricted elective tubal ligation primarily to older married women who had already had children, now young unmarried women are increasingly opting for tubal ligation as a birth control method. While the exceedingly high rate of elective sterilization in the United States cannot be attributed to this fact alone, certainly the frustration of Americans over the lack of other available birth control options is a major reason for the continued and increasing trend.

There are numerous procedures and modifications for contraceptive sterilization, but all basically involve the bilateral cutting, tying, sealing, and/or removing of a small portion of the fallopian tubes (in women) or vas deferens (in men). Electro- or thermocoagulation, mechanical rings or clips, and chemical adhesives may be employed in tubal ligation. Tubal ligation prevents the ovum and sperm from reaching one another at the necessary site of fertilization in the fallopian tube. Va-

sectomy prevents sperm from passing from where they are stored in the epididymis into the vas deferens. Surgical sterilization approaches 100% effectiveness. Generally, there are no major long-term health risks and protection is continuous and lifelong. Although success at reversing the procedures and restoring fertility has increased over the years, both tubal ligation and vasectomy should be considered as permanent and not undertaken unless one is willing to accept termination of reproductive capacity.

IV. FUTURE METHODS OF BIRTH CONTROL

The increasing number of women and men opting for surgical sterilization, the very high rate of unwanted pregnancies (especially among the young and unmarried), the prevalence of abortions, the previously discussed issues of nonuse, underuse, misuse, and method failure, undesirable side effects of current options, and the lack of reversible options amenable to all lifestyles clearly argue that more birth control options are needed in the United States. On a global basis, United Nations' estimates predict a continued increase in worldwide population over the next 100 years, with 95% of the increase occurring in developing countries. Development of new birth control options, improvement of existing methods, and increased efforts in educating people how to properly apply current and potential future options are critical issues for the future.

A. Improvements of Current Methods of Hormone Delivery

1. Oral Contraceptives

Improvements in this now old method of revolutionary contraception has been continuous since the 1960s. In 1989, the FDA recommended removal of all references to age limits for the use of the pill by healthy, nonsmoking women. Newer formulations may now be used by women throughout their reproductive life. The pill may be available as an over-the-counter, nonprescription drug at some future date. While this would likely increase availability and usage and should decrease cost, it has the major potential disadvantage of a woman utilizing a formulation that is not appropriate for her, thereby risking contraceptive failure and adverse side effects.

Research is underway to develop a pill that could be taken orally only once a month rather than daily. This could be targeted either at the time that menses is expected (a potential interceptive action) or on the last day of a woman's period to prevent conception for the following month (a probable contraceptive action). Such a product is probably at least 20 years away from availability.

2. Biodegradable Implants

Biodegradable systems offer many potential advantages over oral, implant, and injectable administration of hormones to control fertility. A major advantage is eliminating the need to remove the delivery system when the supply of the hormone has been exhausted. Biodegradable microspheres that deliver a measured amount of progestagen and provide protection for durations ranging from 3 to 6 months have been investigated in clinical trials and proved effective. A disadvantage of biodegradable microspheres is that they cannot be easily retrieved once administered.

Polymeric membranes allow for superior control of drug release rate compared to any nonbiodegradable system. In addition, the controlled release rate that can be achieved in these products offers the potential to minimize side effects by augmenting the amount and persistence of the drug in the vicinity of target cells, thereby reducing the exposure of nontarget cells. This makes possible the use of lower amounts of hormone and minimizes potential side effects.

One product under investigation, termed Capronor, combines the advantages of retrievable and biodegradable implants. It is a single biodegradable rod containing enough progestagen sufficient to provide contraceptive protection for a number of years, and it will eventually dissolve. However, it would be removable, and hence reversible, for a while following implantation.

3. Transdermal Delivery Systems

This approach is rather similar to the nicotine delivery skin patches now on the market to help wean smokers from cigarettes. The contraceptive hormone is delivered in a skin patch in combination with a substance that enhances skin penetration and absorption. Early trials have been associated with problems in skin irritation due primarily to the effects of the enhancers.

4. Vaginal Rings

The medicated vaginal ring is another "new" method of contraception with a history dating to the 1960s. Vaginal rings, like subdermal implants, are devices made of a biocompatible polymer, usually silastic, loaded with long-acting steroids (either a progestagen alone or a combination of a progestagen and an estrogen) that are slowly absorbed at a predetermined rate. The hormones are absorbed directly into the bloodstream through the vaginal mucus membrane. The ring is placed in the posterior portion of the vagina by the user and left in place for anywhere from 3 weeks to 3 months, or perhaps even for as long as a year. An advantage is that the user herself places and can remove the ring without the need for intervention by a physician. The ring is washable and reusable. One scheme of usage is to insert the ring and leave it in place for three weeks and then to remove it for a week to allow a normal menstrual period to occur. In essence, this is the same concept as the traditional regimen with oral contraceptives, but eliminating the need for daily intervention. The advantage of user control may also be a potential disadvantage since she must ensure its proper placement for the hormones to be appropriately released and absorbed. The position of the vaginal ring may be disrupted during intercourse, so the user needs to be rather constantly vigilant in assuring that it is properly placed.

Effectiveness and side effects of this method of contraceptive hormone delivery are similar to those of the pill. In trials to date, some women have reported localized vaginal irritation and erosion from some models. There also tends to be a disruption of vaginal microbial flora that cause a change in vaginal odor that some women (and their partners) find objectionable. N. J. Alexander predicts that a vaginal ring may be available in the United States by the year 2000 (1995, *Scientific Am.* **273,** 136–141).

5. Vaginal Pills

This is in essence the same concept as the classical oral pill, but is designed to be inserted manually by a woman into her vagina rather than swallowed. It too may be a progestagen alone or a combination progestagen and estrogen. It is of a nature that would be readily absorbed into the vaginal mucosa. Reported theoretical effectiveness is over 99% and it reduces many of the undesired side effects of oral contraceptives, especially nausea. As with oral contraceptives, however, actual effectiveness is dependent upon daily administration by the user. Some women may find the method of daily delivery to be objectionable.

B. Immunocontraception for Women

The 21st century should finally see the realization of several possible options for immunocontraception in women and men. These are vaccines that would stimulate the immune system of the body to make antibodies against the functioning of selected proteins that are involved in various reproductive functions. The immune system is designed to "attack" any protein that is not native to the body. Therefore, injecting a person with a target protein (called an antigen or immunogen) plus other substances to boost the response will result in raising antibodies that would remain in the body for variable periods of time and disrupt the normal functions of the selected antigen.

1. Vaccine against Pregnancy

Beginning around the time of implantation, the placenta of the developing embryo produces a hormone called human chorionic gonadotropin (hCG). Detection of hCG is the basis of pregnancy testing. The major role of hCG is to stimulate the continued production of the hormone progesterone that is necessary for pregnancy to continue. If the normal effects of hCG are negated, progesterone levels fall, implantation is disrupted, and pregnancy fails. Therefore, in women immunized against hCG, the antibodies would counter hCG as soon as it began to be produced and pregnancy would fail.

Several trials have been conducted examining various protocols. In one scenario, immunity against pregnancy was achieved for 1 year by two initial injections and then a booster injection yearly for continued protection. The greatest concern is reversibility. Individuating factors can affect the duration at which antibody protection declines over time. If this decline is too rapid, an unwanted pregnancy may occur due to an unknown lack of protection. If the decline is too slow, there may be a substantial period of time after the last injection before a desired pregnancy may be possible. Since this is an abortifacient approach to birth control, it should only be utilized by women who realize that and accept that mode of action.

2. Targeting the Zona Pellucida

The zona pellucida is a thick membrane that is an outer "shell" around the ovum. Penetration of the zona pellucida by the sperm is essential for fertilization. A specific protein in the zona pellucida called ZP3 has been targeted as the primary immunogenic agent. Numerous studies since the 1970s have shown that ZP3 antibodies interfere with growth of the follicle and maturation of the ovum with contraceptive results. ZP3 is responsible for binding action with the sperm at fertilization. This fact has been exploited by administering ZP3 antibodies that then bind to sperm cells, blocking subsequent fertilization by preventing them from binding to the zona pellucida.

C. Future Options for Men

While birth control options for women are limited, those for men are even more so. Other than withdrawal prior to ejaculation, the only choices available for men are condoms and vasectomy. Given the potential permanence of vasectomy, it is not a viable choice for many. Ignoring the facts that some degree of sexism may be attendant to the history of birth control research and that many men and women simply feel that birth control is primarily a woman's responsibility, there are three major physiological reasons for the inequitable emphasis on female rather than male directed approaches. First, since fertilization and conceptus development occur within women, there are a greater number of reproductive events to manipulate to prevent pregnancy. Second, since women have to centers of "brain control" of the endocrine events controlling the activities of their ovaries and men have only a single brain center of testicular control, it is easier to hormonally interrupt the events controlling female fertility without causing adverse effects on the production of the hormones needed for other body functions. Third, and perhaps most important, it is difficult to argue with the logic of attempting to prevent pregnancy by controlling the fate of a single ovum produced once a month compared with the overwhelmingly more difficult task of interrupting the continuous process of sperm production which yields 20 to 30 million spermatozoa every day of a man's reproductive life, or of trying to control their fate following ejaculation. And the fact that women generally "have more to lose" if an unwanted pregnancy should occur also argues for giving women the greater control.

1. Improved Condoms

A reduction in sexual pleasure is the major reason reported by men for not using condoms. Manufacturers have recently introduced a thin polyurethane condom and are exploring other polymers that provide less interference with sensation while continuing to provide pregnancy and disease protection. Due to the many advantages of the condom in regard to lack of side effects, reversibility, low cost, and widespread availability, improvements in this ancient method of contraception may be the ones of greatest significance for the immediate future.

2. Steroidal Inhibition of Spermatogenesis

One approach involves intramuscular injection of the hormone testosterone. Keeping testosterone at high levels leads to suppression of other so-called gonadotropin hormones necessary for sperm production. Early results are encouraging relative to suppression of sperm numbers. Disadvantages include increased irritability, acne, and the potentially more serious effect of lowering levels of high-density lipoproteins (the "good" kind of cholesterol). Adding a progestagen to the injection may reduce some of the side effects. Despite recent reports of these trials in the popular media, approval of such a regimen is 10–20 years away.

3. Inhibition of Gonadotropin-Releasing Hormone (Gn-RH)

Numerous approaches to blocking the activity of Gn-RH, which is the key controlling hormone produced by the hypothalamic portion of the brain that ultimately leads to production of testosterone by the testes, have been examined for nearly two decades. While sperm production has been decreased, the undesirable side effects of decreased testosterone leading to decreased libido, loss of muscle mass, and alterations of male sexual characteristics have been highly problematic. Replacement testosterone has been the only solution to date of overcoming these negative effects.

4. Inhibition of Sperm Maturation

Rather that directly halting sperm production, this approach would interfere with the maturation of the newly formed sperm in the epididymis where they are housed prior to ejaculation. One reason why disruption of sperm maturation may be preferable to preventing sperm synthesis is that it is easier to deliver products to the epididymis via the bloodstream. There is a natural protective barrier that prevents delivery of potentially injurious products into the testes where sperm production occurs. Also, many products directed at preventing sperm production have had toxic and irreversible effects that caused permanent rather than temporary sterility. Targeting sperm in the epididymis may avoid those problems.

V. CONCLUSIONS

J. Trussell and B. Vaughan examined factors likely to affect contraceptive use and choice to the year 2010 (1992. *Am. J. Obstet. Gynecol.* **167**, 1160–1164). They cited two predictable factors: the changing age distribution of women and the revised upper-age limit for oral contraceptive use. Less predictable factors include the number of women in each age group at risk for pregnancy, the effects of delayed childbearing, and the impact of new birth control methods. Unpredictable factors include adverse publicity about birth control methods (especially as related to health risks), concerns about STDs (especially AIDS), and changes in the availability of legal abortion. Birth control choices are also influenced by a variety of other factors, including socioeconomic status, cultural background, religious beliefs, personal aspirations, health status, and individual moral and ethical values. These (and other) variables argue

for a wide array of birth control options and extensive education about all possible options in order that each woman and man can make an intelligent choice as to the option that best fits their individual situation.

The demand for more effective, safer, easier to use, convenient, readily available, and affordable birth control has never been greater. Current demand will only intensify in the future years. Sadly and perhaps tragically, unless immediate steps are taken to reverse our current situation, the future of birth control in the United States is even dimmer than its present state.

What needs to be done? Government policy must be changed to one that devotes more federal agency research funding to birth control and that encourages private sector research. In our "squeaky wheel gets the grease" political system, there is no constituency pleading to congressmen to increase contraceptive research. According to a 1993 report by the Program for Appropriate Technology in Health, a nonprofit health research organization, executives of 14 major pharmaceutical companies who were surveyed reported that they believed that the market was well served by currently available contraceptives. Concerns of company executives center around the high cost and long time necessary to develop a new product, regulatory hassles, and product liability. Currently only one major U.S. pharmaceutical company has an active contraceptive research program, compared with at least 13 in the 1970s.

Politics and prudery also stand firmly in the path of progress. Political concerns pose a direct barrier to enlightenment when matters of birth control become an issue to be avoided for fear of offending the large segment of citizens (read voters) who either wish to avoid any discussion of sexual matters or believe that "just say no" is the only acceptable policy. Sexual desires, especially among the young, are not amenable to control by social agendas, political positions, or legislative action Our country's history of sexual prudery must bear a significant portion of the blame for the unprecedented high rate of teen pregnancies that politicians love to rail against for political points. Studies indicate no significant difference in sexual activity of American teens and teens in other Western countries. However, there is a dramatically higher pregnancy rate (and higher abortion rate and delivery rate) in the United States. This difference is perhaps attributable to a more open approach to sexuality, more and better preventive sex education, and an easier and freer access to contraceptive services in other countries (Rosenfield, 1994).

In this era of rapid technological progress, and in this country that spawned both the sexual and the technological revolutions, the future of birth control technology, at least for the foreseeable future, is not a bright one. The ability of biomedical science to develop options to control the reproductive process is not the impediment to progress. Dealing, as a society, with the ethical concerns associated with birth control technologies poses the greatest hindrance to future options in controlling reproduction.

Bibliography

Knight, J. W., and Callahan, J. C. (1989). "Preventing Birth: Contemporary Methods and Related Moral Controversies." Univ. of Utah Press, Salt Lake City.

Mastroianni, L., Jr., Donaldson, P. J., and Kane, T. T. (1990). Development of contraceptives: Obstacles and opportunities. *N. Engl. J. Med.* **322,** 482–484.

McLaren, A. (1990). "A History of Contraception: From Antiquity to the Present Day." Basil Blackwell, Cambridge, MA.

Pollard, I. (1994). "A Guide to Reproduction: Social Issues and Human Concerns." Cambridge Univ. Press, Cambridge, UK.

Service, R. F. (1994). Contraceptive methods go back to the basics. *Science* **266,** 1480–1481.

Trussell, J., and Kost, K. (1987). Contraceptive failure in the United States: A critical review of the literature. *Stud. Fam. Plan.* **18,** 237–283.

Brain Death

C. A. DEFANTI
United Hospitals of Bergamo

GLOSSARY

anencephaly A rare condition in which both hemispheres of the brain are not developed. In many cases pregnancy ends with stillbirth. The newborn babies who survive rarely last more than a few days or weeks.

brain death The clinical condition of a patient with massive damage (usually destruction) of the brain, previously defined as **irreversible coma.** This condition was chosen, by the Harvard Committee, as a new criterion of death.

brain life A concept proposed for symmetry reasons as the converse of brain death. The term indicates the beginning of conscious life during fetal development; a controversial construct.

brain stem death A concept proposed by the Royal Colleges of British Physicians and Surgeons, it means the irreversible cessation of the brain stem functions, i.e., the ability to be conscious, to breathe spontaneously, and hence to retain a spontaneous heartbeat. This definition implies that it is neither essential nor possible for the clinician to prove the permanent cessation of hemispheric functions in order to ascertain death.

neocortical (or cortical) death A term meaning the irreversible cessation of the functions of the cerebral cortex; its main clinical feature is permanent unconsciousness.

persistent vegetative state A clinical condition resulting from massive impairment of the brain hemispheres or of the cerebral cortex. The main feature of PVS is permanent unconsciousness coupled with retained functions of the brain stem (i.e., spontaneous breathing and other brain stem reflexes).

whole brain death A refinement of the previous brain death concept and endorsed by the President's Commission (1981). It means the irreversible cessation of *all* brain functions (i.e., the functions supported by the hemispheres and the brain stem).

BRAIN DEATH (BD) may be considered a paradigm of bioethics from many points of view: (a) There was a close temporal overlapping between the proposal of a new criterion of death by the Harvard Committee and the first steps of bioethics as a new, autonomous discipline. (b) The Committee itself, with its multidisciplinary composition, can be viewed, in some ways, as a prototype of the ethics committees which were started afterward. (c) An important point is that both the Harvard paper and the international debate about BD that followed typically involved many different levels of discourse—medical issues relating to the ascertainment of death, purely conceptual issues, and, above all, difficult moral problems (what is the correct approach to severely injured patients dependent on mechanical ventilation and/or permanently lacking consciousness?). This interplay among different conceptual levels, another typical aspect of bioethics, was richly expressed in the relevant literature. (d) The proposal had an enormous practical impact, having been accepted by most legislations in Western countries, and was the necessary condition of the development of modern transplantation medicine. Nevertheless, severe criticisms on the ground of empirical and philosophical reasons and frank hostility in some religious circles and among some areas of the society at large are still in place. So, even though at a first glance it may appear that the concept of BD has been a successful one, no stable consensus has been reached, and that is another typical feature of many bioethical topics. (e) Finally, the concept of BD was linked with the topic of euthanasia, at least during the first years of bioethics, and is still linked with the problem of the beginning of brain life during the development of the embryo. This in turn has important corollaries for the abortion issue. However, the question of whether the concepts of brain death and brain life are symmetrical or not has not been settled so far.

I. THE HISTORICAL FRAMEWORK OF THE HARVARD COMMITTEE PROPOSAL

The historical context in which the paper, "A Definition of Irreversible Coma," by the Harvard Ad Hoc Committee (1968. *J. Am. Med. Assoc.* **205**, 85–88), was published cannot probably be better summarized than in the opening sentences of the paper itself:

> Our primary purpose is to define irreversible coma as a new criterion for death. There are two reasons why there is a need for a definition: (1) Improvements in resuscitative and supportive measures have led to increased efforts to save those who are desperately injured. Sometimes these efforts have only partial success so that the result is an individual whose heart continues to beat but whose brain in irreversibly damaged. The burden is great on patients who suffer permanent loss of intellect, on their families, on the hospitals, and on those in need of hospital beds already occupied by these comatose patients. (2) Obsolete criteria for the definition of death can lead to controversy in obtaining organs for transplantation. (Harvard, 1968, 85)

In fact, this controversy was raging just then in the scientific world and among the general public after the first successful heart transplantation performed by C. Barnard in Capetown (December 1967). It is useful to remember that this historical operation was not only an extraordinary technical achievement, but equally a transgression of the usual practice and of the law: the beating heart of an (irreversibly) comatose individual was removed, so causing the death of this individual, according to the criterion then in force, in order to benefit another person. There is little doubt that this event was the *Primum movens* of the Harvard paper. One clearly recognizes two arguments in this document, the first being a mainly, but not exclusively, empirical point: resuscitation may fail not only in the obvious sense, i.e., not being able to prevent death, but also in another way, insofar as it may lead to a novel clinical state, irreversible coma, whose features are such that they raise the moral question of whether it is sensible to go on with medical treatments in order to maintain it. Secondly, there is the acknowledgment of a need, the need for organs suitable for transplantation.

An earlier draft of the Harvard paper was even more explicit about the second point and was discarded precisely because its wording exposed it to the obvious criticism that the new definition of death was merely instrumental in solving the problem of the scarcity of organs. It is useful to remember how the Harvard Committee was composed: it was a 13-member committee with a majority of doctors, but with the crucial participation of one lawyer, one historian, and one theologian. It was chaired by Henry Beecher, a well-known physician whose seminal papers on the ethical problems of human experimentation were instrumental both in establishing a new legislation for these problems and in stimulating a new sensitivity to these aspects among medical professionals. As such, the Harvard Committee was a truly multidisciplinary body and encompassed those competences (in law, humanities, and theology) deemed necessary to confer on it the authority to make a proposal with a far ranging impact on clinical practice. One does not find any philosophical discussion of the concept of death in the paper; nevertheless, there is an important quotation of a speech given by the late Pope Pius XII to a medical congress, in which two main points are made: (1) the verification of the moment of death is the task of the medical profession and is not within the competence of the Church, and (2) it is not mandatory to continue to use extraordinary means indefinitely in hopeless cases. This is an obvious political move in order to prevent possible criticisms from religious sources.

II. WHY WAS THE HARVARD PROPOSAL SO SUCCESSFUL?

Rereading the Harvard paper long after its publication, it is easy to pick out many obvious flaws. There are both problems of theoretical foundation and of internal consistency. No discussion is attempted of questions like, What is death? Is the definition of death a purely metaphysical, or also a moral, issue? Concerning internal flaws, the reason given in order to equate irreversible coma and death, i.e., permanent loss of intellect, is also valid for persistent vegetative state (PVS). Why then restrict the equation to irreversible coma and why not extend it to this no less troubling condition? Finally, even though there was a rather widespread consensus among neurologists and intensive care physicians on the clinical criteria of brain death (BD), no reliable scientific validation of these criteria (i.e., of their ability to predict true irreversibility) had been reached at that time. Why was the proposal so successful?

One can think of many reasons—the authoritativeness of the Harvard University and of the members of the Committee, its careful consideration of the interests at stake and its clever strategy aimed at preempting juridical and religious opposition, and, above all, the need of organs and the strong support of the community for transplantation medicine. Another political reason for this success was probably the support given to BD by prolife movements in the United States, a seemingly paradoxical support if one thinks of the open opposition of prolife movements in other countries. There is obviously one sense in which a brain-dead individual, whose heart keeps beating, is still alive, and it may seem that the decision to forgo the life support treatments is an antilife move. Probably the main reason of the endorsement by U.S. movements was their hope to relieve the public pressure for euthanasia, as some of them openly acknowledged.

Alternatively, one major reason for this success may have been the far-sightedness of the Committee, i.e., its capacity to accept the revolutionary challenge to old ideas raised by the reversibility of the "traditional," cardiac criterion of death by modern resuscitation techniques and to work out a novel, essentially adequate conceptual response.

III. THE DEFENSE OF THE TRADITIONAL DEFINITION OF DEATH

The implementation of the BD definition in the medical practice was not without opposition. Controversies raged through more than a decade, mainly in the United States, prompting in 1981 a new official statement by another influential body, the President's Commission for the Study of Ethical Problems in Medicine. The Commission issued an important paper in which the conceptual problems of the new definition were discussed at some length (1981. *Defining Death*. Washington, DC). Its conclusion was that a practical agreement had been reached on the concept of BD and that there

was a need for a common legislation. Therefore it proposed a "Uniform Declaration of Death Act" which was subsequently implemented in the legislation of many states.

Many Western countries followed this trend, but not all. In the United Kingdom, for instance, the Royal Colleges of British Physicians and Surgeons proposed the alternative concept of brain stem death (BSD), meaning the irreversible cessation of the brain stem functions, i.e., the ability to be conscious, to breathe spontaneously, and hence to retain a spontaneous heartbeat. The choice of this definition implies, among other things, that it is neither essential nor possible for the clinician to prove the permanent cessation of hemispheric functions in order to ascertain death. Moreover, no legislation was promoted in the matter and the new definition was introduced into clinical practice thanks to the authority of the Colleges.

A remarkable exception among Western countries was Denmark until the late 1980s. In fact, the Danish Council of Ethics still in 1988 had not backed the BD concept and had made instead a different proposition. It suggested considering so-called brain-dead patients not really dead, but having irreversibly entered the dying process. A parallel suggestion was to permit the procurement of organs from these patients if they had previously signed a valid donor card. In this case the removal or organs for transplantation would "become the cause of the conclusion of the dying process, but not the cause of the death of the person." (*Death Criteria*, the Danish Council of Ethics, 1989, 25) However, this interesting proposal did not prevent the Danish government from implementing a BD legislation some time after the Council's report.

Among the Eastern countries, remarkable hostility against BD still exists in the Japanese culture, where no legislation in the matter has passed so far, despite a national framework of well-developed, Western-style medicine.

On a theoretical level, it is important to note the influential attack on the BD concept made by Hans Jonas soon after its proposal. He made a case for sticking to the old concept of death. The main arguments of Jonas were the instrumental character of the Harvard proposal, aimed essentially at making easier the procurement of organs; the obscure, exoteric criteria of BD; and finally the importance of not separating the death of human beings from that of other living creatures.

BD met strong opposition from some religious sources, for instance, from Orthodox Jews, whose strenuous battle against it recently led to the implementation of a law, in the state of New Jersey, recognizing the right to conscientious objection for those not accepting the new definition.

IV. THE MEDICAL INCONSISTENCIES OF THE "WHOLE BRAIN DEATH" DEFINITION

The President's Commission, already discussed, completed in some way the work of the Harvard Committee, confirming, with a few changes, the adequacy of its clinical criteria and backing the BD proposal. However, the Commission addressed more closely the conceptual issues of the definition of death. Different options were analyzed: the "whole brain death" (WBD) formulation (death is the irreversible cessation of all functions of the hemispheres and of the brainstem), the "higher brain" version (death is the irreversible lack of consciousness, due to massive damage of the brain hemispheres), and the "nonbrain" formulations (like the "traditional" definition, based on the arrest of the circulation of body fluids). The choice of the Commission was in favor of the first one on the ground of two complementary arguments, the argument from loss of the primary organ (the brain being the critical system, viz., the integrating center of the organism, complete cessation of which means that the organism is no longer functioning as a whole, i.e., is dead) and the argument from loss of integrated bodily functioning (life is integrated functioning of the organism as a whole, and crucial to this functioning is the interplay among brain, heart, and lungs; hence cessation of brain functions, in the peculiar condition of a ventilated patient in an intensive care unit, is a sign that death has occurred, just as cardiac arrest is a sign of death in ordinary situations). The Commission was aware that both arguments were open to several criticisms, the main ones concerning the critical role given to the brain (in fact many other organs are equally critical to the organism, e.g., the liver and skin) and the supposed irreplaceability of the brain. However, its conclusion was that, "while [it] is valuable to test public policy against basic conceptions of death, philosophical refinement beyond a certain point may not be necessary" (President's Commission, 1981, 36).

The endorsement of the WBD definition by the President's Commission does not mean that it rejected the cardiac criterion of death. In fact this criterion is still valid in most circumstances, when death occurs outside the intensive care context; simply, the validity of the cardiac criterion is dependent on the duration of the cardiac arrest, a duration that has to be such as to allow WBD to follow.

The Commission implicitly rejected the BSD definition—irreversible cessation of the brain stem functions—endorsed by the Royal Colleges of Physicians and Surgeons in the United Kingdom (1976). Moreover, it explicitly criticized the "higher brain" formulation, arguing that there are major problems in defining consciousness and personhood, whose anatomical substrate is poorly known, and that it is very difficult to diagnose irreversible lack of consciousness reliably.

However, the WBD formulation itself is open to many stringent criticisms. In fact, the demonstration of the irreversible cessation of the functions of all parts of the brain is practically impossible, as C. Pallis has shown most convincingly (1983. *The ABC of Brain Stem Death.* British Medical Journal, London). Besides, some recent empirical data, in particular the possibility of keeping biologically alive the body of brain dead people over many weeks or even months, as shown by the famous cases of two pregnant wives so assisted, the second of whom eventually gave birth to a baby, demonstrated unequivocally that modern intensive care is able to replace the integrative functions of the central nervous system, at least during some weeks or months.

Other, recently gathered empirical data show that many individuals in whom a diagnosis of BD is made following the strict criteria suggested by the Consultants to the President's Commission retain some brain activity, e.g., some autonomic reflexes and hormonal secretion. Of course this does not mean that consciousness is preserved in such individuals, much less that they are not doomed, but simply that there is a discrepancy between the clinical criteria for BD routinely used worldwide and the conceptual definition of WBD as irreversible loss of all function of the entire brain.

V. THE MOVE TOWARD A NEOCORTICAL DEFINITION AND ITS PROBLEMS

A. A Remedy for the Medical Inconsistencies of the BD Definition

If such a discrepancy exists, a possible remedy could be to revise the BD concept, equating death with permanent lack of consciousness as does the so-called cortical death theory. Note that the main argument raised by the Harvard Committee in order to back its proposal was that the individual in an irreversible coma suffers from permanent loss of intellect. In fact the present clinical criteria for BD, even though they do not prove the complete loss of all brain functions, are very strong criteria for permanent lack of consciousness.

What are the main criticisms of the cortical death theory?

A first concerns the very term of cortical death. In fact, it is simplistic to affirm that the cerebral cortex is the anatomical substrate of consciousness. Certainly many nervous centers are involved in this phenomenon, such as the thalamus and the brain stem itself; the latter is probably unrelated to the content of consciousness, but is certainly involved in maintaining wakefulness. So the term "cortical death" is an approximate one. An alternative which is less precise but more correct is "higher brain death." However, the major problem with the idea of cortical death is that, would it be accepted, not only BD but other clinical conditions would qualify as death: we are thinking particularly of PVS and anencephaly.

B. Persistent Vegetative State

Let us briefly review the clinical conditions of this state. What is PVS? Due to the development of modern intensive therapy, some patients with an overwhelming damage of the cerebral hemispheres survive, after a more or less prolonged stage of coma, in a state of absence of cortical functions, but with a relative sparing of brain stem functions. This condition is known as PVS and can be reversible, but if it lasts more than a few months, it usually can be considered irreversible. The term *permanent* vegetative state applies to such a case.

Its distinguishing feature is chronic wakefulness without awareness. Individuals in PVS seem to be awake, with open eyes, but they are unable to follow any object moving in their visual field and do not respond appropriately to any kind of stimulation. Breathing, circulation, and regulation of bodily temperature are more or less normal, without artificial support. PVS is clearly a different condition from BD. At variance with brain-dead individuals, those in PVS can survive over months and even years with only careful nursing and artificial hydration and nutrition.

Diagnosis of PVS is much more difficult than recognition of BD. No single test is available to detect PVS, and only by means of a long and skilled clinical observation is it possible to establish the typical picture of a complete dissociation between wakefulness and awareness due to a (demonstrable) massive damage of the cerebral hemispheres. The minimum length of the observation period is still controversial: 3 months are probably sufficient, if the etiology is an anoxic insult, whereas in PVS due to other etiologies 6 or even 12 months are necessary.

C. Anencephaly

What is anencephaly? It is the most severe malformation of the central nervous system, characterized by the lack of development of the cerebral hemispheres and of the cranial vault. There are different types of anencephaly. In the most common (meroanencephaly) the baby has a more or less normal face, but with the absence of a forehead. The eyes may protrude or squint, but the major defect is the lack of a cranial vault: above the eyebrows there is no skin, only a rudimentary brain covered by thin meningeal membranes. These babies have more or less functioning brain stems, and some of them are able to breathe spontaneously. If they are not given artificial ventilation, most of them do not survive longer than a few hours or days, but with mechanical assistance they can be maintained over months or years. In some respects they are similar to individuals in PVS.

A difference, of course, is the fact that an individual in PVS was previously a conscious being with a personal history; in contrast, an anencephalic baby never was nor will be. Another difference is that it is much easier to maintain a PVS individual (by artificial nutrition) than an anencephalic baby, the latter usually requiring mechanical ventilation. Finally, the diagnosis of anencephaly is an easy one and the prognosis of irreversible lack of consciousness can be made without a lengthy observation time.

Under the cortical death definition, both individuals with PVS and those with anencephaly could be considered dead. While it is now usual to pronounce a brain-dead individual as dead and to disconnect her from the respirator (and possibly to retrieve an organ from her), it appears counterintuitive to many to declare a PVS individual or an anencephalic baby as dead. In doing so it would be permissible to bury such individuals (whose bodies lack consciousness but are (usually) able to breathe spontaneously).

D. The Slippery Slope Argument

Finally, a strong criticism of the cortical theory is constructed as a "slippery slope" argument. The opponents to this theory contend that, if the concept of death was to be linked to a psychological property (consciousness) instead of biological data, it would be easy to extend the concept from PVS to other abnormal psychological states, like dementia or severe mental retardation.

Is this objection unanswerable? Certainly not. Although the concept of consciousness is intrinsically difficult to define, the diagnosis of PVS is based on the complete lack of this property, and there is no question of equating simple impairments of consciousness with PVS. In any case, there is no doubt that a change from the BD concept to the cortical one would be a momentous change, no lesser in magnitude than the previous one from "traditional" death to BD.

VI. BACK TO THE "TRADITIONAL" DEFINITION?

Recently some scholars, who formerly were in favor of the cortical definition, have argued for a return to the traditional one. For instance, P. Singer, in his last book, *Rethinking Life and Death* (1995. St. Martin's, New York), wonders if it would not be better to recognize the fact that behind the proposal of BD there was not a true problem of definition, but rather a moral problem, the problem about our duties toward the individual in irreversible coma. After a careful historical reconstruction, Singer emphasizes that the purpose of the Harvard Committee was an essentially practical one and that the Committee chose to avoid this moral problem by introducing a new definition of death.

In this sense, the strategy chosen by the Committee has some analogies with the strategy of the Warnock Committee vis-à-vis the moral status of the embryo. Their distinction between a preembryo and an embryo stage sought to solve a substantial moral question ("What is licit to do to the embryo?") by means of a redefinition, allowing some interventions at the preembryo stage which are not permissible at the following stage.

Singer maintains that Harvard's proposal was widely accepted because of its usefulness: it did not harm brain-dead individuals, it benefited those on a waiting list for organ transplantation, and, above all, it prevented the charge of passive euthanasia being alleged against doctors who were willing to withdraw life support treatments from these patients. However, Singer argues, only a few people really believe that these individuals are truly dead, as a recent sociological inquiry has shown. Even doctors and nurses directly involved in transplantation seem not to take seriously the idea of brain death. The difficulty in understanding the concept of BD, 25 years after the Harvard paper, probably means that something is wrong with this concept. If one adds the fact that it has been proved that some (although weak and clinically not significant) nervous activities are going on in brain-dead individuals, one realizes how fragile the consensus on the BD concept is. Even though the question seems to have been settled for the moment, the need for a thorough reappraisal of the problem will appear again shortly.

In fact, Singer argues that it would have been better to face the moral substance of the question and to recognize that when life is so severely diminished as happens in so-called brain-dead and PVS patients, the respect we owe them is not incompatible with a decision to withdraw life support or possibly to retrieve organs from them for the sake of transplantation. The Harvard Committee, Singer says, was not prepared to propose such a solution, because this would have been tantamount to giving up the traditional sanctity-of-life doctrine. Other scholars partly agree with the historical reconstruction made by Singer and recognize the utilitarian motives that were behind the Harvard proposal, but refute his arguments against the BD concept and the cortical death theory. The argument concerned with misunderstanding from the general public is weak. No doubt it is difficult to comprehend the present concept of WBD, but it would be much simpler to understand the concept of irreversible lack of consciousness (many respondents to the previous inquiry clearly equated BD precisely with cortical death). The argument from repugnance ("It is repugnant to bury a breathing body"), aimed at the cortical death theory, is a typically emotivist one and rather easy to refute. Certainly it seems absurd to bury a body still breathing, but no one maintains an action such as this. Probably the most appropriate behavior, after pronouncing cortical death, would be to stop artificial hydration and nutrition; burial should be performed only after the stoppage of circulation. In any case, there is no logical relationship between ascertainment of death and burial.

VII. BRAIN DEATH AND BRAIN LIFE

A corollary of the BD issue, not thoroughly explored but logically important, is the problem of the beginning of brain life during the development of the embryo. Some scholars have tried to construct, on the ground of a supposed symmetry between the beginning and the end of life, a theory of brain life that would be the complement of the brain death theory. Their argument runs as follows: if we accept identifying human death with the cessation of the brain functions, we can hope to pick out the beginning of these functions during fetal development and so identify this moment with the beginning of the life of the human being. Obviously this beginning does not coincide with conception, but it has to be subsequent. If we manage to pinpoint this moment on purely scientific grounds, we would be able to fix a nonarbitrary boundary for the beginning of a human being. Many questions concerning abortion could be solved that way.

This idea was first suggested by Goldenring, who thought that the 8th week of pregnancy was the critical moment, because at that time brain activity begins to be integrated. Unfortunately, other scholars have proposed different chronologies, generally between the 54th day and the 70th day after conception. More recently, the stress has been laid upon the emergence of consciousness, possible to locate at about 30 to 35 weeks after conception. As one can see, disagreement is great. The reason for this is not obscure. Embryonic and fetal development is a continuous process. It is true that in this process there are some "turning points," but it is not clear at all what we are searching for: are we looking for the beginning of brain activity, whatever it may be?

for the start of the integrating activity of the central nervous system? or for the emergence of consciousness? Is not our inquiry similar to the search of Aquinas for the beginning of the human form of the embryo?

Others deny the usefulness of this search, stressing the continuity of embryonal development; no "natural" boundary can be found in ontogenesis, and, in any case, no moral consequences can be directly derived from empirical data. Moreover, it is controversial that a symmetry exists between "brain life" and "brain death": embryonal development is a lengthy, complex process under genetic control, while death (or, better said, dying), though it too is a process, has many causes and very different durations. Apart from that, BD is not a "natural" fact, but an artifact due to human intervention in the dying process.

VIII. A PARADIGM OF BIOETHICS

A. Some Characterizing Points

From what points of view can BD be considered as a paradigm of bioethics?

The first point to make is that there was a close temporal overlapping among the issue of the Harvard paper proposing the BD concept, the international debate that followed it, and the first steps of bioethics as a new, autonomous discipline (for instance, two major centers of this discipline, the Hastings Center and the Kennedy Institute at Georgetown, were founded in the years 1969–1971).

The Committee itself, with its multidisciplinary composition, can be viewed, in some ways, as a prototype of the ethics committees which have been established subsequently. No doubt multidisciplinarity, and especially the active participation of philosophers and theologians along with physicians and biologists, is a central feature of the bioethical enterprise.

Both the Harvard paper and the related literature typically involved many different levels of discourse: medical issues relating to the ascertainment of death, purely conceptual issues ("What is death?", "What does it mean to be dead?"), and, above all, difficult moral problems (what is the correct approach to severely injured patients dependent on mechanical ventilation and/or permanently lacking consciousness?). This interplay among different conceptual levels, another typical aspect of bioethics, was rather crude in the Harvard paper, but very richly expressed in many contributions to the debate. The most difficult and sensitive point seems to be whether the attempt to give a new definition of death has to be a purely ontological research or a moral inquiry, or both.

Another point to be stressed is the enormous practical impact of the new proposal, which has been accepted, during the 1970s and 1980s, by most legislations in Western countries and has been the necessary condition of the development of modern transplantation medicine. The practical success and the social and political consequences of the Harvard proposal are in many ways a model for the numerous position statements of ethical committees in other fields of medicine that followed. Nevertheless, severe criticisms on the grounds of empirical and philosophical reasons and frank hostility in some religious circles and among some areas of society are still in place. So, even though at a first glance it may appear that the concept of BD has been a successful one, no stable consensus has been reached, and that is another typical feature of many bioethical topics.

B. The Present Status of the Question of the Definition of Death

What can one say about the present status of the question of death? We presently know that no clear-cut concept of death exsisted in the past and that many doubts lingered about its ascertainment, especially during the last three centuries. At the very moment in which reliable diagnostic techniques (i.e., EKG) for documenting cardiac arrest became available, further technical advances (i.e., resuscitation) showed that cardiac arrest, once considered as the central sign of death, no longer was a valid criterion. Now we have, for the first time in history, the opportunity to stipulate a consistent definition of death, taking into account the latest development in medicine. It is not surprising that this definition (both in the WBD and in the cortical version) is in many regards difficult to comprehend and puzzling, but no less surprising and puzzling are other situations created by medical progress (suffice it to think of artificial fertilization). The novelty of situations and the inadequacy of our emotional reactions are probably not a valid argument against the new definition of death. BD (especially in the cortical death version) is a paradigm of the derangements brought about in our culture by the biological revolution.

We have been given, by technological advances, powerful means to alter the dying process, and these means have questioned one of the necessary requirements of the concept of death: irreversibility. What does irreversibility mean today? It seems to be identical to the impossibility of replacing (say, through a transplantation) the cerebral hemispheres while maintaining the memories and dispositions of that particular person. This is not only not technically feasible today, but probably will never be so. How are we to interpret the novel situations? Why should we stick to traditional definitions?

Each socially acceptable death definition has in itself a practical-evaluative component. Even the decision to identify death with the cardiocirculatory arrest was in some way an arbitrary one: it was well known that some biological activities were continuing in the corpse for some time, and only after putrefaction was the process fully completed. No present society accepts complete putrefaction as the criterion of death for exquisitely practical reasons (both moral and hygienic). Quoting J. Lachs, we can say that "death is a biologically based social status" (1988, p. 239. In *Death: Beyond Whole-Brain Criteria* (R. M. Zaner, Ed.). Kluwer, Dordrecht).

Bibliography

Ad Hoc Committee of the Harvard Medical School (1968). A definition of irreversible coma. *J. Am. Med. Assoc.* **205,** 85–88.

Defanti, C. A. (1993). E'opportuno ridefinire la morte? *Bioetica. Rivista interdisciplinare* **1,** 211–225.

Jonas, H. (1974). Against the stream. Comments on the definition and redefinition of death. In "Philosophical Essays: From Ancient Creed to Technological Man." Prentice-Hall, Englewood Cliffs, NJ.

Lachs, J. (1988). The element of choice in criteria of death. In "Death: Beyond Whole-Brain Criteria" (R. M. Zaner, Ed.). Kluwer, Dordrecht.

Mori, M. (1992). Dalla morte cerebrale alla morte corticale: Una breve analisi degli argomenti. In "XI Corso di aggiornamento della Società Italiana di Neurologia." Monduzzi, Bologna.

Pallis, C. (1983). "The ABC of Brain Stem Death." British Medical Journal, London.

President's Commission for the Study of Ethical Problems in Medicine and Biomedical and Behavioral Research (1981). "Defining death." Washington, DC.

Singer, P. (1995). "Rethinking Life and Death." St. Martin's, New York.

Veatch, R. M. (1989). "Death, Dying and the Biological Revolution: Our Last Quest for Responsibility," rev. ed. Yale Univ. Press, New Haven, CT.

Cloning

GREGORY E. KAEBNICK and THOMAS H. MURRAY

The Hastings Center, Garrison, New York

I. Scientific and Technical Background
II. Suggested Uses of Cloning
III. Moral and Religious Issues
IV. Policy Developments
V. Conclusion

GLOSSARY

blastomere separation A technique in which an embryo is split into two parts, each of which can, if implanted into a womb, develop into a fetus and eventually be brought to term.

clone Either a biological copy (of a piece of DNA, a cell, or an organism) or the process by which the copy is produced.

mitochondrial Cellular organelles that provide energy to the cell and that contain their own DNA, derived solely from the ovum.

nucleus A cellular organelle containing the chromosomes.

somatic cell nuclear transfer cloning A technique in which the nuclear DNA from a somatic cell is combined with an ovum from which the nucleus has been removed, creating an embryo having the same nuclear DNA as the organism from which the somatic cell was derived.

CLONING is a term used by scientists as an all-purpose verb that means, in general, to make a copy of a thing. The thing could be a short stretch of DNA (the raw material of

genetics), an entire gene, or a cell. In that broad sense, human cloning has been going on for decades by making copies of genes and cells. Cloning of this sort, though, raises no great ethical concerns. It took the possibility of cloning a person, using either blastomere separation or nuclear transfer, to provoke widespread moral objections.

A person created through cloning would be no different from anybody else in an important respect, namely, in that he or she would be different from anybody else. The clone would be a unique person, and entitled to the respect due persons. This raises a problem for the term "clone," which is used colloquially to suggest a doppelganger—a perfect match in character, temperament, talents, interests, and all physical respects. Someone who is the clone of another in a scientific sense would be no such doppelganger. Nonetheless, this article will typically refer to the person created through cloning as "the clone," for the sole reason that the language necessary to avoid any confusion with doppelgangers would be cumbersome.

I. SCIENTIFIC AND TECHNICAL BACKGROUND

Although it had been used in the 1980s to create animals from embryonic cells, nuclear transfer was made famous when Ian Wilmut and his colleagues employed it to create the sheep they called Dolly. The earlier clones were of early stage embryos, but Dolly was cloned from an adult animal, bringing researchers significantly closer to the possibility of cloning an adult human (Wilmut *et al.* 1997). In the procedure they employed, the genetic material packaged in the nucleus of an unfertilized egg was removed, leaving the remaining contents of the egg—its cytoplasm, mitochondria, and other structures—available to support a new set of genes in a new nucleus. The next step was to fuse the enucleated egg with an adult somatic cell taken from a mammary gland. That cell had been starved temporarily to synchronize its cycle with the egg's. The egg and mammary cell were brought together and a brief electrical shock caused them to fuse into one hybrid embryo. That same shock stimulated the new embryo to begin developing and dividing.

It must be noted that Dolly is not an identical genetic replica of her nuclear progenitor. The mitochondria, the energy factories of the cell, have their own DNA in the cytoplasm of the ovum, not the nucleus. Scientists and ethicists continue to debate the significance of this biological fact. Complete genetic duplication could be achieved, however, by obtaining the ovum and the nucleus from the same animal, or from the nucleus donor's mother.

Dolly was the only success among 277 attempted fusions and 29 implanted sheep embryos. Since the announcement of Dolly's birth, scientists have reported mixed success using nuclear transfer cloning with other species, including mice and cattle. The technique has a high failure rate and even its successes, live-born animals, have been ambiguous; a very substantial proportion of animals created with it appear to have significant abnormalities, while many of the pregnancies leading to their births have likewise been abnormal, and even potentially threatening to the health of the mother.

A. Differentiation and Specialization

Scientists generally have found it easier to use nuclei from embryos and fetuses than from adult animals. The best explanation for this appears to be that as cells differentiate into the various tissues and organs of the human body, it becomes increasingly difficult to return those genes to their infantile, undifferentiated state. The further a nucleus is from a few-celled embryo, the more challenging the task of reawakening all of the genes needed to create a complete, healthy organism.

The newly fertilized embryo, human or otherwise, is a single cell with the capacity to become every type of cell in the adult body, as well as those tissues and organs that support the development of the organism within the uterus—the placenta and related tissues. Through processes only dimly understood at present, that single-celled embryo divides again and again. Particular cells begin to specialize so that they become the precursors of the brain and nervous system, while other cells are destined to make the cellular components of blood, and

still others will be transformed into the heart, liver, intestine, and so on, until all the body's tissues and organs are accounted for.

In the process of division and specialization, many of the nucleus's genes are shut down because the differentiated adult tissue will need only 10% or so of the full cellular complement of genes. Genes are also inactivated through a process known as imprinting, which tags certain genes as coming from the mother or the father and determines whether the maternal or the paternal genes will be active. Proper imprinting patterns seem to be especially important for normal fetal growth. If genes are contributed entirely by one parent, as in the case of cloning, the fetus might turn out to lack acceptable imprinting patterns and so be unable to grow normally. For this and other reasons, many scientists were skeptical that any differentiated adult nucleus could be used to clone a new organism. Dolly suggested that cloning from the cells of an adult was indeed possible. (Wilmut and colleagues did not test the cell used to make Dolly to confirm that it produced the proteins characteristic of a fully differentiated mammary cell, leaving open the possibility that Dolly came from an only partially differentiated stem cell.) Later research has confirmed that cloning from such cells is at least possible, if fraught with difficulty.

B. Cloning of Another Sort: Blastomere Separation

Identical twins are clones produced by nature. Through approximately the eight-cell stage, the human embryo is capable of splitting into two parts, each capable of becoming a fetus and then a child. Cloning by blastomere separation is simply a scientific label for the same sort of division of the early embryo that results in twinning. In the laboratory, human hands and human intentions guide the process. Freezing one or more of the clones thus created opens the possibility of implanting one copy, keeping the others in a deep freeze, and then thawing and implanting them at will. Since the mitochondrial as well as the nuclear DNA will be duplicated, these clones will be identical genetic replicas of each other, unlike Dolly and other clones created via nuclear transfer cloning.

Nuclear transfer cloning and blastomere separation raise many of the same issues, but they are sufficiently different to warrant separate discussions of their ethical implications at certain points in this article.

II. SUGGESTED USES OF CLONING

The potential benefits of cloning are arrayed generally in a rough matrix formed by two simple distinctions—one between animal and human cloning, and another between reproduction and research. The reproductive benefits of animal cloning call attention to the respect in which cloning is a kind of genetic engineering: animal reproductive cloning is attractive chiefly because of the control it gives over reproduction. Animal cloning could allow farmers or scientists, for example, to replicate animals that are valuable for their physical traits. By contrast, human reproductive cloning would be an ineffective way of reproducing the traits we find valuable in people, which have more to do with personality and so are heavily dependent on environmental influences.

Animal and human cloning could contribute to medical research in various ways. Both are believed to be vital for basic research on the way cells differentiate into the myriad tissue types found in the adult body. Such research could lead to the development of new tissue transplantation therapies. Animal cloning could also promote research by allowing researchers to mass produce especially useful research animals.

A. Animal Cloning

1. Agricultural

The agricultural benefit of animal cloning is the prospect of being able to produce very close physical copies of a valuable animal—a prize milk-producing cow, for example. The number of offspring produced by prize animals is already artificially increased through embryo transfer (allowing a prize female to produce more offspring, since surrogate mothers bear all or many of her fertilized ova) and artificial insemination (increasing the reproductive potential of males). Nuclear transfer cloning would provide a more precise way of replicating animals, however, since each product would differ from the original only with respect to the mitochondrial DNA, and would differ not at all if the ovum and nucleus were obtained from the same animal. Embryo splitting could also enable easier replication of genetically desirable animals, with the caveat that the replicas would be created from an embryo of an unknown phenotype.

This use of cloning has the drawback, however, of further restricting stock animals' genetic diversity, likely rendering them more susceptible to epidemics. Because the animals would be physiologically nearly identical, with identical immune systems, it would be easier for a disease to wipe out an entire herd than if the animals were genetically heterogenous.

2. Medical Uses

Related to the agricultural use of animal cloning are some specialized medical purposes. Cloning would allow pharmaceutical concerns to replicate animals that through germline gene transfer have been given properties not naturally occurring in their species. The purpose behind the research that led to Dolly, in fact, was to develop a way of replicating female goats altered so that their milk would contain medically useful substances. Similarly, germ-line gene transfer followed by cloning could enable medical researchers to produce animals altered so as to facilitate xenotransplantation. (Here, the genetic modification would involve altering the animal so that the organ to be transplanted into a human would be less likely to trigger a response from the human immune system, and therefore less likely to be rejected.) These modifications could be transmitted through conventional breeding since the genetic alteration is in germ-line cells, but cloning might be more efficient because of the genetic recombination that occurs in sexual reproduction.

Nuclear transfer cloning might also prove helpful not only in replicating genetically modified animals, but also in initially producing them. Currently, germ-line gene transfer must be accomplished by injecting the new DNA into ova. This step would be easier if the genetic modification could be performed on cultured somatic cells whose nuclei were then transferred into ova. Once an animal expressing that modification was successfully produced, of course, nuclear transfer cloning could also be used to replicate the modified animal. Embryo splitting could also help replicate the animal, if the embryo from which it was created had itself been split and the newly created embryos were then frozen.

3. Medical Research

Animal cloning might be useful in medical research in a variety of ways. First, using the strategies just described, cloning could be valuable both for generating and then reproducing animals whose genetic modifications make them useful research subjects. These animals include the commonly studied "knock-out" animals, in which certain genes have been deleted from the genome so that researchers can study their function. They could also include animals modified to mimic human genetic diseases, or with human genes, such as those that code for cystic fibrosis, inserted into their genome.

Using nuclear transfer cloning to help generate the animals would be especially useful in creating research stocks of large animals. The method typically now used to produce genetically altered animals for research was developed with mice and has not yet been adapted for larger animals, for which it would in any event be very slow because of large animals' longer gestation periods and smaller litters.

Both nuclear transfer and embryo splitting could provide a way of expanding a line of inbred laboratory animals. Inbred mice are commonly used in research because their genetic homogeneity simplifies the analysis. Cloning could be especially useful for propagating genetically homogeneous strains of larger animals because of the longer time necessary to breed them.

Finally, animal cloning could be helpful for basic research on how the DNA in embryonic cells is programmed and how differentiation into the many adult tissue types is begun. Such research might give insights into how partly differentiated stem cells (present in many places in the adult body) or even fully differentiated somatic cells could be directed to generate new tissues. A better understanding of reprogramming factors and basic cell differentiation could thus lead to powerful new medical therapies, which might themselves either involve or circumvent human cloning.

B. Human Cloning

1. Therapeutic Applications

Nuclear transfer cloning of humans holds the promise of a variety of therapeutic applications—distinguished from reproductive cloning because they would not involve the implantation of an embryo into a uterus. In particular, nuclear transfer might make possible the production of tissue lines to repair and replace damaged or defective tissue. Basic research on human cloning might be necessary to learn how somatic cells or partially differentiated stem cells can be directed to generate tissue lines. Human cloning might also itself be part of new therapies. If nuclear transfer were used to produce human embryos, then researchers might be able to extract embryonic stem cells—cells that have the capacity to differentiate into any type of tissue found in the body—and then culture them, direct their differentiation, and produce new tissue to replace diseased tissues. If the somatic cell from which the nucleus was transferred was obtained from the person who needs the tissue, then a fully histocompatible transplant would apparently be produced, surmounting the current problems of graft rejection. Alternatively, if the somatic cell were from some other source, it might be possible to reduce the chances of graft rejection by genetically modifying the nucleus prior to transfer.

Human embryo splitting could also serve therapeutic purposes. In principle, it should be possible to use embryo splitting to create reserve embryos that would be available either for thorough genetic testing (so that parents could ascertain whether an embryo they want to bring to term has genetic diseases) or for later cell-based transplant therapies, using the stem cell techniques just described.

2. Reproductive Cloning

Finally, of course, cloning could also be added to the arsenal of assisted reproductive technologies, in several ways. Indeed, human embryo splitting has been proposed primarily as an aid for those infertile couples who are now treated by removing eggs from the woman, artificially inseminating them, and transferring them to the woman's uterus. Embryo splitting offers a way of increasing the number of available embryos without undergoing multiple cycles of egg extraction.

The reproductive use of nuclear transfer cloning—that is, nuclear transfer in which the resulting embryo is implanted in a uterus with the intention of producing a human fetus—has been proposed for a wide range of cases, although many of these can be set aside at the outset because they are either plainly unethical or, given the complicated contributions to personality of both genes and environment, plainly impracticable. It has been proposed, for example, that cloning could be used to mass produce soldiers or slaves, who presumably would be genetically modified to bring out the right traits, or to make individuals to serve as sources for tissues or organs for transplant. It has also been proposed that cloning might be used to replicate famous or important people in order to preserve their peculiar talents or traits, that people might use cloning to achieve a kind of immortality, or that parents might want to replace a lost child with a duplicate. Given the range of environmental influences on personality, however, none of these reproductive uses could achieve the desired goal. Nothing in Mozart's genotype guarantees that a Mozart clone would be a musical genius, or would even enjoy music. Clone and cloned could not be identical for the simple reason that the clone could not be raised in an environment identical to that in which the original was raised. Indeed, if a clone were intended and regarded as a duplicate of someone else, this alone ensures a difference, realizable in the relationships the clone would have with its parents. The environmental differences could even lead to important physical differences. Perhaps the most basic of the environmental differences concerns the cytoplasm of the egg, which both influences the expression of the DNA and makes its own contribution to a person's genes through mitochondrial DNA, and which in itself might make the difference between excelling and failing in sports.

About a variety of other uses, however, there is ongoing moral debate. In some of these uses precise specification of genotype is, as in the cases above, the motivation for using cloning. For example, cloning is sometimes envisioned as a reliable way of producing a compatible source of biological material (such as bone marrow) for a person who would be seriously and perhaps terminally ill without it. Cloning could also allow a couple at high risk of having a child with a genetic disease to avoid that disease. This would work best if the mother were the carrier: a child could then be produced from one of her eggs and the nuclear DNA from the man.

Some uses of nuclear transfer cloning have been proposed in which cloning serves essentially as a replacement for traditional reproduction rather than as a way to replicate a genotype. In each of these cases, cloning is envisioned as providing a route around physical barriers to reproduction. Most prominently, nuclear transfer cloning is mentioned as a way of treating infertile couples—especially couples in which the male produces no sperm, for if the woman contributed an ovum and the man the nuclear DNA, both would be contributing in important biological ways to the child. Nuclear transfer cloning would also allow lesbian couples to have children to whom both have contributed biologic material—one by contributing the nucleus, and the other by contributing the egg.

III. MORAL AND RELIGIOUS ISSUES

In the debate about whether human cloning ought ever to be permitted, the decisive considerations so far have turned on the possible consequences: research on animal cloning seems to establish that at present, the likely physical harms caused by human cloning outweigh the possible benefits. These consequences are not, however, the focus of either those opposed to human cloning or those who think it might sometimes be permissible. Most of the ethical discussion turns on a range of other values: For those who favor permitting cloning, the individual freedom of researchers and parents is most compelling. Opponents are impressed more by a set of considerations that turn on the intersection between cloning and community. They fear that human cloning is likely to have an unfortunate effect on the family and on the social fabric, and they believe that it is conceptually at odds with our values—values we place on personhood, parent–child relationships, family, or our relationship to nature, for example. The values that move opponents to cloning can be thought of as a handful of the communally oriented values: unlike the value of individual freedom, which establishes some separation between the individual and society, they concern the bonds between people or between humans and nature.

In canvassing these diverse considerations, this section starts with reasons for permitting cloning and moves more or less from individual considerations to social and natural values and from consequentialist considerations to deontological considerations—not to suggest the greater or lesser importance of one or another end of these spectra, but just as a way of setting out the terrain.

A. Arguments for Permitting Cloning

1. Medical Benefit

The case for permitting some kinds of cloning might start by calling attention to cloning's potential medical benefits. This argument typically focuses on the medical advances that basic research on nuclear transfer cloning might make possible. Such research could likely contribute to our understanding of cell differentiation and might foster the development of the cell-based therapies suggested by research on human stem cells.

Reproductive cloning can also be construed as offering, for some people, an important medical benefit, namely, the ability to reproduce. As such, nuclear transfer cloning becomes the latest in a line of assisted reproduction technologies, and as applied in cases of infertility might even be subsumed under the head of "therapeutic cloning." Indeed, the U.S. National Bioethics Advisory Commission (NBAC) lists precisely such uses of nuclear transfer cloning as possible cases of therapeutic cloning, and the nongovernmental National Advisory Board on Ethics in Reproduction (NABER) argues in favor of permitting embryo splitting on grounds that it could prove useful in the treatment of infertility through artificial insemination (NBAC 1997, 30–31; NABER 1994, 267). Embryo splitting could increase the number of embryos available for implantation and reduce the number of egg retrieval procedures necessary for creating embryos.

Nuclear transfer cloning would be different from other assisted reproduction technologies in several important respects, however. For example, the genetic relationships that underlie traditional familial relationships would be replaced with new ones. Chiefly, a clone would be viewed either as having only one genetic parent (excepting the adult who has contributed mitochondrial DNA) or as having been derived from a sibling—as being, in other words, an identical twin of the clonee. The relationship to the person cloned would be different from that of identical twins, of course, in that one twin could be much younger than the other. Also, the usual genetic mixing that is involved both in traditional conception and in other assisted reproduction technologies would be absent; the clone's genetic makeup would be known in advance and possibly would be chosen deliberately.

2. Scientific Freedom

Allied to the argument from medical benefit is an argument concerning the freedom that Western cultures have traditionally accorded scientists in their quest for knowledge. This argument involves both a consequentialist consideration—namely, that the medical and other benefits that scientific progress has provided depend on granting freedom to researchers to pursue their interests and their most promising leads—and a deontological point—that the quest for knowledge is itself an important value, and one that ought not to be restricted without good cause. One might worry that preventing scientists from pursuing human cloning research inappropriately mixes science with religion in a way reminiscent of the restrictions imposed by the Catholic Church on Galileo. It should be noted, however, that scientific freedom is actually championed within some traditions; some Christian, Jewish, and Islamic scholars view scientific discoveries as revelations that are permitted by God, are of God's creation, and afford human beings an opportunity for moral development (NBAC, 45).

The argument for scientific freedom seems to apply primarily to therapeutic rather than reproductive uses of cloning, since reproductive cloning is motivated not by the quest for knowledge but by the quest for children. Some think, however, that the argument does not work straightforwardly even on those limited grounds. The idea that scientific freedom could be justifiably limited by moral considerations is not a completely novel concept. Scientists are commonly restricted in their research on human and animal subjects. In research on human subjects, for example, they are forbidden to perform research on unconsenting subjects, and they are additionally typically required to show that the societal benefits of a research protocol outweigh the risks and harms imposed on subjects.

3. Parents' Right of Reproductive Freedom

The argument most commonly advanced for permitting human cloning is that, at least in some cases, human cloning falls under the range of personal liberties granted in Western societies, specifically under parents' right of reproductive freedom or "procreative liberty." Having a biological connection to one's own children is believed by many to contribute to the most personally satisfying parent–child relationships. Through that link, parents participate in the creation of a person and affirm their love for each other. Raising a child conceived through nuclear transfer cloning techniques could offer an alternative route for realizing these benefits (Strong 1998, 281–282) and cloning techniques in general could be viewed as simply one more of the assisted reproduction technologies. The right to have biologically related children is already believed to permit the use of an assortment of mechanisms for selecting the child's ge-

netic inheritance, currently through prenatal genetic testing and selective abortion.

Whether the right of reproductive freedom generates a right to have a child through cloning might depend on a number of other variables. For example, it may be crucial that the biological parents intend to raise the child. Also, it may be that an especially strong case can be made for permitting cloning through embryo splitting, which is envisioned primarily as a way of helping to treat couples who are using *in vitro* fertilization to overcome gametic insufficiency.

Nuclear transfer cloning can also be envisioned as a kind of infertility treatment. Thus some have argued that the strongest argument for permitting reproductive cloning applies in cases of "reproductive failure," such as when one of the parents is infertile or does not produce enough gametes to make sexual reproduction likely. Such uses could include using either nuclear transfer cloning from embryos to increase the number of embryos available for implantation into the uterus (here, nuclear transfer is essentially an alternative to embryo splitting) or nuclear transfer cloning to completely eliminate the need for gametes from one of the parents. Some "exceptional" cases of nuclear transfer cloning involve especially pressing cases of reproductive failure, suggesting the possibility of even narrower restrictions on the range of cases that could be permitted. A candidate for cloning considered by NBAC involves especially pressing circumstances: suppose a sterile man were the only member of his family to have survived the Holocaust; since without nuclear transfer cloning his family line would be eliminated, to deny him the use of cloning is to compound the tragedy of the Holocaust (NBAC, 54–55).

Such cases contrast with those in which the parents could reproduce sexually but simply prefer to use cloning because it allows them to select their child's genome. These would have to be defended not on the basis of the right to have biologically related children but on the basis of a more general right to select one's child's genome (Robertson 1999, 618).

One might worry, however, whether the distinctions would hold up against the flood of cases that some expect if cloning were permitted. Simply not permitting cloning establishes a bright line, relatively easily enforced. Deciding which cases are "exceptional" is difficult. Permitting certain classes of cases also might generate pressure to widen the range of permitted cases. For example, permitting embryo splitting for infertile couples might generate pressure to permit the freezing of some of the embryos. Once frozen, there might be pressure to permit the parents to bring the "back-up" embryo to term partly in order to save the first, such as by providing tissue. And permitting embryo splitting with cryopreservation of some embryos might generate pressure to permit nuclear transfer cloning, since nuclear transfer provides another way to produce a child that might help save the first. But such use of nuclear transfer would probably be considered justifiable only if the child would also be wanted and loved in and of itself; indeed, some might argue that the *primary* purpose behind such a use of nuclear transfer technology would be the straightforward one of having a second child. Thus permitting nuclear transfer cloning to help save an earlier child might make it difficult *not* to permit using the technique simply to have children (Baird 1999).

B. Arguments for Restricting Cloning

1. Harms to Children

The initial and most telling concern commonly raised about cloning is that it might harm the child created through cloning. The possible harms are of several sorts.

Most obviously, given the uncertain state of science surrounding cloning, the child created through cloning might be physically harmed in important but perhaps as yet unknown ways. For example, it may be that splitting the cells of a blastomere damages those cells in some way undetectable at the early embryo stage. Nonetheless, perhaps since embryo splitting closely parallels the natural process by which identical twins are created and does not involve tampering with the DNA within cells (as in nuclear transfer cloning), NABER did not emphasize in its report the possible harms to fetuses. In contrast, concern about the physical dangers was the primary impetus for NBAC's recommendation that there be no federal funding of research on human cloning through nuclear transfer. "At this time," stated NBAC, "the significant risks to the fetus and physical well being of a child . . . outweigh arguably beneficial uses of the technique" (p. 63).

Fears about the physical harms of nuclear transfer cloning are prompted by the research on animal cloning. A high percentage of the animal embryos created through cloning have not been brought successfully to term, and a high percentage of those that were died shortly after birth. Dolly was the only sheep produced out of 277 attempted fusions and 29 implantations. Attempts to clone cows were more successful but still led to many failures. The 8 calves produced by cloning in Japan were the result of 249 attempts, which led to 38 embryos. Ten of the 38 embryos were transferred to cows, and of these, 8 were brought to term. Four died shortly after birth.

These success rates are cause for concern about possible harms both to the child and to the mother. If the success rates were repeated in human cloning, then pro-

ducing a single child would typically require many attempts, with many children dying during gestation or shortly after birth. Further, if there were only one candidate for egg donor and one candidate for birth mother, then the egg donor would be subject to considerable hormonal manipulation and the birth mother might undergo multiple miscarriages.

Fears about physical harms to children have been exacerbated by the continued research in animal cloning. About half of the animal fetuses created through cloning have exhibited serious abnormalities, including defects in the heart, lungs, and other organs. Some now believe that some of these abnormalities may result from a disruption of genetic imprinting, a poorly understood mechanism that in normal reproduction "tags" genes as coming from either the mother or the father and determines which genes will be active in various parts of the embryo. Embryos are believed to develop normally only when the genes are properly imprinted, and it may be that the genes in the somatic cells used in nuclear transfer cloning are no longer properly tagged (Weiss 1999).

Further concerns have been sparked by research on Dolly. Although Dolly appears healthy to date, the Scottish researchers who produced her report that her DNA may be structurally similar in an important way to the DNA of a significantly older sheep. Apparently her telomeres—strips on the end of each chromosome—are shorter than is normal for a sheep of her age, so that her biologic age in effect may be greater than her chronologic age. Simply put, she may have old DNA. Other scientists have contested this claim, however, and it is also unclear what effect the shorter telomeres would actually have on the ability of Dolly's cells to sustain a sheep's normal life span.

Some have argued that the mere possibility that cloning might harm children bars any attempt to engage in human cloning. Given differences in embryonic development between species, it might be impossible to fully discount the possibility of such harms on the basis of animal studies. Possibly their presence could be determined only through research on human subjects. The central requirement for the moral conduct of human research is that the subjects give their consent to the research before entering it unless the benefit of research seems likely to greatly outweigh its harms or risks. Since the child created through human cloning is unable to consent and the harms and risks cannot be pinned down, the argument goes, the research is impermissible.

Some find this threshold objection unconvincing and even absurd. The fact that the child cannot consent because it does not yet exist raises several responses. Arguably it is unintelligible to ask whether someone who does not exist holds rights. And if a nonexistent person cannot hold rights, then there would be no reason to grant that person the right not to participate in medical research without first granting consent. Alternatively, it is sometimes argued that the objection fails because human cloning is one of the exceptions to the rule requiring consent—it is a case in which the benefit of research greatly outweighs the harms or risks. On the one hand is some possible physical damage, and on the other the good of existence, and it is assumed that it is better to exist, with maybe some disability, than not to exist at all. This response seems at face value, however, to permit subjecting a child-to-be to any degree of harm or risk, as long as that harm is made a condition of existing. Thus this argument may justify too much.

Some other harms to which a child created through nuclear transfer cloning might be subjected are psychological. The central thought here is that because the child would in some sense be a product of somebody else's genetic material, and (depending on the case) might have been produced precisely because of the value placed on that genetic material, the child would have a straitened sense of self. Most children go through a process of self-discovery. They are importantly different from their ancestors, and it is up to them to find out how. A clone would not enjoy the same process of self-discovery, according to some who object to cloning, because it would perpetually be confronted with the example of its genetic precursor. If the clone were created from another child and raised by that child's parents, then there would be an expectation that it would either measure up to its older twin or, if the other child had died, would become what the parents thought the other child could have become. If the child were created from an adult who would also be its parent, then it would have to contend with that parent's accomplishments and beliefs and desires about herself. Whatever the relationship, the child would in effect have a much older twin, different from the relationship that natural twins have with each other partly because of the delay between their births and partly because one child would have been produced from the other rather than emerging simultaneously with it.

This point can be turned around to emphasize the parents' expectations of the child instead of the child's sense of self. Prospective parents who employed cloning might expect that cloning would produce a child with particular characteristics they desired. But since genetic determinism is false—that is, having a particular genome does not guarantee having specific traits, and especially does not ensure a specific personality and character—there is likely to be a mismatch between parental expectations and the child's actual characteristics. A strong belief in genetic determinism, then, could lead to unrealistic parental expectations and deeply troubled relationships between parents and child.

The possible harm of having to follow in the footsteps of one's genetic twin, and perhaps be raised by one's twin, is closely linked although not identical to another possible psychological harm often expressed as a loss of uniqueness or individuality. This claim is that an intrinsic feature of one's self-identity is a sense that one is by nature different from everyone else. The loss of uniqueness differs from the harm of following an older twin in that it need not have an effect on the clone's development; it is simply a good that is lost. While this would be restricted to clones created either through nuclear transfer or through embryo splitting in which one of the embryos had been frozen and used much later, the loss of uniqueness would also affect clones created through embryo splitting and brought to term simultaneously.

The response to these objections is that cloning is not fundamentally different from our current experience. To the concern about a child's self-discovery, one might respond that parents have always tried to control their children's sense of identity, and that the clone's sense of being able to craft her own future is no more disabled than is any other child's. All children have to contend with their parents' hopes and with the examples set by parents and older siblings. A response to the concern about loss of uniqueness takes the form of a dilemma: either what is important about uniqueness is having a unique genome, in which case the experience of homozygous twins simply disproves that uniqueness is as crucial as the objection maintains, or what is important about uniqueness is not threatened by genetic identity. Even natural-born twins vary in myriad ways from each other, after all. Further, a gap between the birth of twins might ameliorate rather than exacerbate the problem of individuality. The fact that a clone of an existing person would be raised differently—in a different environment and perhaps by different parents, or at least by parents with different attitudes—establishes that the clone would not replicate the earlier person. Likewise, it might imply that the clone's *sense* of individuality would not be impaired.

2. Cloning and the Family

Some objections to the reproductive use of nuclear transfer cloning involve either the effect that it might have on the family or the clash between cloning and values instantiated in the family.

One concern, advanced especially from some religious perspectives, is that human cloning would muddy the traditional biological lines within the family and that because these biological connections underpin family members' sense of kinship, cloning would tend to undermine the familial reponsibilities and support networks that are crucial for children's development. The muddying would presumably arise largely because of ambiguity about a clone's biological parentage. Perhaps the clone would have only one parent—namely, the person who donated DNA for nuclear transfer. Others who might be considered biological parents include the donor of the egg, the woman who brings the fetus created through cloning to term, and the biological parents of the person cloned (under the theory that the clone is best thought of as a delayed twin).

The worry, then, is that the ambiguity of the biological connection will attenuate the parent–child bond. For example, if the clone is created as the child of a heterosexual couple in which the male is sterile, then the female, contributor "only" of cytoplasm and mitochondrial DNA, may think she is less a true parent. The male, contributor of the nuclear DNA, may think of himself as partly parent and partly sibling. Further, one might worry that parallel ambiguities in the other biological connections that traditionally delineate the family will debilitate those relationships. The woman's family might not feel that the child belongs to them in the same way that it belongs to the man's family. If brothers and sisters are unsure about which generation and family the clone belongs to, their bond to the clone might also be attenuated.

There are a variety of responses to these worries. It might be that the traditional biological connections are less important than the objection supposes, or that they could and should be less important—indeed are becoming less important as our understanding of the family evolves. Perhaps those with the best claim to being a child's "true" parents are its "social" parents—those who shoulder the burden of actually raising the child. Along this line, it might be claimed that both adoption and established assisted reproduction technologies also confuse or obscure biological connections, but are not widely held to have damaged familial ties. It should be noted, however, that some opponents to cloning also object to other assisted reproduction technologies on precisely this ground.

A second concern about cloning and the family is that cloning jeopardizes "the goods of sexual love and procreation" (NBAC, 51). When a child is conceived sexually, the child is a sine qua non of their physical union, itself a manifestation of their love. If parents conceived through the artifice of nuclear transfer, conception would not be a physical manifestation of their own commitment to each other.

Yet some hold precisely that cloning is best understood as an alternative way of achieving and manifesting just such commitment. Perhaps what is crucial about satisfying parent–child relationships is not that the child is a manifestation of a sexual relationship between the parents, but that there is a biological connection between the parents and the children. Through that link, parents participate together in the creation of a person and affirm their love for each other. Raising a child conceived through nuclear transfer cloning offers a way of maintaining this biological relatedness (Strong).

A third concern, related to the concern about the psychological effect of cloning on the child but centering on a deontological point rather than a consequentialist one, involves the parents' attitudes toward parenting. Cloning would differ from other assisted reproduction technologies in that, at least in some cases, it might be chosen over other ways of creating a child because the parents want precisely that genome, or at least something contained in that genome. Such a goal evinces an attitude toward parenting that many find unacceptable. Having a child is often compared to receiving a gift: the parents are unsure precisely what it is they will get, they try to keep their ideals and expectations reasonable and reasonably vague, and they pledge to love it and want it regardless of what its precise characteristics turn out to be. To specify the child's genome—to gain control of something important about the child and to try to control its development—is to move away from this ideal.

Possibly this objection tells only against certain uses of cloning. Although parents would be de facto specifying their child's genome in any cloning, only sometimes would this be the motivation for choosing the technique. If the parents viewed the specification of the child's genome as an accident, however—a side effect of the only method available to them of rearing a biologically related child—then they are not obviously abandoning traditional parenting ideals.

Another, broader response to this objection is the same as that given to the objection about psychological harms, namely, that cloning would not mark a significant departure from current practices. Parents already strive mightily to determine the kind of people their children will grow up to be, not only by influencing their development but also increasingly by selecting their genetic traits (using prenatal or preimplantation genetic testing followed by abortion). However, those who find the attitudes expressed in cloning anathema to good parenting may dispute the analogy between influencing a child's development through child-rearing and selecting its genome, and may also object to some uses of prenatal genetic testing and selective abortion. Even if raising a child and selecting its genome both have the goal of producing a good child, the different means may make a moral difference, analogous to the difference between achieving athletic greatness through conditioning and training and achieving it through chemical performance enhancers.

3. Effect on Society

Concerns about cloning's effect on children and on the family lead to a broader concern about the effect of cloning on society. This concern was perhaps sharper in earlier discussions of cloning, when the possibility that cloning might be used to create armies of slaves or soldiers, or multiple duplicates of moral monsters or geniuses, was seen as a more legitimate worry. Nonetheless, there might be other reasons to worry about the effect of cloning on society. First, if it is correct that cloning might weaken families, it might also be detrimental to the many socially important values that people acquire primarily through the moral training received in their families. Likewise, if cloning represents a new model of the parent–child relationship, one geared less toward supporting the child's self-awakening and maturation and more toward meeting parents' ideals, a range of values that characterize the traditional model—love, nurturing, loyalty, and steadfastness—might be displaced by values suggested by the second model—vanity, narcissism, and avarice (NBAC, 68).

For some critics, nuclear transfer cloning also produces societal concerns that arise even more sharply with the other genetic engineering technologies to which cloning is linked. To the extent that cloning can be used as a device for specifying a genome and thus avoiding mental and physical disability, and since nuclear transfer cloning lends itself to other techniques for genetic engineering, cloning raises a concern about a kind of eugenics—different from the way eugenics is often conceived because it would not be governmentally directed or organized, but would involve instead a broad-based pressure to conform to societal norms and to compete effectively in a capitalist economy. Further, if the genetic engineering facilitated by nuclear transfer cloning were available only to the wealthy, then cloning might contribute to a deep and self-perpetuating social division between those who can genetically enhance their children and those who cannot.

As many have noted, these objections are highly speculative. Further, the purported effects might have only a marginal societal impact. Given the logistical complexity and likely considerable expense of creating a clone, the appeal of raising a child conceived traditionally, and the possibility that there will be other ways of treating the disabilities that cloning could sometimes help parents circumvent, the use of cloning might turn out to be restricted only to special cases, even in the absence of regulations. And finally, especially if cloning turned out to be used only infrequently, concern about the effects of cloning on society might generate reasons not for banning cloning altogether but for restricting and regulating it carefully.

4. The Value of Personhood

One of the most egregious possible effects on society, and a theme underlying many of the objections so far canvassed, is that cloning might radically change the way we understand personhood. The fundamental thought here is that while no one disputes that clones would be fully human and entitled to the full respect due any human being, permitting and engaging in human cloning would intrinsically violate our conception of human dignity. The claim is entwined with the objection about the possible psychological harms that cloning might give rise to, differing in that this claim emphasizes not the consequences of cloning for the child but its conceptual implications for the value of personhood.

One objection along these lines is that cloning would conflict with human dignity because it denies the uniqueness of personal identity that human dignity seems to imply. That is, cloning might force us to regard people as repeatable, and accepting that people are not one-time occurrences is to allow the value of personhood to be diminished. Or perhaps, more specifically, the problem is not the simple fact that uniqueness could no longer be considered part of personhood, but the particular way in which cloning subverts uniqueness: in nuclear transfer cloning, one person is in some sense the product of the other, and it is this idea of one person originating in another—of a person being replicated or duplicated—that diminishes the value of personhood.

It is not clear that either way of putting this objection can stick, however. While clones would certainly not have unique DNA, natural twins lack genetic uniqueness without confounding our beliefs about personhood. And even if the problem of nonuniqueness is specified as the problem of "having been replicated," whether the clone's uniqueness would be jeopardized in any important sense is unclear. There are other forces—prominently the environment, but also the cytoplasm and mitochondrial DNA of the egg—that join with the nuclear genetic contribution to produce a person and that would differ between clones. In fact, for clones who came into the world at different times, environmental factors would differ more sharply than they do between natural twins. Further, the uniqueness moral philosophers have imputed to each human being may be a function neither of genetic endowment nor of overall personality, but of the nature of rationality itself: each person has a perspective on the world or relationship to the world that can never be anything but solely that person's (NBAC, 65; Jonas 1974, 160). This sense of uniqueness could not be assailed through cloning.

A second objection rooted in the notion of human dignity shifts attention from the nature of what is being acted on to the nature of the act performed on it. Some fear that to replicate people through cloning is to objectify them—to treat them as objects. The worry is that because reproductive cloning would specify the child's genetic makeup, employing it amounts to trying to design and control human identity. Since the purpose of designing human nature would be to achieve certain desired goods, to permit cloning is to condone treating human nature as a means to an end. If design and control were dictated by market forces—so that, for example, the capacity to design children was sold for a profit, the price was set by market demand, and the demand was driven by economic incentives—then human traits would be not only objectified but also *commodified*, that is, treated as a thing to which we can put a price. Such treatment would violate the seeming pricelessness of human nature identified by Kant and by many religious traditions (NBAC, 50–51).

The most common response to this objection is to grant it but limit it. Since it concerns the way people might use cloning, it tells only against certain kinds of uses. Certainly creating the perfect slave, someone who is obedient, fearful, and of limited intelligence, would be ruled out (assuming it is even possible). In general, cloning would objectify human nature only when it is employed in order to try to *specify* human nature.

There are also alternative perspectives on the conceptual relationship between human dignity and such genetic technologies as cloning. One radically opposed view is expressed especially sharply by Joseph Fletcher, according to whom the ability granted by genetic engineering to control human nature is itself a natural product of human nature. Directing a child's development, and escaping from the caprices of chance, is for Fletcher not only consistent with human dignity but a natural way of expressing it (Fletcher 1988).

5. Control over Nature

If human dignity is seen as deriving not from human characteristics themselves—uniqueness, irreplicability, rationality, or pricelessness—but from an exogenous source of those characteristics—God or nature—then the belief that cloning would violate human dignity is related to a concern about human control over nature. This is an objection often associated with some religious traditions. If determining human identity is seen as the domain of God, then cloning—a means of specifying an important factor in a person's identity—may seem to infringe on God's agency. The objection is also some-

times framed in a secular way, however: one might hold that an individual's identity ought to be a natural phenomenon, and that specifying an important factor in a person's identity asserts human agency where it does not properly belong.

Perhaps the best initial response to this sort of objection is simply to observe that medicine is filled with interventions into a person's "natural" physiological characteristics or into the course of nature. Indeed, there are already a range of assisted reproductive technologies that are analogous to cloning in important respects and that have gained broad acceptance. It is relevant, too, that when these technologies were first proposed, there was an initial groundswell of concern about the encroachment of technology into sex and reproduction: witness the press surrounding the first "test-tube baby," so named because fertilization occurred *in vitro*. Even the use of anesthetics during childbirth was initially considered by some to be an unacceptable human intrusion into the natural order.

This line of thought leads to the further point that the objection from control over nature is grounded in a simplistic or misconceived notion of the relationship between humans and nature. Indeed, many in the Western religious traditions—Christian, Jewish, and Islamic—hold views about the human relationship to nature that differ strikingly from the one underlying the objection about excessive human control. The NBAC report identifies various models that theologians have adopted to describe human dominion over nature. One model portrays humans as stewards of God's creation, assigned by God with maintaining and preserving nature. Whether this model generates guidelines for cloning is ambiguous, since what medical and technological interventions are permitted by the mandate to "maintain and preserve" nature is open to debate. In another model, humans are "partners" with God in the act of creation, and perhaps even "created co-creators"— finite beings who are dependent upon God but are intended by God to improve upon creation. Proponents of these views may interpret God's act of creation not as a one-time deed, as the biblical story of creation is often interpreted to suggest, but as an ongoing process. If creation is ongoing, God might intend humans to play their own role in the process. And rather than being restricted to the outlines of God's plan, humans might face an essentially "open future" that they are called upon to develop. At the extreme end of this spectrum, Fletcher—an Episcopalian theologian—has argued that cloning and other genetic engineering technologies ought to be pursued because humans are intended by their creator to act in ways that are deliberate and designed—to exercise dominion over their own nature (1988).

Not all of these models necessarily condone unregulated use of cloning, however. They might be more favorably disposed toward therapeutic cloning than toward reproductive cloning, and if they allow that reproductive cloning is permitted by a right of reproductive freedom, they might grant it only under certain conditions, such as when cloning is the only mechanism for raising a biologically related child, or when the parents intend to raise the children.

C. Moral Arguments and Policymaking

These considerations do not lead neatly to any clear-cut long-term social policy. One reason for this is that most of them do not advance clear-cut, for-or-against conclusions. Although concerns about the physical harms of cloning are widely believed to merit a full ban on human cloning at least in the short term—pending further animal research—most of the other arguments can be construed as advancing reasons for permitting or denying specific kinds of uses, or for permitting but carefully regulating certain uses. Oddly enough, in light of the concern that cloning might be accessible only to the wealthy, perhaps cloning would ideally be simultaneously constrained and supported, in that it would be restricted to certain kinds of uses yet made available to anyone who wanted to put it to those uses.

A second reason these moral considerations might not lead neatly to a straightforward social policy is that it is not clear how or whether they should influence policy at all. Some of the objections, for example, are rooted in some particular metaphysical or religious view and perhaps should not be allowed to set policy for a pluralistic society. Even if the claims that cloning is in some way contrary to human dignity or to a proper relationship with nature were compelling to a majority of people, ought those claims to be enshrined in law against the possibly differing views of the minority? In a libertarian view, a political system should be committed to the principle that an individual's freedom should be constrained only to prevent harm to others, and arguments that are rooted in conceptions of human dignity or humans' relationship to nature should be restricted to the individual, "private" sphere of decision making. Some of the other objections concern the consequences of cloning for society and so may be more likely to generate reasons that can carry weight in public policy making, but because they do not concern the effect of cloning on any identifiable individual but rather on the general good, they may not carry *enough* weight. Democratic political systems do often work to achieve the general good, not merely preventing harm to identifiable people—witness most environmental regulation and much legislation aimed at improving education. But

how such considerations impinge on personal freedom is of course quite controversial.

IV. POLICY DEVELOPMENTS

The political reaction to the possibility of human cloning has been strongly denunciatory around the world. The quickest response has been among a variety of international organizations, including the World Health Organization (WHO), the International Bioethics Committee (IBC) of the United Nations Economic, Scientific, and Cultural Organization (UNESCO), the Council of Europe, the European Commission, the Denver Summit of Eight (comprising the Group of Seven countries plus Russia), and the International Ethics Committee of the Human Genome Organization (HUGO). By and large, these groups have based their opposition to human cloning primarily on the claim that it is contrary to the values attaching to personhood. UNESCO's *Universal Declaration on the Human Genome and Human Rights* declares flatly that reproductive human cloning is "contrary to human dignity"; WHO adopted a resolution declaring human cloning, along with other ways of manipulating human genomes, contrary to human integrity and morality; and HUGO's *Statement on the Principled Conduct of Genetic Research* holds that reproductive human cloning—aimed at producing a child—would violate human dignity and freedom, although it accepts research on human and animal cloning (Eiseman 1999).

Of these international documents, the only legally binding agreement is the Council of Europe's *Convention for the Protection of Human Rights and Dignity with Regard to the Application of Biology and Medicine: Convention on Human Rights and Medicine*. An *Additional Protocol* added to the document explicitly prohibits both nuclear transfer cloning and embryo splitting, although the original document already implicitly prohibited cloning by prohibiting genetic interventions that would modify descendants' genomes (Eiseman, 42). The document has been signed by many European countries, which have thereby committed themselves to enacting legislation that bans cloning. The United Kingdom declined to sign, on grounds that the agreement is contrary to its tradition of scientific freedom.

Some nations have also taken legislative action to ban reproductive human cloning (Eiseman, 32–36). These bans range from the United Kingdom's narrow prohibition of nuclear transfer cloning to produce children—permitting use of the technology to produce embryonic stem cells—to Denmark's broad prohibition on research on any means of producing genetically identical persons. In some countries—Norway, Slovakia, and

Sweden—the legal prohibition of cloning is only implicit in legislation enacted prior to news of Dolly. Other nations that have legally banned human cloning include Australia, France, Germany, Israel, Malaysia, Peru, Spain, and Switzerland.

Policy initiatives in the United States have to date been spearheaded by the federal government, and especially by the National Bioethics Advisory Commission. In its 1997 report, NBAC concluded that human cloning is currently morally unacceptable primarily because it constitutes an unacceptable form of human experimentation. At this early stage of the technique's development any attempt to clone a human child would be research on two parties: the woman carrying the embryo or fetus created by cloning, and the child thus created. Given the risks to both mother and child evidenced by early efforts to clone animals, NBAC concluded that the technique imposes risks far too grave to be justifiably performed on humans. Should cloning in animals eventually be proven safe and reliable, then this risk-based objection to cloning would become less compelling. NBAC thus recommended continuation of the moratorium already established by President Clinton on the use of federal funding for any attempt at human somatic cell nuclear transfer, and called for legislation to prohibit anyone, whatever the source of funding, from attempting human nuclear somatic cell transfer cloning. It cautioned, however, that the legislation should be carefully crafted: it should expire after three to five years to ensure renewed attention to the ban, and it should be written "so as not to interfere with other important areas of scientific research" (NBAC, iv). Animal cloning and cloning of human DNA sequences and cell lines should, in NBAC's view, not be prohibited. Pending a legislative ban, NBAC recommended that the government request private compliance with the intent of the moratorium.

In the United States, existing legislation establishes at least a partial block to attempts at human cloning. In 1996 and 1997, Congress prohibited the use of federal funds for any research that exposes embryos to risk of destruction for nontherapeutic research, and in 1998 and 1999, language was added to similar legislation that defined a human embryo as including organisms created by cloning. These laws have succeeded in severely curtailing research on human cloning. In October 1998, the Food and Drug Administration (FDA) announced that under the provisions of the Public Health Srvice Act it already has regulatory authority over human cloning and that it will prosecute anyone who attempts to clone a human being without first obtaining FDA approval. The basis for this authority has been contested, however.

Legislation that would more explicitly and broadly ban cloning has been proposed at both the federal and the state level. On June 9, 1997, President Clinton introduced a Cloning Prohibition Act that would adhere closely to NBAC's recommendations for federal legislation, and the 105th and 106th Congresses introduced a variety of bills that would restrict cloning in various ways. Some of these bills would have broadened the power of the federal purse to limit cloning by prohibiting federal payments to any organization that engages in cloning. Others would simply have made cloning or some uses of cloning illegal. Some specified that a product of somatic cell nuclear transfer may not be implanted in a uterus—more or less in line with NBAC's recommendations—while others sought to flatly ban human cloning or the use of human somatic cell nuclear transfer technology. None of these were enacted into law. Finally, a number of states have enacted legislation that extends the federal restrictions to privately funded efforts. Some states prohibit research on human embryos, including those created through cloning. As of this writing, five states have enacted legislation that prohibits or restricts human cloning—California, Louisiana, Michigan, Missouri, and Rhode Island—and four other states have legislation pending (Eiseman, 24–26).

V. CONCLUSION

Clearly the moral considerations that have been persuasive in the United States differ somewhat from those that have gained the upper hand in Europe. While at least at the national level the United States has based its restrictions on cloning on the physical dangers involved, in Europe concerns about the violation of human dignity have been at the forefront. These different grounds have led to a further important difference between the U.S. and European responses: the federal moratorium will be periodically reassessed, since the dangers of cloning might diminish with further research on animal models, but the European documents call for permanent bans.

It remains to be seen whether the different reactions people have had to cloning will converge on a common position. Certainly the tone of at least the academic debate about cloning has evolved over the years. Some of the alarmist responses of the 1970s, involving implausible scenarios about mass-produced armies and slaves, have been replaced by a more carefully reasoned discourse in which the possible consequences of cloning reflect the actual science.

Objections to cloning rooted in human dignity and excessive control over nature have undergone less change, however, and it is anyone's guess whether they will eventually fade, following the example of similar objections to *in vitro* fertilization, or, given the growing restlessness about the encroachment of technology into human life and of humanity into the environment, only become sharper and more distinct.

Bibliography

American Medical Association Council on Ethical and Judicial Affairs (1999). *The Ethics of Human Cloning.* CEJA Report 2-A-99. Chicago, IL: American Medical Association.

Baird, P. (1999). Cloning of animals and humans. *Perspectives in Biology and Medicine* **42**(2), 179–194.

"Cloning Symposium" (1997). *Jurimetrics* **38**(1).

Eiseman, E. (1999). *Cloning Human Beings: Recent Scientific and Policy Developments.* Report prepared for the National Bioethics Advisory Commission. Washington, DC: RAND.

Fletcher, J. (1988). *The Ethics of Genetic Control: Ending Reproductive Roulette.* Buffalo, NY: Prometheus Books.

Jonas, H. (1974). *Philosophical Essays: From Ancient Creed to Technological Man.* Englewood Cliffs, NJ: Prentice-Hall.

Murray, T. (1996). *The Worth of a Child.* Berkeley: Univ. of California Press.

NABER (1994). Report on human cloning through embryo splitting: An amber light. *Kennedy Institute of Ethics Journal* **4**(2), 251–282.

NBAC (1997). *Cloning Human Beings,* Vol. 1: *Report and Recommendations of the National Bioethics Advisory Commission.* Rockville, MD: National Bioethics Advisory Commission.

Robertson, J. (1998). Liberty, identity, and human cloning. *Texas Law Review* **76**(6), 1371–1456.

Robertson, J. (1999). Two models of human cloning. *Hofstra Law Review* **27**(3), 609–638.

Strong, C. (1998). Cloning and infertility. *Cambridge Quarterly* **7**, 279–293.

Weiss, R. (1997). Clone defects point to need for 2 genetic parents. *The Washington Post,* 10 May.

Wilmut, I., *et al.* (1997). Viable offspring derived from fetal and adult mammalian cells. *Nature* **385**, 810–813.

Computer and Information Ethics

SIMON ROGERSON

De Montfort University

I. The Uniqueness of Computers
II. Privacy and Monitoring
III. Information Provision
IV. Software as Intellectual Property
V. Organization Structure and the Location of Work
VI. Computer Misuse
VII. Developing Information Systems
VIII. Computer Professionalism

GLOSSARY

information systems A multidisciplinary subject that addresses the range of strategic, managerial, and operational activities involved in the gathering, processing, storing, distributing, and use of information, and its associated technologies, in society and organizations.

intellectual property rights Rights that encompass confidential information, patents, trademarks, and copyright.

software A general term encompassing all programs that are used on computers; it can be divided into **systems software,** which controls the performance of the computer, and **application software,** which provides the means for computer users to produce information.

teleworking Working in flexible locations and at flexible times using computers while ensuring that the needs of the organization and of the individual are catered for.

COMPUTER AND INFORMATION ETHICS came into being as computer technology advanced and people started to become aware of the associated pitfalls that threatened to undermine the potential benefits of this powerful resource. Computer fraud and computer-generated human disasters were indicative of a new set of problems arising from this advancing technology. Perhaps the earliest recognition of this new set of problems was Donn Parker's "Rules of Ethics for Information Processing" (1968. *Communications of the ACM,* **11,** 198–201). By the mid-1970s such issues had been grouped together under the term "computer ethics" (coined by Walter Maner) that represented a new field of applied professional ethics dealing with problems aggravated, transformed, or created by computer technology. Deborah Johnson (D. G. Johnson, 1985. *Computer Ethics.* (1st. ed.). Englewood Cliffs, NJ: Prentice-Hall) defined computer ethics as being the study of the way in which computers present new versions of standard moral problems and dilemmas, causing existing standard moral norms to be used in new and novel ways in attempt to resolve these issues.

This is a narrow scope of computer ethics that focuses on the application of ethical theories and decision procedures used by philosophers in the field of applied ethics. Gradually this scope has been extended, as illustrated by James Moor's definition of computer ethics as the analysis of the nature and social impact of computer technology and the corresponding formulation and justification of policies for the ethical use of such technology (1985. In T. W. Bynum (Ed.), *Computers and Ethics.* Oxford: Blackwell).

The current broad perspective of computer ethics embraces concepts, theories, and procedures from philosophy, sociology, law, psychology, computer science and information systems. The overall goal is to integrate computing technology and human values in such a way that the technology advances and protects human values, rather than doing damage to them (T. W. Bynum, 1997. *Information ethics: An introduction.* Oxford, Blackwell). The term "information ethics" is becoming widely accepted as a better term for this area of applied ethics. This is because, firstly, the computer has evolved into a range of forms including the stand-alone machine, embedded computer chips in appliances, and networked components of a larger, more powerful macro-machine, and so the word "computer" is now misleading. Secondly, there has been an increasing convergence of once-separate industries to form an information industry that includes computers, telecommunications, cable and satellite television, recorded video and music, and so on.

I. THE UNIQUENESS OF COMPUTERS

The case of a company operating a nationwide network of service engineers illustrates what can go wrong if the implications of using computer systems are not carefully and fully investigated. The company had been suffering from several thefts of its service vehicles when parked at night. The attraction was not the expensive, though specialized, service equipment in the vehicles, but the engines of the vehicles themselves, which apparently had a high resale value. The company decided to attach electronic tags to the vehicles, enabling vehicle movement to be monitored from a central office. At night it was possible to place an electronic fence around the vehicles, and should an attempt to move the vehicle beyond the fence occur an alarm was triggered at the central office and the police alerted. The system proved highly successful and thefts were reduced dramatically. The management of the company then realized that this system could be used to monitor indirectly the movements of the sales engineers throughout the working day, providing information about abnormal activity instantaneously and without the knowledge of the engineers. The computer manager was briefed to develop this spin-off system, and therein lies the problem—the legitimate use of technology giving rise to the opportunity of questionable unethical action by the company which would affect every service engineer and ultimately anyone who used a company vehicle. The computer manager was placed in a very difficult position because of the conflict in professional responsiblity to the company on one hand and to the employees as members of society on the other.

This situation arose because of the uniqueness of computers. While spying on employees can be done without the use of computers, it is the power of computers that makes such activities viable in this situation. According to Walter Maner (1996. *Science and Engineering Ethics,* **2**(2), 137–154), the characteristics of the computer's uniqueness include storage, complexity, adaptability and versatility, processing speed, relative cheapness, limitless exact reproduction capability, and dependence on multiple layers of codes. This uniqueness has resulted in a failure to find satisfactory noncomputer analogies that might help in addressing computer-related ethical dilemmas. Indeed this is an area that raises distinct and special ethical considerations that are characterized by the primary and essential involvement of computers, exploit some unique property of computers, and would not have arisen without the essential involvement of computers.

There is a need to discover new moral values, formulate new moral principles, develop new policies, and find new ways to think about these distinct and special ethical considerations, particularly in the organizational context (Maner). The sections that follow consider the major issues within information ethics with the exception of issues specifically related to the Internet.

II. PRIVACY AND MONITORING

Privacy is a fundamental right because it is an essential condition for the exercise of self-determination. Balancing the rights and interests of different parties in a free society is difficult. Problems of protecting individual privacy while satisfying government and business needs are indicative of a society that is becoming increasingly technologically dependent. Sometimes individuals have to give up some of their personal privacy in order to achieve some overall societal benefit.

Organizations are increasingly computerizing the processing of personal information. This may be without the consent or knowledge of the individuals affected. Advances in computer technology have led to the growth of databases holding personal and other sensitive information in multiple formats, including text, pictures, and sound. The scale and type of data collected and the scale and speed of data exchange have changed with the advent of computers. The potential to breach people's privacy at less cost and to greater advantage continues to increase.

There are two important types of privacy: consumer privacy and employee privacy (R. A. Spinello, 1995. *Ethical aspects of information technology.* New York: Prentice Hall). Consumer privacy covers the information complied by data collectors such as marketing firms, insurance companies, and retailers; the use of credit information collected by credit agencies; and the rights of the consumers to control information about themselves and their commercial transactions. Indeed the extensive sharing of personal data is an erosion of privacy that reduces the capacity of individuals to control their destiny in both small and large matters. Organizations involved in such activities have a responsibility to ensure privacy rights are upheld. Consumer privacy focuses on the commercial relationship. Expanding this concept to client privacy includes consideration of non-commercial relationships where privacy is equally important. For example, medical, penal, and welfare relationships have, without doubt, serious privacy relationships. According to Spinello the issues that need to be addressed regarding movement of consumer data (and client data) can be summarized as:

- Potential for data to be sold to unscrupulous vendors
- Problems with ensuring the trustworthiness and care of data collectors
- Potential for combining data in new and novel ways to create detailed, composite profiles of individuals
- The difficulty of correcting inaccurate information once it has been propagated in many different files

Employee privacy deals primarily with the growing reliance on electronic monitoring and other mechanisms to analyze work habits and measure employee productivity. Spinello explains that employees have privacy rights which include the rights to control or limit access to personal information that he or she provides to an employer; to choose what he or she does outside the workplace; to privacy of thought; and to autonomy and freedom of expression.

In the modern workplace there are increasing opportunities to monitor activity. It is important to ensure that the use of monitoring facilities does not violate employee privacy rights. Some of the potential problem areas are:

- Personal computer network management programs that allow user files and directories to be monitored and to track what is being typed on individual computer screens
- Network management systems that enable interception and scrutiny of communications among different offices and between remote locations
- E-mail systems that generate archives of messages that can be inspected by anyone with authority or the technical ability to do so
- Electronic monitoring programs that track an employee's productivity and work habits
- Close circuit television surveillance systems that are computer controlled and have extensive archiving facilities and digital matching facilities

A modification of the data protection principles within the United Kingdom's Data Protection Act (1984) provides a framework that can be used to address the issue of privacy, develop a reasonable privacy policy, and ensure that the development and operation of information systems (IS) are sensitive to privacy concerns. The modified principles are as follows (E. France, 1996. *Our answers: Data protection and the EU Directive 95/ 96 EC.* Wilmslow: The Office of the Data Protection Registrar): (i) Personal data shall be processed fairly and lawfully. (ii) Personal data shall be collected for specified, explicit, and legitimate purposes. (iii) Personal data shall not be further processed in a way incompatible with the purposes for which they are collected.

(iv) Personal data shall be adequate, relevant, and not excessive in relation to the purposes for which they are collected or further processed. (v) Personal data shall be accurate and, where necessary, kept up to date. (vi) Personal data shall not be kept longer than is necessary for the purposes for which they are collected and further processed.

(vii) An individual shall be entitled, at reasonable intervals, without excessive delay or expense and under no duress, to be informed by any controller when he or she processes personal data of which that individual is subject and to certain information relating to that processing; to access the personal data of which the individual is subject and to any available information as to their source; and to knowlege of the logic involved in any automatic processing of data concerning him or her involving certain automated decisions; and, where appropriate, to have such personal data rectified, erased, or blocked, and to have details of such rectification, erasure, or blocking available to third parties to whom personal data have been disclosed.

(viii) Appropriate security measures shall be taken against unauthorized access to, or alteration, disclosure, or destruction of, personal data and against accidental or unlawful loss or destruction of personal data.

III. INFORMATION PROVISION

Information has become one of the most valuable assets, for it is through information that people gain knowledge that can then be used in both current and future decision-making activities. Information is concerned with communicating a valuable message to a recipient. Thus information must be clear, concise, timely, relevant, accurate, and complete. A message which has no value to its recipient is simply termed data. The majority of information provision is likely to use computer-based IS. The integrity of information is reliant upon the development and operation of these systems. The responsibility for these activities is a complex issue. For example, it is not clear whether IS provision is a service or the supplying of a product, nor is it possible in the case of a large IS for a single individual to fully understand it, and therefore no single individual can be held responsible for the whole system. It often turns out that an organization together with several individuals within that organization have a shared responsibility.

It is important to understand the nature of responsibility, which, according to D. G. Johnson, comprises four concepts (1994. *Computer ethics* (2nd. ed.). Englewood Cliffs, NJ: Prentice-Hall):

1. Duty—a person has a duty or responsibility by virtue of the role held within the organization
2. Cause—a person might be responsible because of undertaking or failing to undertake something which caused something else to happen
3. Blame—a person did something wrong which led to an event or circumstance
4. Liability—a person is liable if that person must com- pensate those who are harmed by an event or action

Specific responsibility issues often include several of these concepts. For example, a computer programmer knowingly reduced the testing procedure for a program in order to meet a deadline by not using the supplied test data that were for very infrequent cases. This resulted in a major operation failure several months after implementation. In this situation the programmer was to blame because failure to complete the specified testing had caused the program malfunction, and it was the programmer's duty and responsibility to ensure adequate testing. In this circumstance the programmer may be found legally liable.

One practical way of dealing with responsibility is to assign it to individuals involved in information provision within an organization. Individuals can be grouped into three broad categories: development, implementation, and maintenance of IS; collection and input of data; and output and dissemination of information. Responsibility clauses should be included in each job specification within these three areas of organizational work. Each clause should explain the extent of responsibility. Both management and nonmanagement jobs should be covered. Individuals should be adequately briefed on their responsibilities regarding the authenticity, fidelity, and accuracy of data and information. They should be encouraged to accept such responsibilities as part of their societal responsibilities. Should an undesirable event occur it should be considered on its own merits, and responsibilities can be identified using the responsibility framework already in place.

IV. SOFTWARE AS INTELLECTUAL PROPERTY

Intellectual property rights (IPRs) raise complex issues which organizations have to address. IPRs related to software and data are particularly difficult to assign and protect, and require careful deliberation before executive action occurs. Society has long recognized that taking or using property without permission is wrong. This extends not only to physical property but also to ideas. It is generally accepted that software is a kind of intellectual property and that to copy it or use it without the owner's permission is unethical and often illegal.

Ownership might not be clear. Johnson (1994) argues that a consequentialist framework is best for analyzing software IPRs because it puts the focus on deciding ownership in terms of affecting continued creativity and development of software. Software may be developed by a number of people, each making a contribution. Individuals might have difficulty determining which elements belong to them and to what degree they can claim ownership. Individuals may be employees or contractors. The development of software on behalf of a client raises fundamental IPR issues. It is important that agreement concerning the ownership of IPRs is reached at the onset before any development commences.

If an organization or group of individuals invests time, money, and effort in creating a piece of software they should be entitled to own the result by virtue of this effort and be given the opportunity to reap an economic reward. For the sake of fairness and equity, and to reward initiative and application, one should have the right to retain control over intellectual property and to sell or licence the product. However, the extent of these rights is debatable. Parker, Swope, and Baker explain that there is a responsibility to distribute software that is fit for the purpose for which it was developed, so the owner does not have the right to distribute software that is known to be defective and that has not been thoroughly tested (D. B. Parker, S. Swope, & B. N. Baker, Eds., 1990. *Ethical conflicts in information and computer science.* Wellesley: QED Information Sciences). Software embodies ideas and knowledge that can often benefit society as a whole. To have unrestricted rights may curtail technological evolution and diffusion, which will disadvantage the consumer and society. For example, there may be a piece of software that is deemed to be societally beneficial but which is withheld on commercial grounds. It is questionable whether the owner has the right, simply on grounds of optimizing economic gain, to withhold distribution. Some reasonable limit must be placed on the IPR so an equitable balance is struck. For example, currently copyright legislation in the USA protects the expression of an idea and not the idea itself. This constraint appears to achieve a balance between the right to private property and the furthering of common good.

There is reasonable agreement in countries of the West that individuals or groups of individuals have intellectual property rights regarding software. The law in many countries recognizes that computer software is worthy of protection because it is a result of a creative process involving substantial effort. The principal instrument of protection is copyright. However, the interpretations in other countries and situations are sometimes different. For example, IPR safeguards in countries of the Far East are minimal mainly due to a different philosophy that tends to treat intellectual property as communal or social property. In economically poor developing countries the view often taken is that the right to livelihood takes precedence over other claims on which IPRs are based. It is only when prosperity increases that there is a shift from a social well-being interpretation of IPRs to one with more emphasis on the individual. Such differences will have an impact on organizations involved in international trade and must be considered carefully.

V. ORGANIZATION STRUCTURE AND THE LOCATION OF WORK

With the advent of computers there has been a shift from traditionally stable organizational structures toward more flexible working arrangements. New computer-enabled working practices are creating more dynamic structures that are highly flexible and capable of responding to environmental uncertainty. For example, with the advances in telecommunications and IS, many jobs can be redefined as telework, which involves working remotely via a computer link. Many organizations are now using teleworking, communal office desks and computers, and geographically dispersed virtual teams to reduce organizational operating costs, but there may be serious disadvantages in this. For example, teleworking might result in the breakup of social groups in the workplace and the disenfranchising of those without the resources to participate. This may detrimentally affect organizations and society in the long term.

The impact of computer-enabled work will continue to grow. Work that is capable of being transformed into computer-enabled work must have a low manual labor content, be undertaken by individuals rather than teams, require minimal supervision, be easily measurable, and not depend upon expensive or bulky equipment. This means that there are many activities that might be organized as flexible computer-enabled work, including:

- Professional and management specialists such as accountants, design engineers, graphic designers, general managers, and translators
- Professional support workers such as bookkeepers, proofreaders, and researchers
- Field workers such as auditors, sales representatives, insurance brokers, and service engineers
- Information technology specialists such as software programmers and systems engineers
- Clerical support workers such as data entry staff, telesales staff, and word processor operators

Without doubt there are opportunities for benefit gains through the use of computer-enabled work for both organizations and individuals. However, this change in work practice raises many ethical dilemmas, and as computers evolve the dilemmas will continue to change. The following list illustrates some of the dilemmas and questions that may arise:

- The ability to employ people and sell goods and services globally through technological support may result in localized ghettos comprising people who have redundant or overpriced work skills and people who cannot afford the goods and services produced. Does an employer have a responsibility to the local community to ensure such ghettos do not exist or are minimized?
- Is it right to exploit low labor costs in the economically poor areas of the world, ignoring the injustice of wage differentials and an employer's responsibility to the community in which its employees live?
- Given the access to a global workforce and an increased need for flexibility to respond to the dynamic needs of the marketplace, the permanency of jobs and job content are likely to change. Is this acceptable to individuals, and how might organizations support individuals in coping with this often stressful situation?
- Computer-enabled communication only supports some of the elements of human communication. The loss of non-verbal communication or body language and the creation of electronic personalities could have an impact on the way people interact. Will this have a detrimental effect on individuals and the way they work?
- The workplace provides a place for social interaction at many levels. Individuals cherish this interaction. Commuting provides psychological space that separates work from home, which is important to some people. The move to teleworking radically changes this situation, potentially causing social isolation and disruption in home life. How can organizations safeguard individuals when adopting teleworking?

VI. COMPUTER MISUSE

As computers become more widely used, the risk of misuse and abuse increases, and the impacts of such acts are likely to be greater. For example, in the United Kingdom there was a threefold increase in the number of computer abuse incidents reported in 1993 compared with 1990, with virus infection, fraud, and illicit software accounting for 40% of the total incidents. Computer abuse covers a wide spectrum of activity, as summarized as follows (Audit Commission, 1994. *Opportunity makes a thief.* London: HMSO):

- Fraud through unauthorized data input or alteration of data input; destruction, suppression, or misappropriation of output from a computer process; alteration of computerized data; and alteration or misuse of programs, but excluding virus infections
- Theft of data and software
- Use of illicit software by using unlicensed software and pirated software
- Using computer facilities for unauthorized private personal work
- Invasion of privacy through unauthorized disclosure of personal data and breaches of associated legislation, and disclosure of proprietary information
- ''Hacking'' through deliberately gaining unauthorized access to a computer systems, usually through the use of telecommunication facilities
- Sabotage or interfering with the computer process by causing deliberate damage to the processing cycle or to the equipment
- Computer virus infections by distributing a program with the intention of corrupting a computer process

Spinello (1995) argues that organizations and individuals are ethically obliged to protect the systems and information entrusted to their care and must strive to prevent or minimize the impact of computer abuse incidents. He suggests that those stakeholders at greatest risk from a computer abuse incident might be party to decisions made concerning security arrangements. He argues that computer abuse offenses should not be treated lightly, even if the detrimental outcome is negligible, because, at the very least, valuable resources will have been squandered and property rights violated. Spinello also points out that a balance has to be struck regarding stringent security measures and respect for civil liberties. There is a dual responsibility regarding computer abuse. Organizations have a duty to minimize the temptation of perpetrating computer abuse, while individuals have a responsibility to resist such temptations.

VII. DEVELOPING INFORMATION SYSTEMS

Developing a computer-based IS is frequently a complicated process requiring many decisions to be made. As well as economic and technological considerations, there are ethical and social issues that need to be taken into account, but these are sometimes overlooked. It is generally accepted that IS development is best undertaken using a project team approach. How the project is conducted will depend heavily upon the perceived goal. The visualization of this goal should address many questions, including:

- What will the goal of the project mean to all the people involved in the project when the project is completed?
- What will the project actually produce? Where will these products go? What will happen to them? Who will use them? Who will be affected by them and how?

These types of questions are important because through answering them an acceptable project ethos should be established and the project's scope of consideration defined, so that consideration of ethical and societal issues is included, as well as that of technological, economic, and legal issues. The problem is that in practice these fundamental questions are often overlooked. It is more likely that a narrower perspective is adopted, with only the obvious issues in close proximity to the project being considered. The holistic view promoted by such questioning requires greater vision, analysis, and reflection. However, the project manager is usually under pressure to deliver on time and within budget, and so the tendency is to reduce the scope and establish a close artificial boundary around the project.

Within computing there are numerous activities and decisions to be made, and most of these will have an ethical dimension. It is impractical to consider each minute issue in great detail and still hope to achieve the overall goal. The focus must be on the ethical hotspots where activities and decision making include a relatively high ethical dimension because they are likely to influence the success of the particular information systems activity and promote ethical sensitivity in a broader context. The scope of consideration is an ethical hotspot and is influenced by the identification and involvement of all stakeholders both within and outside the organization.

The widespread use of and dependence upon IS within organizations and society as a whole means that the well-being of individuals may be at risk. It is therefore important that in establishing the scope of consideration of an IS project the principles of due care, fairness, and social cost are prevalent. In this way the project management process will embrace, at the onset, the views and concerns of all parties affected by the project. Concerns over, for example, deskilling of jobs, redundancy, and the breakup of social groupings can be aired at the earliest opportunity. Fears can be allayed and project goals adjusted if necessary.

An IS project is dynamic and exists in a dynamic environment. Appropriate information dissemination is essential so that the interested parties are aware of occurring change and assignments can be adjusted accordingly. Being over-optimistic, ultra-pessimistic, or simply untruthful about progress can be damaging not only to the project but also to both the client and the supplier. This is true whether the supplier and client are in the same or different organizations. Typically, those involved in this communication would be the project team, the computer department line management, and the client. An honest, objective account of progress which takes into account the requirements and feelings of all concerned is the best way to operate. Thus the second project management ethical hotspot has to do with informing the client. No one likes to get shocking news, so early warning of a problem and an indication of the scale of the problem are important. The key is to provide factual information in non-emotive words so the client and project manager can discuss any necessary changes in a calm and professional manner. Confrontational progress meetings achieve nothing. The adoption of the principles of honesty, non-bias, due care, and fairness help to ensure a good working relationship.

Turning to the overall development process, there are numerous methodological approaches to information systems development. Few deal adequately with the ethical and societal dimensions of the development process, instead tending to stress the formal and technical aspects. Consideration of the human, social, and organizational consequences of system implementation must not be overlooked during the development process. Management should encourage systems developers to adopt the principles of non-bias, due care, fairness, and consideration of social cost and benefit. In particular they should include the social design of computerized systems and work settings in the overall systems development project; build systems that are attractive to those whose work is most affected by them; and undertake information systems development in parallel with any necessary reorganization of work, taking into account changed responsibilities, relationships, and rewards.

VIII. COMPUTER PROFESSIONALISM

In discharging their professional duties, computer professionals are likely to enter into relationships with employers, clients, the profession, and society. There may be one or several of these relationships for a given activity. Quite often there will be tensions existing between all of these relationships, and particularly between the employer and societal relationships. There are three skills that a computer professional should possess so that professional duties might be undertaken in an ethically sensitive manner: (1) the ability to identify correctly the likelihood of ethical dilemmas in given situations; (2) the ability to identify the causes of these dilemmas and to suggest appropriate, sensitive actions for resolving them, together with an indication of the probable outcomes of each alternative action; and (3) the ability to select a feasible action plan from these alternatives.

Codes of conduct can be useful in helping computer professionals discharge their duties ethically, because the code provides a framework within which to work, and indicates to the new professional what are acceptable work practices. An excellent example of a code of conduct is that adopted by the Association for Computer Machinery (ACM) in 1992. The extract in Box 1 shows one of the 24 imperatives and its associated guideline that sets the overall tenor of the code and relates to many issues of computer and information ethics, for example, those raised in this article about the location of work and privacy.

Focusing on obligations makes it possible to consider carefully the implications of advancing computing technologies. There are four types of obligations for computer professionals: those as a supplier, those as a client, those as an end user, and those as a member of the community (W. R. Collins, K. W. Miller, B. J. Spielman, & P. Wherry, 1994. *Communications of the ACM,* **37,** 81–91). These obligations can be summarized as follows:

Obligations as a supplier to the client are to provide a reasonable warranty and be open about testing processes and shortcomings. Those to the end user are to provide clear operating instructions, give reasonable protection from, and informative responses to, use and abuse, and offer reasonable technical support. Obligations to the community are to ensure reasonable protection against physical, emotional, and economic harm from applications, and to be open about development processes and limits of correctness.

Obligations as a client to the supplier are to negotiate in good faith, facilitate adequate communication of requirements, and learn enough about the product to make an informed decision. Those to the end user are to provide quality solutions appropriate to the end user's needs within reasonable budgetary constraints, be prudent in the introduction of computing technology, and

Extract from the ACM Code of Ethics and Professional Conduct[1]

Imperative: 1.2 Avoid harm to others.

Guideline: "Harm" means injury or negative consequences, such as undesirable loss of information, loss of property, property damage, or unwanted environmental impacts. This principle prohibits use of computing technology in ways that result in harm to any of the following: users, the general public, employees, employers. Harmful actions include intentional destruction or modification of files and programs leading to serious loss of resources or unnecessary expenditure of human resources such as the time and effort required to purge systems of "computer viruses."

Well-intended actions, including those that accomplish assigned duties, may lead to harm unexpectedly. In such an event the responsible person or persons are obligated to undo or mitigate the negative consequences as much as possible. One way to avoid unintentional harm is to carefully consider potential impacts on all those affected by decisions made during design and implementation.

To minimize the possibility of indirectly harming others, computing professionals must minimize malfunctions by following generally accepted standards for system design and testing. Furthermore, it is often necessary to assess the social consequences of systems to project the likelihood of any serious harm to others. If system features are misrepresented to users, co-workers, or supervisors, the individual computing professional is responsible for any resulting injury.

In the work environment the computing professional has the additional obligation to report any signs of system dangers that might result in serious personal or social damage. If one's superiors do not act to curtail or mitigate such dangers, it may be necessary to "blow the whistle" to help correct the problem or reduce the risk. However, capricious or misguided reporting of violations can, itself, be harmful. Before reporting violations, all relevant aspects of the incident must be thoroughly assessed. In particular, the assessment of risk and responsibility must be credible. It is suggested that advice be sought from other computing professionals.

[1]Reprinted with permission of the ACM, 1515 Broadway, New York, NY 10036-570
Email: acmhelp@acm.org

represent the end user's interest with suppliers. Obligations to the community are to acquire only products having reasonable public safeguard assurances and be open about product capabilities and limitations.

An obligation as an end user to the supplier is to respect ownership of rights. Those to the client are to make reasonable requests for computing power, communicate needs to the client effectively, and undertake to learn and use the products responsibly. Obligations to the community are to make a conscientious effort to reduce any risk to the public and encourage reasonable expectations about computing technology capabilities and limitations.

Obligations as a community member are to become aware of the limitation of computing technology, encourage effective economic and regulatory frameworks, support societally beneficial applications, and oppose societally harmful applications.

Organizations are an essential part of society. Those in charge of organizations have a responsibility to ensure that when computers are applied in pursuit of business objectives it is done so in a balanced manner that accounts for the needs of both individuals and society, as well as those of the organization. Senior executives must strategically manage computer usage to ensure that issues such as privacy, ownership, information integrity, human interaction, and community are properly considered. Computer professionals and their managers must be trained so that they are sensitive to the power of the technology and act in a responsible and accountable manner. The adoption of a broader approach that addresses economic, technological, legal, societal, and ethical concerns will help to realize a democratic and empowering technology rather than an enslaving or debilitating one, both now and in the future.

Bibliography

Berleur, J., & Brunnstein, K. (Eds.) (1996). *Ethics of computing: Codes, spaces for discussion and law. A handbook prepared by the IFIP Ethics Task Group.* London: Chapman & Hall.

Bynum, T. W., & Rogerson, S. (Eds.) (1996). Global information ethics, special edition. *Science and Engineering Ethics,* **2**(2).

Huff, C., & Finholt, T. (Eds.) (1994). *Social issues in computing: Putting computing in its place.* New York: McGraw–Hill.

Johnson, D. G., & Nissenbaum, H. (Eds.) (1995). *Computer Ethics and Social Values.* Englewood Cliffs, NJ: Prentice-Hall.

Langford, D. (1995). *Practical computer ethics.* London: McGraw–Hill.

Rogerson, S., & Bynum, T. W. (Eds.) (1997). *A reader in information ethics.* Oxford: Blackwell.

Consequentialism and Deontology

MATTHEW W. HALLGARTH
United States Air Force Academy

GLOSSARY

altruism The view that egoism is not enough for morality, and that taking into account other persons' interests, for their own sake, is a necessary condition for morality.

a priori Means "before." In philosophy, the fact of knowing a proposition prior to experience, that is, without referring to experience to verify its truth. It has been hotly debated over the centuries whether *a priori* knowledge is even possible.

categorical imperative Kant's phrase for an absolute moral obligation, of the unequivocal form, "Do X," always. He proposes three tests to ascertain what these are.

consequentialism Any ethical theory that argues fundamentally that right action is an action that produces good results or avoids bad results.

deontology Literally means the "science of duty." It refers to any moral theory that emphasizes that some actions are obligatory irrespective of the pleasurable or painful consequences produced.

egoism The view that actions are right that satisfy self-interest.

foundational A fundamental assumption or axiom of a particular theory. A foundational assumption of Kant's ethics is that humans are autonomous.

libery (harm) principle Mill's principle that utility is maximized in societies where the guiding hand of the state is restricted to intervening in one's personal life only to prevent harm, one to another, but not to prevent you from harming yourself. It prescribes minimum limits to human and government sovereignty consistent with his principle of utility.

maxims What Kant calls a "subjective principle of action." These are the rules people operate by when they perform actions. Maxims that are indeed moral ones have to meet certain criteria.

prima facie On the face of it, at first glance, out of context. For example, killing human beings intentionally is *prima facie* wrong, although it is actually permitted in a just war.

summum bonum Latin for "highest good." A good that is an end in itself, and not a means to a higher order good. In Artistotle's theory, health is not the *summum bonum* because it is a means to the *summum bonum,* which is *eudaimonia* (happiness).

teleological Emphasizing design, goals, ends, that is, purposiveness in nature. Teleological ethical theory grounds moral obligation in observations about the design, goals, ends, and purposes of human beings. All consequential moral theories are teleological.

utilitarianism An altruistic variety of consequentialism that holds that good results are results that maximize benefits and minimize harms, even if this entails self-sacrifice. Usually, benefits is translated "pleasure," and harm is translated "pain."

utility principle Foundational moral principle espoused by Mill and Bentham. Acts are right if they maximize happiness for the greatest number of people. By happiness is meant maximizing pleasure and minimizing pain, unhappiness is vice versa.

CONSEQUENTIALISM refers to any of a class of normative theories that will argue that morally right action is action that produces good results. Theories of this type are teleological, in that they assume first an empirically grounded, natural theory of human good as a prelude to deriving moral obligations. Consequential theories in various ways always subsume moral obligations under the higher umbrella of a question best answered through observation, "What, given our environment and what is obvious about human nature, is good?" By understanding what our design suggests constitutes the ultimate goal of human action, our moral obligations logically follow as the "right" way to achieve that goal.

The word *deontology* originates from the Greek words *deon* (duty), and *logos* (science). Hence, it means the science of duty. In everyday reasoning the notion of duty is not a particularly divisive concept. When a person makes a decision, he normally chooses based on a common-sense assessment of his interests and the interests of others in light of his other long- and short-term commitments, a job, the offices he holds, previous promises, and various other obligations. However, a theoretical approach to the concept of duty is often very technical and the subject of much debate.

Deontology refers to any of a class of moral theories, the most noteworthy of which comes from the influencial philosopher Immanuel Kant, which argue that there are some moral obligations which obtain absolutely, irrespective of the consequences produced. Whereas the teleological moral theories of the ancient Greeks and the modern utilitarians emphasized the instrumentality of moral obligation as conducive to individual and corporate happiness, or the good life, answering the question "What is good?" deontological moral theories ground moral prescriptions in terms of the question, "What is right?" For deontological theories, the moral law is absolute and supreme. If considerations for one's well-being or the well-being of others contradicts a rationally acceptable moral law, obedience to the law is obligatory and must prevail. This is doing your duty for the sake of the duty alone. Kant even went so far as to say that lying to save the life of a friend is wrong. Unlike the observationally grounded consequential theories, Kant's position is that moral laws are understood *a priori* and then implemented by the practical reason

humans use to regulate action. Although we can go back as far as the Stoics to see a budding concept of deontology mentioned as fundamental to the moral life (e.g., their avowed duty to live according to nature), we find our fullest, most influencial spokesperson for deontological morality in the person of the eighteenth-century philosopher Immanuel Kant.

I. CONSEQUENTIALISM

Variations among consequential theorists are nearly always rooted in different assessments of human nature. If, generally speaking, consequential theories argue that right action produces good results, consequentialists remain a good deal removed from consensus. Just what *is* the good? What is the good result we are obligated to seek? How a consequentialist answers this question depends on the different answers given to foundational teleological questions such as these: Are humans necessarily selfish? Is it in their capacity to act autonomously? and, Are they able to put reason above inclination? Consequential theorists posit a variety of answers to these questions, and their answers imply vastly different normative obligations, both individually and collectively. Each, however, is consequentialist in consistently grounding obligations in the view that right action is that which produces good results. If there is any consensus among consequential theorists, it is a very general sort of agreement, based on observation, that humans are naturally driven to live a full life, to, in some sense, flourish as a human being in a community. This flourishing is what Plato calls justice, Aristotle calls *eudaimonia,* Hobbes calls peace or security, and the utilitarians call happiness.

Aristotle skillfully captures what each of these theorists generally mean by the natural human drive to live a full life, to flourish. He observed that there are many human goods that people strive for, such as health, money, friendship, power, and fame. A middle-aged man cuts down on his fat intake for health reasons, a young man saves to buy a motorcycle, a rich person creates a foundation to support philanthropic causes, and a young music student diligently practices the guitar. However, Aristotle was astute enough to ask if, among all good things, there is a highest human good, or *summum bonum.* It would be something all people want, not as a means to other goods, but as an end in itself. This ultimate goal would be the final desire of all human striving, that is, justice, security, *eudaimonia,* or happiness. Each of these posited "ends in themselves" serve as proposed theoretical definitions of what it means to flourish as a human being.

Consequential theorists will universally agree that

the *summum bonum,* this ultimate end of human striving, is not amenable to *a priori* logical proof. Nevertheless, a close observation of human behavior shows that all human pursuits are a means to, or constitutive of, living a full, flourishing life. This inability to logically prove an ultimate end of human striving, is, Mill argues, the case with all questions of ultimate foundations. Foundational principles are, at their best, generalizations inferred from empirical observation. At their worst, they are ad hoc assumptions used to derive pet obligations a particular theorist feels strongly about.

So, consequential theorists generally agree that right action is that which produces good results, but disagree on the details of what the nature of this *good* is because they disagree on fundamental questions about human nature. Additionally, consequential theorists disagree on a deeper practical issue concerning the good. Who's good gets precedence when a moral agent makes decisions about alternatives that each produces good results, contributing to this variously defined idea of what constitutes human flourishing? Should the good of the self come first? Are we obligated to sacrifice for the sake of a greater good of the community? Are the criteria for human flourishing objective (e.g., good job, talents, utilized, healthy family) or subjective (maximize pleasure/minimize pain)? What is the state's proper role in creating conditions where activity that produces good results are maximized? What is the proper integration between concerns for the common good and my private good? Practical considerations like these are the source of much of the debate in consequentialist circles. Now it is time to survey the chief varieties of consequential theory. For the sake of brevity, the quintessentially famous consequential theory called "utilitarianism," championed by Jeremy Bentham and John Stuart Mill, will be most substantively addressed.

A. Egoistic Consequentalism

Egoistic versions of consequentialism argue that when moral agents consider courses of action, they will be or should be motivated by self-interest, irrespective of aggregate consequences produced for the community, that is, unless those community consequences best satisfy self-interest. Self-interest is the egoist consequentialist's answer to the scope question, Who's good counts when I have to decide between competing actions that produce good results. Notice the operative words "will be" or "should be."

The egoistic consequentalist who holds that agents "will be" motivated by selfish considerations must defend the view that persons are motivated by self-interest, necessarily. This is a descriptive theory, and normative talk about "should and ought" loses most of its moral

force. On this account, it makes no sense to call for sacrifice when sacrifice, unless it produces something the self wants (like honor), is impossible. In this view, the only two normative issues at stake are those of method and the level of community intervention. The first suggests that agents ought to be taught to think clearly and critically so as to best calculate what the self wants and the most effective method to achieve it. This might, as Hobbes argues, entail cooperation given some assurance to personal security enforced by the state. The second issue concerns the community's need to structure an adequate system of rewards and punishments in order to steer its citizens' necessarily hedonistic drives in directions that effectively balance self-interest with the larger community's interest.

An egoistic consequentialist who argues that humans "should" pursue personal interest admits that humans are not determined by self-interest considerations in moral decision-making, but they should be. This variety of egocentric consequentialism is prescriptive about its hedonism, arguing that beneficial consequences are maximized for society and the individual when persons pursue their own ends, that is, mind their own business. Ann Rand would fit in here. This view defends a belief in human autonomy but criticizes personal and social altruism as more harmful than good, and hence wrong by consequential standards. The prohibition (twentieth) amendment to the U.S. Constitution serves as a good example of an idealistic and socially atruistic law that was morally wrong in contradicting the individual and hence the common good. Illegalizing liquor made more people feel oppressed on a personal level and eroded respect for law on the public level. It produced bad results. Prescriptive egoistic consequentialists have the same concerns over method and community intervention as the descriptive egoistic consequentialist, although the prescriptive variety has, in accepting a notion of human autonomy, a broader range of alternatives than personal and social conditioning.

It should be noted that egoism as a set of consequentialist theories vaguely defines a range of views that might, depending on interpretation, include a large host of positions. The character-centered theories of the Greeks, most notably Plato and Aristotle, are often referred to as enlightened egoist positions, because they argue, I think quite well, that the path of virtue and excellence is, given human nature, *the* way human beings are fulfilled in this life. In Plato's case, this virtue produces the good result of harmony of soul and harmony in the community, which Plato essentially defines as personal and corporate justice. The social contract positions of the ancient Hebrews (via Abraham), Thomas Hobbes, and lately John Rawls serve as popular religious and rational models for how the human person can achieve the good results of right relationship with God, security from the brutality of anarchy, and distributive justice. These can also be interpreted as egoistic positions, although based on diverse metaphysical assumptions about the origin of the universe and what motivates human nature and why. With each of these theories, note the agreement on the premise that right action is action that produces good results. What this good is and who's good is most important generates the disparity in these positions.

B. Altruistic Consequentialism

Altruistic consequentialism is really a euphemism for utilitarianism. These theories argue that in situations where self-interest and the interests of the community clash, the self is obligated to sacrifice its interests for the good of the community, assuming of course that the benefits obtained by the larger community exceed those to be gained by the self. Utilitarianism typically answers the question, "What is the good?" as actions that promote happiness, with happiness defined as acts that maximize pleasure or minimize pain. This is a prescriptive formulation with greater normative force, because it defends the human capacity to freely choose the greater good even to the detriment of self-interest. Both Bentham and Mill fall into this camp. Mill even goes so far as to say that personal sacrifice for the greater good of the community is the highest virtue that can be found in human beings. "Though it is only in a very imperfect state of the world's arrangements that anyone can best serve the happiness of others by the absolute sacrifice of his own, yet so long as the world is in that imperfect state, I fully acknowledge that the readiness to make such a sacrifice is the highest virtue which can be found in man." However, Mill argues, unless the good results are produced, the sacrifice is intrinsically worthless. "The utilitarian morality does recognize in human beings the power of sacrificing their own greatest good for the good of others. It only refuses to admit that the sacrifice is itself a good. A sacrifice which does not increase, or tend to increase, the sum total of happiness, is considered wasted." Mill insists that the need for these sacrifices is symptomatic of an imperfect world, although in the best social conditions this dilemma infrequently presents itself. A society properly educated and ordered according to consequentialism (his principle of utility) is to Mill the best framework for harmonizing the interests of the individual with those of society. More on how Bentham's and Mill's positions cash out in a moment.

C. Idealistic Consequentialism

This version of consequentialism is the brainchild of G. E. Moore. Also called ideal utilitarianism, it likewise makes good results the criteria for rightness and wrongness in moral decision-making, but argues that the attempt to define what the good is is futile, because good itself is indefinable. Hence, it makes sense that he refuses the strivings of humanity to a specific good such as, for example, gaining pleasure and avoiding pain. Moore espoused instead a richer view of human consciousness in his *Ethics.* By Moore's account, goodness is an intrinsically indefinable quality or property that may issue from many experiences such as contemplation, gaining new knowledge, or aesthetic enjoyment. The real debate here is whether these human experiences are actually reducible to gradations of something like pleasure and pain. For it seems Mill's emphasis on quality in his consequential scheme can and does absorb the intellectual and the aesthetic into a higher order of pleasure. One's view on this issue depends on the efficacy of Moore's argument for good as an indefinable quality. It has exerted considerable influence on the philsophical community. A detailed account of Moore's position would require another essay.

D. Consequential Decision-Making

Let us assume for a minute that some variety of consequentialism in its descriptive or prescriptive forms is true. Each theory makes some good arguments. How are moral agents to calculate the possible good results of their moral decisions so as to determine what course of action is morally right in a situation? And what is the larger community's role in facilitating the best possible accomodation between competing claims of maximizing self-interest and maximizing the common good? To restate an earlier formulation, "If good results indeed determine what action is right, what makes those results good, and who's good counts?" Jeremy Bentham and John Stuart Mill, the quintessentially famous consequential thinkers, serve as good examples to illustrate how moral obligation using a results model can be worked out. For a full view of different ways consequential morality is cashed out, familiarize yourself with the ancient Greeks, the social contract theorists, in addition to the following utilitarians.

1. Bentham

As an example of disputes over consequential moral decision methodology, consider the altruistic consequentialists, that is, utilitarians, Jeremy Bentham and John Stuart Mill. Bentham was a philosopher and legal reformer interested in changing the British legal system into what he argued was a fairer model that meted out rewards and punishments according to the beneficial or deleterious consequences produced by the questionable act. Bentham was frustrated by the fact that a hungry peasant in Britain would get 20 years in prison for stealing an apple from a street vendor, while a more affluent white-collar criminal (e.g., Michael Milken), who caused much greater harm to society, would get a relatively mild sentence.

Bentham, like many other minds in the nineteeth century, was enamored with the accomplishes of science and held out high hopes for the scientific method's ability to also solve moral and social problems, in addition to practical problems of health, food, and industrial production. To support his self-avowed legal consequentialism, Bentham wrote *Introduction on the Principles of Morals and Legislation,* a carefully written guide on his method for correct consequential moral decision-making. In this book, Bentham developed what he called a "hedonic calculus," a quantifiable scheme whereby levels of pleasure and pain could be rigorously and accurately assessed to determine the consequential effects for the various alternatives in a moral decision. The choice that produced the most pleasure or the least pain, as evaluated using various criteria like intensity, duration, and propinquity, identified the correct moral course of action. Fairer laws would be established, and social rewards and punishments would be meted in proportion to quantifiable results verified by this calculus.

Now this might seem, *prima facie,* how we do in fact make many of our moral judgments. It seems to work in simple cases, like what game to play with friends. And yet, Bentham's method is vulnerable to criticism using some simple case situations. For example, suppose a person is a compulsive peeping tom because that person enjoys voyeurism. In Bentham's calculus, if the victims are ignorant, the agent does not get caught, and his actions generate intense pleasure, it seems the act is not only acceptable, but a moral obligation according to the calculus. Examples like these defy common-sense perceptions about justice and rights. The same difficulty would emerge in other cases as well, for example, the justifiability of carefully framing an innocent person for a crime to satisfy a community's desire for justice, and hence, to maximize corporate pleasure.

2. Mill

Mill's famous brand of consequentialism owes much of its substance to Bentham's ideas, for Bentham knew Mill as a child and was a friend and colleague of Mill's father. Mill figured out Bentham's vulnerability to the aforementioned criticism and thus offered some cogent amendments to Bentham's views, while at the same time retaining the position that consequences, in terms of levels of pleasure and pain, is nonetheless the foundational principle of moral action.

Mill argued that restricting pleasure to quantifiable pleasures ignored quality as a vital consideration in the moral situation. This concern for quality in the agent's attempt to maximize personal happiness is not a novel idea in Mill. Epicurus supported a similar position in suggesting that pleasures of the mind are superior (purer) than bodily ones. They are not, as Epicurus says, "mixed." Epicurus argues that bodily pleasures are usually accompanied by associated evils as well, for example, hangovers for alcohol and jealousy for physical love.

Regarding this quality/quantity issue, Mill asks a now-famous question, "Is it better to be a pig satisfied than Socrates dissatisfied?" According to Bentham, it seems one ought or will (Bentham seems at times to equivocate between descriptive and prescriptive statements) choose the quantifiably pleasurable life of the pig, while, as Mill asserts, no one who has experienced both sides of this dilemma would choose the more vulgar existence. Figuratively speaking, only the pig would. This qualitative factor has plenty of mundane practical applications too. For example, should I read pulp fiction or a classic? Listen to Mozart or to pop music? Watch *Citizen Kane* or championship wrestling? If you ask people who have experienced each, who know and understand each, the higher quality pleasure is preferred. Mill puts it plainly, "Of two pleasures, if there be one to which all or almost all who have experience of both give a decided preference, irrespective of any feeling of moral obligation to prefer it, that is the more desirable pleasure."

Another way the quantity/quality issue is broached in contemporary discourse is in the distinction between act utilitarianism (consequentialism), and rule utilitarianism. Bentham's emphasis on the quantifiable assessment of acts placed him in a position to have to accept problematic examples to remain consistent. Mill tried to overcome this by arguing that a holistic determination of beneficial consequences required asking this general question, "What general types of acts, as a rule, tend to maximize beneficial consequences over time?" When you take counterintuitive cases like the peeping tom example and generalize them into a social norm of behavior, what, *prima facie,* seems at first to support the consequentialism, it turns out, thwarts it when viewed as an acceptable pattern of behavior over time. The common good is dimished. Hence, peeping tom behavior is wrong, in spite of exceptional cases.

What Mill has done through his emphasis on consequence maximizing rules of behavior is to salvage the possibility of making respect for human rights consistent with consequentialism. Let us suppose that researchers are very close to discovering a cure for sudden infant death syndrome, or SIDS, and the discovery of this cure will save 10,000 infant lives per year. Also suppose that the only way this cure can finally be discovered is through painful and fatal experimentation on 10 healthy newborn infants. Although Bentham, as an act consequentialist, would have to justify this experimentation, Mill as a rule consequentialist would not. For what would a society be like if acts of this type, as a rule, were generally accepted? This rule, by ignoring a right to life, would make people insecure and afraid, perhaps rebellious, and thus, would significantly diminish the collective happiness of that society.

Therefore, according to Mill's richer, rule-oriented, quality-emphasizing consequentialism, persons should take the responsibility for developing an enlightened notion of self-interest that emphasizes seeking pleasures of quality. Likewise, the state should structure educational policy and the broader system of social rewards and benefits according to higher concerns for quality in maximizing the common good. This includes respecting rules that honor human rights in ways that cultivate the tenuous balance between personal and community happiness.

Mill also amends Bentham's consequentialism by more cogently broaching the difficult task of harmonizing personal autonomy in pursuing one's own ends (pleasures) with the state's mandate to maintain a civil community life (corporate pleasure). He does this through his proposed "liberity principle," or "harm principle," as it is also called. Based on astute observations about what makes human beings happiest as members of a community, the "harm principle" places a (happiness maximizing) limit on the autonomy that adults may exercise in pursuing their ends in a community, while at the same time limiting the state's prerogative to intervene in the personal lives of its citizens. Persons may do anything free of government intervention to the extent that their behavior does not harm someone else. The government may not intervene in someone's life to protect that person from himself. Mill argued that the happiness of individuals and society is maximized when persons are at liberty to partake of even potentially self-destructive behaviors. Thus, on Mill's account, smoking should remain legal, but may be banned in restaurants. Mill articulated it this way, "Mankind are greater gainers by suffering each other to live as seems good to themselves, than by compelling each to live as seems good to the rest." By Mills account, this "harm principle" is therefore a necessary condition for harmonizing beneficial consquences for person and society.

E. General Criticisms

Following are several general criticisms of consequential moral theory. Many more specific criticisms that apply to specific consequential theorists, that is, the Greeks, Hobbes, Rawls, Moore, and so on, deserve further study but are beyond the scope of this essay.

1. A common criticism of all versions of consequentialism concerns measuring levels of pleasure and pain produced in a moral decision. Exactly how do we measure it? Whether you focus on quantity or integrate quality considerations, it is still nearly impossible to accurately predict which act over time will maximize good consequences. And in Mill's argument that we should concern ourself with act types as a rule, we at best can only claim to have a general idea as to which act types tend to produce the best results. But specifically in practice, how can I possibly determine accurately what impact by moral decision might have on, say, my yet unborn grandchildren? To calculate, it seems I am forced to construct arbitrary boundaries on the range of calculation. At worst I am pressured by time considerations to make a decision without adequate time to make a thorough assessment. These considerations can and often do countermand the purpose of consequentialism. Mill answers this criticism merely by appealing to man's expanding wisdom over time, arguing that we have had thousands of years of collective human experience to inform us as to which decisions to make. In a limited sense, this is true.

2. Another criticism of consequential morality concerns the fact that, to be consistent, it is forced to be exclusively forward looking. This is problematic. For example, suppose a criminal violates Mill's harm principle and murders an innocent civilian in cold blood. The culprit is convicted of the crime beyond a reasonable doubt. If only consequences matter, the sole, justifiable concern of the judge sentencing the convict must be future concerns, for example, satisfying the public's need for security from this person's maladjustment, quenching their justifiable desire for justice, or the need to make the punishment harsh enough to deter other potential felons. But what about backward-looking reasons? What about the appropriateness of punishing the culprit for some asocial behavior done in the past, such as murder? We naturally think of the past as critically relevant to the moral domain, especially situations of this type. This problem of needing to be forward looking also works with less heinous moral issues like promise keeping. A promise made in the past is morally important despite the fact that breaking it for some future enjoyment might increase the total of pleasure. These issues demonstrate some problems consequentialists have with (intrinsically valuable?) concerns for justice.

3. In still another criticism of consequentialism, R. M. Hare in his book, *Freedom and Reason*, argues

that the supposed distinction between act and rule consequentialism collapses into the act variety. In the aforementioned peeping tom example, Hare would argue that if act consequentialism justifies peeping tom behavior, it follows that rule consequentialism will justify it too. For in all situations with the same conditions, one can easily say that, as a rule, one ought to act this way. Hare argues that when rule consequentialists reject these types of exceptional cases, they do so by altering the parameters of the situation. In the peeping tom example, the rule consequentialist would have to assume the behavior will become public knowledge with the instigation of the rule, and this redefines the situation being analyzed. If Hare is right, consequentialism's tenuous connection to human rights remains.

4. A final criticism of consequentialism concerns the common-sense notion that motives have intrinsic moral value, something the consequentialist denies. A rule consequentialist will instead extol the virtues of good motivation as an instrumentally valuable state of mind that tends, more often, to produce good results. Suppose someone is drowning in a lake at a public park. One person knows the victim to be rich and famous, and hopes for both publicity and wealth from rescuing the individual. He is a good swimmer and he successfully rescues the struggling swimmer. In a similar case suppose the victim is neither rich nor famous. In this case a different person comes along who recognizes a duty to protect life regardless of personal gain and risk. But this person is a weak swimmer. He attempts the rescue and they both drown. In these cases, the consequentalist will ascribe greater moral value to the first scenario. To Mill, the latter rescuer's good intentions are "wasted." There is an inherent virtue and a nobility about the latter situation that most recognize, but that consequentialists must dismiss to remain consistent. A later contemporary of Bentham and Mill, Henry Sigwick, cogently addresses this last criticism by drawing an important distinction between notions of right and wrong and praise and blame. In the drowning example, Sidwick would argue that this agent's action was morally right in that it produced good results. However, the agent still merits blame for acting from a blameworthy motive. Likewise, the latter agent did something wrong in not producing beneficial results, though he nevertheless is of praiseworthy character for owning pure motives in the situation. The consequentialists' reluctance to ascribe intrinsic value to motives and difficulty in accounting for criticisms based on justice and rights introduces the discussion of deontological ethical theories, a wholly different class of moral theories. They argue that obedience to moral principles is an absolute obligation, and not instrumentally contingent on the production of beneficial results.

II. DEONTOLOGY

A. Kant

We will confine the lion's share of the discussion of deontology to the well-articulated, rigorously defended duty ethics of Immanuel Kant, with some time later devoted to the deontology of W. D. Ross. It was Kant's contention that if moral obligation was going to be universally binding on humanity in the truest sence, it had to be grounded in bedrock, logically, consistent, *a priori* rational principles. To pull this project off, Kant works out a deontic scheme not dependent on empirical observations about human behavior or the world humans inhabit. In this way, Kant salvages, he thinks, the right to assume the truth of human autonomy.

1. Moral Law versus the Law of Nature

Why does Kant insist that genuine moral living consists of a rational dedication to do one's duty for duty's sake alone, independent of the consequences produced? One way to look at Kant's thinking on this position is to elucidate his emphasis on the importance of making moral obligation something that is ontologically above the system of nature.

Kant lived in the optimistic excitement following Newton's scientific discoveries. During this period, many eminent philosophers genuinely thought it was a matter of time before science would discover the causal relationships that explain every phenomenon that occurred in the universe. To Kant, this was a troubling hypothesis for moral responsibility. Consequential morality, Kant argues, grounds moral obligation in observations about the desires, strivings, and ends of human action. If this were true, Kant showed, then moral obligation was confined to the natural order, and being confined thus, would inevitably be predictable in terms of cause and effect. If this were true, the deliberations and strivings of humanity would ultimately reduce to a descriptive behaviorism, and hence destroy the deeper meaning of words like "ought" and "should," "praise" and "blame." Genuine human morality, Kant argues, must be grounded in rational principles beyond the predictable causality of laws of nature, in an authentic, autonomous human will dedicated to duty for duty's sake, in spite of the predictable vicissitudes of natural inclination.

An example will help. Envision three persons watching the Jerry Lewis Telethon. The first person watching gives money because he needs a tax deduction for that fiscal year. The second individual sees a crippled little child on the television, is moved with pity, and, in tears, sends in a donation. The last person has indifferent emotions about the need but recognizes a rational obligation to help if she can. Most people, consequentialists included, would say that the second person's decision to give has moral value. Kant disagrees. In his deontological view, only the third person's motives have moral worth, because they are rationally dedicated to dutiful obedience to moral principle. The second person's motives lack moral worth because they are grounded in a contingent, behavioral law of nature, not in an autonomously chosen will. It just so happens that this second person's particular physical constitution is susceptible to pity, which is ultimately no different than the way a lion's particular constitution makes it inclined to ruthlessly kill and eat zebras for food. In the case of the pitiful giver and the lion, praise, and blame are irrelevant, even though the constitution of the second person is, no doubt, socially useful. The second person, moved by the same predictable class of forces that motivate a serial killer or a compulsive gambler, does not give a genuine moral response.

2. Intentions

Kant's contrast between the moral law and the law of nature explains why deontological morality is grounded in human motivation and not in consequences. Ascription of moral worth rests totally with the agent's intentional state. Kant says, "There is only one thing that is good without qualification, and that thing is the good will." There are many things, Kant states, that are viewed as good, for example, intelligence, good fortune, character traits, and so on. But these things can be used for evil ends unless they are under the control of a person with the "good will." Only this is intrinsically valuable. "Considered in itself, [the good will] it is to be esteemed beyond comparison as far higher than anything that it could ever bring about merely in order to favour some inclination or, if you like, the sum total of inclinations." Kant takes this still further. "Suppose the good will is "impotent" and wholly incapable of bringing about whatever it tries to accomplish. He says that "even then it would still shine like a jewel for its own sake as something which has its full value in itself." To transcend the predictable system of nature, the "good will" must be autonomous and thus rationally generated, because it is reason alone that enables the human person to overcome myriad variations of inclination and desire.

3. Good Will and Maxims

Kant sets humans apart as unique beings endowed with a special capacity to make genuine moral decisions grounded in a rationally governed, autonomous will obedient to moral laws, irrespective of natural desires

or inclinations to the contrary. But what are the true moral laws? And how does the error-prone person assess the true ones? In Kant's terminology, reason communicates to the mind things it should do according to rules he calls maxims. How does this process work? To illustrate, suppose I am leaving the grocery store and I notice that an elderly lady is encumbered with a load of groceries by the store exit. I decide it is my duty to open the door for her. In this case, Kant argues, my mind is operating according to a subjective principle of action, a maxim, of the following form, "When I infer that a person needs help and I have the capacity to help, I have a duty to help." Acting according to maxims is not necessarily a conscious mental activity, although the mind is continually making such judgments about alternatives of action according to these principles for action. Naturally, not all maxims necessarily cash out absolute moral laws. The aforementioned person moved by pity to give to the Jerry Lewis Telethon is acting according to a maxim of action, but the maxim is not a moral law and the intentions have no moral worth; it does not express a "good will." In this case, the person acts according to a contingent maxim grounded in a predictable and natural cognitive dissonance, "When something moves me with pity or sympathy, take action to assuage guilt or restore homeostasis." This is not an autonomous, rational response.

4. Categorical Imperatives

Kant's deontological morality, as you might suspect, offers a rational test the agent can use to determine if the maxims the mind and will deliberate on are in fact maxims of genuine moral worth. Maxims of supreme moral worth take the form of what Kant calls "categorical imperatives," that is, they are necessary, and of the unconditional form, "Do X!" not of the contingent form, "If you want Y, do X."

Human lives encounter scores of conditional maxims too, and by not being categorical, are not necessarily wrong. These are the prudential judgments that make up most of life. For example, we know that physical fitness is desirable as conductive to a long, vigorous life. Hence, an exerciser might operate according to the hypothetical maxim, "If you want to be healthy, get some regular exercise." Kant calls these types of maxims hypothetical imperatives. Thus, although we can argue with some success that people have a duty to exercise, we cannot argue that they have an absolute moral duty to do so. Categorical imperatives as a different species, are optionless.

Kant's criteria for determining whether a maxim for action is a genuine universal moral principle, remember, must be grounded in *a priori* principles to avoid begging a key question of objective rational foundations. His three criteria rely on the principle of contradiction, and each is a necessary condition to ascribing categorical moral value of the maxim at stake.

1. Universalizability: Act only on that maxim through which you can at the same time will that it should become a universal law.
2. Means/ends: Act in such a way that you always treat humanity, whether in your own person or in the person of any other, never simply as a means, but always at the same time and end.
3. Autonomy: Act so that you treat the will of every rational being as a will which makes universal law.

To briefly illustrate how these criteria (tests) work, let us look at Kant's example of promise keeping. A person in desperate need is driven to borrow money to get by. He reasons that he will promise to pay back the money without ever really intending to do so. Could breaking a promise ever be a universal moral law? It would contradict itself. For if every promise were broken, the purpose of the principle would become meaningless and inert, for no one would promise anything, knowing beforehand that each promise would be meaningless. Truth is, the capacity to break promises is paristic on the moral law of promise keeping, and requires the assumption of the morality of promise keeping to sustain itself. Because breaking promises cannot be universalized without contradiction, the issue is settled. The promise must be kept because keeping one's word is an absolute moral law.

And yet when a maxim cannot be universalized without contradiction, it will also fail the other criteria as well. In the breaking of a promise, the person also violates Kant's second criteria for categorical assessment, for in the breaking of the promise, one is treating someone as a means only, and not at the same time an end. The word "only" is of practical importance here, because in our many personal interactions we usually rely on other persons and groups as a means without violating a moral rule. If the destitute person were to pay back the money as promised, he nevertheless would still use the giver as a means to get out of a tough situation, but he would not be using that person as a means "only." As Kant states, "For then it is manifest, that a violator of the rights of man intends to use the person of others *merely* as a means without taking into account considerations that, as rational beings, they ought always at the same time to be rated as ends, that is, only as beings who must themselves be able to share in the end of the very same action." Finally, if a maxim for action fails to meet the first two criteria for categorical imperatives, it will fail the third one also, to "treat the will of every rational being as a will which makes universal law." In

short, because each rational agent is an autonomous will given the capacity to know and obey moral laws, they are entitled to the respect and dignity due to a being of that type. Hence, in the case of promise breaking, the maxim contradicts itself, treats humanity as a means only, and fails to accord the dignity and respect due to the lender as a rational, autonomous agent.

5. Human Nature

The concept of duty as primary in deontological moral systems entails a unique view of morality in light of true facts about human nature. Kant offers a rational methodology to determine whether the maxims a moral agent acts upon are in fact absolute moral rules. And yet, the notion of duty is important in that human beings are a species that often chooses to follow other inclinations in spite of its knowledge of moral obligations. This is an important difference between the teleogical position of the consequentialists and the deontological thinkers. The Greeks and the utilitarians were concerned to understand what, for human beings, is moral given reliable observations about their specific nature as creatures. Deontological morality, as evidenced in Kant, is concerned with what is moral *in spite* of the nature of human beings. Genuine morality applies to all rational beings, of which humans are an imperfect and easily distractable example. The concept of duty would be unnecessary, in Kant's view, for perfectly rational beings, for those beings would always understand the purely reasonable moral law and act on it. Humans though, as imperfectly rational beings, require the concept of duty as a practical motive for action that is necessary to subordinate conflicting human desires to the rational, autonomous, law-abiding will.

This analysis again suggests why intentions (good will) are morally valuable while consequences are not. Right and wrong motivations can obtain in situations where the correct action is executed by both parties. A shopkeeper can be honest in his business practices because he wants his business to be successful, or he can be honest because he recognizes a rational obligation to dutifully obey the moral law. The first motive merely coincides with what duty requires and lacks genuine moral value, while the latter motive expresses obedience to an absolute moral maxim that expresses a good will, and hence, has genuine moral worth. Such examples are not meant to be complicated and Kant himself argued that this perspective on duty was neither original nor esoteric. Everyone, he argues, knows that doing something you are morally obligated to do and doing something you want to do are two different things. And we all know that the former is meritorious in a unique way; the latter is not.

6. Perfect and Imperfect Duties

Thus far in the discussion of deonology, particularly Kant, one might think that duties are duties, period, and hence must be obeyed. This is true of categorical imperatives. But if you reflect a minute, you can no doubt think of situations where different duties seem to conflict and a decision must be made as to which one to fulfill. In these cases, Kant recognizes an important distinction between what he calls perfect and imperfect duties. This distinctive allows us to resolve many of our moral dilemmas.

In short, perfect duties are defined as negative duties, that is, associated with the moral requirement not to cause harm. In many, if not most cases, this implies a right to redress, including many times, state intervention, if the violation of a perfect duty harms you. Perfect duties are phrased as maxims that lend themselves easily to verification as categorical imperatives by Kant's three tests for moral maxims. If someone punches you in the nose for no apparent reason, they have violated a perfect, categorical duty. The rule, "Everyone ought to punch anyone in the nose whenever they are so inclined," cannot be universalized without contradiction, treats humanity as a means only, and violates the dignity and respect due to rational, autonomous moral agents. The same analysis holds for most other negative duties, such as the duty not to steal, to lie, to murder, to molest, and so on.

Imperfect duties refer to positive duties, that is, prudentially bringing a positive consequence to a situation. When these duties are phrased in the form of maxims, they most often do not cash out as categorical moral obligations. They are hypothetical, and hence are of the form, "If you want X, do Y." With imperfect duties, we can only say with assurance something like the following: "It is praiseworthy to do acts of this *type.*" However, in specific situations, the failure to fulfill the imperfect duty does not entail a right to compliance or enforcement by the receiver. If you are driving down the highway and you notice a motorist stranded with a flat tire, do you have a categorical duty to stop and help? No, and yet it would be nice to help, and furthermore, we praise people who do acts of this type when they can. Likewise, you would laugh if the stranded motorist claimed a right to your help in fixing the flat, and expected the state to fine you if you did not stop to assist. The duty to give to charity would also classify as an imperfect duty. You can argue generally that persons should give to charity, but as a member of the Boy Scouts of America you cannot expect that any individual is categorically obligated to give to your organization, nor could you expect the police to arrest those who did

not give to your cause. And yet, persons who have a "good will" will help in similar ways.

7. The Virtues of Kant's Deontology

This has been a rough-and-ready recount of Kant's complicated rational scheme for the real possibility of a genuine human morality grounded in an autonomous will that is above the predicatable system of nature. His views derive from an exhaustive metaphysical and epistemological philosophy that attempted to overcome David Hume's incisive arguments against all attempts to demonstrate the possibility of certain knowledge. Kant's deontology does have practical virtues that deserve mention.

1. Kant was wise to emphasize the valuable role that intention plays in the ascription of praise and blame to moral agents.

2. His desire to ground morality in an autonomous will that somehow can "rise above" the predictable system of nature is foundationally crucial to our belief in moral responsibility. This is just as important today as it was in Kant's time. For a universe in which human behavior is reducible to merely a higher order of animal behavior, minimalizes morality to a need for conditioned social control and abdicates a rich view of personal accountability.

3. His view respects contemporary notions of human rights. The tests for assertaining categorical moral rules omit any opportunity for violating the dignity and respect due all rational agents, regardless of perceived social benefit. Rational, autonomous persons who "can do otherwise" deserve respect due free agents who possess genuine moral responsibility.

4. His deontology correctly requires that persons never treat themselves as unique or special, deserving consideration as exceptional cases. Such a consistency requirement entails that if I have reasons for doing something categorically moral, those reasons must be adequate for anyone reasoning in that or similar cases.

5. Finally, Kant's deontology, in defending a rich view of human freedom, lends itself to support for the classic notion of justice, that is, to get what you deserve. Moral agents who, through decisions subject to possible discipline by a rational, autonomous will, forsake performing their perfect duties, forfeit their rights to be treated with the dignity the categorical imperative mandates. Hence, a thief, in violating an absolute moral law, loses his right the similar treatment under that law, and may be incarcerated. This perspective supports the common-sense notion that we punish people precisely

for injustices they committed in the past, and not just to deter the behavior in the future. There is genuine accountability here.

8. Criticisms of Kant

Kant's deontology is not without its opponents. Here are some of the criticisms commonly raised.

1. One criticism of Kant's ethics is that he ultimately assumed the freedom of the will without proof. Given our observations about the human condition, it is viewed by many as more tenable to postulate determinism over freedom, regardless of what that does to human morality. In Kant's case, he seems to make a pragmatic decision that moral responsibility is so important that this justifies the assumption of human freedom.

2. Many reject Kant's contention that genuine moral worth is grounded in a purely rational activity of the will married to an impersonally calculable logical consistency. Kant does not try to hide this. And yet, his view of intentionality, though rigorous, fails to capture much of what we admire as moral motivation, such as the admiration of maternal instinct judiciously applied.

3. Another criticism of Kant concerns the issue of maxims. When humans act, Kant says that we act according to subjective principles of action called maxims. What he never does is argue for a consistent way to form these maxims prior to assessment via the three tests for categorical imperatives. Once again honesty serves as a good case in point. Suppose I am unemployed and tempted to steal to feed my family. If I phrase the maxim generally, "It is OK to steal," I am breaking a categorical moral law by the three tests. But Kant does not really specify that I cannot phrase the maxim anyway I wish, such as "I may steal to feed my starving family when I know I will not get caught, and when I can reasonably assume that others in the same position are not doing the same." Universalize this maxim, and there is no apparent practical contradiction. Here we seem to have a case where a maxim fails the first test but passes the second and third ones.

4. Kant insists that moral situations never, if thought out carefully, cash out as genuine dilemmas over conflicting categorical imperatives. Otherwise, the status of absolute moral rules as a class would be in jeopardy. During the Second World War, Dutch fishermen hid Jewish fugitives in their boats and ferried them to safety over in England. Often, SS patrol boats and submarines would stop these fishing boats. When Nazi captains asked the fishermen if there were Jews on board, what were the fishermen to do? If the maxims,

"It is wrong to lie," and "Permitting the murder of innocent people is wrong," are both viewed as categorical imperatives, as I think they must be, then we have a genuine categorical dilemma. In these cases, most Dutch captains viewed it as permissible to sacrifice absolute honesty to save innocent life. Kant has trouble with these types of scenarios.

B. W. D. Ross's Twentieth-Century Deontology

The persistence of problematic dilemmas for Kantian deontological thinkers is, in many respects, accounted for by the amendments made to deontological thought in the twentieth century by W. D. Ross. Consequentialist thought dominated ethical debate in the nineteenth and early twentieth centuries until Ross rearoused philosophical interest in the notion of duty. An influential gadfly that prompted Ross in this area was an article by H. A. Prichard entitled, "Does Moral Philosophy Rest on a Mistake?"

1. "Plain Man" Intuition

Prichard argues that this "mistake" of moral thought is to base our ethical obligations on the good to be achieved as compared to alternative courses of action. As a matter of fact, Prichard argues, we do not appreciate an obligation because of arguments that support it, but merely by a sense of direct awareness of the importance of the obligation. This analysis, directed at the consequentialists, also criticizes the rational, calculative methodology of Kant. Prichard grounds obligation in human intuition, and this intuition often contradicts reasoning about results, and also rejects dry recourse to rational speculation based on the principle of contradiction.

Ross accepts Prichard's argument for intuition and uses it to ground his deontological morality in the principle of *prima facie* duty. According to Ross, the terms "right" and "good" are distinct, irreducible, indefinable objective qualities, consistent with G. E. Moore's position. An act is either right or wrong, but it is motives that are good or bad. On this account, there are four possible conclusions in assessing the moral quality of an agent's response in a situation. Ultimately, and this is where Ross is most easily subject to criticism, a person by intuition "knows" which course of action is right and which motive connotes a "good" one. But because right and good are indefinable, there is no need to analyze further, as the consequentialists do in making good an instrumental quality.

How does Ross's view work in practice? Well, he argues that the "Plain Man" understands his duty in a situation according to Prichard's notion of immediate awareness. So, if a murderer stops at my front door and asks, "Is your child here, I'm going to kill him?" the "Plain Man" knows intuitively that this situation justifies a lie to save life. This is how Ross tries to overcome exceptional cases that plague the Kantian scheme. The aforementioned Dutch fishermen example is another case in point. The "Plain Man" or average Joe will just "know" that his duty to be truthful may be suspended for a more stringent duty, in this case to save Jews from concentration camps.

Ross addresses an obvious concern closet critics should have about the nature of intuition. Obviously, many people intuit very badly, and they often rationalize things they want into morally acceptable decisions. Ross's answer to this is to point out that the intuition he has in mind is developmental. Hence, education and practical training are vital to molding the intuitive faculty into one that makes good moral decisions. Is this still what we would call intuition? Well, in a way, yes. Here is an example that illustrates how it should work. As a adolescent, many children go through a period where they do not respect their parents, (although they are good parents by all normal standards), they think their parents owe them more than they receive, or they surmise that their parents are overly conservative and have habits that are distinctly passé. But if the parents are patient, they remain hopeful that the children will "grow up" and eventually "understand" why they are like they are. In the mind of the maturing adolescent, and this comes at different times for different children, it will just "dawn" on the child that his parents are not as unfair, dumb, or passé as he thought. He will just "see" it, usually as a response to understanding the good reasons behind the parent's actions through encountering adult situations himself, such as having children of his own, or working out a personal budget. In this case, Ross would say the intuition of the maturing person has positively developed. Upon "understanding" the reasons behind the parents' actions, the adolescent might recognize a *prima facie* duty of gratitude to his parents, perhaps even a duty of reparation for past injuries inflicted out of ignorance. Although it is true that as humans, we use reason to make moral deliberations, the point of understanding is intuitive in Ross's sense. Remember how you have worked and worked to figure out a difficult math problem, only to have the whole solution just "dawn" on you. This is what Ross means by intuition.

2. *Prima Facie* Duty

Prima facie duty (duty on the face of it) refers to a duty one must perform, unless it conflicts with a more important duty intuitively recognized. In a situation where two or more *prima facie* duties conflict, the one my "Plain Man" intuition tells me I must perform is what Ross calls the "actual duty". In the absence of a conflict of duties in a moral situation, any *prima facie* duty is the agent's actual duty. Thus, in the case of the Dutch fishermen, the duty to be truthful is the actual duty in cases where no other exceptional circumstance warrants overriding it for a more important obligation, such as saving innocent life.

Ross argues that there are seven *prima facie* duty types, and each is an obligatory actual duty unless there is a conflict with a greater *prima facie* duty. This list is the duty types Ross observes, though he is humble enough to say that this list could be incomplete. The seven *prima facie* duties are as follows:

1. Fidelity: Keep promises and commitments.
2. Reparation: Correct past wrongs.
3. Nonmaleficence: Duty to prevent harm.
4. Beneficence: Duty to increase general pleasure.
5. Justice: Duty to prevent unfair distribution of benefits.
6. Gratitude: Duty to repay kindness, and
7. Self-improvement: Duty to better oneself.

In this list, Ross admitted the complexity of the moral domain in human affairs. Another example will help demonstrate how he cashes these duties out in practice. Suppose a nurse makes an appointment to give a lecture on Friday at 6 P.M. at a gathering of oncology nurses for the Nurses' Association. En route to the engagement, she witnesses an automobile collision with life-threatening injuries. If she stops and renders assistance, she will miss her engagement, but if she makes her appointment, someone will likely die. Ross would say this nurse has a conflict between the duties of fidelity and nonmaleficence. Irrespective of consequences, (making the engagement might create more general utility), the nurse will know intuitively that in this case the duty of nonmaleficence overrides the duty of fidelity. She has an actual duty to stop and render assistance. However, after the assistance is complete, her failure to fulfill her *prima facie* duty of fidelity obligates her to a duty of reparation, that is, to call and explain the situation to the Nurse's Association and probably offer to speak again at a later date. Likewise, the Nurse's Association should recognize a *prima facie* duty of gratitude for the call and demonstrate an understanding of her predicament as a duty of beneficence. This intuitive weighing of our various duties, Ross, argues, is in fact how we work out our morality in practice. Though it has its weaknesses, it does offer a practical explanation for the moral complexity humans encounter day to day.

3. Criticisms of Ross

The main criticisms of Ross are as follows:

1. An intuitive method for moral decision-making is unreliable and often inconsistent
2. Much of his notion of *prima facie* duty is left undefined, that is, the "Plain Man" must figure out on his own which of the seven duties apply and which one overrides which.

Regarding the first criticism, Ross argues that by intuition, he means self-evident in the sense in which mathematical axions are self-evident to those familiar with them. Also, if the system of *prima facie* duties seems unsystematic, that is because that is how we, in fact, reason about moral issues. In this sense it is descriptive. Ross argues that though his deontology is apparently less systematized than consequentialist positions, they are in reality no less precise in practice. In practice, the principle of utility is at least if not more difficult to cash out in actual decision situations, and is based on a principle that is not how humans really think about moral issues. Besides, Ross does have a governing principle of action implicit in his deontology. An act is mor-

ally right (i.e., an actual duty), if and only if it is a *prima facie* duty, and no other conflicting act represents a more stringent duty.

Bibliography

Aristotle. (1925). *Nicomachean ethics.* (W. D. Ross, Trans.). London: Oxford University Press.

Abraham, E., Flower, E., & O'Connor, F. (Eds.). (1989). *Morality, philosophy, and practice: Historical and contemporary studies:* New York: Random House.

Bentham, J. (1948). *Introduction on the principles of morals and legislation.* W. Harrison (Ed.). London: Oxford University Press.

Brennan, J. G. (1973). *Ethics and morals* New York: Harper and Row.

Hudlin, C. W. (Ed) (1949). *Moral traditions in moral philosophy.* Dubuque, IA: Kendall/Hunt.

Kant, I. (1948). *Groundwork of the metaphysics of morals.* (H. J. Paton, Ed. and Trans.). London: Hutcheson & Co, Ltd.

Mill, J. S. (1979). *Utilitarianism.* G. Sher (Ed.). Hacket Publishing.

Mill, J. S., (1947). *On liberty.* A. Castell, (Ed.) Arlington Heights, IL: AHM Publishing Co.

Rachels, J. (1993). *The elements of moral philosophy.* (2nd ed.). New York: McGraw-Hill Inc.

Ross, W. D. (1930). *The right and the good.* Oxford: Oxford University Press.

Solomon, R. C. (1984). *Ethics: A Brief introduction.* New York: McGraw-Hill.

Embryology, Ethics of

SØREN HOLM

University of Copenhagen

I. A Short Outline of the Development
of the Human Embryo
II. Embryology and *in Vitro* Fertilization
III. What Is the Connection between Embryology
and Applied Ethics?

GLOSSARY

abortion The termination of a pregnancy. Abortion can be spontaneous or induced.

conceptus, zygote, preembryo, embryo, fetus Developmental stages of the developing human being; see Section I.

embryology The science that studies the development of the embryo, fetus, etc., from just prior to conception to birth.

gamete A reproductive cell; sperm and eggs are gametes.

IVF *In vitro* fertilization, a technique developed to treat infertility. IVF involves the fertilization of ova outside of a woman's body. IVF is one of a range of techniques usually called the "new reproductive technologies" or "assisted reproductive technologies."

EMBRYOLOGY is the science that studies the development of the embryo, fetus, etc., from immediately before conception to birth (or hatching in the case of birds, fish, reptiles, etc.). Although the terms "embryo" and "fetus" denote different stages of the developing being, this is not reflected in the name of the science dealing with all these stages.

There is no separate science of "fetology." The science of embryology contains both descriptive and explanatory elements; i.e., it describes the various development stages (their anatomy, physiology, etc.), but it also tries to unravel the mechanisms which control the development.

For the purpose of the present article embryology will be used as synonymous with human embryology.

Embryology is a scientific area where there are still large lacunae in our knowledge, but it is also an area of very rapid development. This means that some of the things we think we know today may very well be superseded by new discoveries in the coming years.

I. A SHORT OUTLINE OF THE DEVELOPMENT OF THE HUMAN EMBRYO

The first step in the process leading from fertilization to birth is fertilization itself; i.e., the process whereby the male and female gametes (i.e., the ovum (egg) and spermatozoon (sperm)) unite to form a new entity. This step in the process is also sometimes referred to as conception.

There is no moment of fertilization; it is a process which is extended in time. When a spermatozoon penetrates the outer membrane (the zona pellucida) around the ovum, the membrane changes configuration and becomes impenetrable to further spermatozoa. The head of the spermatozoon then penetrates the cell membrane of the ovum, and its genetic material is injected into the cytoplasm of the ovum where it forms the male pronucleus.

Over a period of hours the male and female pronuclei fuse and form the nucleus of the newly formed entity called the zygote. When one nucleus has been formed we talk of syngamy. The zygote then begins its first division, the single cell becoming two. Genetic studies have shown that the genetic material from the father is not activated until 24–48 hr after the beginning of the fertilization process, that is, after the occurrence of syngamy and after the first division.

In normal fertilization, after sexual intercourse the fertilization usually takes place in the fallopian tubes (the tubes leading from the ovaries to the uterus (the womb)) and the initial divisions take place while the zygote is transported through the fallopian tubes.

After the first division the zygote continues to divide, forming a 4- and an 8-cell stage. At the 8-cell stage all the cells are still totipotential (or pluripotential); i.e., they can all form a new complete zygote if separated. At this stage the zygote looks like a mulberry and is called a morula. It is possible to induce a human ovum to begin its divisions without having been fertilized by a spermatozoon. This activation can be done with a range of different chemical and electrical stimulations. However, such parthenogentically activated ova never develop further than the 8- to 16-cell stage. True parthenogenesis (virgin birth) is impossible in humans.

As cell divisions continue the mass of cells develops a central cavity and is called a blastula or blastocyst, and later the embryo proper begins to develop as a localized mass of cells (the embryonic disk) protruding into the central cavity. If development goes wrong at this stage it may in some cases lead to the creation of a tumor known as a mola which consists only of the kinds of cells which would have formed the placenta in normal pregnancy.

It is at the stage of the formation of the embryonic disk that monozygotic (identical) twinning can occur by a division of the embryonic cell mass. As far as we know, monozygotic twinning is a random process; i.e., the embryo is not genetically predestined to twin (dizygotic (nonidentical) twins arise when two ova are fertilized by two different spermatozoa).

The reverse process of twinning, where two embryos fuse and form a chimera consisting of two genetically different lines of cells, is also possible.

After the appearance of the embryonic disk the zygote is called a preembryo (preembryo may also be used for the entire period prior to the embryonic stage). At about 11 days gestational age (g.a., gestational age, is counted from fertilization) the embryonic disk begins to develop a central depression (the primitive streak) which marks the position of the future brain and spinal cord. When the primitive streak is fully formed at 13–14 days g.a., twinning is no longer possible and the preembryo is now called an embryo.

The term preembryo was first used in the early 1980s and there has been considerable debate over the use of the term. It has been claimed by opponents of embryo research that it was deliberately invented to make research on embryos seem more palatable by redescribing them as preembryos. The term is, however, now in common usage in embryology.

The process of nidation runs simultaneously with the development of the zygote. At 6–7 days g.a. the embryo starts embedding itself in the lining of the uterine wall, and this process is finished at about 14 days g.a. when the first primitive placenta is also formed.

More than 50% of all fertilized eggs do not develop properly in the early stages or do not nidate properly and are expelled with the mothers next menstural flow.

From 2 to 8 weeks g.a. the embryo develops very rapidly, and at 8 weeks all major organs are formed (although very few are functioning), and the embryo is clearly recognizable as a human embryo.

The first brain waves appear at 8 weeks g.a., but the cerebral cortex, which is presumably the part of the brain supporting consciousness, is only developed around 20 weeks g.a.

From 8 weeks g.a. until birth we talk of a fetus. Given present techniques, a fetus is viable outside its mothers womb from 22 to 24 weeks g.a. Prior to this time the lungs of the fetus are not sufficiently developed to sustain extrauterine life.

The mother is able to feel the movement of the fetus from about 16 to 18 weeks g.a. The time when she first feels fetal movement is called "quickening."

The exact time at which the fetus can feel pain is not known, and there is still disagreement on this point among embryologists.

The period from 24 to 40 weeks g.a. is characterized by growth and final maturation of the organs of the fetus. Birth usually takes place at 40 weeks g.a.

II. EMBRYOLOGY AND *IN VITRO* FERTILIZATION

In vitro fertilization (IVF) is a technique originally developed as a treatment for infertility. In IVF a number of ova are removed from a woman's ovaries and fertilized outside of her body. When they have developed to the eight-cell stage they are transferred back into her uterus, or the uterus of another woman. A number of variations of the IVF technique have been developed, each with its own acronym. These include GIFT (gamete intrafallopian transfer), ZIFT (zygote intrafallopian transfer), ICSI (intracytoplasmatic sperm injection), and SUZI (subzonal sperm injection). As a group these techniques are usually called the "new reproductive technologies" or the "assisted reproductive technologies."

The development of IVF opened new possibilities for embryological research, and new possibilities for intervention at the embryonic stage of human life. Some of these possibilities are briefly outlined here.

Zygotes can be frozen and stored indefinitely, although 25–50% are lost in the process. Zygotes can be split to form two individuals (cloning), or two or more zygotes can be fused. These two possibilities have not yet been tried in humans. The genetic makeup of the zygote can be determined by preimplantation diagnosis where one cell is removed and the genetic material in this cell amplified through modern gene-amplifying techniques. Genetic engineering can be performed on the zygote.

Ova can be donated from one woman to another, or one woman (a so-called surrogate mother) can bear a child for another couple. Ova from aborted female fetuses can be matured and used in IVF procedures. The normal reproductive span of women can be extended by hormonal treatment and egg donation.

These possibilities can furthermore be combined in various ways and can give rise to quite exotic scenarios. By cloning and embryo freezing a woman could, for instance, give birth to her own identical twin.

It is fairly obvious that some of these possibilities raise ethical questions.

III. WHAT IS THE CONNECTION BETWEEN EMBRYOLOGY AND APPLIED ETHICS?

Embryology is important for applied ethics in three ways: (a) it creates new ethical problems by being the basis of a range of new reproductive techniques, (b) it delivers empirical premises which are part of many of the arguments in biomedical ethics concerning the old problem of abortion and the new problems concerning the reproductive techniques, and (c) the early developments of the fertilized egg have proved a fertile field of examples for developing and testing more general theories of personhood and individuation.

Confusion may be created in the mind of the reader of articles of embryology and ethics if the authors do not make it clear whether they are trying to "solve" the ethical problem at hand or to develop a new ethical theory. The reader looking for answers to real world ethical problems is not always satisfied with philosophical analysis performed purely for the sake of its intrinsic philosophical interest.

A. The Moral Status of the Embryo

The moral status of the human embryo and fetus is not a new subject in philosophy. It has a history going back to Aristotle who wrote about the development of the human fetus and identified three consecutive ensoulments with a vegetative, an animal, and a rational soul. In scholastic philosophy in the middle ages this idea was further developed in connection with discussions of the proper punishment for abortion, and some scholastic philosophers maintained that male fetuses were finally ensouled with a rational soul 40 days after conception whereas female fetuses were similarly ensouled 80 days after conception.

In modern applied philosophy the question of the moral status of the embryo again emerged in the early 1970s in connection with discussions about the morality of abortion, and the question has continued to be prominent in discussions about ethical problems in the new reproductive techniques. Knowledge about embryology has been used extensively in these discussions. The question has often been stated in terms of whether or not embryos are persons, where a person is taken to be an entity with the moral status of a normal adult human being, including rights not to be harmed and not to be killed.

It is possible to discern two basic positions on the moral status of the human embryo: (1) The human embryo has no intrinsic moral status, but its status depends on the value conferred on it by others (e.g., its mother). (2) The human embryo has intrinsic moral status, independent of how others value it. The third possible position, that embryos begin with very little or no moral status and then gradually acquire more and more moral status as they develop, is probably the common sense view. This gradualist view has, however, been rejected by most ethicists as unsustainable. This may in part be because it shares the problem of many gradualist theories—it is often easier to argue for the extreme positions than for the middle ground. The gradualists will be attacked from both sides and will be vulnerable to the claim that their distinctions are arbitrary. Some versions of the argument from potential do support gradualist conclusions (see the last part of Section III.A.2).

1. Embryos Have No Moral Status

The position that embryos have no intrinsic moral status or value can be reached in a number of different ways. The two most common are via (a) a theory of rights, or (b) a preference consequentialist theory. The first approach usually proceeds by arguing that interests are a necessary condition for the ascription of rights, and that an entity has interests only if it is possible for the entity to know that these interests have been harmed. An entity therefore has an interest in not experiencing pain if and only if it can experience pain, and it has an interest in not being killed if and only if it is conscious of its life as a life, and not just as a series of unconnected experiences. In the context of the embryo this leads to the conclusion that it does not have any rights until it develops sentience in the fetal stage, and that it never develops a right not to be killed, because it does presumably not develop a conception of its own life as a life before well after birth.

The second (preference consequentialist) approach proceeds by arguing that what really matters ethically is preference satisfaction, and that the wrongness of doing specific acts can be located in the degree to which they thwart the preferences of the entities concerned. Since embryos and fetuses do not have a preference for going on living (or anything else), it is not wrong to kill them.

One problem for views of this sort is that they necessarily lead to the conclusion that full moral status is not attained until well after birth, since it is highly unlikely that infants have any idea of themselves as existing over time (i.e., of having a life). Infanticide is therefore not intrinsically wrong.

Proponents of the view that embryos have no moral status also usually argue that intrinsic value can come only from intrinsic properties; i.e., if an entity has intrinsic value (value in itself), this value must derive from some intrinsic property of the entity, and not from its relationship to other entities. The main problem with this argument is whether it is possible to provide a compelling account distinguishing intrinsic and extrinsic properties of entities.

2. Embryos Have Moral Status

The claim that embryos do have moral status has been supported by a number of different arguments.

The traditional Christian view, which is presently used primarily by Catholic moral theorists and by some evangelical groups, claims that an embryo has full moral status from the point of conception/fertilization. Similar views are held by some Orthodox Jewish and Islamic scholars. This view can be based on a theory of immediate ensoulment, or by reference to the fact that at fertilization a human being is created with a unique genetic makeup.

The main problems for this view are: (a) that it relies on a theological premise if it makes reference to ensoulment, and such premises are not generally accepted in a secular context; (b) that there is a possibility of twinning until 14 days after fertilization, which would require either two souls in one entity prior to twinning or the infusion of a new soul at the point of twinning; (c) that there is a possibility of chimera formation where two zygotes fuse, and this would seem to lead to an entity with two souls; (d) that the argument referring to the embryo as a human being may be guilty of speciesism (i.e., relying on the mere fact that something is human as an argument for giving it moral status); and (e) that there are other human cells with a unique genetic makeup which we do not accord the same status.

A different argument for the full moral status of the human embryo proceeds by localizing the core wrong in killing adult humans in the fact that we deprive them of "a life like ours"—we deprive them of all the experiences and other things which their life would have contained if we had not killed them.

This is also true of any embryo we might kill, and thereby makes killing embryos wrong. There may be other wrong-making characteristics involved in killing adult human beings (the pain, the fear, etc.) which play no role when we discuss killing embryos. This indicates that it may be more wrong to kill adults than embryos, but it does not show that killing embryos is morally innocuous. The main problem in this argument is that it makes killing wrong in all cases, even in cases where a person might want to be killed. This is not a problem if we are only concerned with embryos or fetuses, but it could be a problem if the account is intended to be a general account of the wrong done in killing human beings.

A third way of arguing for the proposition that an embryo has moral status is through an argument from potentiality. Arguments of this kind acknowledge that embryos do not have present conscious interests or present preferences, but they then proceed to the claim that an entity with a potential for possessing such interests or preferences does have moral status.

Such arguments rely on a clarification of what "potential" really means. Potential cannot be the same as logical possibility, since it is not logically impossible for most things to turn into other things (e.g., it is not *logically* impossible for the egg of a hen to develop into a human fetus). Potential also cannot be mere material possibility, i.e., the possibility that a certain piece of marble could turn into a statue of David in the hands of Michelangelo. It must, in the present context, entail that the entity having the potential also is responsible in some sense for the development leading to the fulfillment of the potential. Finally, the notion of potential must rely on some background notion about a stable environment. If this stable environment requirement is not brought in it would be the case that the potential of an embryo would depend upon whether or not its mother wanted to abort it.

If a coherent notion of potential can be established, there still has to be an argument for the move from "entity X has the potential to be Y" to "entity X now has the moral status which it will have when it becomes Y." It is often mentioned in the criticism of the argument from potential that the fact that someone is a potential president of the USA does not give him or her the same powers and prerogatives as the incumbent of the position. This is obviously correct, but it is equally possible to find examples where someone being a potential incumbent of a position does give special privileges, although these privileges may not be exactly the same as those of the actual incumbent (think of the role of the heir to the throne in monarchies).

Another problem for the potentiality argument is that two gametes also have the potential to become a human being. After all, the gametes have a potential to become a zygote, and the zygote has a potential to become a human being. The proponent of the argument from potential therefore seems commited to a prohibition of contraceptives in order not to frustrate the potential of the gametes. This problem has been dealt with either by arguing that potential is a property of entities such as zygotes and not of assemblies of entities such as sperm and ova, or by arguing that gametes do not possess the same degree of control over the developmental process as does the zygote and the later stages of the human embryo.

B. Replacement Arguments and Personal Identity

A specific class of arguments based on an intriguing observation about personal identity have become a standard feature of discussions about ethics and embryology. These are the so-called replacement arguments which all proceed from the observations that (a) if a woman conceives a child this month, the child will be different from the child she could have conceived next month, because they will come from the union of different gametes and be genetically different, and more generally (b) any change in reproduction which entails a change in the timing or manner of conception leads to the production of different children (children with different identities).

This has implications for many arguments of the type, "It is not good for children to grow up in condition X, Y, or Z, therefore it is wrong to procreate in condition X, Y, and Z because it harms the future children." Proponents of the replacement argument argue that there is an underlying conceptual confusion at play here. We may believe that we are comparing the welfare of the child growing up in adverse conditions with the welfare of the *same* child growing up in better circum-

stances and deciding which would be better, but this is not true. What we are doing is comparing two *different* children, the child growing up in adverse conditions and *another* child growing up in better circumstances. The life of the child in adverse conditions is the only life this child can have, and what we have to decide is not whether there are better lives, but whether the life of this child is so bad that it would be better not to have it. This is an unlikely proposition in most cases, so the argument that it would be better for the child not to be born than to be born disadvantaged is in most circumstances false.

The argument thus maintains that children born to single mothers, to lesbian mothers, or after IVF using matured eggs from aborted fetuses could only have been born this way, and that it makes no sense to prohibit or discourage such pregnancies for the sake of the children.

The replacement type of argument is just one of a larger class of arguments concerning the connection between the adult human being and the zygote. This class of arguments are concerned with the question of whether or in what way the adult and the zygote can be said to be identical. Common sense seems to indicate that I, the adult, am identical with the zygote from which I developed, but there are arguments against this view.

First of all it can be questioned whether the important thing is biological or personal identity. I may be biologically identical with the zygote without having any kind of personal identity with the zygote. This follows, for instance, if we accept a view of personal identity in which it is exclusively a function of psychological connectedness. Since the zygote has no psychology I cannot be psychologically connected to it.

Secondly the possibility of twinning and chimera formation makes it questionable whether I am even numerically identical with the zygote from which I developed. If I am a monozygotic twin both my brother and I developed from the same zygote, and it is logically impossible that both he and I can be identical with the same entity, since we are undoubtedly at the present time distinct and numerically nonidentical. It is simply logically impossible for A and B to be nonidentical at the same time that they are both identical to C. This assertion of logical impossibility is, however, only valid in all cases in a nontemporal logic like the usual first-order predicate logic. A full discussion of the possibilities of resolving the logical problem within a suitably enriched temporal logic is, however, beyond the scope of the present article.

Arguments about personal identity and individuation have been important in discussions about time limits on embryo research. Many countries now have a legal regulation stating that destructive embryo research is permitted only prior to 14 days g.a. This similarity in the legal regulations is most likely the result of a direct transfer of the 14-day limit suggested by the British Warnock committee which, with the moral philosopher Mary Warnock in the chair, published a report in 1984 on the ethical issues created by the new reproductive techniques (republished commercially as M. Warnock, Ed., 1985. *A Question of Life: The Warnock Report on Human Fertilization and Embryology*. Basil Blackwell, Oxford). The committee based its recommendation of a 14-day limit on the following argument (reconstructed from the text):

1. Prior to 14 days g.a. the primitive streak is not fully formed and there is a possibility of twinning
2. If there is a possibility of twinning the embryo cannot be an individual
3. Only individuals can have moral status
4. It is not wrong to perform destructive experimentation on entities without moral status

Therefore, the embryo becomes an individual at 14 days g.a. (the time of individuation). Prior to this time it has no moral status, and it is not wrong to perform destructive embryo experimentation.

There must obviously be a hidden premise stating that being an individual is not only necessary but also sufficient for some kind of moral status, otherwise the argument would not justify a limit at 14 days g.a. as the committee proposed, but only be an argument to the effect that no limit lower than 14 days g.a. could be justified. All the nonempirical premises of the argument have been criticized, but as could be expected proponents and opponents of embryo research cannot agree on which premises are wrong, or in what way they are wrong.

C. Is the Moment of Fertilization Important?

The simple answer to the question in the head of this section is "no," since there is no moment of fertilization but a process extended over time from the first penetration of the sperm to the occurrence of syngamy. A proponent of the view that fertilization is important could, however, without losing very much, move to either syngamy or the first activation of paternal genes as the important step marking the creation of a unique human being. What is important for the proponent of the view that fertilization matters is presumably not the exact point in time but the fact that at a specific point in time a genetically unique entity is created which is undeniably human.

There is, however, also a more radical challenge to

the view that fertilization is important. A challenge which casts doubts on any specific moment in human life as carrying special significance, be it fertilization, birth, sexual maturity, childbearing, or death. This view would point to the way life is presently conceptualized in biology.

If we look at life not as something bound to specific identifiable individuals or entities, but as a process encompassing the whole species and all its members, we find that sexual reproduction is just one of many subprocesses necessary for the continuation of life, but a subprocess which is neither more nor less important than others. Cooperative food gathering may be equally, or more, necessary for the continuation of life in some societies. The same point can also be illustrated if we look at species which reproduce both sexually and asexually (like the common hydra), where it seems odd from a biological point of view to say that sexual reproduction is more important than asexual reproduction, or that entities (new hydras) created by sexual reproduction are more important than entities (new hydras) created by asexual reproduction.

D. Is Viability or Birth Important?

On most views, both those claiming that a fetus has moral status and those claiming that it does not have moral status, there is no direct change in this status just because the fetus is viable outside its mother's womb or just because it has actually been born. The fetus is the same, it is just its possible or actual relationship to its mother which has changed.

This means that a view not according moral status to the fetus would necessarily entail that there is nothing intrinsically wrong in killing viable fetuses or neonates and infants. There may be social reasons not to allow such killing, but it is not wrong in itself. It has been claimed that this consequence of the main theories depriving the fetus of moral status amounts to a *reductio ad absurdum* of these theories.

The fact that the viable fetus is potentially independent of its mother, and that the neonate is independent of her, does, however, entail a difference in decision making. The strength of any reasons to allow the mother to control her own body, and have the embryo or fetus killed in the process, diminishes considerably in a situation where the question is no longer only a question about bodily integrity, but also a question about decision making for incompetent individuals. The mother may have a right not to have the embryo implanted into her body or a right to have the fetus dispelled from her body, but these rights do not necessarily entail that she has a right to have the embryo or fetus destroyed.

E. Are There Specific Problems in Creating Embryos for Research?

Embryos can be used in research for a number of purposes ranging from basic embryological research to improvements of IVF techniques. It is often claimed that embryo research is necessary for the further development of IVF techniques, and that the information produced about the earliest development of the human being may prove useful in the understanding of aging and of cancer. The embryos used in embryo research can either be so-called "spare" embryos that are surplus to requirements in the context of IVF treatments (e.g., because the treatment is a success and there are stored embryos which the parents do not need anymore), or they can be created specifically for research. There has been some debate about whether it is ethically acceptable to create embryos specifically for research.

The arguments produced to show that it is not acceptable to create embryos, but acceptable to use spare embryos, usually refer to Kant's categorical imperative and its prohibition against using someone merely as a means and not at the same time as an end. It is claimed that embryos created specifically for research are used merely as means to further the researchers project, and that this is wrong. The problem with this argument is that it does not really distinguish between spare and specifically created embryos. All embryos are created as means to somebody else's project (the "progenitors" or the "researchers"), and even if we want to deny this, it seems incontrovertible that even if a spare embryo was created as an end in itself it is transformed into a mere means as soon as it is donated to research. The Kantian argument may, however, be more applicable in the context of custody or inheritance disputes about frozen embryos. In many of these cases it is obvious that the embryos in question are even further reified than they were as part of the IVF procedure.

There is also the further empirical problem that the number of spare embryos is not a natural given. Any number of spare embryos can be produced by manipulating the number of eggs retrieved and the number of eggs fertilized. Since there is usually no clearcut distinction between the people performing IVF treatment and the people performing embryo research, the researcher may, in his guise as physician, ensure that a sufficient number of spare embryos are produced.

Bibliography

Chadwick, R. (Ed.) (1987). "Ethics, Reproduction and Genetic Control." Croom Helm, London.

Evans, D. (Ed.) (1996). "Conceiving the Embryo—Ethics, Law and Practice in Human Embryology." Nijhoff, The Hague.

Harris, J. (1993). "Wonderwoman and Superman." Oxford Univ. Press, Oxford.

Hursthouse, R. (1987). "Beginning Lives." Basil Blackwell, Oxford.

Kamm, F. M. (1992). "Creation and Abortion: A Study in Moral and Legal Philosophy." Oxford Univ. Press, New York.

Sadler, T. W. (1995). "Longman's Medical Embryology," 7th ed. Williams & Wilkins, New York.

Steinbock, B. (1992). "Life before Birth: The Moral and Legal Status of Embryos and Fetuses." Oxford Univ. Press, New York.

Environmental Impact Assessment

MARJA JÄRVELÄ and KRISTIINA KUVAJA-PUUMALAINEN

The University of Jyväskylä

GLOSSARY

ecosocial morality The principle or practice of adjusting values or way of life according to the principle of sustainable development.

environmental impact assessment A tool used in environmental management that examines the environmental consequences of development actions in advance.

nature contract An idea of a symbiotic relationship between humanity and nature to be taken as a parallel of the social contract, or to reform this contract.

policy of small action A policy of making everyday life choices in accordance with ecosocial morality.

social impact assessment A tool used to complement environmental impact assessment by predicting and evaluating the impact of planned actions on social and human aspects of changes in the living environment.

sustainable development Development that meets the needs of the present without compromising the ability of future generations to meet their own needs.

way of life The process of everyday life that is based upon its essential social conditions and people's perspectives.

ENVIRONMENTAL IMPACT ASSESSMENT (EIA) is one of the main tools used in current environmental management to produce sustainable development. Sustainable development has most often been defined in terms of economic development and biological reproduction. In its narrow sense it refers to the balance of economy and environment, but new formulations also emphasize free intellectual activity and development of cultural diversity. The authors of this entry perceive that the wider interpretation of sustainable development includes a moral obligation to transmit a viable living environment to future generations. This moral obligation can be treated as part of the social impact assessment (SIA) related to EIA processes. This involves basically the problem of promoting and integrating a layperson's perspective into the experts' activity in EIA. The policy of small action by the common citizens and other activities referring to the new norms of ecosocial morality should be considered seriously as social resources for sustainable development and for EIA processes.

I. INTRODUCTION

Environmental impact assessment (EIA) is one of the main tools used in current environmental management to produce sustainable development and to prevent the accumulation of environmental risks. "EIA is a *process*, a systemic process that examines the environmental consequences of development actions, in advance" (J. Glasson, R. Therivel, & A. Chadwick, 1995. *Introduction to environmental impact assessment* (p. 3). *The natural and built environment*, series 1. London: UCL Press). Taking a technical stance, the EIA process can be analyzed by its steps (e.g., identification of key impacts and prediction of impacts on natural environment) in terms of its efficiency in physical and biological risk minimizing. Yet a broader view of EIA refers to the comprehensive idea of sustainable development and even to the capacities of the civil society to act as a social agent in building a sustainable heritage for future generations.

This entry does not describe in detail the process and potentialities the EIA has in technical risk minimizing; neither do we argue that the current EIA processes are "unethical" or valueless. Our aim is merely to propose that the concept of ecosocial morality be considered as an integrative ethical element referring to the social aspect of EIA and to the concept of sustainable development, which is the ultimate aim of EIA processes.

II. ENVIRONMENTAL ETHICS

Sustainable development has most often been defined in terms of economic development and biological reproduction. In its narrow sense it refers to the balance of economy and environment. The concept of sustainable development was used already at the time of the Cocoyoc declaration on environment and development in the early 1970s. The concept gained wide popularity after the Brundtland Commission's 1987 report *Our Common Future* (Oxford: Oxford Univ. Press), in which sustainable development was defined as a tool for economic growth and equity among the world's nations. The United Nations Conference on Environment and Development (UNCED) in Rio de Janeiro in 1992, in turn, gave a new emphasis to this concept. *Agenda 21*, which was formulated during the conference, comprised three operational dimensions: an ecological dimension,

a social dimension, and a cultural dimension. Hence, in addition to the Brundtland Commission's definitions of ecological and socially just economic development, the new formulation of the concept also emphasized free intellectual activity and development of cultural diversity.

As the authors of this entry perceive it, the wider interpretation of sustainable development refers to the totality of environmental and social policies and their practical implications. The basic argument we propose is that there is a moral obligation inherent to the concept of sustainable development: we should do at least as well for our successor generations as our predecessors did for us. We refer to the responsibility to transmit the common patrimony to an unspecified number of generation as understood, e.g., by François Ost.

Discussion on environmental ethics has previously led to a dramatic division of anthropocentric and biocentric interpretations (B. Norton, 1987. *Why preserve natural variety?* Princeton, NJ: Princeton Univ. Press). According to the anthropocentric approach, humans are the sole moral and valuable entity, and therefore nature is considered valid only when it is related to our goals and aims. Based on this premise, all values connected to nature are derived from human beings. Biocentric environmental ethics, in turn, argues that nature has internal values in itself which make nature equal to humans. Furthermore, our attitudes toward nature should create a more respectful behavior in relation to nature.

Humans, as individuals and communities, assume values and behavior in everyday life. Ways of life can be understood along anthropocentric and biocentric lines of reasoning. An anthropocentric position is represented by ways of life pursuing the well-being and happiness of human kind as ultimate aims. The biocentric position is represented by ways of life dedicating intrinsic value to nature and to populations of various species, as exemplified in the discourse of Deep Ecology.

III. WAYS OF LIFE AND THE ENVIRONMENT

In a study of self-regulation in modern society, there are at least two levels at which we can meaningfully use the conceptual tools of the social sciences to examine the relationship of changing ways of life to the environment. (The concept of the way of life is considered here as a process of everyday life which is based upon its essential social conditions and people's perspectives. People's perspectives refer here to the subjective goals, aspirations, and visions attached to their activities (see P. Ahponen & M. Järvelä, 1987. In J. P. Roos and A. Sicinski (Eds.), *Ways and styles of life in Finland and*

Poland (p. 71). Gower: Avebury).) On the one hand, there is the question of what kind of environment individual people can and will produce for themselves through everyday life choices. On the other hand, it is important to understand how the individual, as an integral part of nature, adapts and adjusts to changes in nature, and what sort of impacts and interactions are caused by human activity on and in different ecosystems.

The first level of recognition refers to the human–environment relationship, which starts with considering how people can interact with their environment without causing serious harm to themselves, to others, or to nature. Promoting sustainable development implies recognition of the ecosocial point of view in all development efforts. Robert Tessier refers to a shift in motivation that is expected to favor the community over individual survival or interest. In analyzing the concept of sustainable development Tessier even refers to collective instead of individual salvation, because he finds structural similarity in the rational action inspired by "sustainable development" and Max Weber's protestant ethic.

As regards the second level, the key thing is to look at the foundations of the ecological self-organization of way of life. Edgar Morin and Anne-Brigitte Kern have said that one relevant measure for a human society is whether it operates like a living machine rather than a built machine. A living machine is capable of self-corrective action, in spite of its functional deficiencies, whereas a built machine needs external intervention in an event of any malfunction. The same principle of ecological self-organization could be thought to apply to people's everyday way of life.

The discussion on environmental ethics has only recently been connected seriously to the discussion on Western ways of life and consumption. In the 1980s the concept of sustainability referred more to the environmental stress and catastrophes resulting from increased population and industrialization in the developing world or from such accidents as Chernobyl. The present discussion focuses more on the effects of daily life of an ordinary Western citizen and refers to the aggregated environmental and social impacts of the consumption culture.

As far as large-scale environmental risks are concerned, a major problem lies in the difficulty of predicting the ecological impacts of human activity. The environmental sociologist faces the additional difficulty of predicting human activity in modern society. We still know fairly little about how and why people move about in their daily labyrinths: we know much more about the restrictions that limit the scope of our choices and about what members of modern society can do and what they leave undone. For many people it is difficult not to take

the car out of the garage every morning, even though they are well aware that the catalyzer does not completely eliminate the environmental impacts of driving. On a very general level, then, we can conclude that ways of life are structured by modern economic and social systems that tend to mobilize interests and rational action that are not fully conscious of environmental risks inherent to them. Yet each citizen may, as a moral being, take a critical position on the risks induced by his or her choices, or on the reproductive policies of one's community as regards the environmental and social impact of daily decisions.

IV. THE NATURE CONTRACT

Ervin Laszlo's caricature of modern Man describes him as an accumulation of false beliefs about his chances of controlling nature (1989. *The inner limits of mankind*. London: Oneworld). In his critique Laszlo says that the idea of us being able to externalize the environment derives from the belief that nature is an inexhaustible reserve of raw materials. Laszlo is also sharply critical of urbanization and the urban way of life, drawing attention to the pompous aspects of a consumer society's way of life, which are products *not* of the independent logic of economy, but of human greed, short-sightedness, and oversized desires.

The French philosopher Michel Serres (1990. *Le contrat naturel*. Paris: Bourin) goes even further when describing the unbalanced relation between humans and nature. Serres argues that our relation to the environment is parasitical; we use environmental resources, reaping all the rewards without giving any returns. Modern social structures formulate their own justification: the fundamental misbelief that human rights be accorded "every human being." According to Serres this is the very essence of the parasitical relationship—there exists no justification which includes the whole global system as a value in the final assessment.

Serres argues that in order to establish a new balance in human societies' relation to their environment, the "social contract" should be replaced by the "nature contract," in which the environment is an equal counterpart to humans. In *Le Contrat Naturel*, Serres presents the idea of this symbiotic relationship, which is based on the ecosocial morality of the unique human being.

The policy-making processes of modern societies reproduce elements of the social contract that refer back to the constitutional history of modern nation states. In those contracts nature is not included as an inherent entity of vital interest or of intrinsic value. Traditionally, in modern societies social norms refer to the social contract, where nature remains an external entity to be conquered and exploited. From the sociological point of view, Serres' idea of a symbiotic relationship between man and nature, the nature contract, implies development of new norms by the ecosocial dynamics of the present civil society. This, in turn, addresses the final moral responsibility of the social community to develop the natural contract. This moral call is universal in the sense that each community has a common heritage, including the environment, to be maintained or restored not only for the benefit to us, but also for that to the future generation.

V. ECOSOCIAL MORALITY IN THE COMMUNITY

For the common person, environmental changes appear as global and comprehensive entities that are often difficult to reduce into a chain of causes and effects. If a community is to aim at a critical review of the effects of citizens' everyday life at the community level, it has to evaluate its common heritage of ecological and social ways of life. In addition, the community should be able to project impacts of any major development on the living environment, including the impact of the human–nature relation in that particular community. According to Mary Douglas and Aron Wildavsky, communities actually have a strong tendency to define impacts of change as risks that are morally loaded. However, in self-protective communities risks may become "village secrets" if no procedural evaluative practice is created.

The process of perceiving risk and of taking responsibility of the common heritage by the community is specific to particular cultural forms. Anthony Giddens refers to the culture of modern societies as post-traditional (1994. In U. Beck, A. Giddens, and S. Lash (Eds.), *Reflexive modernization.* Cambridge: Polity). In post-traditional cultures risks are not defined "domestically," but in the situation where impulses from the outside have a generative function. Giddens compares today's experts to traditional guardians who formulated local wisdom. But expertise, from the angle of the community, is a much more situational role that has to be tested not only in terms of knowledge but also in respect to the moral values it represents. In the final analysis, then, what is important is the trust between expert and layperson which may or may not be confirmed in any process of "environmental impact assessment." Thus, the assessment should include as its consituent element a building up of a common ecosocial morality that generates new patterns of reflexive modernity in societies which seem to have lost much of their traditional capacity for guarding the common heritage.

Even in a post-traditional society, there is a sense of solidarity with future generations. This aspect has been underlined by the discussion of sustainable development, even if it is often neglected in experts' assessments finding shortcuts in practical technologies to deal with environmental problems and "development." Particularly in families, children represent the potential victims of environmental risks. Their children's fate make parents reflect on the threat of environmental risks and consider how long "we (our descendents) can be safe."

VI. GLOBALIZATION AND ECOSOCIAL SOLIDARITY

Many manifestations of environmental risks (such as depletion of ozone layers, the accelerated greenhouse effect, and scarcity of potable water) have led to the conclusion that risks of human impact on nature are not only local but also global by character. Correspondingly, there has been strong agreement that principles of environmental policy should be agreed upon at the level of international politics or even defined by the "global community." The critical point for globalization of environmental risks and the call for sustainability and the nature contract is whether ecosocial solidarity may be raised high enough to result in a new moral comprehension at the world level. According to this comprehension, factors which were considered as "development" in the past may not be considered as such any longer (see also Laszlo, 1989).

In relation to poverty, the controversy between economy and ecology is of vital interest. Particularly, the "developing countries" have had difficulties in finding sustainable options that serve their own interests and that meet with critical Western projections (as manifested, e.g., by "green" movements). Nevertheless, there has appeared a new kind of a solidarity concerning "distant others," not only the members of future generations, but also individuals who live far away. This kind of solidarity has increased simultaneously with the globalization of environmental risks, but has also facilitated the globalization of culture (mass media, exported cultural products, etc.). Nowadays many people are aware that "their day-to-day actions are globally consequential" (Giddens, 1994, 57). Social morality thus acquires a new dimension of global solidarity that may restructure the way of life of the individual or even of the community.

VII. POLICY OF SMALL ACTION

Citizens' response to the challenge of changing their way of life through their own choice is in most cases a policy of small actions. Apart from those who choose a radical alternative of asceticism, citizens show their solidarity toward distant others through small actions that do not reverse their own way of life (such as recycling, rejecting harmful products and packaging, and saving electricity and gas). Yet, no matter how minute the acts may appear even to the actors themselves, they lead to the formulation of a new ecosocial norm in civil society. The formulation of ecosocial norms in the daily patterns of action indicates that the personal assessment of environmental changes may lead to new ecological adaptation—or defense mechanisms—from the bottom up and not only by top-down experts' strategies. For example, in Finland citizens commonly believe that they may affect their own ways of life and, to a certain extent, even introduce to others ecosocial morality. New ecosocial norms are, therefore, increasingly a part of daily discussions in families, workplaces, and schools, as well as of other everyday interactions (M. Järvelä & M. Wilenius, 1996. In *The Finnish Research Programme on Climate Change, final report*. Publications of the Academy of Finland).

The application of the personal nature contract and the policy of small actions creates a new ecosocial moral community which emphasizes frugality as a civil virtue. The interpretations of frugality may vary from conserving a household's electricity to choosing second-hand shopping. The main idea is that everyone is able to make some of these choices and become a member of the new ecosocial community. Some acts are practiced very consciously as symbolic manifestations of an ecosocial orientation. In Finland, two of these widely adopted representations are giving up the use of plastic bags (used by the majority) in daily shopping or using the bicycle when going to work. Based on these symbolic actions, the members of the ecosocial community can recognize each other in the middle of daily routines.

The policy of small actions reflects citizens' orientation to a new possible future in which society has chosen a more ecological direction of development. It is considered more valuable as an act of human solidarity to "do even something" in the face of an unsecure future than just to wait and see what will happen. Even if there is no radical change of the frames or patterns of ways of life, ecosocial norms gradually generate new environmental awareness. Instead of the externalized threat of nature expressed by the agonizing relation, "nature versus me," as many come to realize the globalized risks of living as we have, people will start to think in terms of "I and nature" to prepare more or less consciously a reformed basis to pursue community-level sustainability. Consequently, the policy of small actions and acts of frugality can be seen as elements of a more general ecosocial self-regulation of everyday practices. This social development seems to be crucial for the transition

of awareness and action that will bring the issue of environmental reproduction and impact assessment to the community level.

VIII. LEGITIMACY AND POLITICAL EFFICIENCY

There is, however, a basic problem of legitimacy and of political efficiency inherent to citizens' policy of small actions. The everyday norms of ecosocial morality and their strong attachment to the daily policy of small actions keep citizens' world of living (*Lebenswelt*) at a considerable distance from the society's environmental politics, which remains to a great extent an experts' "playground" (Järvelä & Wilenius, 1996). It can also be argued that the members of ecosocial moral communities do not necessarily identify themselves with environmental movements which have become familiar through the media. Frugality and prudence as commonly appreciated virtues of civil society lead to a situation where, for example, Greenpeace's acts—if they are considered to be legal at all—gain only passive support by members of the ecosocial community. Legality is considered not only in social movements' activities, but is also in personal norms. Environmental actions ought to meet the demands of legality. Consequently, ecosocial moral norms do not, in fact, replace traditional social norms referring to the social contract. Rather, the norms deduced from the nature contract, as perceived by laypeople of the post-traditional society, should be added to the "contract basis" that sets norms and values in social policy and in everyday life.

IX. ENVIRONMENTAL IMPACT ASSESSMENT AS RATIONAL ACTION

The formulation of the ecosocial moral community follows a certain reasoning in which the principles for personal action are formulated through an individual implementation of the "nature contract." Personal frugality and a policy of small actions may be based on a generalized moral obligation without any further development of the rational strategy that is usually called for in societal environmental policy and expert discourse. In a highly technological modern society with complex policy-making patterns it is, however, important to ask in what social arena and how the values and goals can be mutually understood by experts and laypeople. How is the civil society mobilized to protect nature or to secure a sustainable living environment?

According to our view the responsbility of sustainable environmental policy cannot simply be transferred to experts or environmental movements from the community level. Civil society should create particular rational models of self-regulation to guarantee a social arena for risk perception, for dealing with values and preferences, and for democratic choice of policy action. This would be an arena where experts and laypeople could meet on an equal basis. Personal impact assessment and ecosocial profiles built and represented in everyday life would, then, serve as social resources to increase the self-regulative capacities of the society.

The personal mode of action commonly follows the model of teleological reasoning which may be composed as follows (M. Järvelä & M. Wilenius, 1993. *Climate change, living environment and ways of life*, working paper 9/1993. Tampere: Research Institute for Social Sciences, Univ. of Tempere): (1) The actor recognizes a factual situation that has to do with a highly appreciated object. (2) Having realized the factual situation, the actor chooses another situation in relation to the appreciated object. The more preferred situation is set as the ultimate goal of action. (3) The actor then pursues this goal by specific means that are considered to be essential for achieving the goal. For example, in the case of the accelerated greenhouse effect, the members of the ecosocial moral community can choose appropriate means in their daily lives to prevent the increase of carbon dioxide emissions and to join in this way a more global strategy of rational action to mitigate environmental risk. This kind of reasoning is structured in the way that usually is recognized as an ideal model of expert discourse, no matter how much the actual mitigation may differ from the ideal model. As an ideal model, it may, however, serve as an analysis of the gap between the expert and the lay perspective on the issue of sustainability of the environment and environmental risk.

As said before, the formulation of the ecosocial community in civil society is usually actualized through a policy of small actions: there is rarely involved any direct structural changes at the level of state policies due to these actions. However, communication between experts, professionals, and laypeople can be considered a desired outcome of the development of ecosocial communities. The personal impact assessment and ecosocial profiles of individual families provide a channel of information which is likely to offer new resources for environmental planning and administration, if this agency of the civil society can be channeled to the societal management processes, such as environmental impact assessments.

X. MODEL OF SOCIAL ACTION

There can be drawn a crude distinction between (1) a strategy of rational political influence and (2) a mythical

survival strategy. It is commonly assumed that experts and professionalists are inclined to follow a strategy of rational environmental influence and that laypeople are inclined to resort to a mythical environmental survival strategy. The baseline assumption is that there is an important difference in the strategic approach or in the form of rationality between experts and laypeople (Järvelä & Wilenius, 1993). It is likely that a pure division of social values and action orientation, such as presented by the model of "theoretical teleologies" (see Figure 1), has never been found historically. A more reasonable approach is to argue that there is a large scale of variation around the ideal types. Both experts and laypeople can and do combine elements of both rational and mythical analysis in their strategic approaches. Yet it is more probable that we find the ideal types when analyzing the strategic behavior in different social fields, as dominating strategies on each side (Järvelä & Wilenius, 1993). To facilitate communication between two relatively distant partners, it is of general interest to develop methods of auditing and participation by civil society such as the environmental impact assessment. There is, however, no technological solution to the problem of social distance between experts and laypeople. This is why environmental impact assessment needs to be completed and reformed by social impact assessment.

XI. ENVIRONMENTAL IMPACT ASSESSMENT

Environmental impact assessment was institutionalized rapidly in the 1980s as an internationally recognized strategy for environmental policy. Since the 1980s EIA procedures have developed into multiple strategies and vary by country (see A. Gilpin, 1995. *Environmental impact assessment* (pp. 16–35). Hong Kong: Cambridge Univ. Press). As a process, EIA may, however, be defined by its phases. According to Glasson *et al.* (1995, 4), the main steps of the EIA process can be defined as in Table I.

Public consultation is a formal part of the EIA, which may be attached to several phases of the process. More important here is that this involves the official recognition of the "parties interested" in each EIA process *separately*. From the point of view of sustainable development, however, a good environmental standard may be defined to be a *general interest*. Consequently, it is important to discuss in what framework this general interest can be presented.

From a sociological point of view, EIA has recently been introduced with reference to reflexive modernization, society risk, and communicative planning. It is assumed to develop into a tool of environmental management in order to promote ecological modernization of

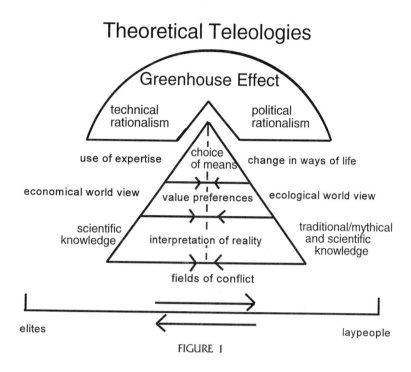

Theoretical Teleologies

FIGURE 1

a society. A critical view of EIA is concerned with the introduction of an effective shift from scientific-data-based impact assessment to a value-based assessment of environmental change. It also calls strongly for the mobilization of the social action resources of the civil society, and, furthermore, it emphasizes developing linkages between laypeople's perceptions and expert knowledge.

TABLE I

Step	Tasks
I.	Project screening (is an EIA needed?)
	Scoping (which impacts and issues to consider?)
	Description of the project/development
	Description of the environmental baseline
	Identification of key impacts
II.	Prediction of impacts
	Evaluation and assessment of significance of impacts
	Identification of mitigating measures
III.	Presentation of findings in an environmental impact statement (EIS)
IV.	Review of the EIS
V.	Decision making
VI.	Postdecision monitoring
	Auditing of predictions and of mitigation measures

XII. SOCIAL IMPACT ASSESSMENT

While *environmental* impact assessment has focused mainly on the physical and biological properties of natural environments, *social* impact assessment (SIA) concentrates on the distinctively human side of particular environments. The very heart of the SIA is to examine the social impact of formulating policies, instituting programs, and building projects. As does the environmental impact assessment, SIA aims to predict and evaluate the impact and outcomes of planned actions. This is, however, approached from the individual, family, or community point of view. SIA methods have been developed since the 1970s in parallel with EIA methods, but often with a distance and distinctiveness from the development of EIA expertise. Today the problem of the (re)integration of EIA and SIA appears to be one of the basic questions not only from a pragmatic policy point of view, but also when considering ethical aspects of the more indirect human impacts related to development acts.

According to its basic idea, SIA uses local citizens' environmental awareness as an input to the societal process of environmental impact assessment. The SIA process should bring forth the shared or controversial values of the community and create a value-based process of mitigation in the interest of the local community. There are several alternatives of practical strategies for the SIA. Auditing may be part of the institutionalized process of EIA, citizens may be involved in the debate concerning EIA in mass media, or a local future workshop can be created. We suggest one more action idea that could be practical as a more permanent participative appraisal for ecosocial communities.

There have been, e.g., in Finland, some recent tentative experiences of integrating social work in local communities with the zoning processes of the cities in a way which we find to be a very interesting example of a new community-level policy of sustainable development. The ecosocial approach to social work pursues in this case an empowerment strategy, trying to match city planning to the ecosocial profiles and projection of the local population. The conventional experts of the EIA process are involved, of course, but there is an additional contribution by the local social workers to mediate between experts and laypeople. In this case social workers are developing capacities as negotiating professionals in ecosocial mitigation. By outlining community values to the experts they help the experts to determine the relevant criteria of a good living environment. On the other hand, the experts' knowledge and technical resources can be translated to the members of the community. The application of this two-way process of advocacy and empowerment calls for an open professionalism where traditions of social closure may be avoided among social workers. The capacities of the reflexive and open professionalism are tested particularly in controversial cases such as in construction of new roads or supermarkets at the next neighborhood of the ecosocial community.

We mention the example of the Finnish social workers as a particular application of SIA mainly because their experiences have further clarified some basic elements of the lay perspective on the immediate living environment. The social workers' activity as SIA agents also increased our understanding of what kind of social mechanisms are applicable to channel laypeople's ethical purposes to the EIA process in order to promote sustainable development. The most crucial qualities or basic values of the urban living environment were brought to the fore through the process of reciprocal participation between citizens and experts. The SIA interpretation by the social workers of the vital value basis refers to the preference for three qualities of an urban environment: (1) good quality or state of the physical and biological environment, (2) social and ecological diversity of the environment, and (3) facility of quotidian life. The ecosocial community is, of course, expected to appreciate highly qualities (1) and (2), whereas the third element is not as easily assumed as part of the quality of a sustainable environment. From the community's point of view, this aspect may, however, be very decisive in everyday life when people choose the partic-

ular acts of their ecosocial policy of small actions. Thus, the final outcome of the SIA process should be balanced as an assessment of the man–nature relationship and should include even the feedback the man-to-man relationship has on the former relation.

XIII. CONCLUSION

Environmental problems are inherently social problems. They have been generated mainly by the industrial societies and by consumers' ways of life. Hence, it is consistent that particularly these societies create strategies of social action aiming at mitigation of these problems. Environmental impact assessment has been developed into a legitimate tool for experts to define the impacts of environmental change. In a modern society ecological development is, however, a major social issue in the everyday life of the common people. Daily choices by individuals and communities do have global consequences. This is why environmental impact assessment ought to be reformed and complemented by social impact assessment.

Social impact assessment involves communication not only from expert to expert but also between experts and laypeople. It is ultimately the mission of the civil society to create a culture of self-control in relation to the living environment. In highly individualized societies—and in the world of globalized risk for the environment—the social resources of sustainability can be restored to a great extent to the moral obligations of the solidarity the citizens feel for the "distant other," whether these others are represented by faraway people or by people of future generations. The new ecosocial morality of small steps of environmental protection in everyday life is not to be minimized but to be recognized as an element of a new culture promoting ecosocial morality in civil society. The more communities are empowered by the new ecosocial morality, the stronger the capacities they have in generating political impulses to the process of environmental impact assessment in order to support future prospects for sustainable development.

Bibliography

Bourdieu, P. (1987). *Choses dites*. Paris: Les Éditions de Minuit.

Douglas, M., & Widavsky, A. (1983). *Risk and culture, an essay on the selection of technological and environmental dangers*. Berkeley: Univ. of California Press.

Finsterbusch, K., Llewellyn, L. G., & Wolf, C. P. (Eds.) (1993). *Social impact assessment*. Beverly Hills, CA: Sage.

Hofbeck, J. (1991). La deep ecology: Essai d'évaluation éthique. In J. A. Prades, J.-G. Vaillancourt, et R. Tessier (Eds.), *Environnement et développement: Questions éthiques et problèmes sociopolitiques*. Québec: Éditions Fides.

Morin, E., & Kern, A. (1993). *Terre-Patrie*. Paris: Seuil.

Norton, B. (1987). *Why preserve natural variety?* Princeton: Princeton Univ. Press.

Ollikainen, M. (1992). Kestävä kehitys: Talouden ekologisointi eettisten ja taloustieteellisten periaatteiden nojalla (Sustainable development: Ecological perspective on economy with reference to principles of ethics and economy). Yhteiskunta ja ympäristö, Ympäristöpolitiikan arvo-ongelmia M. Järvelä ja Y. Uurtimo (Eds.), Julkaisuja Yhteiskuntatieteiden tutkimuslaitos, Tampereen yliopisto 2/1992, 101–123.

Redclift, M. (1987). *Sustainable development.* London: Methuen.

Sairinen, R. (1993). Environmental impact assessment in Finland. In A. Haila (Ed.), *Cities for tomorrow—Directions for change.* Espoo: Helsinki Univ. of Technology, Centre for Urban and Regional Studies.

Tessier, R. (1991). L'éthique du développement durable: Quels fondements? Une comparaison avec l'ascétisme séculier chez Weber. In J. A. Prades, J.-G. Vaillancourt, et R. Tessier (Eds.), *Environnement et développement. Questions éthiques et problèmes sociopolitiques.* Québec: Éditions Fides.

Wolf, C. P. (1983). Social impact assessment: A methodological overview. In K. Finsterbusch, L. G. Llewellyn, and C. P. Wolf (Eds.), *Social impact assessment.* Beverly Hills, CA: Sage.

Feminist Ethics

ROSEMARIE TONG
Davidson College

GLOSSARY

ethics A systematic attempt to understand moral concepts such as right, wrong, permissible, ought, good, and evil; to establish principles and rules of right behavior; and to identify which virtues and values contribute to a life worth living.

feminism A political, economic, cultural, and social movement aimed at eliminating systems, structures, and attitudes that create or maintain patterns of male domination and female subordination.

gender A social term denoting the traits, behaviors, and identities that a culture associates with being female or male; females exhibit "feminine" traits and males exhibit "masculine" traits.

sex A biological term denoting the chromosomal structure of a human being. An XX chromosomal structure is female, whereas an XY chromosomal structure is male.

sexual preference A choice made by individuals to have sexual relations with members of the other sex (heterosexual), the same sex (homosexual or lesbian), or both sexes (bisexual).

FEMINIST ETHICS is an attempt to revise, reformulate, or rethink those aspects of traditional Western ethics that depreciate or devalue women's moral experience. Among others, feminist philosopher Allison Jaggar faults traditional Western ethics for failing women in five related ways. First, it shows little concern for women's as opposed to men's interests and rights. Second, it dismisses as morally uninteresting the problems that arise in the so-called private world, the realm in which women cook, clean, and care for the young, the old, and the sick. Third, it suggests that, on the average, women are not as morally developed as men. Fourth, it overvalues culturally masculine traits like independence, autonomy, separation, mind, reason, culture, transcendence, war, and death, and undervalues culturally feminine traits like interdependence, community, connection, body, emotion, nature immanence, peace, and life. Fifth, and finally, it favors culturally masculine ways of moral reasoning that emphasize rules, universality, and impartiality over culturally feminine ways of moral reasoning that emphasize relationships, particularity, and partiality.

Feminist ethicists have developed a wide variety of approaches to ethics, including those labeled "feminine," "maternal," "political," and "lesbian." Each of these *feminist* approaches to ethics highlights the differences between men's and women's respective situations in life, biological and social; provides strategies for dealing with issues that arise in private as well as public life; and offers action guides intended to undermine rather than bolster the present systematic subordination of women (A. Jaggar, 1991. In *Feminist Ethics* (C. Card, Ed.). Univ. of Kansas Press, Lawrence). Considered together, the overall aim of all these feminist approaches to ethics is to create a gender-equal ethics.

I. FEMININE APPROACHES TO ETHICS

Biological and/or social differences between men and women are the foundation of men's and women's respectively different styles of moral reasoning, behavior, and identity, according to those formulating feminine approaches to ethics. For example, moral psychologist Carol Gilligan maintains that because women have traditionally focused on others' needs, they have developed a language of care that stresses the importance of creating and maintaining intimate human relationships. In contrast, because men have traditionally devoted themselves to the world of enterprise (business, medicine, and law), they have developed a language of justice that emphasizes the use of mutually agreed upon rules. Gilligan also maintains that widely accepted scales of moral development—for example, Lawrence Kohlberg's Six-Stage Scale—are attuned to the voice of justice but not of care. For this reason, those who speak the language of care (primarily women) rarely climb past Kohlberg's Stage Three ("the interpersonal concordance or 'goodboy–nice girl' orientation"), while those who speak the language of justice (primarily men) routinely ascend to Kohlberg's Stage Five ("the social contract legalistic orientation") or even Stage Six ("the universal ethical principle orientation").

Gilligan's ethics of care is feminist because she insists that women's morality of care is no less valid than men's morality of justice: all human beings should be both caring *and* just. Thus it concerns Gilligan that men seem far less willing than women to embrace the moral values associated with the opposite sex. She suspects that this reluctance on men's part is rooted in Western culture's continuing disrespect for so-called effeminate men.

Even more than Gilligan, Nel Noddings, a philosopher of education, argues that ethics is about caring relationships between individuals. There are, she says, two parties in any relationship: the one caring and the cared for. In the ideal, the dynamic of care consists of the one caring being emphatically engrossed in the cared for as a person, and the cared for recognizing and gratefully responding to the one caring's attention by "turn[ing] freely towards his own projects, pursu[ing] them vigorously, and shar[ing] his accounts of them spontaneously" (N. Noddings, 1984. *Caring: A Feminine Approach to Ethics and Moral Education*, p. 9. Univ. of California Press, Berkeley).

Noddings' feminine ethics of care is feminist because it underscores the differences between men's and women's typical styles of moral reasoning. Although women are just as capable of deductive-nomological moral reasoning as men are, Noddings claims that, far more than men, women prefer to consult their "feelings, needs, impressions, and ... personal ideals" when they make moral decisions (Noddings, 1984, 5). Another reason to view Noddings' feminine ethics of care as feminist is that she provides strategies for dealing with many private or domestic issues, including the care of children, animals, plants, possessions, and ideas.

Whether Noddings' feminine ethics of care is also *feminist* in the sense of subverting the systematic subordination of women is less clear. Although Noddings insists that care giving is a fundamental *human* activity, virtually all of the caregivers Noddings describes are women, some of whom seem to care too much—that is, to the point of imperiling their own identity, integrity, and even survival in the service of others. Moreover, although Noddings claims that the one caring needs to care for herself, she often conveys the impression that self-care is morally legitimate only insofar as it enables the one caring to be a better carer. Finally, Noddings suggests that relationships are so important that "ethical diminishment" is almost always the consequence of breaking a relationship—even a destructive one. For all these reasons, some feminist critics of Noddings urge her to take seriously Sheila Mullett's warning that women should engage only in the kind of care that "takes place within the framework of conscious-raising practice and conversation" (S. Mullett, 1989. In *Feminist Perspectives: Philosophical Essays on Method and Morals* (L. Code, S. Mullett, and C. Overall, Eds.), pp. 119–120. Univ. of Toronto Press, Toronto).

II. MATERNAL APPROACHES TO ETHICS

Closely related to feminine approaches to ethics are maternal approaches to ethics. Virginia Held, Sara Ruddick, and Caroline Whitbeck stress that, in the course of their ethical deliberations, all moral agents should use the concepts, metaphors, and images associated with the practice of mothering. They claim that, as mothers proceed in rearing their children, they typically manifest a type of "maternal thinking" that constitutes human moral reasoning at its best. In Ruddick's estimation, if we all thought like mothers think about their children's individual well-being and social acceptability, we could help each other transform our adversarial and competitive public world into a harmonious and cooperative one.

Maternal approaches to ethics are feminist not only because they value maternal thinking but also because they provide strategies for dealing with issues that arise in the private or domestic realm—for example, how to avoid the kind of psychological and physical exhaustion that contributes to violence in the home. It is not as clear, however, that maternal approaches to ethics help subvert the systematic subordination of women. Ruddick herself notes that her emphasis on mothers could be misread as (1) an unwarranted idealization of mothers as saints who attend to their children's needs no matter the costs to themselves; (2) a privileging of *biological* mothers as the "best" kind of mothering persons; and (3) a presentation of a certain kind of mothering—the kind white, heterosexual, middle-class women exhibit—as somehow better than the kind of mothering some minority women are forced to do in extremely oppressive circumstances. Unless maternal thinkers carefully stipulate their novel use of the term "mother," the mother–child relationship might lead all human beings, but especially women, down some disempowering rather than empowering moral pathways.

III. POLITICAL APPROACHES TO ETHICS

More than feminine and maternal approaches, political approaches to ethics offer action guides aimed at subverting rather than reinforcing the present systematic subordination of women. Liberal, Marxist, radical, socialist, multicultural, global, and ecological feminists have offered different explanations and solutions for this state of affairs. Likewise have existentialist, psychoanalytic, cultural, and postmodern feminists. Proponents of these varied schools of feminist thought maintain that the destruction of all systems, structures, institutions, and practices that create or maintain invidious power differentials between men and women is the necessary prerequisite for the creation of gender equality.

Liberal feminists charge that the main cause of female subordination is a set of informal rules and formal laws that block women's entrance and/or success in the public world. Excluded from places such as the academy, the forum, the marketplace, and the operating room, women cannot reach their potential. Women will not become men's equals until society grants women the same educational opportunities and political rights it grants men.

Marxist feminists disagree with liberal feminists. They argue that it is impossible for any oppressed person, especially a female one, to prosper personally and professionally in a class society. The only effective way to end women's subordination to men is to replace the capitalist system with a socialist system in which both women and men are paid fair wages for their work. Women must be men's economic as well as educational and political equals before they can be as powerful as men.

Disagreeing with both Marxist and liberal feminists, *radical feminists* claim that the primary causes of women's subordination to men are women's sexual and re-

productive roles and responsibilities. Radical feminists demand an end to all systems and structures that in any way restrict women's sexual preferences and procreative choices. Unless women become truly free to have or not have children, to love or not love men, women will remain men's subordinates.

Seeing wisdom in both radical and Marxist feminist ideas, *socialist feminists* attempt to weave these separate streams of thought into a coherent whole. For example, in *Women's Estate*, Juliet Mitchell argues that four structures overdetermine women's condition: production, reproduction, sexuality, and the socialization of children. A woman's status and function in *all* of these structures must change if she is to be a man's equal. Furthermore, as Mitchell adds in *Psychoanalysis and Feminism,* a woman's interior world, her psyche, must also be transformed, for unless a woman is convinced of her own value, no change in her exterior world can totally liberate her.

Multicultural feminists generally affirm socialist feminist thought, but they believe it is inattentive to issues of race and ethnicity. They note, for example, that U.S. "white" culture does not praise the physical attractiveness of African-American women in a way that validates the natural arrangement of black facial features and bodies, but only insofar as they look white with straightened hair, very light brown skin, and thin figures. Thus, African-American women are doubly oppressed. Not only are they subject to gender discrimination in its many forms, but racial discrimination as well.

Although *global feminists* praise the ways in which multiculturalist feminists have amplified socialist feminist thought, they nonetheless regard even this enriched discussion of women's oppression as incomplete. All too often, feminists focus in a nearly exclusive manner on the gender politics of their own nation. Thus, while U.S. feminists struggle to formulate laws to prevent sexual harrassment and date rape, thousands of women in Central America, for example, are sexually tortured on account of their own, their fathers', their husbands', or their sons' political beliefs. Similarly, while U.S. feminists debate the extent to which contraceptives ought to be funded by the government or distributed in public schools, women in many Asian and African countries have no access to contraception or family planning services from any source.

Ecofeminists agree with global feminists that it is important for women to understand how women's interests can diverge as well as converse. When a wealthy U.S. woman seeks to adopt a child, for example, her desire might prompt profiteering middlemen to prey on indigent Asian or African women, desperate to give their yet-to-be-born children a life better than their own. Ecofeminists add another concern to this analysis: In wanting to give *her* adopted child the best that money can buy, an affluent woman might not realize how her spending habits negatively affect not only less fortunate women and their families, but also many members of the greater animal community and the environment in general.

Departing from these inclusionary ways of understanding women's oppression, *existentialist* feminists stress how, in the final analysis, all selves are lonely and in fundamental conflict. *In The Second Sex,* Simone de Beauvior writes that, from the beginning, man has named himself the Self and woman the Other. If the Other is a threat to the Self, then woman is a threat to man, and if men wish to remain free, they must not only economically, politically, and sexually subordinate women to themselves, but also convince women they deserve no better treatment. Thus, if women are to become true Selves, they must recognize themselves as free and responsible moral agents who possess the capacity to perform excellently in the public as well as the private world.

Like existentialist feminists, *psychoanalytic* and *cultural feminists* seek an explanation of women's oppression in the inner recesses of women's psyche. As they see it, because children are reared almost exclusively by women, boys and girls are psychosocialized in radically different ways. Boys grow up wanting to separate themselves from others and from the values culturally linked to their mothers and sisters. In contrast, girls grow up copying their mothers' behavior and wanting to remain connected to them and others. Moreover, because of the patriarchal cues they receive both in and outside the home, boys and girls come to think that such "masculine" values as justice and conscientiousness, which they associate with "culture" and the public world, are more fully *human* than such "feminine" values as caring and kindness, which they associate with "nature" and the private world.

In the estimation of many psychoanalytic and cultural feminists, the solution to this dichotomous, women-demeaning state of affairs rests in some type of dual-parenting arrangement. Were men to spend as much time fathering as women presently spend mothering, and were women to play as active a role in the world of enterprise as men currently do, then children would cease to associate authority, autonomy, and universalism with men, and love, dependence, and particularism with women. Rather, they would identify all of these ways of being and thinking as ones that full persons incorporate in their daily lives.

Finally, as *postmodern feminists* see it, all attempts to provide a single explanation for women's oppression not only will fail but should also fail. They *will* fail because there is no one entity, Woman, upon whom a label may be fixed. Women are individuals, each with a unique story to tell about a particular self. Moreover, any single explanation for Woman's oppression *should* fail from a feminist point of view, for it would be yet another instance of so-called "phallogocentric" thought, that is, the kind of "male thinking" that insists on telling as absolute truth *one* and *only* one story about reality. Women must, in the estimation of postmodern feminists, reveal their differences to each other so that they can better resist the patriarchal tendency to center, congeal, and cement thought into a rigid "truth" that always was, is, and forever will be.

IV. LESBIAN APPROACHES TO ETHICS

Lesbian approaches are to be distinguished from feminine and maternal approaches to ethics on the one hand and from political approaches to ethics on the other. Lesbian ethicists, like Sarah Lucia Hoagland, generally regard feminine and maternal approaches as espousing types of caring that contribute to women's oppression. They insist that lesbians should engage only in the kind of caring that does not bog down in a quicksand of female duty and obligation from which there is no escape. Lesbian ethicists also take exception to those political approaches to ethics that represent heterosexual relationships as ethically acceptable even in a patriarchal society. As they see it, separation from men is the only course of action for women who wish to develop themselves as truly free moral agents.

Although lesbian ethicists believe that heterosexual women, and even some men, can, and perhaps should, learn from lesbian ethics, they do not believe it is their responsibility to share their insights with men or even with heterosexual women. Rather, they believe it is their calling to create moral meanings and values for lesbians only, leaving it to nonlesbians to create their own moral meanings and values.

V. APPLICATIONS OF FEMINIST ETHICS

Feminist ethicists do not propose uniform answers to the kind of moral problems women typically face. However, there are some key concepts that most feminist ethicists do employ. Whether they are analyzing issues related to sexuality and sexual preference, reproduction, family structure, or work, feminist ethicists are apt to stress one of the following concepts: choice, control, or connection.

A. Sexuality and Sexual Preference

Many feminists claim that, in order to secure men's approval and support, women have struggled to shape their bodies to fit the contours of an androcentric ideal of female beauty. Women have dieted to the point of developing eating disorders like anorexia (self-starvation) and bulimia (binging and purging). They have had multiple cosmetic surgeries, often risking their health in the process. Worse yet, mothers have participated in binding their daughters' feet, or in breaking their ribs with whalebone corsets, or in mutilating their genitals—all in the name of male sexual pleasure. Saddened by the ways in which women have harmed themselves and each other to meet men's sexual needs, these feminists urge women to refuse to forsake their naturally given bodies and faces.

Other feminists claim that cultural ideals of beauty are not necessarily androcentric, and that just because a woman diets, wears make-up, or has a face-lift does not mean she is obsessed with looking like a top fashion model. On the contrary, she might simply be a woman who wishes to look her personal best, and sees no reason not to do so provided that she does not harm herself or others in the process. These feminists insist that women should feel free to perfect their bodies as well as their minds if doing so increases their self-esteem and self-confidence.

To the degree that feminists have expressed considerable disagreement among themselves about how healthy or pathological it is for women to focus on their physical appearance, they have tended to agree that male sexual violence against women is not only widespread but also almost always about power. Despite their uniform opposition to male sexual violence against women, however, feminists have expressed differing views about what it is and how best to combat it.

For example, some feminists insist that whenever persons with unequal power in the workplace or academy have sexual relations the term "sexual harassment" is applicable. As they see it, it is virtually impossible for the less powerful party not to feel that she (he) *has* to have sex with the more powerful person. Other feminists disagree. They fear that an important distinction between sexual harassment and sexual attraction is dissolving—that there can be genuine love between some employers and employees and some professors and students, for example.

Feminists also debate the meaning of the word "rape." Some feminists stress that acquaintances or dates that force themselves upon women are no less deserving of the label "rapist" than strangers who do so. The fact that a woman knows her rapist, has consumed drugs or alcohol with him, or has previously permitted him certain "liberties" with her body does not change the meaning of her "no, stop" to "yes, go." Other feminists, in contrast, believe that women have to be clearer about their sexual needs or wants. As they see it, when a woman gives a man mixed signals about her desire for sexual intercourse, he might *justifiably* interpret her hesitation to mean "yes" after all. These same feminists are also willing to consider that what some women call "date rape" is really only their morning-after reaction to a regretted act of consensual sexual intercourse.

Another sexual violence issue feminists debate is how to best end the cycle of violence that energizes woman battering. Some feminists believe that because women are socialized to blame themselves for everything that goes wrong in their relationships and to "forgive and forget" harms perpetrated upon them, it is crucial that women not give their batterers "another chance." These feminists think that the law should deal with woman batterers in the same way that it deals with anyone who batters another person. In contrast, other feminists maintain that battered women should be the ones to decide whether punishment or treatment is the most appropriate remedy for the men who beat them. They reason that, provided a battered woman is receiving the kind of strengthening counseling she needs for proper self-esteem and self-respect, she should have the option of giving her batterer a chance to salvage his relationship with her.

Given the extent of male sexual violence, it is not surprising that feminists have often questioned whether sexual relations between men and women serve women's best interests. Some feminists claim that heterosexuality is an institution that not only gives men a "right" to women's bodies but also deprives women of the possibility of being lesbians. As these feminists see it, given the intensity of the mother–daughter bond and the tendency of women to establish close emotional friendships with women, there is reason to think that had men not used their power to compel women to love them first and foremost, many women would have chosen women as their primary love objects.

Although it would seem that women can be what lesbian Adrienne Rich terms "woman-identified" without being lesbian, some feminists maintain that, in order to be fully feminist, a woman must be a lesbian even if she is not "naturally" sexually attracted to women. Therefore, if a heterosexual woman wishes to be fully feminist, she must try not only to refrain from sexual relationships with men but also to orient her sexual desires toward women. Other feminists believe that heterosexuality and feminism are compatible, provided that heterosexual women engage in heterosexual relations on their own terms. As these feminists see it, many women have male partners who treat them as full human persons. Not only do these men view their female partners as their social, political, and economic equals, they also view them as their sexual equals—that is, as persons whose sexual needs and desires are just as important as their own.

B. Reproduction

Feminists have expressed a wide range of opinions on the reproduction-controlling technologies (contraception, sterilization, and abortion) and the reproduction-assisting technologies (artificial insemination, *in vitro* fertilization, and surrogate motherhood). Many feminists emphasize that unless a woman can control her reproductive density, she will not be able to pursue her personal and professional interests as successfully as a man can. Thus, laws that prohibit contraception, sterilization, and abortion must be fought, and policies that limit access to these technologies must be challenged. To tell women that they have a right to abort their fetuses, for example, but that no one has a duty to fund or perform abortions is to provide women with a hollow "right."

Conceding that the reproduction-controlling technologies have benefitted many women, other feminists note that these same technologies have probably harmed at least an equal number of women. A woman may "choose" to have an abortion, but only because she lacks the means to raise a child she would very much like to have. Or a woman may "choose" to be sterilized simply because she cannot tolerate the physical and psychological side effects of the few contraceptives currently on the market.

Feminists also disagree about women's use of the new reproduction-assisting technologies. Some feminists believe that these technologies are beneficial because they increase women's reproductive choices. As they see it, provided that they know the risks as well as the benefits of these technologies, women, especially infertile women, should feel free to use the services of gamete donors and even surrogate mothers. They should not be judged on how much of their funds they spend on costly fertility treatments, including those offered by *in vitro*-fertilization clinics.

Other feminists believe that the new reproduction-assisting technologies are harmful because they segment and specialize reproduction as if were just another mode of production: One woman provides the egg; another woman gestates the embryo; and still another rears the child once it is born. The same feminists note that, increasingly, pregnant women are instructed to eat healthy diets, to exercise, to forsake licit and illicit drugs, and, if necessary, to submit to cesarean sections so that

their children will be born mentally and physically healthy. They are also urged to screen their fetuses for a variety of genetic diseases and defects, and to abort or genetically alter the unhealthy ones. Finally, they are told that women who, because of the way they conducted their lives during their pregnancies, give birth to a diseased or defective child risk being charged with fetal abuse or neglect.

C. Family Structure

Many feminists resent the kind of family structure that consists of a working father, a stay-at-home mother, and one or more children. They argue that in addition to confining women to activities associated with cooking, housecleaning, and child care, this type of family structure alienates men from their children. Men are banned from the nursery and women are locked out of the worlds of medicine, business, law, and so on. As a result, men and women become half-persons. Women are either not permitted or not encouraged to develop their intellectual skills, and men are either not permitted or not encouraged to develop their emotional skills. Convinced that the only effective way to enable women and men to become full persons is to eliminate the family structure as it has been constituted under patriarchy, many feminists have pressed for the development of family structures other than the traditional "nuclear family" consisting of a heterosexual couple and their biological children. These feminists maintain that society should bolster rather than undermine single-parent families (especially female-headed ones), lesbian (and gay) families, and extended families in which several generations of biologically related adults and children live together.

Another group of feminists claim that the families just described are not genuine *alternatives* to the nuclear family since they mirror the values of the traditional nuclear family. Thus, in the estimation of these feminists, society should support the growth of so-called family communities in which a number of adults (some heterosexual and others lesbian or homosexual; some married and others single; some with children and others without children) intentionally develop a lifestyle that eliminates child-rearing inequalities between men and women, the sexual division of labor, the sexual exclusivity of couples, and parental possessiveness of children. In such family communities, children would have as much power as adults, homosexuals and lesbians would have as much sexual freedom as heterosexuals, people of color would be accorded as much respect and consideration as white people, and everyone would share their economic resources communally.

A final group of feminists remain unconvinced that the traditional nuclear family is unreformable. As they see it, *less* than two parents is not necessarily good for children. If gender equality is the aim of feminism, then boys and girls should probably be raised by both a man and a woman. Moreover, *more* than two parents is not necessarily good for children as anyone who has reared an adopted child or a step-child knows. Although it is wrong to view children as putty, to shape in one's own image and likeness, these feminists suggest that intense parent–child bonds—a sense of belonging to each other on account of nature (genetics) as well as nurture (environment)—can be a source of great strength. Finally, these feminists maintain that sexual exclusivity, especially within marriage, probably serves women's interests better than sexual libertarianism.

D. Work

Feminists have mixed feelings about women's increasing success in the workplace. Some feminists point with pride to the fact that affirmative action policies have enabled women in large numbers to enter previously male-dominated professions and occupations. Other feminists note that the women who do the best in the public world do so because they adopt "male" ways of thinking and acting there. Thus, they express disappointment that women have not used "female" ways of thinking and acting to transform the competitive world of enterprise into a cooperative one in which people play to each other's strengths and minimize each other's weaknesses. Yet other feminists observe that it is not enough for women to enter the public world of work; men need to assume their fair share of domestic and parental responsibilities in the private world. Without household and child-caring assistance from their partners or hired domestic help (a definite luxury), many women find themselves working an exhausting "double day": one at the workplace and one in the home. Thus, these feminists conclude that if men can have it all—career and family—without exhausting themselves in the process, justice demands the same state of affairs for women.

A final group of feminists emphasize that not all women *want* to work outside the home. These feminists note that work means different things to different kinds of women. For rich, well-educated women, work is a creative outlet or a source of power, in some cases satisfying enough to substitute for an intimate family life. For poor women, who have always had to toil, work means sweating in factories or doing rich women's laundry. Such women dream of being suburban housewives—of having time to play with their children and to chat with their friends. Moreover, many women, irrespective of their class status, think that housework, and particularly child care, is a responsibility that demands the *full-time* attention of at least one parent. Women who wish to assume this responsibility should feel free to do so. Their choice is just as legitimate as the choice of a full-time career woman or a woman who combines career and family responsibilities.

VI. CONCLUSION

Although feminist approaches to ethics are all women-centered, they do not impose a single normative standard on women. Rather they offer to women a variety of accounts that validate woman's moral experience, but in a way that points to the weaknesses as well as the strengths of the values and virtues culture has traditionally labeled "feminine." In addition, they suggest to women a variety of ways to work toward the one goal that is essential to the project of feminist ethics, namely, the elimination of gender inequality.

Although feminists' different interpretations of what constitutes a voluntary and intentional choice, an illegitimate or legitimate exercise of control, and a healthy or a pathological relationship reassure the intellectual and moral community that, after all, feminism is not a monolithic ideology that prescribes one and only one way for *all* women to be, this variety of thought is also the occasion of considerable political fragmentation among feminists. Asked to come to the policy table to express *the* "feminist" perspective on a moral issue, all that an honest feminist ethicist can say is that there is no such perspective. Yet, if feminists have no clear, cogent, and unified position on a key moral issue, then a perspective less appealing to women may fill the gap. Although it is crucial for feminist ethicists to emphasize, for example, how a policy that benefits one group of women might at the same time harm another group of women, it is probably a mistake for feminist ethicists to leave the policy table without suggesting policies that are able to serve the *most* important interests of the *widest* range of women. For this reason, many feminist ethicists believe that, over and beyond their commitment to eliminating gender inequality, feminists need to develop a mutually agreeable methodology that will permit them to achieve a consensus position on many, if not all, the moral issues related to women. Feminist ethicists have a moral duty first to listen to each others' differing points of view, and then to develop policy recommendations that, despite their shortcomings, will nevertheless help inch as many women as possible toward the goal of gender equality with men.

Bibliography

Gilligan, C. (1982). "In a Different Voice." Harvard Univ. Press, Cambridge, MA.

Held, V. (1987). Feminism and moral theory. In "Women and Moral Theory" (Kittay E. and Meyers, D. Eds.). Rowman & Littlefield, Savage, MD.

Hoagland, S. L. (1989). "Lesbian Ethics." Institute of Lesbian Studies, Palo Alto, CA.

Jaggar, A. (1991). Feminist ethics: Projects, problems, prospects. In "Feminist Ethics" (Card, C. Ed.). Univ. of Kansas Press, Lawrence, KS.

Mullett, S. (1988). Shifting perspectives: A new approach to ethics.

In "Feminist Perspectives: Philosophical Essays on Method and Morals" (Code, L. Mullett, S. and Overall, C. Eds.). Univ. of Toronto Press, Toronto.

Noddings, N. (1984). "Caring: A Feminine Approach to Ethics and Moral Education." Univ. of California Press, Berkeley, CA.

Rich, A. (1970). "Of Woman Born." Norton, New York.

Ruddick, S. (1989). "Maternal Thinking: Toward a Politics of Peace." Beacon, Boston.

Tong, R. (1982). "Feminine and Feminist Ethics." Wadsworth, Belmont, CA.

Whitbeck, C. (1984). The maternal instinct. In "Mothering: Essays in Feminist Theory" (Trebilcot, J. Ed.). Rowman & Allanheld, Totowa, NJ.

Fetal Research

ROBERT H. BLANK

University of Canterbury

GLOSSARY

embryo Developing human organism prior to body definition at approximately 6 to 8 weeks after fertilization.

ex utero Outside of the uterus.

fetal tissue transplantation Use of fetal tissues for therapy through transplantation to affected organs.

fetus Developing human organism after body definition from approximately 9 weeks to term.

in utero In the uterus.

neural grafting Transplantation of tissue into the brain or spinal cord to treat neurological disorders.

preimplantational embryo Embryo after fertilization but prior to implantation and formation of placenta.

FETAL RESEARCH encompasses a broad array of research and potential clinical applications. Research can be conducted on the preimplanted embryo, on fetuses in the uterus prior to elective abortion, on previable fetuses after abortion, or on dead fetuses. Clinical applications using fetal or embryonic matter such as tissues, cells, or organs represent another form of experimentation incorporated under the definition of fetal research. Issues arise concerning the moral status of the fetus or embryo and the use of aborted fetuses or preimplantation embryos for research or experimentation. Questions also center on who may consent for the use of fetal materials, in what circumstances the abortion procedure can be modified to meet the needs of the research or transplantation procedures, and what type of compensation, if any, should be allowed for fetal tissues. As a result of these issues, fetal research has been elevated to the public agenda and has become a highly volatile moral and political issue.

I. TYPES OF FETAL RESEARCH

One of the problems in analyzing fetal research is that it encompasses a broad variety of types. One important distinction is that between investigational type research that is not beneficial to the fetus and therapeutic research that is possibly beneficial to the fetus but more likely to be beneficial to future fetuses. Another key distinction centers on the stage of development of the human organism when the research is conducted, from preimplantation to late fetal stages. The general sources of fetal material include tissue from dead fetuses; previable or nonviable fetuses *in utero*, generally prior to an elective abortion; nonviable living fetuses *ex utero*; fetal tissue transplantation research using tissue from dead fetuses; and embryos, *in vitro* and preimplantation. Although the variety of potential uses of fetal tissue is virtually unlimited, five areas are summarized here.

A. Fetal Development Studies

The first category of research deals with investigations of fetal development and physiology. The purpose of this research is to expand scientific knowledge about normal fetal development in order to provide a basis for identifying and understanding abnormal processes and, ultimately, curing birth deformities, or treating them *in utero*. These studies primarily involve autopsies of dead fetuses. However, studies of fetal physiology include the fetus *in utero* as well as organs and tissues removed from the dead fetus. In some instances, this research requires administration of a substance to the woman prior to abortion or delivery by caesarean section. This is followed by analysis to detect the presence of this substance or its metabolic effects in a sample of umbilical cord blood or in the fetal tissue. This research also advances knowledge of fetal development *in utero* by monitoring fetal breathing movements. Fetal hearing, vision, and taste capabilities have been documented by applying various stimuli to the live fetus *in utero*. In addition, some physiological studies utilize observation of nonviable but live fetuses outside the uterus to test for response to touch and for the presence of swallowing movements.

B. Fetal Diagnosis Research

A second type of research focuses on the development of techniques such as amniocentesis to diagnose fetal problems. The initial research was conducted primarily on amniotic samples withdrawn as a routine part of induced abortion to find the normal values for enzymes known to be defective in genetic disease. Once it was demonstrated that a particular enzyme was expressed in fetal cells and normal values were known, application to diagnosis of the abnormal condition in the fetus at risk was undertaken. Recent prenatal diagnostic research on fetuses involves extension of diagnostic capacities to additional diseases, development of chorionic villus sampling, and attempts to detect fetal cells in the maternal circulation.

Research has also been directed toward techniques that allow the identification of physical defects in the developing fetus. Ultrasound, alpha-fetoprotein tests, amniography, tests for fetal lung capacity, and a variety of techniques for monitoring fetal well-being or distress are recent products of this category of fetal research. In each case, following animal studies that indicated the safety and efficacy of the procedure, human fetal research was conducted in a variety of settings. Fetoscopy, for example, because of the potential risk to the fetus, was developed selectively in women undergoing elective abortion. The procedure was performed prior to abortion and an autopsy was performed afterward to determine its technical success.

C. Pharmacological Studies

A third area of fetal research involves efforts to determine the effects of drugs on the developing fetus. These pharmacological studies are largely retrospective in design, involving the examination of the fetus or infant after an accidental exposure. For instance, all studies on the influence of oral contraceptives or other drugs on multiple births or congenital abnormalities have been retrospective, as were most studies of the effects on the fetus of drugs administered to treat maternal illness during pregnancy. In these designs, no fetus was intentionally exposed to the drug for research purposes. However, some pharmacology research involves intentional administration of substances to pregnant women prior to abortion in order to compare quantitative movement of these agents across the placenta as well as absolute levels achieved in fetal tissues. These studies serve as guidelines for drug selection to treat intrauterine infections such as syphilis by examining the dead fetuses after abortion and demonstrating the superiority of one drug over another.

D. Embryo Research

The availability of human embryos for research purposes followed the development of *in vitro* fertilization (IVF). In 1982, Steptoe and Edwards, who 4 years earlier reported the first birth through IVF, announced plans to freeze "spare" embryos for possible clinical or laboratory use. Theoretically, fresh or frozen spare embryos could be augmented by the deliberate creation of embryos for research where donors consent to have their gametes or embryos used in this way. There is considerable hesitancy to move to the deliberate production of embryos for research even though this might be the only way to satisfy expanding research needs.

Although assisted reproduction technologies are all experimental at some stage and thus might be described as research, there are many nonclinical applications that clearly fit a research paradigm. Among the many nonclinical uses of human embryos are to (1) develop and test contraceptives; (2) investigate abnormal cell growth; (3) study the development of chromosomal abnormalities; (4) conduct implantation studies; and (5) initiate cancer and AIDS research. Potential genetic uses of the embryos include (1) attempts at altering gene structures; (2) preimplantational screening for chromosomal anomalies and genetic diseases; (3) preimplantation therapy for genetic defects; and (4) development of characteristic selection techniques including sex preselection. At some stage research on artificial placentas will be dependent on the availability of human embryos.

E. Fetal Tissue Transplantation

A final area of fetal research involves the use of fetal cells for transplantation. Unlike other areas of fetal research where the tissue is used to develop a treatment that might help future fetuses, in fetal transplantation research the tissue is the treatment used to benefit an identifiable adult patient. Although some persons see no ethical difference between the transplantation of adult organs and that of fetal tissues to benefit individual recipients, others contend that the use of fetal tissue for this purpose is ethically questionable because there are no possible benefits to particular fetuses or to future fetuses.

Despite the ethical controversy surrounding the use of fetal tissue discussed in Section IV, many scientists consider it to be well-suited for grafting. In contrast to adult tissue, fetal tissue cells replicate rapidly and exhibit tremendous capacity for differentiation into functioning mature cells. This capacity is maximal in the early stages of fetal development and gradually diminishes throughout gestation. Moreover, unlike mature tissue, fetal tissue has been found to have great potential for growth and restoration when transplanted into a host organism. Nutritional support provided by blood vessels from the host is readily accepted and likely promoted by fetal tissue.

In animal experiments, fetal tissue has displayed a considerable capacity for survival within the graft recipient. Fetal cells also appear to have increased resistance to oxygen deprivation, which makes them an especially attractive source of transplant material. Furthermore, fetal cells are easily cultured in the laboratory, thus allowing development of specific cell lines. They are also amenable to storage via cryopreservation, and because of their immunological immaturity fetal cells are less likely than adult cells to provoke an immune response leading to rejection by the host organism.

Although most applications of fetal tissue transplantation are highly experimental, the potential use is significant (Table I). The only proven effective treatment to date is the use of fetal thymus transplants for the rare DiGeorge's Syndrome, which has been standard treatment for over 20 years. Also, despite less than encouraging success to date, fetal pancreatic tissue transplants for juvenile diabetes are seen as promising. Fetal liver cells have been transplanted to patients with aplastic anemia with reasonable success, although clinical trials and more research are needed to ascertain the mechanism for recovery of these patients. Other poten-

TABLE I Clinical Conditions and the Types of Human Fetal Tissue Transplanted

Condition	Tissue
Acute leukemia	Liver
Addison's disease	Adrenal
Aplastic anemia	Liver
Bare lymphocyte syndrome	Liver
DiGeorge's syndrome	Thymus
Huntington's disease	Neural
Juvenile diabetes	Pancreas
Metabolic storage disorders	
Fabry's disease	Liver
Fucosidosis	Liver
Gaucher's disease	Liver
Glycogenosis	Liver
Hunter's disease	Liver, fibroblasts
Hurler's disease	Liver, fibroblasts
Metachromatic leukodystrophy	Liver
Morquio Syndrome, type B	Liver
Niemann-Pick types A, B, C	Liver
San Filippo syndrome, type B	Liver
Parkinson's disease	Neural, adrenal
Radiation accidents	Liver
Schizophrenia	Neural
Severe combined immunodeficiency	Liver and thymus

Source: Vawter, D. E., (1993). "Fetal Tissue Transplantation Policy in the United States." *Politics and the Life Sciences, 12,* p. 80.

tial applications of fetal liver tissue transplants include bone marrow diseases such as severe combined immunodeficiency and acute leukemia, where HLA-matched donors are unavailable; an array of inherited metabolic storage disorders; and radiation accidents.

II. STATUS OF THE FETUS

At the center of the controversy over fetal research is disagreement over the moral and legal status of the fetus. There is evidence that abortion politics has had a stifling effect on many areas of this research. Because in the United States federal funding of basic research, largely through the National Institutes of Health, is important, recent restrictions on specific types of fetal/embryo research have raised questions concerning their long-term impact on reproductive science and health.

Because the fetus is unable to consent to being a research subject, there is concern over what type of consent and by whom is sufficient. There are also questions as to what types of fetal research ought to be pursued, how to balance research needs with interests of the pregnant woman, and the proper ends of such research. Fetal research has surfaced periodically on the political agenda for two decades and has been addressed in the United States as an issue by national and presidential commissions, Congress, and many state legislatures.

Although the context of fetal research results in various shadings of support and opposition for particular applications, in general there are two major sides in the debate. On the one hand, the research community and its supporters argue that research using human fetal and embryonic materials is critical for progress in many areas of medicine. By and large they recognize no moral objections to using fetal tissues from electively aborted fetuses for basic research and transplantation, and they cite an array of areas where significant contributions can be made.

On the other hand, critics argue that research using aborted fetuses will give abortion greater legitimacy and encourage its use in order to aid research and/or individual patients. They contend that use of embryos and fetuses for research exploits them as the means to another person's ends and reduces them to biological commodities. Moreover, other interests not fully opposed to such research express concern over potential commercialization and payment questions, the ownership of fetal materials, and the need to delineate boundaries for acceptable uses of fetal material.

III. THE POLITICS OF FETAL RESEARCH

A. National Commission

Fetal research first appeared on the national policy agenda in the early 1970s after widely publicized exposés on several gruesome experiments conducted on still-living fetuses. In response Congress passed the 1974 National Research Act (Public Law 93-345) which established the National Commission for the Protection of Human Subjects of Biomedical and Behavioral Research, whose first charge was to investigate the scientific, legal, and ethical aspects of fetal research. The statute also prohibited all federally funded research on fetuses prior or subsequent to abortion until the commission made its recommendations and regulations were adopted.

In 1975 the commission issued its recommendations, which set a framework for the conduct of fetal research. In 1976 regulations for the federal funding of such research were promulgated (45 CFR 46.201-211) and are still in effect. Under the regulations, certain types of fetal research are fundable, with constraints based on parental consent and the principle of minimizing risk to the pregnant woman and the fetus. With respect to

cadaver fetuses the regulations defer to state and local laws in accordance with the provisions of the Uniform Anatomical Gift Act. Fetal cadaver tissue is to be treated the same as any other human cadaver.

With respect to research on live fetuses, the regulations provide that appropriate studies must first be done on animal fetuses. The consent of the pregnant woman and the prospective father (if reasonably possible) are required, and the research must not alter the pregnancy termination procedure in a way that would cause greater than "minimal risk" to either the pregnant woman or the fetus. Moreover, researchers must not have a role in determining either the abortion procedure or the assessment of fetal viability. Where it is unclear whether an *ex utero* fetus is viable, that fetus cannot be the subject of research unless the purpose of the research is to enhance its chances of survival *or* the research subjects the fetus to no additional risk and its purpose is to develop important, otherwise unobtainable, knowledge. Research on living, nonviable fetuses *ex utero* is allowed if the vital functions of the fetus are not artificially maintained. Finally, *in utero* fetal research is permissible if it is designed to be therapeutic to the particular fetus and places it at the minimal risk necessary to meet its health needs, or if it imposes minimal risks and produces important knowledge unobtainable through other means.

B. Federal Regulations

In 1985, Congress passed a law (42 U.S.C. 289) forbidding federal conduct or funding of research on viable *ex utero* fetuses with an exception for therapeutic research or research that poses no added risk of suffering, injury, or death to the fetus *and* leads to important knowledge unobtainable by other means. Research on living fetuses *in utero* is still permitted, but federal regulations require the standard of risk to be the same for fetuses to be aborted as for fetuses that will be carried to term.

In 1985, Congress also passed legislation (42 U.S.C. 275) creating a Biomedical Ethics Board whose first order of business was to be fetal research. In 1988, Congress suspended the power of the secretary of the Department of Health and Human Services (HHS) to authorize waivers in cases of great need and great potential benefit until the Biomedical Ethics Advisory Committee conducted a study of the nature, advisability, and implications of exercising any waiver of the risk provisions of existing federal regulations. However, in 1989 the activities of the committee were suspended, thereby leaving the question of waivers unresolved.

Federal regulations on fetal research, then, appear to be quite clear in allowing funding within the stated boundaries. However, in two key areas—embryo research and fetal tissue transplantation research—federal funding was in effect prohibited. The de facto moratorium on embryo research existed between 1980 and 1995. One of the provisions of the 1975 regulations prohibits federal funding of research involving the embryo entailing more than minimal risk unless an Ethics Advisory Board (EAB) recommends a waiver on grounds of scientific importance. The EAB was chartered for this purpose in 1977, first convened in 1978, and in May 1979 recommended approval for federal funding of a study of spare, untransferred embryos. However, in September 1980, Department of Health, Education, and Welfare (HEW) Secretary Patricia Harris allowed the charter of the EAB to expire. Although some observers have speculated that Harris did so to avoid overlap with the planned presidential commission, others concluded that she instead did so out of opposition to federal funding of IVF research. According to them, Harris was fully aware that the EAB was the only lawful body that could recommend waiver of minimal risk in research.

Although NIH directors throughout the 1980s called for a recharter of the EAB, no HHS secretary took action until 1988. Under pressure from Congress, Robert Windom, assistant secretary for Health, announced that a new charter was to be drafted and a new EAB appointed. The draft charter was published in the *Federal Register* as required by law and the charter was approved by HHS secretary Otis Bowen shortly before he left office. The incoming Bush administration, however, never acted and the EAB was not reestablished. As a result, the moratorium on federal funding of all human embryo research, including IVF and other assisted reproduction techniques, continued. According to the Institute of Medicine, this moratorium severely hampered progress in medically assisted reproduction. In 1994 the Human Embryo Research Panel, set up by the National Institutes of Health, recommended federal funding of certain types of embryo research. The recommendations have not yet been acted upon, in part because of resistance in the U.S. Congress.

C. State Regulations of Fetal Research

As with federal activity in fetal research, state regulation was largely a response to the broad expansion of research involving legally aborted fetuses after *Roe v. Wade* [410 U.S.113 (1973)]. Table II shows those states with current statutes restricting fetal research. Many of these laws were enacted by conservative legislatures as an effort to foreclose social benefits that might be viewed as lending support to abortion. Of the 25 states with laws specifically regulating fetal research, 12 regu-

late only research concerned with fetuses prior or subsequent to an elective abortion, and most of the statutes are either part of or attached to abortion legislation. Moreover, of the 13 states that apply more general regulations, 5 impose more stringent restrictions of fetal research in conjunction with an elective abortion.

Under state law, research on fetal cadavers is regulated through the Uniform Anatomical Gift Act (UAGA), which has been adopted by all 50 states. However, some states have excluded fetuses from the UAGA provisions, and others regulate it through fetal research statutes. Although a total of 45 states permit the use of tissues from elective abortions, 14 have provisions regulating research involving fetal cadaver tissue that deviate from the UAGA either in consent requirements or in specific prohibitions on the uses of such tissue. Five states currently prohibit any research with fetal cadavers except for pathological examinations or autopsies. Of these, 4 apply prohibitions exclusively to electively aborted fetuses.

State laws regulating research on live fetuses (*in utero* or *ex utero*) generally constrain research that is not therapeutic to the fetus itself. Because these state laws were adopted in the context of the abortion debate, the primary focus is on research performed on the *ex utero* fetus. Twenty states regulate research on *ex utero* fetuses, while 14 regulate research on *in utero* fetuses. Although the specifics of the prohibitions and the sanctions designated differ by state, most would appear to prohibit research involving transplantation and nontherapeutic research.

Another restriction imposed by some of the fetal research statutes addresses concerns over remuneration for fetal materials or participation in research. At present at least 16 states prohibit the sale of fetal tissue, 7 for any purpose and 9 specifically for research purposes. Importantly, some of these restrictions apply only to elective abortions, not spontaneous abortions or ectopic pregnancies. In some states the penalties for violation are quite high. For instance, selling a viable fetus for research in Wyoming is punishable by a fine of not less that $10,000 and imprisonment of 1 to 14 years.

IV. THE CONTROVERSY OVER FETAL TISSUE TRANSPLANTATION

Although fetal research has been embroiled in controversy for two decades, the recent debate has focused on the use of fetal tissue for transplantation research. As noted earlier, while the case can be made that much fetal research leads to knowledge and treatment that may benefit fetuses, transplantation of fetal tissue will benefit primarily adult patients. Because of the impor-

TABLE II   State Statutes Regulating Fetal Research

State	Regulates use of fetal cadavers	Prohibits nontherapeutic research on fetus		Prohibits sale of fetal tissue	May restrict preembryo research
		live *ex utero*	live *in utero*		
Arizona	×	×			×
Arkansas		×		×	×
California		×		×	×
Florida		×	×	×	×
Illinois	×[a]			×	×
Indiana	×	×			
Kentucky		×		×	×
Louisiana		×[b]	×	×	×
Maine		×	×	×	×
Massachusetts	×	×	×	×	×
Michigan	×	×	×	×	×
Minnesota		×	×	×	×
Missouri		×	×	×	
Montana		×			
Nebraska		×		×	
Nevada				×	
New Mexico		×	×	×	
North Dakota	×	×	×	×	×
Ohio	×	×		×	×
Oklahoma	×	×	×	×	×
Pennsylvania	×	×	×	×	×
Rhode Island	×	×	×	×	×
South Dakota	×[c]	×[c]	×		
Tennessee	×[c]	×[c]			
Utah			×	×	×
Wyoming		×		×	×

Source: Adapted from OTA (1990). "Neural Grafting: Repairing the Brain and Spinal Cord," pp. 133–134. Government Printing Office, Washington, DC.

[a] Federal District Court ruled this law unconstitutional.

[b] Statute found unconstitutional in *Margaret S. v. Edwards* (1986).

[c] Requires consent.

tant political implications and ethical dimensions raised by transplantation and because it so vividly raises questions of consent and rights, this section examines in more detail this one area of fetal research and the political context of the debate.

A. Controversy over Source of Fetal Tissue

The major policy issue surrounding neural grafting using fetal central nervous system (CNS) tissue centers on the use of fetal tissue in transplantation research. Four possible sources of fetal tissue are:

1. Spontaneous abortions
2. Induced abortions on unintended pregnancies
3. Induced abortions on fetuses conceived specifically for research or therapy
4. Embryos produced *in vitro*

A dependence on spontaneously aborted fetuses for research is impractical because of the limited number available, the inability to control the timing, and the fact that fetal tissue is fragile and deteriorates quickly after the death of the fetus. The major supply of fetal tissue, therefore, is likely to come from induced abortions on unwanted pregnancies. Instead of discarding or destroying the tissue, it is retrieved for research or transplantation. If even a small percentage of women elected to donate fetal tissue for research, the million and a half elective abortions performed in the United States each year would appear to be more than adequate to meet most research needs.

Unlike the use of fetuses spontaneously aborted, however, dependence on elective abortions as the primary source raises vehement objections on moral grounds by groups opposed to abortion. Other serious questions are raised when the research or therapy needs affect the timing and method of abortion, or if pressures are placed on the women undergoing the abortion to consent to donation. To ensure against the latter, most observers recommend that the decision to terminate a pregnancy be made separately from the consent for fetal research.

An even more troublesome source of fetal material might arise if a human fetus is conceived specifically for the purpose of aborting it for research or therapy. Although there is no documented evidence that this has happened, in at least two cases women have asked if they could become pregnant to produce tissues or organs for another person. In one case, the daughter of an Alzheimer's patient asked to be inseminated with the sperm of her father and, at the appropriate stage, to abort the fetus to provide her father with fetal neural tissue for transplantation. Although there is no evidence at present that this is technically possible, and the women's request was denied, this case demonstrates a possible demand for such applications. Similarly, another woman requested that her midterm fetus be aborted and the kidneys be transplanted to her husband who was dying of end-stage renal disease. There is also concern that increased pressures for fetal tissue could lead to a marketplace for this scarce resource which, in turn, could lead to exploitation of poor women (in the United States or elsewhere) paid to conceive solely to provide fetal materials, even if the practice were illegal.

The fourth source of human cells or tissue comes from embryos produced through IVF. Again, the situation is clouded by the need to distinguish between those embryos deliberately created for research purposes and those untransferred embryos remaining after IVF of multiple ova. Although persons who believe that life starts at conception are likely to oppose any use of human embryos for research purposes, many supporters of the use of spare embryos find the production of human embryos specifically for research unacceptable. Questions of consent, ownership, and payment are common to both categories.

B. Interim Moratorium on Fetal Tissue Transplantation Research

In March 1988, in response to a growing political controversy surrounding the publicity over Mexican and Swedish attempts at grafting fetal CNS tissue into Parkinson's patients, Robert Windom, assistant secretary for Health, imposed a moratorium on federal support for fetal tissue transplantation applications, pending a report from a special panel he directed NIH to convene in order to answer ethical and legal questions posed by a proposal for transplanting fetal brain tissue into patients with Parkinson's disease. In September 1988,

the Human Fetal Tissue Transplantation Research Panel recommended that funding be restored and concluded that such research is "acceptable public policy." In order to protect the interests of the various parties, the panel recommended guidelines to prohibit financial inducements to women; prohibit the sale of fetal tissue; prevent directed donations of fetal tissue to relatives; separate the decision to abort and the decision to donate; and require consent of the women and nonobjection of the father.

In December 1988, the NIH Director's Advisory Committee unanimously approved the special panel's report without change and recommended that the moratorium be lifted. The committee concluded that existing procedures governing human research and organ donation are sufficient to regulate fetal tissue transplantation. In January 1989, James Wyngaarden, the Director of NIH, concurred with the position of the Advisory Committee and transmitted the final report to the assistant secretary for Health. The report languished in the department without action until November 1989, when Secretary Louis Sullivan announced, in direct conflict with the recommendations, an indefinite extension of the moratorium.

C. Congressional Action

In April 1990, the U.S. House of Representatives Committee on Energy and Commerce's Subcommittee on Health and the Environment held hearings on human fetal tissue transplantation research. In July 1991 the House passed HR 2507, which would have limited the authority of executive branch officials to ban federal funds for areas of research. In May 1992, the House voted to approve the language of the conference committee, which included privacy and consent provisions added by the Senate, and it sent the bill to the president for a certain veto. In anticipation of a likely override of his veto, President Bush issued an executive order directing NIH to establish a fetal tissue bank from spontaneously aborted fetuses and ectopic pregnancies, even though there was little evidence that such a bank could provide sufficient amounts of high-quality tissue for transplantation.

The political controversy continued in June 1992 when a compromise bill was introduced in both houses. This bill in effect gave the administration 1 year to demonstrate that the tissue bank would work. If researchers were then unable to obtain suitable tissue from the bank within 2 weeks of a request, they could obtain it from other sources, including elective abortions. In October 1992, a Senate filibuster by opponents of fetal research ended attempts at passage, leading majority leader Mitchell to vow that the bill would be the first order of business when the Senate reconvened in 1993. On January 23, 1993, on his second day in office, President Clinton as expected issued an executive order that removed the ban on the funding of fetal tissue transplantation research.

D. Issues in Fetal Tissue Transplantation

Despite the President's action opening the way for public funding of fetal tissue for transplantation research, many legal and ethical issues remain. These issues include:

1. The determination of fetal death
2. The distinction between viable and nonviable fetuses as tissue source
3. The suitability of federal and state regulations and guidelines on the use of fetal tissues
4. Issues surrounding the procurement of fetal tissue, specifically who consents, the procedures for consent, the timing of consent, what information must be disclosed, and who seeks consent
5. Questions as to permissibility of altering routine abortion methods or timing in the interest of increasing the yield of fetal tissue suitable for research or transplantation
6. Quality control of fetal tissue, including screening of tissue and storage procedures
7. Issues surrounding the distribution system, including financial arrangements between physicians performing abortions and researchers using tissues, payment for cell lines, and the legality of designated recipients by the tissue donors

Contention over these issues is bound to intensify as the amount of transplantation research increases and the need for fetal tissue expands, particularly for issues 4 and 5.

1. Consent for Use of Fetal Tissue

The question of individual rights is clearly present in the debate over procurement of fetal tissue. Does a woman who decides to abort the fetus maintain any interests in the disposal of the fetal cadaver? In those states that include fetal cadavers under UAGA, proxy consent is required. The most logical proxy is the pregnant woman. Not only is she the next of kin, but the privacy argument suggests the woman's right to control her own body and its products. Observers argue that to deny the woman the opportunity to veto the use of fetal remains for transplant research or therapy denies her autonomy and that consent, therefore, must be obtained to protect the interests of the women. Following this approach, the NIH fetal tissue transplantation research

panel concluded that maternal consent is essential prior to the use of fetal cadavers for research.

In contrast, others argue that through abortion the woman abdicates responsibility for the fetus and that as a result there is no basis for seeking her consent concerning the disposition of the fetus. Once abortion has taken place the tissue is no longer part of the woman's body; therefore her claim to the use of tissues from her body carries little weight. Furthermore, since the woman clearly does not intend to protect the fetus it is inappropriate for her to act as a proxy.

Another argument against requiring maternal consent is that such a requirement may be an unwelcome intrusion upon a woman already facing a tortuous decision in the abortion. For instance, in *Margaret S. v. Edwards* [794 F.2d 994 (5th cir. 1986)] the U.S. Fifth Circuit struck down a Louisiana regulation that required the woman's consent for the disposal of the dead fetus. The court noted that informing a woman of the burial or cremation of the fetus intimidated pregnant women from exercising their constitutional right or created unjustified anxiety in the exercise of that right (at 1004). For the sake of reducing a woman's emotional burden and preventing harm to the woman, the court ruled that the woman need not be informed of, nor consent to, the disposal of the fetus.

One might ask whether informing a woman that her aborted fetus may be donated for research is potentially unwelcome information. Although some women might find it an intrusion, others may welcome the opportunity to specify donation for that purpose. The dilemma is that while the UAGA and state statutes require the woman's informed consent in order for her fetus to be a legal donation, *Margaret S.* suggests that this requirement may intrude upon her abortion decision and thus be unconstitutional.

Robinson [The Moral Permissibility of In Utero Experimentation. *Women and Politics* 13, 19–31 (1993)] raises an interesting issue surrounding a woman's decision regarding abortion once she has consented to *in utero* research on the fetus is whether the woman should be able to change her mind about the planned abortion once the experiment on the fetus has began. If yes, the woman's autonomy is ensured at the cost of potentially severe harm to the fetus. If no, limits on her autonomy are accepted in order to prevent possible harm to the fetus. Robinson argued that in this situation, the woman should not be permitted to change her mind once she has begun participation in the experiment—the woman has freely chosen to limit her own autonomy when she agrees to *in utero* experimentation and should be clearly told that she is making an irrevocable decision. This is the case in any abortion where there is at some time a point of no return—this point simply comes earlier in the process when the experiment starts. In either case, even if the fetus were to survive, it would have likely suffered great harm needlessly. Whether one agrees with Robinson's conclusion or not, this scenario reiterates the need for full and informed maternal consent prior to any experimentation on the fetus.

2. Modifications of Abortion Procedure

Another concern in fetal research, especially in grafting procedures where the pressures to obtain tissue of the most appropriate gestational age and optimal condition for transplantation are strongest, is whether the pregnant woman's medical care can be altered in order to meet research purposes. Although some observers approve of modifications in the abortion procedure if they pose little risk for the woman and she is adequately informed and consents, no one has publically advocated changing abortion procedures that entail a significant increase in the probability of harm or discomfort to the woman. At this stage there is general agreement that the means and timing of abortion should be based on the pregnant woman's medical needs and not on research needs. The Fetal Tissue Transplantation Research Panel made this requirement a high priority.

However, the difficulty of ensuring cooperation from abortion clinics and obstetricians in making fetal tissue available for research, attributed in part to reluctance to meet the additional time and resource requirements, has raised concerns for maintaining an adequate supply of fetal tissue. Moreover, the availability of RH-486 and other abortifacients in the near future might diminish the supply of usable fetal tissue at a time when demand might increase should fetal tissue prove to be a successful treatment for a common disease. Despite near consensus that fetal tissue procurement should not pose significant risk to the pregnant woman and that procedural protections must be in place, pressures for an expanding supply of usable fetal tissue demands vigilance to minimize abuses.

V. FETAL RESEARCH AND ABORTION

The ongoing debate over whether the use of fetal tissue from elective abortion encourages or legitimizes abortion will continue to be unresolved. Persons opposed to abortion will continue to oppose the use of fetal tissue altogether or from all but spontaneous abortions and ectopic pregnancies. The arguments that fetus research devalues the fetus and amounts to use of the fetus as the means to another's ends are strongly rooted in the antiabortion position, and arguments of the good derived from such research are unlikely to sway strong

opponents. There is no evidence that the use of fetal tissue will encourage women who are ambivalent about the abortion choice to have one, and most experts believe it will not do so.

It is reasonable to establish procurement procedures that remove to the maximal extent possible any likelihood of encouragement or pressure on the women to abort in order to provide fetuses for research or transplantation. Precautions to this end include distancing the abortion choice from the consent for research by prohibiting abortion clinic personnel from even discussing donation until after the woman has formally consented to abortion. The separate consent for fetal donation might even follow the abortion procedure in those situations where possible. Procedures should also require that the woman be fully informed of any risks to her privacy or well-being associated with donation of the fetal tissue and of any interest the abortion provider might have in her donation.

Furthermore, in order to reduce the presence of any coercion of women into conceiving and aborting in order to provide fetal tissue to benefit identifiable others, specification of tissue recipients, including the woman herself, should be prohibited. Payments to women who abort, including compensation for the cost of the abortion procedure, should be prohibited. To further reduce the profit motive in the disposition of fetal remains, payments to doctors, clinics, and other parties involved in the abortion procedure should be prohibited. Fetal materials for research should be distributed through public registries established to distribute them on a nonprofit basis once fetal tissue transplantation needs expand. As noted by Vawter: "Appropriate guidelines for fetal tissue procurement are necessary to replace the inadequate practices currently in widespread use so that fetuses and the women asked to donate tissue after elective abortion receive the respect and protection owed them" (Vawter, D. E. (1993). "Fetal Tissue Transplantation Policy in the United States." *Politics and the Life Sciences, 12,* 79–85.). Although this will not defuse the debate over fetal research nor eliminate the need for continued vigilance, it seems to represent the best balance of the competing interests.

VI. CONCLUSIONS

Clearly some important areas of fetal research have been explicitly constrained on political rather than scientific grounds. The presence of abortion politics continues to exert strong influence on research funded by the government across a wide range of substantive areas. In the process, long-term scientific goals are being compromised by immediate, pragmatic political objectives.

Given the sensitivity of human embryo and fetal research and its interdependence on abortion, this should not be surprising. This research will continue to elicit intense opposition and support, thus placing it well within the political agenda throught the 1990s. Fetal research policy raises moral red flags for many persons. In contrast, research on human embryos and fetal tissue transplantation promise significant advances in our understanding of the human condition and represent potentially revolutionary treatment options.

Bibliography

Andrews, L. B. (1993). Regulation of experimentation on the unborn. *Journal of Legal Medicine,* **14,** 25–56.

Donovan, P. (1990). Funding restrictions on fetal research: The implications for science and health. *Family Planning Perspectives,* **22,** 224–231.

Fletcher, J. C. (1993). Human fetal and embryo research: Lysenkoism in reverse—How and why? In "Emerging Issues in Biomedicine" (R. H. Blank and A. L. Bonnicksen, Eds.), Vol. 2. Columbia Univ. Press, New York.

Mullen, M. A. (1992). "The Use of Human Embryos and Fetal Tissues: A Research Architecture." Royal Commission on New Reproductive Technologies, Ottawa.

Office of Technology Assessment (1990). "Neural Grafting: Repairing the Brain and Spinal Cord." Government Printing Office, Washington, DC.

Stein, D. G., and Glasier, M. M. (1995). Some practical and theoretical issues concerning fetal brain tissue grafts as therapy for brain dysfunction. *Behavioral and Brain Sciences,* **18,** 36–45.

Vawter, D. E., Kearny, W., Gervais, K. G., *et al.* (1990). "The Use of Human Fetal Tissue: Scientific, Ethical, and Policy Concerns." Center for Biomedical Ethics, Minneapolis.

Gene Therapy

ADAM M. HEDGECOE
University College London

GLOSSARY

DNA The basic material of inheritance.
genome The total set of genes in any one person's body.
germ-line gene therapy Interventions on a person's genome that will be passed onto future generations.
in vitro **fertilization (IVF)** The fertilization of an ovum outside the body before reimplantation ("test-tube" baby).
somatic gene therapy Genetic interventions that do not affect future generations.

GENE THERAPY is the use of genetic information to intervene in the DNA of a human cell to relieve the symptoms and prevent the causes of diseases with a genetic component. The obvious candidate diseases are those which run in families, such as sickle cell anemia and Huntington's disease, although many other diseases have been cited as possible candidates for treatment, including complex disorders such as cancer and even HIV/AIDS. The ethical concerns about gene therapy center around the fact that such treatment can have direct and permanent effects on future generations, as well as on the fact that changes are made at a very "basic" biological level.

I. INTRODUCTION

Gene therapy is one of the areas of greatest ethical concern in modern genetics. It is also novel in that it is one of the few cases where large-scale debate of the ethical considerations has taken place well before the technologies concerned have become available on anything but a very experimental level. Although debate about other issues in genetics, such as screening and counseling, has taken place, this has occurred either after technologies have been developed or at the same time. Gene therapy is one of the first chances to allow ethical debate to shape and contribute to a technology from the very first stages of its development.

Despite its highly novel nature, there has been a great deal of commercial and public interest in gene therapy because it is seen as having such great potential for curing disease. It is suggested that because it attacks the genetic cause of diseases, gene therapy is far more beneficial than current treatments which only deal with the symptoms and which are far less specific. Gene therapy holds out the promise of permanently eradicating diseases which have plagued humans for millennia.

The aim of this entry is to present the various types of gene therapy and to outline the ethical arguments both in favor of and against its development and use. This means that this entry will *not* address ethical issues associated with gene therapy but which are not specific to it; these include objections to embryo experimentation, abortion, and biotechnology as a whole. All these areas are relevant, especially to the experimental development of gene therapy, but to include them would overlap with other entries. An example of one relevant area of discussion is that of cloning. At the end of February 1997, it was announced by scientists in Scotland that they had successfully cloned an adult sheep, producing an identical (though younger) copy, called Dolly. Although cloning of mammals had taken place at the embryonic stage (by splitting an embryo—inducing identical twins in short), the cloning of an adult, i.e., taking cells from an adult mammal and producing a clone from them, had never been done before. As a result, a great deal of interest and concern was generated worldwide over the possible (future) application of these techniques to humans. Although these issues have concerns in common with gene therapy (such as interfering with nature or affecting future generations againt their will), it is important that they be separated from gene therapy. We do not need clones to have effective gene therapy (though they might conceivably be useful in experimental tests), nor do we need gene therapy to produce clones.

II. DEFINITIONS AND DISTINCTIONS

A. Somatic Gene Therapy

This form of therapy has been carried out already and is the simplest and least controversial form of gene therapy. It involves acting on the DNA contained in a person's body, or "somatic" cells; this means that any changes produced by the therapy will be limited to that patient, i.e., changes will not be transferred to the patient's children.

In 1993, experimental trials were run in London on a somatic gene therapy for cystic fibrosis (CF). This involved patients inhaling a fine spray which was composed of fatty packets called liposomes. Inside each of these was a copy of the gene which cystic fibrosis sufferers lack. This gene was inserted into the lining of the nasal passage by the liposomes. Once in the body, it was hoped that the gene would start to produce copies of the protein which is missing in CF patients and which contributes to their breathing difficulties. This trial followed the now standard approach to somatic gene therapy, which is to use a vector, in this case liposomes, to carry the target gene into the body where it is hoped that it will produce the missing protein. In trials carried out at the same time in the United States, scientists used an adenovirus (the same type of virus as the common cold) to insert genes into the lungs of CF patients. The viral vector was more efficient at transferring the genes into the patient than the liposome method, but led to more side effects.

The first ever effective gene therapy, carried out in 1990 in the United States, involved the immune deficiency disease "ADA deficiency." In this case, scientists removed some of the white blood cells from a young girl's body, inserted new copies of the ADA gene into the cells, and then put the cells back into the girl's body with a transfusion. The modified cells were able to manufacture the missing chemical, ADA, and the girl's immune system improved remarkably. In many ways, these descriptions of somatic gene therapy make it sound very much like any advanced medical procedure, and indeed, that is the way it is viewed by the majority of those people working in the field and those who are concerned with the ethics of such experiments. Somatic gene therapy, because changes are limited to the initial patient, and because they are not passed down to later generations, is not generally seen as ethically problematic.

Since 1989, trials of somatic gene therapy have been run for various forms of cancer, familial hypercholesterolemia, hemophilia, and even AIDS. Despite the advances that have been made in understanding, it is now clear that effective gene transfer is a much more complex proposition than originally thought. The early high hopes for this form of treatment have given way to a realistic, long-term view that clinical somatic gene therapy may be many years away [T. Friedmann "The Origins, Evolution, and Directions of Human Gene Therapy" in T. Friedmann (ed.), *The Development of Human Gene Therapy* (New York: Cold Spring Harbor Press, 1999)].

B. Germ-Line Gene Therapy

Germ cells are those cells found in the ovaries and testes of humans that give rise to sperm and ova. These are the cells that germ-line gene therapy would act upon. The theory is that at a very early stage of an individual's development, perhaps even before conception, changes could be made to a person's genome that would have an effect on every single cell of their body (since all other cells would develop from those early ones) and upon children that they have (and upon all subsequent generations). Any change made to a person's genome using germ-line gene therapy would have long-term, and possibly unpredictable, effects. At the moment, germ-line gene therapy is not even at an experimental stage, there is a worldwide ban on its development, and ethical debate about its acceptability has been increasing.

Very closely linked to germ-line gene therapy, in fact one might almost call it passive germ-line therapy, is the technique of selective implantation in *in vitro* fertilization (IVF). This technique does not directly act on the genome like germ-line gene therapy, but produces the same results. By using genetic screening, it is possible for couples undergoing IVF to select embryos to implant which are free from disease genes (for example, if their families have a history of muscular dystrophy). This would produce a child without the gene for that disease, just as germ-line gene therapy would. This technique is already available and being used by some hospitals, yet although there has been ethical discussion of it, it has not really been of the same tenor as that around germ-line therapy. The point is that if both selective implantation and germ-line therapy produce the same results, are they ethically different, and if they are different, then there must be something ethically significant about directly acting on and manipulating a person's DNA, which is not the case in simple selection.

C. Therapy vs. Enhancement

The final distinction to draw is between genetic technologies used as a form of medical therapy and those same techniques used to enhance the abilities and characteristics that persons already have. For example, gene therapy might be used to increase the height of children with growth hormone deficiency to that of the norm in the population. This would count as therapy. But the same (or similar) techniques could be used to increase the height of a normal child so that they were of above average height. This would be enhancement of characteristics. These possibilities have raised fears of "designer babies" and are felt to produce some of the strongest arguments against gene therapy, or specifically, germ-line gene therapy, since the general view is that effective enhancement treatment would have to act on the germ cells.

As the above example suggests, the line between therapy and enhancement is far from clear, and for some this has led to the conclusion that any form of germ-line therapy, however well intentioned initially, will inevitably lead to the development of enhancement technologies. Such "slippery slope" arguments are often used to oppose germ-line therapy as a whole.

III. PRAGMATIC CONCERNS

When one looks at the ethical debate surrounding germ-line gene therapy, two distinct types of argument can be distinguished: the practical difficulties and the categorical objections. The former are often presented as if they are a final and definite proof of the ethical wrongness of germ-line gene therapy, but in effect, they are merely the normal pragmatic concerns that surround any new technology. For example, the committee set up by the UK government to look at gene therapy (the Clothier Committee) offers only practical, safety-based

objections to germ-line therapy: "there is at present insufficient knowledge to evaluate the risks to future generations … [and] therefore … gene modification of the germ line should not yet be attempted" (C. Clothier *et al.,* 1992. *Report of the Committee on the Ethics of Gene Therapy* (p. 18). London: HMSO). The practical problems stem from the fact that gene therapy deals with systems in the body that are still largely unknown and unpredictable, and in the case of germ-line therapy, many of the effects of the treatment will not be known for a long time (many generations perhaps). But these are the sorts of problems associated with most new medical treatments. Before we make widespread use of heart surgery, we would want to make sure that the techniques used were as safe as possible, that there were no unforeseen side effects, and that we understood as much as possible about the parts of the body that might be effected by such treatment. The same holds true for gene therapy (of whatever type). The U.S. cystic fibrosis trials were stopped because of patient reaction to the virus used as a vector; this does not necessarily mean that such treatment is unethical, merely that all such experiments must be carried out with great caution. This is what we would expect from responsible medical science.

The pragmatic difficulties with somatic gene therapy are manifold. Even if safe vectors can be discovered (and there is some concern over the use of viruses in gene therapy, because of their tendency to mutate and change), the actual insertion of the new DNA into the patient's cells is still very haphazard. At the moment, there is no sure way of getting the new DNA to "slot into" the correct place in the patient's genome. There is the possibility of disrupting the working parts of the genome, and even of triggering cancer. With germ-line gene therapy any effects on an individual treated in this way would transfer into their children and even later generations. But of course the possibility is that many of the side effects might never be revealed *until* those later generations are born. Does this place an impossible burden of proof on germ-line gene therapy? Is it possible that we will never know, to a reasonable, acceptable level of doubt, whether such treatment is actually "safe"? Such doubts should certainly be considered. Perhaps germ-line gene therapy will never be safe to use, but it is important to note that this is not a categorical ethical objection to germ-line gene therapy, but depends upon empirical results of its efficacy and safety. There may come a time in the future when we have to accept that germ-line therapy is safe, or at least as safe as some other medical techniques. It may be unethical to use a treatment that is not tested and safe, but when such a treatment becomes safe, can we still call it unethical? The answer from the second type of objection raised against germ-line therapy, categorical objections, is that it remains unethical.

In the following discussion of gene therapy, most of the issues will be concerned with germ-line therapy and the categorical objections to it that have been raised. As has already been suggested, the ethical problems associated with somatic gene therapy are similar to those issues related to organ transplants and other current medical procedures; i.e., they stem from practical concerns about the treatment's safety. Germ-line therapy is far more ethically problematic, whatever the state of its practicality and safety. It must also be borne in mind that gene therapy may turn out to be safe, but ineffective, or even safe, but too expensive to realistically develop into a standard technique. These points have their own ethical issues (such as distribution of limited resources), but these will not be addressed here.

IV. ARGUMENTS IN FAVOR OF GENE THERAPY

A. Beneficence

This is the idea that doctors should not just "do no harm" (nonmaleficence) but should also actively strive to benefit their patients. Thus, if a doctor knows that by using germ-line therapy he or she can ensure that the child who is born does not have a particular disease, the duty of beneficence would seem to require that the doctor carry out this form of therapy. Put another way, if the doctor knows that there is a high chance of any child born to a particular couple suffering from a disease (for example, if they are both cystic fibrosis sufferers), then we might hold the doctor responsible in some way if he or she allows a child to be born with that disease, when the means of preventing it are available.

Although this might appear a clear case of the duty of beneficence coming into play, it has to be noted that it extends the limits of this duty a little more than is normal. Doctors are normally required to "do good" to current patients, but germ-line gene therapy involves doing good to a person who is not yet even in existence (even if one counts personal existence as starting at conception, the fact that germ-line therapy may operate *before* conception still means that there is no person in existence at the time of action). Thus the responsibility of doctors is spread to people who do not exist. For some this might not be problematic. Perhaps we are all felt to have some sorts of duties toward future generations (with regard to natural resources, for example), and this extension of beneficence might been seen in a similar light. There is even a precedent in that doctors inoculate women against German measles/rubella to prevent them catching it during pregnancy and harming

their child. This could be seen as analogous to germ-line gene therapy. Others may not be convinced, perhaps feeling unease at the shift in emphasis in medicine, which has traditionally is concerned with treating the sick in the population. Germ-line gene therapy could be seen as a way of, instead, preventing the sick from joining the population in the first place.

B. Consequences

This broader position holds that ethical actions are those that produce the best consequences (however best is measured; traditional utilitarian calculations use "happiness"). The argument would follow that the use of germ-line therapy would produce an increase in happiness in the future, with less people suffering from disease, and therefore it should be used. Against these obvious benefits are questions about whether people in the future, if they know they are the product of genetic engineering (of some form), might not suffer psychologically or feel less human because of the action that was taken at such a early stage in their life. This then has to be measured against how "happy" they would feel if they were left alone and developed the disease that germ-line therapy could have prevented. These are speculative questions, hindered by difficulties of comparing people with nonexistent persons ("persons they might have been"), and as such are very hard to answer one way or another. But these are the sorts of issues that need to be considered before one can decide whether germ-line gene therapy is ethical or not.

C. A Duty to Enhance?

John Harris has suggested that if the technology exists, then we should use germ-line therapy to enhance future generations. His analogy is with schooling. If we are prepared to pay large sums of money to ensure that our children get the best education possible, to make them as intelligent as possible, then why should we not use germ-line treatments to enhance their intellectual capabilities as well? "If the goal of enhanced intelligence ... is something that we might strive to produce through education ... why should we not produce these goals through genetic engineering?" (J. Harris, 1992. *Wonderwoman and Superman* (p. 142). OUP). One should note that as in the case of selective implantation of fetus, this is a case where germ-line therapy reaches the same results as another, less controversial technique. If there is something ethically wrong with germ-line therapy, then it must intrinsically have to do with its direct action upon human DNA, rather than its end results.

Possible objections to enhancement include the idea that it is cheating in that it short circuits the admirable human activity of working toward a specific goal through effort, and it undermines the social value of those activities (test-taking or athletic achievement, for example). In addition, genetic enhancement can be viewed as an abuse of medicine—seeking to solve, through medical intervention, problems which are in essence, social (E. T. Juengst and L. Walters "Ethical Issues in Human Gene Transfer Research" in Friedmann op.cit.).

V. ARGUMENTS AGAINST GENE THERAPY

A. Right to an Unaltered Genome

It has been suggested that every human being has a right not to have their genome altered by other humans. Germ-line gene therapy would alter human genomes (though somatic therapy would not), and therefore germ-line gene therapy is unethical. This argument has the advantage of being rooted in the language of human rights, which is widely recognized as voicing valid ethical concerns (we think of the UN Universal Declaration of Human Rights as outlining ethically ideal behavior for governments, for example). But this right is certainly not one of the "traditional" rights debated over time (for example, the American right to "life, liberty and the pursuit of happiness"). On what basis do we gain such a right? It is not at all clear. Perhaps it is on the basis of human dignity (see next section). Another difficulty is the idea of a nonexistent person having rights (which is very controversial). While people who complain about their being treated with germ-line gene therapy would be rational beings, the same cannot be said for them when their right was infringed, i.e., just after (or before) their conception. They did not actually have that right at that time, since they did not exist; therefore it is hard for them to complain about infringements of rights they did not have.

The right to an unaltered genome is problematic in other ways since, following the ideas of Derek Parfit, we can say that almost everything we do alters our offspring's genome, the most obvious being our choice of sexual partner. When we choose the person we want to be a parent with, we are automatically making decisions about the genome of our offspring. How is germ-line gene therapy different? If it is different, then this difference must lie in the actual process of the therapy, rather than in the fact that the offspring's genomes are different to how they might otherwise be. The problem is, can we even describe this as "altering" a person's genome? is it not better described as affecting the characteristics that person will have? The same description could be applied to germ-line gene therapy, certainly that type which operates on sperm or ova (i.e., before conception, before any possible person exists), and perhaps, if we do not count a fertilized ovum as a person, germ-line therapy as a whole. Can a person describe themselves as having an altered genome, when in fact what happened was that their parents chose the characteristics that they would have? Does this make an ethical difference? It certainly makes germ-line gene therapy closer to selective implantation and IVF in ethical terms.

B. Human Dignity

A related objection to germ-line gene therapy is that it offends against human dignity, an idea derived from Kant, specifically his categorical imperative that we should always treat people as ends in themselves and not merely as means to an end. Yet it is hard to see how seeking to cure people, or more accurately, prevent them from developing disease, is treating them as means to an end. What end is it that we are treating them as a means to? Perhaps it could be claimed that the end is one of a healthy society, and using the therapy is treating people as merely a means to this end. But then surely this is true of all medical treatment. Rather, medicine is administered to treat individuals, to make individuals better—a side effect is the improved health of society as a whole. Proponents of the human dignity argument have to show why trying to prevent someone getting a disease is not treating them as an end in themselves.

Of course, such a case can be made for *experimental developments* of gene therapy, where the patients are, to some extent, human guinea pigs. In this case, it could be claimed that use of (experimental) germ-line gene therapy treats people as means to an end (that end being experimental results) rather than ends in themselves and could thus be seen to act against human dignity. Perhaps such arguments could be used against the development of germ-line therapy; even if the use of such therapy is ethical, its development is not. The trouble with this position is again the similarity to standard medical treatments. Were these considered as acting against human dignity when they went through experimental testing?

C. Naturalness

Is gene therapy "unnatural," and if so, does this make it unethical? It is difficult to know exactly what is meant when it is claimed that gene therapy is unnatural or "against nature." Certainly it does not occur in the natural world, but then this is true of much that human beings do and build. Perhaps it is rather that such claims of unnaturalness express a deep unease with technologies that act at such a very basic level. The actual DNA that is inserted into a patient's body or genome during gene therapy is as natural as the DNA that it is replacing or overriding. While it may be true that gene therapy thwarts the "natural course of events," this is also the case with taking aspirin, stitching a wound, or administering the kiss of life. All medical treatment effectively acts against the natural course of events. In some ways, that is what medicine is for.

D. Playing God

Ruth Chadwick has suggested that when the term "playing god" is used in moral arguments, it can mean two separate things, depending on the context. In the first case, it is a warning to decision makers that they are not infallible, that they can make mistakes, and that they are not omniscient and divine. The second way is as a warning of unease and disquiet about a particular course of action being taken—that human beings might be going too far and risk unforeseen consequences and disaster. Both of these interpretations are useful for informing our decisions about germ-line gene therapy, but neither of them provides a categorical opposition to it. The first meaning brings us back to the practical problems associated with gene therapy. We must know that it is safe (however that is defined in this context), and that decisions we make about its use are always open to question. The second meaning is useful as counsel, in warning us that we must be cautious in our use of the new technology, not just in terms of physical risk but the changes that it may bring about in our perception of ourselves and our children (and their self-perception). "The playing-God objection is more than mere rhetoric, but less than an argument against a particular course of action. As a counsel it has some value" (R. Chadwick, 1989. Playing God. *Cogito,* Autumn, p. 193).

E. Consent of Future Generations

This argument rests on the premise that the use of germ-line gene therapy involves making decisions about the lives of future generations that they have no say in (for obvious reasons, they do not exist yet). We may be doing things to them that they, if they had the choice, would not want done to them. We cannot tell what future generation, perhaps thousand of years hence, would want us to do for them. Who knows what sort of people or society might exist then or what its values will be. It is unethical for us to make decisions on their behalf.

This argument can be used with quite some strength against those who favor allowing genetic *enhancement* for future generations, perhaps for things such as height, body shape, and hair color; these are things we cannot really make anything more than a vague guess about with regard to their desirability for future generations. It is less effective against the use of germ-line gene *therapy* since we already make decisions concerning the health of future generations, and these are not regarded as very difficult to make. The most obvious example is the one of toxic or nuclear waste. When there are fears about deep-sited nuclear waste dumps, these are rarely about current populations. Such claims are usually made on behalf of future generations, and so they should be. It will not effect me if highly radioactive plutonium is placed at the bottom of a mine shaft in containers that allow radioactivity to leak out into the atmosphere and water table in minute quantities. It may not even effect my grandchildren, but it could possibly effect later generations that are born long after my death. The same argument can be made for genetic risks. The fact that I am a carrier for a lethal genetic disorder may not affect me (since I am only a carrier), but it could, at some unspecified time in the future, effect future generations. Just as we think we should protect future generations from the foreseen harm resulting from nuclear waste, the same argument can be made that we should protect them from the foreseen harm resulting from bad genes. It is hard to argue that we do not really know what future generations want with regard to health, certainly in terms of serious genetic disease.

E. Eugenics

One concern of opponents of germ-line gene therapy comes from the historical precedents that have been set by people trying to affect human inheritance and health. The word that is used is "eugenics," which literally means "well born" and which was adopted by a movement in the United States and United Kingdom (and whose ideas were later picked up by the Nazis in the 1930s Germany) which at the turn of the century tried to improve society by breeding out of the population undesired traits. Much of the rhetoric used was racist, and many of the actions carried out in the name of eugenics, such as the forcible sterilization of thousands of people in 1920s America, are now looked on with horror. The argument goes that germ-line gene therapy is eugenics by other means, or, at the very least, it has the potential to descend into the same sort of mistakes as the original eugenics movement.

This argument has a great deal of emotional impact due to the distaste in which the original eugenics is held, but there are important differences between eugenics and the widespread use of germ-line gene therapy. The most obvious is the involvement of the state and the involuntary nature of much of eugenics. But germ-line therapy is conceived of within the framework of modern medicine, with its emphasis upon individual decision making and patient autonomy. It is not suggested that germ-line gene therapy would be applied against parent's wishes, to force them to have healthy children, though of course it is possible that a subtle form of social pressure and drive to conform could arise if germ-line therapy became freely available. Supporters of germ-line therapy have pointed out that eugenics is concerned with preventing the disabled from reproducing, while germ-line therapy is concerned with preventing couples from producing sick and disabled children. This leads into a related claim which is often voiced by disabled people and their support groups, that the use of germ-line therapy to prevent disablement and disease is in effect devaluing those people who currently have those disabilities and diseases, a way of saying they are less important, what has been called the "expressivist objection" (A. Buchanan, 1996. Choosing who will be disabled: Genetic intervention and the morality of inclusion *Social Philosophy and Policy,* **13**(2), p. 28). While this is an understandable reaction, it is hard to make this accusation stick.

> In advocating the use of genetic science to reduce disabilities one *is* saying that avoidable disabilities ought to be avoided, and that in that sense one *is* saying that our world, in the future, should not include the existence of so many people with disabilities. But it is not the people who have disabilities that we devalue; it is the disabilities; and one need not and should not wish to reduce the number of people with disabilities by taking the life of any person who is disabled.
>
> (BUCHANAN, 1996, 32–33)

Some people have argued that we have to accept the fact that selective implantation and germ-line gene therapy will lead to a form of eugenics, what has been labeled as "utopian eugenics" (P. Kitcher, (1996). *The lives to come.* New York: Simon & Schuster).

F. Broader Consequences

The previous objections are all put in categorical terms in that they attempt to show that germ-line gene therapy is wrong for a single specific reason. Although all these objections have their points, it is not clear that they are overly persuasive. The best case against germ-line theory may in fact lie in looking at the possible consequences of germ-line therapy on future society. The most obvious area of concern is access to these new technologies. They are likely to be very expensive, and certainly when first available. Is there not a real risk that germ-line therapy will reinforce economic differences within society? If such treatments are only available to

the very wealthy, huge disparities in wealth could be added to by disparities in health. While this may already be the case to some extent, it is valid to ask whether we would want to introduce technology which will increase such differences even further. Such concerns are outside the simple risk assessment of whether gene therapy is safe in purely medical terms and questions of whether these technologies will have a detrimental effect on society. As consequencialist arguments, these would of course have to be considered against similar arguments in favor of germ-line gene therapy, with its effect on individual health.

VI. CONCLUSIONS

Gene therapy is one of the areas of greatest ethical interest in genetics, but there is a great deal of speculation about the actual practicality of these technologies. Nevertheless, it seems likely that at least some forms of somatic if not germ-line gene therapies will be developed in the future. In some ways the ethical issues associated with these technologies are nothing new, with comparisons being drawn to other novel medical techniques. In other respects germ-line gene therapy presents us with new problems, because of its long-term, unpredictable impacts on future generations, and its involvement with very basic elements of life. Many of the objections to germ-line gene therapy which are normally made rely on categorical opposition, saying that there is something uniquely wrong about germ-line gene therapy because of the actual direct intervention in the genome. It is not clear how germ-line therapy differs from other medical treatments currently available. Stronger objections to germ-line therapy rely on analysis of the possible consequences on society of such technologies and the wider dangers that exist. What is clear is that the issues raised by gene therapy will continue to be debated as genetic technologies advance.

Bibliography

Anderson, W. A. (1990). Genetics and human malleability. *The Hastings Center Report,* **20**(1), 21–24.

Munson, R., & Davis, L. (1992). Germ-line gene therapy and the medical imperative. *Kennedy Institute of Ethics Journal,* **2**(2), 137–158.

Reiss, M., & Straughton, R. (1996). *Improving nature? The science and ethics of genetic engineering.* Cambridge: Cambridge Univ. Press.

Wachter, M. (1993). Ethical aspects of human germ-line gene therapy. *Bioethics,* **7**(2/3), 166–177.

Waters, L., and Palmer, J. G. (1997). *The Ethics of Human Gene Therapy.* New York: Oxford University Press.

Genetic Counseling

ANGUS J. CLARKE

University of Wales College of Medicine

GLOSSARY

carrier testing The identification of healthy (unaffected) carriers of a genetic condition who may transmit the disorder to their children. This applies particularly to carriers of autosomal recessive diseases, who have one faulty copy and one intact copy of the relevant gene; to female carriers of sex-linked disorders, to whom the same applies; and to carriers of balanced chromosome rearrangements, who are generally unaffected themselves but who may transmit to their children an unbalanced set of chromosomes (with a deficiency of some and an excess of other chromosomal material).

chromosome A physical structure that carries genes through the processes of cell division and that is involved in the coordination of gene expression in the cell nucleus. There are 23 pairs of chromosomes in each cell of the human body, with one copy of each chromosome being derived from each parent. Twenty-two of the 23 pairs of chromosomes are the same in males and females—the autosomes. The twenty-third pair consists of the sex chromosomes, X and Y. The X chromosome contains many genes that relate to a wide range of body functions.

directiveness A style of genetic counseling which leads clients to make specific decisions, often in relation to reproduction. Directiveness may be a conscious policy, where the counselor frankly recommends what she thinks is best for the client, or it may be the result of subtle, implicit attitudes and behaviors on the part of the counselor.

DNA Deoxyribonucleic acid is a linear molecule composed of a backbone of alternating ribose (a 5-carbon sugar) and phosphate groups with one of four bases attached to each ribose element. The molecule adopts a helical configuration, and the bases—adenine, guanine, cytosine, thymine—project inwards toward the longitudinal axis. Two such linear molecules, oriented in opposite directions, interlock to form a double helix molecule in each chromosome. The bases on one strand form stabilizing hydrogen bonds with the bases on the other strand of the double helix. Adenine can only pair with thynine and guanine can only pair with cytosine, so that knowledge of the sequence bases on one strand is enough to determine the sequence of bases on the other strand. Knowledge of the sequence of bases in the coding regions of a gene determines the sequence of amino-acids in the corresponding protein specified by that gene.

gene A message encoded in DNA that instructs the

body in how to develop and function. Genes generally come in matching pairs, one of each type being inherited from each parent. There are many different genes, each present in every cell of the body. Genes are organized along the length of the chromosomes, rather like beads on a length of string. If one copy of a gene is faulty—is disrupted by a mutation—this will often cause no problems, as long as the other copy of the same gene is intact; this type of gene defect is termed **recessive,** and an individual will only be affected if they have inherited a faulty copy from both their parents (who may be affected by the condition or who may be unaffected carriers). If a single faulty copy of a gene is sufficient to cause disease, then the faulty gene is said to be **dominant.** An individual will have this type of gene defect either if one of their parents is affected and has transmitted to them their faulty copy of the gene, or if a new mutation occurred in that gene in the production of either the egg or the sperm that went to create them.

multifactorial conditions Also known as polygenic or complex traits, these are characteristics or diseases influenced by multiple genetic factors and often also by other, nongenetic circumstances, i.e., environmental or chance. A single genetic factor may modify susceptibility to a multifactorial disease but could not be said to cause it or to determine its development.

predictive genetic testing The identification of healthy individuals as likely or inevitably to develop a specific disorder, usually in relation to a well-defined, Mendelian genetic disorder. Such testing may be carried out to identify those likely to benefit from surveillance for the early detection of tumours (in family cancer syndromes), or it may be carried out at the request of individuals at risk of developing an untreatable, late-onset condition (such as Huntington's disease), primarily in order to clarify their genetic status and thereby enable them to make important life decisions with full information.

screening test A test applied to an entire population or subpopulation, rather than to individuals or specific families, and which may not be of very great sensitivity or specificity. Screening tests are usually offered pro-actively by health professionals to large groups of people who individually have a small chance of being identified as affected or at risk of problems—e.g. to all newborn infants, to all pregnant women.

single-gene (Mendelian) disorder A disorder that is determined genetically by mutations at one particular genetic locus. The transmission of the condition through families will usually demonstrate the mode of inheritance as autosomal dominant, autosomal recessive or sex-linked.

susceptibility testing Testing individuals for their inherited susceptibility to a multifactorial disorder. Such testing is not able to predict whether a person will develop a condition, but will merely indicate their relative risk.

GENETIC COUNSELING is a process of communication between clients and professionals. Clients have questions or concerns about a disorder in their family that is, or may be, genetic in origin; they put these to genetic counselors, health professionals who attempt to clarify the situation as far as possible.

The focus of a consultation with a genetic counselor may be on:

1. Establishing the diagnosis of the condition in the family

2. Finding out more about the condition and what the future may hold for affected individuals

3. Considering who else in the family may also come to be affected

4. Considering who could have affected children, and how this may be avoided

5. Discussing the treatments available for affected individuals, and whether or not complications can be treated more successfully if they are diagnosed early by a program of surveillance

6. Considering the likely emotional impact on the client and on other family members of the various possible outcomes of any genetic test that the client may take, whether it be a predictive test, a prenatal diagnostic test, or a test of carrier status

The extent to which the concerns of a client will focus on any one of these areas will vary enormously, depending upon the prior experience and knowledge of the family members, the nature of the disorder in question, and numerous other factors. There are certain tasks, however, that will frequently be required of any genetic counselor, and these can be summarized as (i) listening to the client, (ii) establishing or confirming the diagnosis of the condition in the family, (iii) communicating information to the client, (iv) scenario-based decision counseling, and (v) providing ongoing support to the client and family. It may be helpful next to put some of these genetic counseling activities and tasks into context.

I. THE SCOPE OF GENETIC COUNSELING

Listening to the client has to be the first task of genetic counseling—to find out what questions and concerns the client has, and what she already knows or understands. It is otherwise impossible to provide the appropriate information for the client in a comprehensible form that the client can incorporate into her decision making. Furthermore, the doctor who has referred the client may have an expectation of genetic counseling that differs from the client's, and the genetic counselor should not fulfill the doctor's expectations at the expense of the client, for example, if the referring doctor expects the genetic counselor to instruct a couple to have no more children.

Diagnostic questions will often predominate when a child has multiple unexplained developmental difficulties and/or physical anomalies. Do the child's problems amount to a specific condition—a pattern or "syndrome"—or have they occurred together in the child by chance? It may be possible to identify a genetic cause for such a child's problems, although often this is not the case. Clinical (medical) experience and expertise will be required for this element of the clinical genetic assessment, which may be performed by the same individual who provides the genetic counseling—but these activities are sometimes performed by different members of the same multidisciplinary team.

It is important to note that the naming of a genetic condition in a child carries its own burden of responsibilities. There is the process of labeling the child with a diagnosis, which can be performed well or not so well, and how those around the child respond to this process. This involves professional skill and judgment and requires careful consideration of complex issues. Sensitivity to the needs of the parents is required when, for example, parents have been unable to accept the serious nature of their child's developmental difficulties—they may not be ready to accept certain genetic diagnoses. Or, it may be necessary to consider the likely effect on the child's education of giving the diagnostic label to his or her school—will it be helpful? There is also the separate question of the name that is given by professionals to a specific condition; this name must be chosen with care if it is not to offend the families and individuals to whom the label is attached.

Some clients wish to discuss the consequences of the condition that has been diagnosed in their family. What does the future hold for the affected person? How will the condition progress? And what implications are there for other family members? Who else may come to be affected in the future, or who may have affected children? Would the client's future children be at risk of developing the same condition? If so, is it possible to perform prenatal diagnosis with the selective termination of affected fetuses? May other members of the family be unaffected carriers of the condition, with a risk of having affected children? For some clients, a simple explanation of the mode of inheritance of the condition may answer most of their concerns, while others may wish to have much more detailed information about the underlying mechanisms of the disease process or the likely course of the illness.

If affected individuals are likely to develop serious complications from their genetic condition, are there useful methods of surveillance that can identify these complications at an early stage and improve the outcome through improved treatments? In family cancer syndromes, can tumors be detected early and treated better as a result, or can surgery be offered to reduce the risk of malignancy? Under conditions that predispose one to cardiovascular disease, can a catastrophe be prevented or deferred by control of blood pressure, by medication, or by a surgical intervention?

The emotional impact of the condition on members of the family will often need to be considered in a genetic counseling consultation. Where reproductive decisions or predictive testing decisions are to be made, however, these may need to be considered with special care and in a systematic way. For example, in the context of someone weighing whether or not to proceed with predictive testing for a disorder with a bleak outlook—e.g., for a neurodegenerative disorder such as Huntington's disease (HD)—it may be helpful for them to be invited to consider the likely impact on themselves and on other members of their family for each possible outcome of the counseling process. How will they and others respond to the following scenarios?

1. They choose not to proceed with testing, and thereby retain uncertainty but also hope
2. They proceed with testing and are shown to have the disease-causing HD gene allele
3. They are tested and are shown not to have the disease-causing HD gene allele

Although this last outcome is obviously the preferred one, such favorable results may still cause much distress to the individual and serious difficulties within the family. A favorable result may leave the unaffected family member feeling guilt and/or detachment from the family, and other family members may feel resentment or envy. Comparable considerations of the various possible outcomes for the client and for the family may also be helpful when a couple are deciding whether or not to embark upon prenatal diagnosis or whether to continue with a pregnancy in which the fetus may be affected by a serious genetic condition. This type of discussion—in any disease context—can be termed "scenario decision counseling."

Finally, it may be appropriate to maintain occasional contact with a family over a period of years. If the condition that affects a child has not been identified, it may be helpful to reassess the diagnosis at intervals. If an adult at risk of developing a late-onset condition has decided not to undergo predictive genetic testing, then occasionally he may find it helpful to reassess his family situation—and the reasons that counted for and against testing. It may also be appropriate to maintain contact with those at risk of developing such diseases, so that they have an opportunity to discuss the difficulties of their situation—living under the cloud of a genetic disease—and so that appropriate interventions can be offered if they develop signs of the condition. Those clients at risk of carrying a recessive or sex-linked disorder may also wish to maintain contact so that they can discuss the issues when it feels most relevant, and so that they can be offered the most appropriate carrier or prenatal diagnostic testing available. In families where several generations have carried or been affected by a genetic condition, it may be helpful to maintain contact so that the younger generation can be offered counseling and testing when they judge it to be relevant to themselves.

II. GENETIC TESTING AND GENETIC SCREENING

Genetic testing may be carried out in many different contexts. It is helpful to distinguish these contexts because the ethical issues that arise can be very different. Genetic testing can be offered to individuals because of their specific family histories or other circumstances, or it can be offered to large groups, populations, or subpopulations (e.g., all newborn infants, all pregnant women, or all adults of a specific racial or ethnic group). The first type of testing is specific to the individual or the family context; the second type of testing is termed "population screening."

Genetic testing (of the individual in a family context) and genetic screening (offered to a large population group) can also be categorized as predictive or reproductive. Reproductive testing can take the form of carrier testing—to identify healthy carriers who may have children affected by genetic disease in the future—and prenatal testing—carried out in a pregnancy to identify those in which the fetus is affected by a genetic condition. Genetic testing of the individual can also be diagnostic; this context arises when patients have signs or symptoms of disease, and thus is not applicable to population screening programs.

A. Predictive Testing

Predictive (presymptomatic) genetic testing is carried out on individuals known to be "at risk" for a serious genetic disease because other members of their family are affected. There may be definite health benefits from such testing if complications of the disorder can be man-

aged better when they are identified early. In some family cancer syndromes, for example, the early detection of tumors improves the clinical outcome, and so the identification of those at risk of the tumors helps in coordinating the program of surveillance. Furthermore, those who are at risk of being affected but are shown not to carry the relevant faulty gene are spared the costly and troublesome surveillance process or prophylactic surgery.

For other diseases, however, testing may give knowledge and hence relieve uncertainty, but without leading to any improvement in the course of the illness. Particularly for these conditions, such as the neurodegenerative diseases where presymptomatic diagnosis does not improve prognosis, it is important to ensure that those who seek testing have thought through the practical and personal consequences. To this end, many centers have adopted protocols for counseling individuals seeking such tests.

B. Diagnostic Testing

Diagnostic genetic testing is performed when an individual presents to a doctor with signs or symptoms of a disease that may be genetic in origin, and genetic tests are employed to clarify the cause of the illness. Genetic disease may be diagnosed by more traditional tests, such as muscle biopsy (for Duchenne muscular dystrophy) or renal ultrasound scan (for polycystic kidney disease), but molecular genetic tests are being used increasingly to diagnose (or exclude) symptomatic genetic disease. Such tests are often cheaper and less invasive than the traditional diagnostic tests.

When a patient is found to have a genetic condition, there may well be implications for other members of the family. When the condition comes to light inadvertently there is little chance to prepare the patient or the family for this possibility. But if the possibility of a serious genetic disorder is acknowledged early in the diagnostic process, it may be helpful to raise this with the affected individual and/or other members of the family before the tests are carried out.

If the doctor knows that the patient has a family history of a genetic disorder, and if this could account for the patient's problems, then the tests may be focused more rapidly on the relevant inherited condition. This may be more efficient, although there is also a chance that a treatable, nongenetic condition may fail to be diagnosed in someone whose problems are attributed too readily to the familial disorder. In someone at risk of Huntington's disease, for example, a complaint of forgetfulness, personality change, or lethargy could be caused by another illness (depression, perhaps) and may respond well to appropriate treatment—whether or not the patient carries the HD gene mutation.

C. Newborn Screening

Presymptomatic genetic screening is carried out in many countries on newborn infants. This is not usually controversial from an ethical perspective because the goal of most such screening programs is to identify infants who will benefit very substantially from early diagnosis and treatment. This is known to be the case with respect to phenylketonuria (PKU), congenital hypothyroidism, and sickle cell disease, and may also be true for certain other metabolic disorders and cystic fibrosis (CF).

Screening to identify infants affected by Duchenne muscular dystrophy (DMD) is more controversial because there is no effective treatment for DMD and so its early diagnosis provides little benefit to the affected boy. DMD is a lethal, sex-linked disorder affecting 1 in 3500 infant males and is often not diagnosed until 3–6 years of age. The potential benefits of early diagnosis include the avoidance of a prolonged and distressing diagnostic process and providing the opportunity for the extended family to seek genetic counseling and, perhaps, prenatal diagnosis in future pregnancies. There are potential problems too, however, including the difficulty of ensuring adequate informed consent for such testing, the distress caused by the imposition of such unwelcome knowledge some years earlier than would otherwise have been the case, and the possible disruption of family relationships as a result. This is a good example of an ethical issue that cannot be resolved by a priori reflection but requires a careful evaluation of such a screening program in operation.

The concerns about possible harmful effects of newborn screening for DMD apply a fortiori to screening for fragile X syndrome, the commonest familial form of learning difficulties and mental handicap. This affects about 1 in 4000–5000 children. In addition to the problems suggested above is the potential danger that labeling a child as being affected by fragile X will lead to his educational problems being exacerbated by diminished expectations from his earliest days—a self-fulfilling prophecy.

D. Susceptibility Screening and Commercialization

Screening for genetic susceptibility to the common, complex disorders is not yet a reality, but it is expected to become feasible over the next 5–10 years. In the past, most genetic research focused on the understanding of conditions inherited as simple, Mendelian traits (i.e., single-gene disorders). Over the past few years, the genetic dissection of the more complex, polygenic or multifactorial diseases has begun. Already, the localization of many loci contributing to juvenile and maturity-onset diabetes, raised blood cholesterol level, hypertension, Alzheimer's dementia, and several types of cancer has been achieved. The search for these genetic factors, and others involved in psychiatric disease, alcohol tolerance and dependence, personality traits, and "intelligence," is being pursued by numerous academic and commercial research groups.

While the primary scientific motivation for this research is to improve our understanding of the disease mechanisms and thereby open up new possibilities for therapy, there will be strong professional and commercial pressures to apply this knowledge so as to identify individuals at high or low risk of these important, common diseases. This may be justified when it leads to health benefits from useful clinical interventions or lifestyle changes, but there are real possibilities of causing harm by such susceptibility screening in the absence of useful treatments or prevention. Individuals identified as being at high risk of a disease will often feel distress and anxiety. They may also respond fatalistically by indulging in a high-risk lifestyle (e.g., smoking or consuming more cholesterol), while those given a low risk of disease may behave in the same "unhealthy" way if they feel themselves to be invulnerable. Such paradoxical responses that worsen the outlook for individuals in the high- and low-risk groups have been observed after cholesterol screening programs. Other possible difficulties include the consequences of labeling a healthy person as "sick," as when asymptomatic individuals identified as hypertensive adopt the sick role. Regulation of the commercial promotion of such tests and the development of minimum standards for the associated counseling will be necessary, but may not be sufficient to prevent these problems altogether.

One particular concern about susceptibility screening is that it may undermine collective public health measures to improve the welfare of society, such as attending to public policy on housing, food, education, transport, and safety in addition to health. Because each person will be able to tackle the risk factors peculiar to him- or herself, the political will to tackle these issues collectively may dwindle, but such a collective approach may be the most efficient way of improving the whole population's health and quality of life. Furthermore, collective approaches may be the only means of helping the most vulnerable groups in society. It is these processes that focus on the genetic factors predisposing one to health problems (and which lead to proposals for individualized rather than collective solutions), rather than focusing on social or environmental factors, that have been referred to as "geneticization."

E. Cascade Carrier Testing

When one member of a family has been identified as a carrier of a genetic disorder, the question arises as to whether any other members of the family may similarly be carriers. Such testing of the relatives of a gene carrier for some genetic disorder is termed "cascade testing." This situation arises in families with late-onset Mendelian dominant disorders and also, especially in the context of reproductive decisions, in relation to autosomal recessive disorders, sex-linked (X chromosome) disorders, and balanced chromosome rearrangements that do not cause disease in the carrier but which may cause serious problems in the carrier's future children. A child will only be affected by an autosomal recessive disease if both parents are carriers, but may be affected by a sex-linked disease (if male) when the mother is a carrier or by an unbalanced chromosome translocation if either parent carries the balanced translocation. The carrier status of one family member may come to light because an affected child has been diagnosed in the family, if the genetic testing was carried out as part of a population screening program, or if the genetic tests were being performed for some other reason (e.g., prenatal diagnostic testing for some other condition).

The term "cascade testing" can also be applied in the context of autosomal dominant disorders in which serious complications of the disease may develop in a carrier of the faulty gene without any prior indication of the person being "at risk." This applies in some of the familial cancer syndromes where family members may be at risk of malignant tumors in the colon, breast, thyroid gland, or other organs. Knowledge of a person's genetic status allows one to take preventive action or to be screened regularly to detect complications and optimize treatment. Tracking the condition back through the family in a cascading (or snowball) fashion allows other family members at risk of complications to take appropriate action.

The attitude of family members to the offer of cascade testing will vary greatly, depending upon their experience of the disorder, the chance of their developing the disease or of having an affected child themselves, and other factors. The healthy sibs of individuals with cystic fibrosis can find the process of carrier testing to be disturbing; the best age at which to offer carrier testing in these circumstances is controversial and is likely to depend upon factors specific to each family. Members of an extended family may have lost touch with each other, and it can be both difficult in practice

and emotionally distressing to reestablish family contact under the cloud of a serious genetic disease.

To what lengths should families be expected to go to pass on genetic information to their relatives? This will depend in part upon the probability of the relatives being carriers of the disease gene in question and upon the chance of this being significant for them (the risk of a future child being affected, or of a gene carrier developing a complication of the disease), as well as on the feasibility of establishing contact. The nature of the obligation to share some types of genetic information with relatives is discussed further below.

F. Population Carrier Screening

This is genetic testing to identify healthy carriers of autosomal recessive diseases from among the general population or a specific subgroup. This type of testing is offered to individuals with the standard, population risk of carrying the disorder in question rather than to those with a family history of the condition. Thus, testing the general public in Britain would identify about 1 person in 25 as a healthy carrier of a disease-causing allele at the cystic fibrosis gene locus. Testing individuals with a history of CF in their close family would identify a much higher proportion of carrier individuals than this, and would amount to cascade testing.

One important difference between family based genetic testing and population-based genetic screening is that population screening is offered proactively by health professionals (or commercial organizations) to individuals who had not previously decided to seek testing. They will often have no personal knowledge of the disease in question, and a positive test result—showing that they are carriers—may never have any implications for their own health or the health of other family members, because few of their partners will be carriers.

Many more individuals will be identified as carriers than there are carrier couples, so for most identified carriers the process of counseling and testing may complicate their family lives and reproductive decisions without it yielding any "benefit." The screening may introduce anxiety into the lives of many without delivering clear health benefits. There is evidence that some carriers of recessive disorders identified in population screening programs can have misunderstandings about their carrier status, with inappropriate concern about their own future health and confusion about the implications of their test result. What benefits to some carriers would justify these concerns among others? How should society decide its collective response to such questions?

It is known from pilot studies of CF population carrier screening in Britain, Europe, and North America that the way in which the offer of screening is made has a great influence on the proportion of people who accept the offer. As many as 70 to >95% of individuals will agree to testing if it is offered in person to those attending a health center or antenatal clinic and if it is available on the spot without inconvenience; if the offer is made by mail, or if a specific visit to the clinic must be arranged, then the rate of uptake of the test is much reduced (to 10–15%). This suggests that the high uptake rate that prevails in some circumstances reflects the compliance of an apathetic public in the face of professional enthusiasm. If health professionals, society, or commercial interests decide it would be advantageous for population screening to be introduced on a wide scale, it seems that it would be possible to adjust the numbers of those "choosing" to be tested to suit the interests of those offering the test. Given this, to what extent can people's decisions to accept the offer of such screening be considered to reflect their "true" wishes and their "adequately informed" consent?

Another important fact relating to population screening is that the frequency of carriers for many recessive disorders varies between populations (ethnic groups). It may therefore seem appropriate for carrier screening programs in multiethnic societies to be targeted toward those of particular ethnic backgrounds. This can cause problems, however, if one racial group (more prone to a specific disease) is subject to discrimination and stigmatization at the hands of a different, socially dominant group. Identification as a carrier of sickle cell disease, for example, added to the burden of racial discrimination already being experienced by black people in the USA during the 1970s when legislation to promote carrier screening among black people was abused in relation to employment and insurance.

G. Prenatal Diagnostic Testing

This is offered to women or couples whose pregnancies are known to be at increased risk of a genetic disorder or congenital anomaly because of specific grounds for concern about the health of their fetus. This will be based either on a family history of a genetic condition or on some prior evidence from the pregnancy suggesting the possibility of a problem. Such evidence might include an ultrasound scan suggestive of a chromosomal anomaly or an increased probability of Down syndrome calculated from maternal age and serum screening. Where there is a family history of the disorder in question, a couple are likely to have some experience or knowledge of the relevant condition. The genetic counselor will still need to explore their level of knowledge and understanding, but this is likely to be much more substantial than that of those entering routine population screening programs.

Diagnostic testing usually provides a definitive result, confirming that the fetus either does or does not have a specific disorder. Molecular genetic (DNA) and cytogenetic (chromosome) testing usually achieve this, although a small proportion of results may be of uncertain significance. If a few cells with a cytogenetic anomaly are identified from amniocentesis or chorionic villus sampling (CVS) while most cells are normal (chromosomal mosaicism), then this may indicate a problem in the fetus, may represent an anomaly confined to the placenta or fetal membranes, or may have arisen as an artifact in the cells cultured in the genetics laboratory. A further invasive (and therefore hazardous) diagnostic procedure may then be needed if the significance of the finding is to be established.

Another type of difficult result is the finding of an unanticipated anomaly—an incidental finding rather than the condition that was being sought. This often results in difficult counseling scenarios, as regularly happens when a case of sex chromosome aneuploidy is diagnosed (any sex chromosome constitution other than the standard XX or XY). In these conditions, when there is only one sex chromosome (Turner syndrome, with one X chromosome) or when there are additional sex chromosomes, the outlook for the future child may be much less serious than with many other chromosomal disorders. For example, a child with Turner syndrome is likely to be of short stature and to have delayed puberty, but these problems can be treated. A minority of affected girls will require cardiac surgery but the outlook is usually very good. The affected girl will be infertile, but not mentally or physically handicapped.

Despite the relatively lesser seriousness of these conditions, many fetuses with Turner syndrome are terminated. Factors influencing the decision include the specialty of the doctor who discusses the chromosome results with the woman or couple, with fewer pregnancies being terminated if the woman discusses the results with a clinical geneticist than with an obstetrician. Being told that the chromosome result is abnormal can lead some women or couples to make a rushed decision to terminate their pregnancy even though the condition diagnosed may not be the condition for which prenatal diagnosis was being performed. Such women can regret their decision when they subsequently learn more about the condition. Being given more information about the chromosome anomaly at the time of the initial giving of the results can help some couples to pause and reflect instead of proceeding immediately to a termination.

Some of these problems can be avoided if the possibility of unusual results, either of uncertain significance or indicating relatively minor anomalies, is explained to the woman or couple in advance of the test being performed. This may make it easier for the woman to defer a decision about termination in the event of such a result until she has sought further information or advice. These considerations serve to emphasize the importance of women being offered full information and the opportunity to consider the implications of testing before they make any decisions about prenatal screening or diagnostic testing. For consent to prenatal testing to be meaningful, it must include an adequate delay between the offer of testing and the test itself to permit the woman to reflect, to discuss the issues with friends and family, and to request further information.

H. Antenatal Screening

Antenatal screening for genetic disorders and malformation is offered to most pregnant women in developed countries, and also to many women in developing countries. There are five types of genetic screening tests to mention in this article.

1. Amniocentesis entails taking a sample of the fluid from around the fetus. This is usually performed at about 16 weeks' gestation and permits the chromosomal analysis of cultured cells from the fluid. This may be carried out in the context of a couple's high risk of genetic disease or as part of a screening program aiming to identify (and terminate) Down syndrome fetuses. Amniocentesis can result in an increased risk of miscarriage in the pregnancy (the increase in risk being of the order of 1 in 200 procedures, but varying with the center and the operator).

2. CVS is a similar invasive test with a slightly higher associated risk of miscarriage (about 1 in 50). It entails removing a sample from the developing placenta, and can be performed from about 10 weeks' gestation.

3. Maternal serum screening is carried out on a blood test from the pregnant woman; the level of particular materials in the serum (notably alpha fetoprotein and beta-HCG) gives some indication of the probability that the fetus has Down syndrome; this is interpreted along with the age-dependent risk of Down syndrome to give a combined probability. Women whose pregnancies have a chance of having Down syndrome of greater than 1 in 200 to 250 are often then offered amniocentesis as a diagnostic test.

4. Fetal ultrasound scanning may be offered at several stages in pregnancy, but fetal anomaly scans are frequently performed at 18 weeks' gestation. These scans are very effective at identifying structural anomalies such as spina bifida, and scans earlier in pregnancy will also identify many pregnancies with chromosomal anomalies.

5. There are newer techniques of prenatal screening being devised, such as the sorting of fetal cells from maternal blood samples. The sorted fetal cells can be subjected to genetic tests without putting the pregnancy at risk. The development of such techniques will clearly be a great benefit to high-risk families who want prenatal diagnosis for serious genetic disorders. They may result in less positive consequences for some women, however, such as those who would prefer not to have prenatal diagnosis at all but who decline the offer of the current, invasive tests on the grounds of the associated risks of miscarriage. Such women may find it harder to decline testing in the future.

Because ultrasound scans may also be used for other purposes in the management of a pregnancy, fetal anomaly scans are often performed without the woman fully appreciating the purpose of the scan. Worse, scans may be regarded as a means of promoting maternal "bonding" with the fetus. The irony of this becomes apparent when such "bonding" is rudely shattered if a congenital anomaly is identified and staff suggest that the mother should consider a termination of the pregnancy.

The major ethical issues arising in the context of prenatal screening concern the social nature of the screening program as opposed to the individual focus of prenatal diagnosis arising within known high-risk families. First, because such screening programs are offered to the whole population of pregnant women, inadequate attention may be paid to the nature of the consent obtained before women participate in the screening. The offer of screening may be routinized and staff may exert pressure on women to comply with the general policy of the clinic instead of encouraging them to consider the issues and weigh the potential advantages and disadvantages of screening for themselves as individuals. There is ample empirical evidence that such unsatisfactory practices are widespread.

Second, there is the question of equity of access to screening tests. Access to, and uptake of, most health services is greater in the professional and middle classes than in the working class and in some ethnic minority groups. Is this also true for prenatal screening programs? If so—as often is the case—this could lead to a higher incidence of genetic disorders being found in working class communities—those sections of society with the least independent resources to cope with such problems. The higher mean maternal age in the higher social classes may in part counter this effect because Down syndrome births will accordingly be more frequent, but insofar as prenatal screening is a benefit, it should be equally accessible to women from all social groups.

Third, there is the question of how society comes to decide which possible screening test should be offered. There have been two routes to the introduction of new prenatal screening tests. Fetal ultrasound scanning, although unevaluated, was introduced piecemeal by enthusiastic professionals. The technology—the ultrasonic scan machines—are readily available because they are used for many other purposes than solely fetal anomaly scans, and these scans are popular with many clients and professionals, so that it would now be difficult to provide antenatal care without offering scans routinely.

Maternal serum screening to identify pregnancies with an increased risk of Down's syndrome has been more controversial, but has become effectively entrenched in much of Britain and North America through the concerted efforts of professional and commercial interests to persuade pregnant women (service users), obstetricians (service providers), and health authorities or insurance corporations (the purchasers of health services) of its worth. Indeed, in California it is mandatory for physicians providing antenatal care to offer serum screening to every pregnant woman. Such serum screening programs are backed up by amniocentesis as a diagnostic test, with the specific goal of reducing the birth incidence of Down's syndrome. These efforts have largely succeeded, despite objections that the problems caused by these screening programs have largely been ignored. The principal argument put forward to justify such screening for Down syndrome is that screening costs the health services less than providing care for liveborn affected children. This justification would be acceptable in a business venture, but is not usually accepted as the standard justification for items of health care.

This discussion leads on naturally to the central ethical question arising in prenatal diagnosis of genetic conditions: how "severe" does a disorder have to be for it to justify the termination of an otherwise wanted pregnancy? It is possible for a society to legislate a solution—to draw up a list of conditions that are acknowledged as being sufficiently severe to permit prenatal diagnosis and the termination of an affected pregnancy (as in Norway). It is also possible for society to permit whatever a woman and her physicians agree together as appropriate (as effectively is the case in Britain). Practice in the USA differs between the private sector and the public sector supported by federal funds: there are constraints in the public sector that prevent even the discussion of the termination of pregnancy in many publicly funded clinics.

From the perspective of a clinical geneticist, it is possible to respond differently in the two distinct contexts of prenatal diagnostic testing and population screening. In the case of a family with intimate knowledge of a genetic condition, who have sought out genetic counseling and testing because they fear a recurrence of a specific disorder, it would be difficult for a profes-

sional, or for society, to refuse the request for prenatal diagnosis when the family has agonized over this very issue, perhaps for years. Complying with the family's wishes will probably lead to fewer "casualties" than any other policy, as long as the counselor ensures that the family has a good understanding of their genetic situation and of the testing process. But screening programs are very different, because they touch the lives of so many individuals—not just those few at very high risk of problems. Accordingly, even a low incidence of iatrogenic complications resulting from the screening program may lead to vastly more problems than would be avoided through screening. And when one considers that there is real disagreement within society over whether or not Down syndrome or spina bifida—the two principal conditions identified by antenatal screening—is sufficiently serious to justify the termination of pregnancy, it can be appreciated that a permissive approach to prenatal diagnosis can reasonably coexist with a skeptical approach to prenatal screening programs.

I. Preimplantation Genetic Diagnosis (PGD)

This entails the application of genetic technologies to a very few cells removed from an early embryo—often the 8–10 cell stage. The techniques are either molecular genetic, based on the polymerase chain reaction to amplify the small number of copies of the relevant genes, or molecular cytogenetic, employing fluorescence *in situ* hybridization carried out during interphase to assess the number of specific chromosomes or subchromosomal regions. Genetic counseling prior to such methods entails making sure that the couple understand the limitations of the *in vitro* fertilization (IVF) technology and procedures in addition to the potential errors of the tests themselves.

PGD is attractive to couples who have concerns about their risk of a child with genetic disease in addition to problems with fertility. It is not so often sought by those who have no difficulty achieving a conception, partly because of cost and accessibility, and partly because of the difficulty in complying with the IVF procedures while also using contraceptive measures to ensure that no unassisted conception occurs. Clients who have strong feelings against terminating a pregnancy may consider PGD as worthwhile despite these problems because it allows reassurance that the child will be unaffected without having to consider the use of pregnancy termination. Those who believe that a fertilized egg is already of infinite value, equal in worth to an independent, living man or woman, will have objections to the discarding of those embryos that are not implanted—but most people who cannot face terminations of pregnancy can consider PGD as an option for them, at least in principle.

Although PGD may appear to have a number of advantages, at least for some couples, it does raise some problems. In particular, it can be seen as a step toward the technology of the "designer baby"—the methods could be applied to select an embryo on the basis of nondisease criteria if genetic testing for such traits were feasible. Selecting infants on the basis of sex or carrier status for autosomal recessive traits is already possible—concerns have been raised that the application of PGD in such ways may lead down a slippery slope to selection on the basis of cosmetic criteria, intelligence, or personality in the future.

Objections to prenatal diagnosis and screening can be made on the grounds that such practices devalue those individuals with such genetic conditions as Down syndrome, muscular dystrophies, spina bifida, and cystic fibrosis, and lead to their stigmatization and to social discrimination against them and other people with disabilities. Any increase in genetic testing that resulted from the use of PGD would raise even more concern that individuals with genetic conditions and disabilities would be devalued by society at large—it might lead society to be even less welcoming to individuals with such problems. The fact that PGD allows prenatal selection between embryos independent of questions about the termination of pregnancy presents the issues around choosing among viable, future people in a starker and clearer light than before.

There are also questions raised by PGD for the professionals involved—what type of responsibility do they have for children born as a result of their involvement in the process of fertilization and selection? Some of these questions are similar to those raised by any *in vitro* fertilization process, but others have not had to be faced in practice before. The UK Human Fertilisation and Embryology Authority (HFEA) has ruled against the use of PGD to select on the grounds of fetal sex, except in relation to specific sex-linked diseases. Whether or not this decision will be supported in other countries remains to be seen. It is certainly the case that prenatal sex selection is practiced widely, especially in Asia, using genetic technologies to identify fetal sex and then terminating pregnancies in which the fetus is of the undesired (usually female) sex. In such circumstances, PGD to achieve the same ends may be portrayed persuasively as the lesser of two evils.

The comparison between fetal sex selection and antenatal screening to detect pregnancies in which the fetus has Down syndrome is also instructive. If sex selection is regarded as unacceptable because it is offensive to women and perpetuates a society in which they are devalued, why does the same argument not apply to Down syndrome and the sex chromosome aneupolidy conditions (including Turner syndrome, Klinefelter syndrome, and the XYY syndrome)? This is a powerful argument used by advocates of disability rights, and can be answered more easily by libertarians—who maintain that any form of selection is acceptable as long as the couple are properly informed about the potential hazards of their chosen course—than it is by the HFEA and others, who would like to prevent sex selection but who endorse PGD and prenatal diagnosis as currently practiced.

III. THE GOALS OF GENETIC COUNSELING

There are many possible goals of genetic counseling, which can be approached from the perspective of the individual client or of the population served by a genetic counseling service. In examining the effectiveness of a service, both perspectives may need to be evaluated, but it is important to decide which is dominant. If the principal goal were to be the avoidance of suffering, how should this be assessed? Using quantified measures of the quality of life, by in-depth interviews with affected individuals and families, or by counting the births of affected children? The aim of defining the goals of genetic counseling is to enable the effectiveness and quality of genetics services to be examined, and this will have important implications for the nature of the services provided.

Measures of individual client outcomes may assess information retained by the client (e.g., the mode of inheritance, likely medical complications, or the risk of recurrence of the condition), the influence on reproductive plans or (with a longer timescale) reproductive behavior, or client satisfaction with the service provided. None of these measures is very suitable, however, because using these frameworks of evaluation entails prejudging the reasons for seeking genetic counseling and how information gained in such counseling should be used. Thus, information about reproductive risks may well be sought by the client, but may not be remembered in the form in which it was given to them but in a simpler, binary form of the risk being either inevitable or negligible. Again, some clients are seeking a diagnostic label for the condition in their family, or an explanation for the problems faced by their child, and have no plans for further children—the focus on reproductive plans or behavior is inappropriate for them.

A focus on client satisfaction may at first seem more attractive, but it too is problematic because many clients will be given unwelcome information in genetic counseling, and their dissatisfaction with the information may lead them inappropriately to express dissatisfaction with the service. Also, they may have unrealistic expectations of the service, e.g., diagnostic certainty may simply be unattainable in their family circumstances. Another approach to the assessment of genetics services is to develop a measure of satisfaction that will not be so distorted by this problem of "blaming the messenger"; this could take the form of a retrospective assessment of satisfaction with the service provided once the client has sufficient experience to know what expectations of the service would have been reasonable.

Measures of population outcomes for genetic counseling and screening have also been proposed. In effect, is the birth incidence of children affected by serious genetic disorders decreased when genetic counseling services are provided? This approach to measuring the impact of genetics services ignores the impact of the service on individual clients and is really seeking to justify expenditure on genetic services on the grounds that they spare resources (i.e., save money)—that there are fewer individuals born who will be a burden on the taxpayer or insurance corporation through provision for their health, social, and educational needs.

There are two principal grounds for rejecting this approach. The first is because it is profoundly disrespectful to so many people with genetic conditions, suggesting that their lives are not worthwhile and that they are merely a burden to their families and to society at large. Many individuals with genetic disease and congenital malformations do support the availability of prenatal diagnosis for "their" conditions, but the widespread promotion of screening to prevent the birth of people like themselves can still have very negative effects on their feelings of self-worth—and there are many individuals with Down syndrome or spina bifida, for example, who are advocates for the rights and worth of people with these conditions.

Second, if society decides to provide genetic counseling services simply on the grounds that it saves money in the long run, then any process of audit of the service that seeks to maximize the "benefits" to society—the savings—will lead to pressure on service providers to pay less attention to individuals who are seeking only explanations, and to present the reproductive options to those considering further children in such a way that they minimize the chance of an affected child being born. In such circumstances, genetics services may be led toward a frankly eugenic role in which they would be held accountable for the births of "avoidable" individuals affected by the conditions identified as undesirable. This may appear to be alarmist fantasy, but it is already happening in China; market forces could lead to a similar result in other countries if the state or private health insurance corporations refused to make adequate provision for children with "potentially avoidable" con-

ditions. It could even be argued that the provision of care for individuals with genetic disease and disability is currently inadequate even in many developed countries, so that families already have to pick up many of the costs of caring for children with disabilities.

An attempt has been made by Modell and Kuliev to bridge the gap between measures of individual and population outcomes in genetic counseling through a focus on "informed reproductive decisions." This approach is attractive in many ways, but depends upon the notion of informed reproductive decision being both easily measured and intellectually coherent. Unfortunately, many informed reproductive decisions would escape measurement in practice (e.g., decisions to have no further children), and many such decisions would have little relevance to medical care but would rather reflect social pressures. For example, there are many informed reproductive decisions facilitated by medical technology around the world each year which lead to the termination of female fetuses on social, not medical, grounds. To define "informed reproductive decisions" as the goal of genetic counseling is to evade the value-laden question as to which of the possible reproductive decisions that could be made are the desirable ones.

In summary, the goals of genetic counseling can be viewed from an individual or population perspective. The goals of genetic screening programs must almost inevitably be viewed from the population perspective, and can be considered under three categories. Screening programs may be justified by appeal to

1. The avoidance of human suffering, which is not controversial unless it also leads to the avoidance of particular types of people (those with Down syndrome, for example, or females)
2. The promotion of informed reproductive choices
3. The most effective use of limited resources

As discussed above, the goal of a screening program is important because it has implications for the ethos of service delivery, and hence for the personal experiences of those individuals offered screening.

IV. AUTONOMY, INFORMED CONSENT, AND "NONDIRECTIVENESS"

The eagerness of clinical geneticists to escape from the aura of negative emotions surrounding their past involvement in the eugenic movements in Germany, the USA, Britain, and other countries has led us to reject any suggestion that we might encourage, persuade, or lead our clients to make specific reproductive decisions. As a professional group, genetic counselors have accordingly espoused "nondirectiveness." It is our task to present clients with the information they seek, thereby assisting them in the process of making their personal reproductive decisions, but it would be unacceptable for us to impose a decision on our clients—however subtly. Indeed, any suggestion that we might be even unintentionally directive, as a profession, can provoke fierce objections.

The eugenic past doubtless accounts for the vehemence of the feelings in support of nondirectiveness, but there are other grounds for the adoption of this slogan. First, there is the current enhanced respect for autonomy as the cardinal principle of health care ethics, with less support for the other acknowledged principles, including beneficence (also known as "paternalism"). Second, for counselors there is some utility in a degree of detachment from the decisions made by their clients. This affords relief from the emotional burden of clients' decisions—whether in predictive testing or decisions relating to prenatal diagnosis and the termination of pregnancies—and it also simplifies the medicolegal aspects of the genetic counseling practice. As long as "the relevant facts" have been presented, then it is the clients who carry the emotional and legal burdens.

What are the relevant facts here? How "informed" is "informed consent"? The quantity of technical information conveyed is only one aspect of this question, and can usually be resolved in discussion between the client and an adequately informed counselor. It would not be possible for genetic counselors to describe in detail every possible outcome of a prenatal diagnostic test, but it is feasible for them to outline the range of possible outcomes and the range of possible consequences. Thus, it is possible to explain that results may not be simply "normal" or "affected," but also may fall into a "gray area"—unclear or unanticipated. The possible consequences of the decisions being made by the client can be explored in a technical sense, as well as in terms of their personal meaning for the client and the emotional impact of the test results on her and on her relationships with family and friends. At entry to maternal serum screening for Down syndrome, therefore, the possibility of the test leading toward a termination of the pregnancy should be mentioned, and the client's likely reactions to this prospect can be considered. To gloss over these aspects of an important decision could result in the client making a decision that was inadequately informed.

The claim that genetic counseling should be nondirective is comprehensible; the claim that it is so in practice, however, is somewhat ingenuous. While the genetic counseling of individuals and families may often contain no "directiveness"—it may not be related to decision making at all—there is a clear message conveyed to society at large from the existence of genetic counseling

clinics and—a fortiori—from the existence and operation of antenatal genetic screening programs. The existence of genetic counseling clinics conveys the message that health professionals, and perhaps society at large, consider it reasonable—even desirable—for individuals to make rational reproductive decisions in light of information concerning the possible outcomes of future pregnancies. Society, in a sense, has set up genetic counseling to combat genetic fatalism as well as the fear of genetic risks. Most genetic counselors would defend this stance, and would willingly argue against fatalism with respect to genetic disorders. But genetic counseling is not neutral in relation to this issue.

This argument applies more powerfully in relation to genetic screening programs because the very existence of a screening program amounts, in effect, to a recommendation that the testing thereby made available is a "good thing." Health professionals and society would hardly establish and promote antenatal screening for Down syndrome unless they wanted people to make use of it—the existence of such a program is an implicit, but powerful, recommendation to accept any screening offer made. Screening programs, therefore, simply cannot be nondirective. Health professionals may respect the decisions of those who decline the offer of screening, not coercing them into compliance, but those who decide against participation will carry a label of social deviance unless great efforts have been made to avoid this. In practice, at least some of the personnel offering such tests are likely to regard those who decline screening as irritating, if not irresponsible, and to make this clear to their "clients."

Another issue is whether or not clients want genetic counseling to be nondirective. There are some clients who would actively prefer their genetic counselor to provide guidance as to what course of action to follow. Given that many genetic decisions are difficult and involve choosing between two or more options, all of which entail unhappy long-term consequences, this is thoroughly understandable. It is the client, however, who will have to live with the emotional consequences of their decision in the long term—perhaps with regret at having had a termination of pregnancy, or with an affected and suffering child—and most genetic counselors are unwilling to accept the burden of explicitly sharing such responsibilities, quite aside from medicolegal considerations.

Finally, there are circumstances in which genetic counselors are expected at least to attempt to persuade their clients to take one course of action rather than another. This applies in particular to two sets of circumstances: (1) situations where the genetic counselor can recommend a course of action as being medically advantageous for the client, as with surveillance for some types of familial cancer, and (2) where the counselor recommends that the clients should share information about the genetic condition in their family with other family members. Genetic counselors usually find that improving family communication is helpful to all concerned; attempts to keep information back from family members to whom it is relevant are often damaging—besides being ultimately unsuccessful in most cases. In these situations, the strict adherence to nondirectiveness is clearly inappropriate. This is discussed further below.

V. CONFIDENTIALITY AND GENETIC PRIVACY—WITHIN THE FAMILY

Genetic information is intensely personal, relating as it does to a person's very biological existence as an individual. It would therefore seem reasonable to treat it with the greatest respect as private and confidential, not to be divulged to others without consent except in the most extraordinary of circumstances. On the other hand, genetic information is usually—by its very nature—a family concern. We share our genes with other members of our biological family. Should we not then share information about our genes with them, at least when it might be of any relevance to decisions they might make about their health care, reproductive plans, or other life decisions?

Even when family members have lost affection for each other or have strayed apart geographically, they are usually willing to share personal information with other family members when it could be helpful to do so. If individuals are at first reluctant to contact other family members, the thought that they could be held morally responsible for future problems in the family is usually enough for them to change their minds. Few would wish to be blamed for another member of their family suffering from an avoidable cancer or preventable heart disease, for example, or for the birth of a child affected by a severe genetic disease or set of malformations.

Occasionally, however, an individual will refuse to make contact with family members about a condition that could well be relevant to their welfare. What should health professionals do in such circumstances? One response, exemplified in several reports from the Anglo-Saxon world, has been the suggestion that professionals should breach confidentiality and pass on the information to other family members without the consent of the first individual. While it is difficult to rule out this course in all conceivable circumstances, there are good grounds for caution about even drawing up proposals to define the circumstances in which confidentiality can be breached in this way.

First, the Hippocratic origin of medical confidentiality has served the public and the profession well, and any grounds for recommending systematic breaches of this must be very strong; any harm to be prevented in this way would have to be a very serious harm, not just a potential concern, and this harm would have to be clearly avoidable if the relevant information were passed on.

Second, the mere existence of guidelines for breaching respect for confidentiality will have an effect on the trust placed in genetic counselors by clients. Many clients may decide not to seek a referral to the genetics service at all, or might give misleading information about their family history that would make the interpretation of their genetic situation much more difficult. This could interfere in many consultations because many clients initially have doubts about the motivation and integrity of the genetics staff and only gain confidence in genetic counseling services over time. Damage could result from this interference in many more families than the very few where respecting confidentiality appears to be a problem now.

Third, families in which information is shared without the application of persuasion or moral coercion may be strengthened by the process; the opportunity for this will diminish if it is generally known that the relevant family member had no choice about disclosure of the information concerned.

Fourth, the existence of guidelines for breaching confidentiality will lead to an expectation that genetic counselors should trace the members of their clients' families with whom the clients have lost contact. This could evolve all too easily into an unrealistic legal obligation on the counselor to trace families, with genetic services becoming liable for damages if they do not go to great lengths to do so.

VI. CONFIDENTIALITY AND GENETIC PRIVACY—THIRD PARTIES

In what other contexts does the issue of genetic privacy arise? Apart from family members, the principal other parties who might wish to acquire genetic information about an individual are employers, insurance companies, and the state.

Employers may wish to select employees who are likely to remain healthy and to have suitable personalities for their work; some employers may also wish to select employees who would be less likely to develop specific occupation-related illnesses. Insurance companies will want to have access to any information about their clients' chances of developing ill health that is available to their clients—otherwise those at high risk of ill health will take out larger policies (leading to the feared phenomenon of "adverse selection"). The state may wish to gather genetic information on its population for a variety of purposes—such as for planning health services, for restricting immigration to blood relatives of citizens, and for forensic purposes.

What is it reasonable to permit these third parties to do?

It could be argued that employers should be free to assess the current health of an employee or a prospective employee by conventional clinical examination, but that tests of genetic susceptibility to future illness should not be imposed as a condition of employment except when public safety depends upon the good health of the employee. Such testing would have implications of much wider scope than just employment, such as health insurance schemes, and might have repercussions for other family members, and so should not be imposed by employers. For such testing by employers to be both fair and rational, there would have to be some intervention that could be offered to improve the outlook for an employee identified as being at increased risk of a health problem; the mere knowledge of risk would not justify the imposition of testing. If testing could identify those at increased risk of an occupational health hazard—such as exposure to dust or chemicals—then it could be worthwhile as long as it did not substitute for attempts to minimize exposure to the hazard for all employees, and as long as an employee at high risk was offered equivalent work elsewhere, without the hazard. Such testing, to identify susceptible employees, should not replace safety measures for all; even those at below average risk of an occupational disease will still be at risk from exposure to the hazards.

If insurance companies could use genetic information to predict the future illnesses and the likely ages and causes of death of most individuals, then the whole rationale of health and life insurance would be destroyed. Instead, each individual would need to pay in advance for their likely future health care needs. Only accidents would be insurable.

Because the future is unknown, it makes sense now for many individuals to contribute to a common fund to meet the costs of the few individuals or families who have to cope with unforeseen premature death or disease. Insofar as our knowledge of the future becomes more certain, the system of social solidarity represented by the insurance industry will be undermined. Our present collective approach to confronting disease and death will be fragmented and individualized. Arguably, we will all be the poorer as a result.

At present, such accurate foreknowledge is not available, and in principle it is unattainable because the complexity of our body systems, and the effects of environ-mental factors and chance, will conspire to prevent it. But it is possible to identify those at high risk for the uncommon single-gene disorders, and it may become possible to assess individuals' risks for common diseases such as breast and colon cancer, heart disease, diabetes, hypertension, stroke, and Alzheimer's dementia. There are reasons for doubting some of the claims made about the impending genetic revolution in health care, but even a crude risk stratification applied to large numbers of people could have serious adverse social consequences in limiting the availability of health care resources to some groups in some societies. How should insurance companies use such data? How could they be regulated so that their use of such data has few adverse effects?

While a strategy for society's collective response to the future availability of susceptibility testing for common diseases is being considered, it has been suggested that we must adopt a policy for dealing with the genetic testing issues that are already being confronted in practice—problems with the single-gene disorders. It is necessary for insurance companies to be protected from adverse selection, while not denying the social benefits of insurance to those individuals destined to develop the serious Mendelian disorders. This is particularly true in the USA in relation to health insurance, and in Britain in relation to life insurance, because housing in Britain is often obtained through a life insurance policy. One interim proposal put forward by Harper has been that insurance companies should not demand that genetic tests for Mendelian disease be performed, and that they should not even have the right to request the results of such tests as have been performed unless the life insurance policy being sought is for an unusually large sum. A similar policy has been implemented in The Netherlands.

A discussion of the right of the state to gather genetic information on its citizens falls outside the scope of this article. Suffice it to say that many geneticists are skeptical about the wisdom of entrusting such information to the state when it is understood how easily such data could be obtained illicitly for commercial applications, or how data could even be abused by the state itself.

VII. GENETIC TESTING OF CHILDREN

It is as easy to carry out genetic testing on children as on adults. Sometimes this will be perfectly appropriate, as when a child has presented with symptoms or signs of a disorder that could be genetic in origin. Equally, if a child is at risk of complications from a disease that runs in her family, and if early (presymptomatic) diagnosis may be to her advantage—improving the surveillance for complications if she does have the faulty gene and avoiding the need for it if she does not—then genetic testing will be seen by everyone as helpful. There are circumstances, however, where the issues are not so clear. These arise when parents or others request genetic testing for a child as either (1) a predictive test for a late-onset (usually adult-onset) disorder where there is no question of useful medical interventions in childhood, or (2) a test of the child's genetic carrier status, which is of possible relevance to the child's future reproductive plans but not to her health in childhood.

We are concerned here with the imposition of testing on those children who are not sufficiently mature to request testing in their own right. Although maturity is more difficult to define than chronological age, it is this that is important in ethics and in law. Predictive tests for late-onset genetic disorders have several potential disadvantages. Identifying children as destined to develop a serious, late-onset disorder such as Huntington's disease or familial breast-and-ovarian cancer could affect their upbringing and emotional development. Expectations of children's future relationships or educational attainments may be altered in such a way as to damage their progress, and these altered expectations could act as self-fulfilling prophecies; the children's adult lives would then be blighted even more thoroughly than living at risk of the disease would achieve. In addition, the future right of children to decide for themselves about testing—the right to autonomy—and the right of the future adults to confidentiality of the test result would both have been abrogated. This is a particular concern when we know that many adults at risk of such disorders choose not to be tested, and that some tested individuals choose not to inform other members of their families about the test results or even about the fact of being tested.

Against these considerations, the case can be made that children may adjust better to unwelcome genetic information if they find out about it at an early age, and that there is no firm evidence of any harm being done by such testing. On the other hand, while children may adjust well to the knowledge of genetic risk, it may be that individuals cope with unwelcome genetic information more readily if they are actively involved in seeking testing; genetic testing could then be likened to the search by an adopted child for his or her biological parents.

Given that young children at risk of Huntington's disease and other such disorders have not been tested, it will of course be difficult to obtain evidence about the long-term consequences of such testing without performing the tests, having decided that the ethical con-

cerns—the abrogation of the child's future rights to autonomy and confidentiality—can be ignored, and then waiting for 20–30 years until the consequences can be assessed.

The arguments in relation to carrier testing in childhood are similar, but the stakes are not so high. The child's future autonomy and right to confidentiality are both lost, but his or her own future health and existence are not affected. What may be affected is the child's emotional development if the parents seek to raise the child in such a way as to influence his or her future pattern of relationships and reproduction. Quite apart from the possibility of such attempts in practice being counterproductive, they may also be emotionally harmful. There is a better chance of obtaining evidence on this point than with predictive testing, because carrier testing for cystic fibrosis has been carried out routinely in many centers since the isolation of the CF gene in 1989, and children in families with balanced chromosomal translocations have been tested for carrier status since the early 1970s.

Another context in which the possibility of inappropriate genetic testing could arise is in relation to susceptibility screening. There is some evidence that children can be harmed by dietary restriction after cholesterol screening, and similar problems could arise after molecular genetic testing for susceptibility to many common diseases. There may be contexts in which testing for predispositions could be helpful, but inappropriate testing could be a problem if commercial promotion of such tests was targeted toward the testing of children.

VIII. CONCLUSION

Genetic counseling brings together traditional medicine with the "new genetics," and applies them both to the difficult circumstances of those individuals and families who are confronted by the reality of genetic disease. It is therefore not surprising that the ethical issues that arise in this context are so complex. There are other ethical issues generated by the new genetics that have not been considered here, such as the conduct of genetic research, the implications of the Human Genome Diversity Project for the members of oppressed ethnic minority groups, and the genetic dissection of individual variation for nondisease traits such as personality attributes and intelligence (but note that a few references to these issues have been included in the Bibliography). These topics were outside the scope of this article, but if (or when) the future application of such new knowledge to the genetic testing of adults, children, fetuses, and embryos becomes possible, it may be that genetic counselors will have to confront even more, and more difficult, ethical problems.

Bibliography

Abramsky, L., and Chapple, J., Eds. (1994). *Prenatal Diagnosis: The Human Side.* London: Chapman & Hall.

American Society of Human Genetics (1996). ASHG Report. Statement on informed consent for genetic research. *Am. J. Human Genetics* **59,** 471–474.

American Society of Human Genetics Social Issues Subcommittee on Familial Disclosure (1998). ASHG Statement. Professional disclosure of familial genetic information. *Am. J. Human Genetics* **62,** 474–483.

Andrews, L. B., Fullarton, J. E., Holtzman, N. A., and Motulsky, A. G., Eds. (1994). *Assessing Genetic Risks: Implications for Health Policy.* Washington, DC: National Academy Press.

Annas, G. J., and Elias, S. (1992). *Gene Mapping: Using Law and Ethics as Guides.* New York/Oxford: Oxford Univ. Press.

Baumiller, R. C., Cunningham, G., Fisher, N., Fox, L., Henderson, M., Lebel, R., McGrath, G., Pelias, M. Z., Porter, I., Seydel, F., and Wilson, N. R. (1996). Code of Ethical Principles for Genetics Professionals: An explication. *Am. J. Medical Genetics* **65,** 179–183.

Bekker, H., Modell, M., Denniss, G., Silver, A., Mathew, C., Bobrow, M., and Marteau, T. (1993). Uptake of cystic fibrosis testing in primary care: Supply push or demand pull? *British Medical Journal* **306,** 1584–1586.

Biesecker, B. B., and Marteau, T. M. (1999). The future of genetic counseling: An international perspective. *Nature Genetics* **22,** 133–137.

Billings, P. R., Kohn, M. A., Cuevas, M. de, Beckwith, J., Alper, J. S., and Natowicz, M. R. (1992). Discrimination as a consequence of genetic testing. *Am. J. Human Genetics* **50,** 476–482.

Bosk, C. L. (1992). *All God's Mistakes: Genetic Counseling in a Pediatric Hospital.* Chicago: Univ. of Chicago Press.

Boston, S. (1994). *Too Deep for Tears.* London: Pandora.

Clarke, A., Ed. (1994). *Genetic Counselling: Practice and Principles.* London/New York: Routledge.

Clarke, A., Ed. (1998). *The Genetic Testing of Children.* Oxford: Bios Scientific.

Clarke, A., and Parsons, E. P., Eds. (1997). *Culture, Kinship and Genes.* Basingstoke: Macmillan.

Clinical Genetics Society Working Party (1994). The genetic testing of children. *Journal of Medical Genetics* **31,** 785–797.

Conrad, P., and Gabe, J., Eds. (1999). Special Issue on the New Genetics. *Sociology of Health & Illness* **21**(5), 505–706.

Davison, C., Macintyre, S., and Smith, G. D. (1994). The potential social impact of predictive genetic testing for susceptibility to common chronic diseases: A review and proposed research agenda. *Sociology of Health & Illness* **16,** 340–371.

Duster, T. (1990). *Backdoor to Eugenics.* London/New York: Routledge, Chapman and Hall.

Green, J. M., and Richards, M. P. M., guest Eds. (1993). Psychological aspects of fetal screening and the new genetics (special issue). *Journal of Reproductive and Infant Psychology* **11**(1), 1–62.

Harper, P. S. (1993). *Practical Genetic Counseling,* 4th ed. Oxford: Butterworth-Heinemann.

Harper, P. S., and Clarke, A. (1997). *Genetics, Society and Clinical Practice.* Oxford: Bios Scientific.

Kessler, S. (1989). Psychological aspects of genetic counseling: VI. A critical review of the literature dealing with education and reproduction. *Am. J. Medical Genetics* **34,** 340–353.

Koch, L., and Stemerding, D. (1994). The sociology of entrenchment: A cystic fibrosis test for everyone? *Social Science and Medicine* **39,** 1211–1220.

Lock, M. (1994). Interrogating the Human Diversity Genome Project. *Soc. Sci. Med.* **39,** 603–606.

Macintyre, S. (1995). The public understanding of science or the scientific understanding of the public? A review of the social context of the "new genetics." *Public Understanding of Science* **4,** 223–232.

Marteau, T., and Richards, M., Eds. (1996). *The Troubled Helix: Social and Psychological Implications of the New Genetics.* Cambridge: Cambridge Univ. Press.

Modell, B., and Kuliev, A. M. (1993). A scientific basis for cost–benefit analysis of genetics services. *Trends in Genetics* **9,** 46–52.

Nuffield Council on Bioethics (1993). *Genetic Screening—Ethical Issues.* London: Nuffield Council on Bioethics.

Pelias, M. Z. (1992). The duty to disclose to relatives in medical genetics: Response to Dr. Hecht. *Am. J. Medical Genetics* **42,** 759–760.

Pokorski, R. J. (1995). Genetic information and life insurance. *Nature* **376,** 13–14.

Richards, M. P. M. (1993). The new genetics: Some issues for social scientists. *Sociology of Health & Illness* **15,** 567–587.

Rothman, B. K. (1988). *The Tentative Pregnancy.* London: Pandora.

Royal College of Physicians Committees on Ethical Issues in Medicine and Clinical Genetics (1991). *Ethical Issues in Clinical Genetics.* London: Royal College of Physicians of London.

Schrander-Stumpel, C. T. R. M. (1998). What's in a name? *Am. J. Medical Genetics* **79,** 228.

Welch, H. G., and Burke, W. (1998). Uncertainties in genetic testing for chronic disease. *J. Am. Medical Association* **280,** 1525–1527.

Wertz, D. C. (1992). Ethical and legal implications of the new genetics: Issues for discussion. *Social Science and Medicine* **35,** 495–505.

Williamson, R., and Savulescu, J. (Eds.) (1999). Special Issue on Genetics, *Journal of Medical Ethics* **25**(2), 75–214.

World Federation of Neurology (1989). Research Committee Research Group: Ethical issues policy statement on Huntington's disease molecular genetics predictive test. *Journal of the Neurological Sciences* **94,** 327–332; (1990). *Journal of Medical Genetics* **27,** 34–38.

Zallen, D. T. (1997). *Does It Run in the Family? A Consumer's Guide to DNA Testing for Genetic Disorders.* New Jersey: Rutgers Univ. Press.

Genetic Engineering

MATTI HÄYRY and HETA HÄYRY

University of Helsinki

GLOSSARY

biotechnology The application of biological advances such as genetic engineering to industry.

DNA Deoxyribonucleic acid. Any of the nucleic acids that are localized especially in the cell nuclei and that form the molecular basis of heredity in organisms.

gene The unit of hereditary information, which is in most organisms composed of DNA.

genetic engineering The directed alteration of genetic material, particularly by recombinant DNA techniques.

genome The totality of the chromosomes of a species, including all genes and their connecting structure.

recombinant DNA DNA that is produced technologically by breaking up and splicing together DNA from different species of organisms.

GENETIC ENGINEERING is the science of altering genetic material by intervention in genetic processes to produce new traits in organisms. Humankind has manipulated the qualities of living beings for thousands of years through, for instance, breeding, domestication, and training, but since the 1970s, biologists have been able to bring about considerably more radical and rapid changes by employing recombinant DNA techniques—that is, by isolating and combining the genetic materials of different organisms, frequently across the boundaries of natural biological species.

The practical uses of genetic engineering are expected to be abundant. The manipulation of plants, animals, and bacteria can generate improved crops, inexpensive food products, and more efficient pharmaceuticals, while the genetic alteration of the human constitution can be useful in medicine, especially when it comes to the treatment and prevention of hereditary diseases.

The ethical rightness or wrongness of genetic engineering can be examined from four different philosophical perspectives. From the viewpoint of *positive utilitarianism,* the central question is one of whether the good consequences of gene technology outweigh its costs and undesired side effects. For *teleological* ethicists, the important point is whether the genetic alteration of living beings is regarded as natural or unnatural. *Deontological* moralists, in their turn, stress the necessity of inflexible rules and limits in the regulation of human actions. And *negative utilitarians* emphasize the need of humankind to prevent and to remove unnecessary suffering whenever this is possible.

I. THE CONSEQUENCES OF GENETIC ENGINEERING

A. The Advantages

The advantages of genetic engineering, as seen by its proponents, include many actual and potential contributions to medicine, pharmacy, agriculture, the food industry, and the preservation of our natural environment.

1. Genetic Medicine

Within medicine and health care, the most far-reaching consequences are supposed to follow from the mapping of the human genome, an enterprise which holds the key to many further developments. Accurate genetic knowledge is a precondition for many foreseeable improvements in diagnoses and therapies as well as an important factor in the prevention of hereditary diseases, and possibly in the general genetic improvement of humankind. Provided that such knowledge becomes available, potential parents can in the future be screened for defective genes and, depending on the results, they can be advised against having their own genetic offspring, or they can be informed about the benefits of prenatal diagnosis.

One of these benefits is that an adequate diagnosis may indicate a simple monogenic disease in the fetus which can be cured by somatic cell therapy at any time during the individual's life. Another possibility is that an early diagnosis may reveal a more complex disorder which can be cured by subjecting the embryo to germ-line gene therapy—this form of prenatal treatment also ensures that the disorder will not be passed down to the offspring of the individual. Even in the case of incorrigible genetic defects the knowledge benefits the potential parents in that they can form an informed choice between selective abortion and deliberately bringing the defective child into existence.

Prenatal checkups and therapies do not by any means exhaust the medical applications of future genetic engineering. Somatic cell therapies are expected to help adult patients who suffer from monogenic hereditary diseases, and the increased risk of certain polygenic diseases can be counteracted by providing health education to those individuals who are in a high-risk bracket. In addition, the purely medical benefits of genetic engineering are extended by the fact that gene mapping will probably prove to be useful to employers and insurance companies as well as to individual citizens. Costly mistakes in employment and insurance policies can be avoided by carefully examining the applicants' tendencies toward illness and premature death prior to making the final decisions.

2. Biotechnology

Gene therapies and genetic counseling are practices which require advanced knowledge concerning the human gene structure. But the genetic manipulation of *nonhuman* organisms in biotechnology can also be employed to benefit humankind. The applications of genetic engineering to pharmacy, for instance, can in the future produce new diagnosing methods, vaccines, and drugs for diseases which have to date been incurable, such as cancer and AIDS. The agricultural uses of recombinant DNA techniques include the development of plants which contain their own pesticides. In dairy production, genetically engineered cows give more milk than ordinary ones, and with the right kind of manipulation the proteins of the milk can be made to agree with the digestive system of those suffering from lactose intolerance. As for other food products, gene technologies can be applied to manufactured substances like vanilla, cocoa, coconut oil, palm oil, and sugar substitutes. And biotechnology can even provide an answer to the growing environmental problems: genetically engineered bacteria can be employed to neutralize toxic chemicals and other kinds of industrial and urban waste.

B. The Disadvantages

The disadvantages of biotechnology, as seen by its opponents, are in many cases closely connected with the alleged benefits. An efficient strategy in opposing genetic engineering is to draw attention to the cost and risk factors which are attached to almost all inventions and developments in the field.

1. Costs and Dangers

One general critique can be launched by noting that the applications of recombinant DNA techniques are enormously expensive. Millions and millions are spent every year by governments and multinational corporations in biotechnological research and development. These resources, opponents of the techniques argue, would do more good to humankind if they were allocated, for instance, to international aid to the Third World.

Another problem is that, despite the undoubtedly good intentions of the scientists, the actual applications of genetic engineering are often positively dangerous. Consider the case of plants which are inherently resistant to diseases, or which contain their own pesticides. Although there are no theoretical obstacles to the production of such highly desirable entities, corporations—who also sell chemical pesticides—might prefer to market another type of genetically manipulated plant, which is unprotected against pests but highly tolerant to toxic chemicals. The result of this policy would be an increase in the use of dangerous chemicals in agriculture, particularly in the Third World—which is to say that the outcome is exactly opposite to the one predicted by the proponents of biotechnology. Besides, it is quite possible that genetically engineered grains are less nourishing than the grains which are presently grown. If this turns out to be the case, then the employment of biotechnology will intensify instead of alleviating famine in the Third World. And to top it all, genetically manipulated plants may contain carcinogenic agents, and thus contribute to the cancer rates of the developing countries.

An oft-used criticism against agricultural biotechnology is that the introduction of altered organisms into the natural environment can lead to ecological catastrophes. Scientists working in the field of applied biology have themselves noticed this danger, and set for themselves ethical guidelines which are designed, among other things, to minimize this particular risk. But as the opponents of genetic engineering have repeatedly pointed out, not all research teams follow ethical guidelines if the alternative is considerable financial profit.

2. Injustice

Apart from the excessive expenses and the increased risk of physical danger caused by biotechnology, its opponents can appeal to yet another disadvantage, which is related to widely shared moral ideals rather than to straightforward estimates concerning efficiency. Genetic engineering, it can be argued, is conducive to economic inequity and social injustice, both nationally and globally. Even in the most affluent Western societies, gene therapies are too expensive to be extended to members of all classes and age groups. Subsequently, these therapies are likely to become the privilege of an elite, and they will drain scarce resources from the more basic areas of health care provision. In the developing countries, the situation is even more absurd. Medical problems which originally stem from lack of democracy, lack of education, shortages of fresh water, population explosion, archaic arrangements of land ownership, and the like cannot possibly be solved by high-tech Western innovations which can barely be made to work in the most affluent and democratic of countries.

Another type of injustice emerges from the fact that the natural national products of many developing countries are superseded in the market by the biotechnological products of multinational corporations. Genetically engineered substitutes for sugar, for instance, could adversely affect the lives of nearly 50 million sugar workers in the Third World. Biotechnological vanilla could increase the unemployment figures by thousands in Madagascar, Reunion, the Comoro Islands, and Indonesia. And plans to produce cocoa by genetically manipulating palm oil threaten the current export market of three poverty-striken African countries, namely, Ghana, Cameroon, and the Ivory Coast. In all these cases, the profits of multinational Western corporations are clearly and directly drawn from the national income of the developing countries.

Finally, as regards medical biotechnology, there are those who believe that advanced knowledge concerning the human genome will inevitably become an instrument of genetic programming, which in its turn leads to subtle forms of genocide and general injustice. The opponents of gene splicing argue that the development begins inconspicuously with attempts to eliminate hereditary diseases. This practice of what is called "negative" eugenics will, however, soon be followed by more "positive" efforts toward altering the human genome: the inborn qualities of future individuals will first be improved in their own alleged interest, and then, later on, in the interest of society at large. When this development has gone far enough, scientists will also be asked to design special classes of subhuman beings who can do all those occupations which are too dangerous or too tedious for ordinary people. The outcome, according to the opponents of biotechnology, will be something like Aldous Huxley's *Brave New World.*

C. Optimism and Pessimism

Although the expected advantages and dreaded disadvantages of genetic engineering are fairly well publicized, it is difficult to assess objectively what the actual consequences of employing the techniques would be. The outcome depends, namely, upon the social and political setting in which the application takes place. As different groups of people hold different views concerning the structure and dynamics of social and political life, these groups inevitably also disagree upon the consequences of the development of recombinant DNA techniques.

The optimism of the proponents of genetic engineering can be defended by noting that during the last decades nothing particularly alarming has happened in genetics. No bizarre life-forms have been created, and nobody seems to have any major plans to try to improve the human race. Thus, at least the worst-case scenarios involving genetically determined social classes seem, at the moment, farfetched.

On the other hand, it would be naive to believe that the multinational corporations which presently control biotechnological development would voluntarily undertake to further general welfare and global justice. Pessimism in this sense is therefore justified. Doubts are also warranted when they prevent people from believing uncritically what genetic engineers claim about the advantages of the new techniques—applied biologists do, after all, have many vested interests in biotechnology.

The upshot of the disagreement regarding the consequences of genetic engineering is that it is not possible either to justify or to condemn all applications of recombinant DNA techniques in a straightforward positive utilitarian analysis. The benefits are in many cases indisputable, but they are almost always counterbalanced by equally considerable disadvantages. The ultimate moral rightness or wrongness of genetic engineering must, therefore, be determined in the light of other ethical theories.

II. GENETIC ENGINEERING AND UNNATURALNESS

A. The Teleological View

The core idea of teleological ethics is that all beings have a natural *telos*—a goal toward which they are ideally inclined to move or to develop. The telos of human beings is, according to the original Aristotelian reading of the view, a good life in a just society, or, beyond that, intellectual contemplation. Later, Christian interpretations of the theory have stated that the ultimate goal of human beings is an afterlife of everlasting joy.

It is not easy to apply the teleological model in its more philosophical—or theological—forms to the questions of genetic engineering. The link between recombinant DNA techniques and the human good is obscure, and it seems that an accurate view of the consequences of genetics would be required to support an adequate analysis of the connection. On a less philosophical level, however, traces of teleological thinking can be detected in many popular objections to biotechnology, especially in the claim that genetic engineering is unnatural.

B. The View of the German Enquete Commission

The best attempt to formulate and employ the argument of unnaturalness against human genetic engineering can be found in the report of the Enquete Commission to the German Bundestag. In its 1987 report the Commission tackled three questions which are fundamental to the issue, namely, the definition of the natural as opposed to the unnatural, the reasons for preferring naturalness to unnaturalness, and the division of different kinds of biotechnology according to their natural and unnatural characteristics.

As for the question of definition, the development of individual human beings is regarded in the report as natural only if it is not determined by technical production or social recognition. Technological and social processes can, according to the Commission's view, produce only unnatural artifacts.

The value of promoting naturalness and avoiding artificial elements in practices which concern human development is linked in the report with the need to protect the humanity and dignity of human beings. Our humanity, so the Commission asserts, "rests at its core on natural development," and our dignity "is based essentially on the naturalness of our origins." (1988. Prospects and risks of gene technology: The report of the Enquete Commission to the Bundestaq of the Federal Republic of Germany. *Bioethics* **2,** 256–263.) If technological or social interventions are allowed, then the result is that people will be created by other people, and the Commission regards this possibility with extreme suspicion. Human beings whose existence and personal qualities depend on the planning or caprice of other human beings are not free persons in the full meaning of the term, and their lives lack the individual worth of naturally developed human lives. It is the pure chance of nature that secures our independence from other people, our personal freedom, and our individual worth as human beings.

These considerations lead in the report to the following normative views regarding different kinds of human genetic engineering. First, somatic cell therapies performed on fetuses, infants, and adult human beings are, at the moment, justifiable as experimental treatments, since they are intended to cure only the individuals who are actually being treated. Whether or not such treatments should be abandoned or condoned in the future remains to be judged by their practical success. But the humanity of individuals is not threatened by the use of genetic medicine when the individuals in question have already developed into the beings that they "naturally" are.

Second, the mapping of the human genome is legitimate as long as it is employed to diagnose the need for somatic cell therapies. The potential use of gene maps for other (eugenic) purposes is more controversial. Third, cloning and large-scale eugenic programs must, according to the report, be banned as gross instances of manufacturing people. And fourth, if a strict interpretation is given to the Commission's ideas concerning naturalness, germ-line gene therapies, which are expected to rectify hereditary disorders both in the patients themselves and in their descendants, must also be prohibited. All interventions in the germ lines of individuals diminish, according to the foregoing argument, their independence, uniqueness, and worth as human beings.

C. A Disagreement among the Commission

The opinions among the Enquete Commission diverged, however, regarding the legitimacy of germ-line gene therapies. Only some members of the Commission upheld the strict interpretation of naturalness, while others advanced a more moderate view. The core of the latter, moderate interpretation is that the medical corrections of obvious defects are not unnatural, as they do "not manufacture the human genome capriciously, but measure it against nature, that is, good health." (1988. Prospects and risks of gene technology: The report of the Enquete Commission to the Bundestaq of the Federal Republic of Germany. *Bioethics* **2,** 256–263.) Illness and suffering can be a part of a person's identity, but if they are prevented before the person even exists, there is no point in maintaining that her or his individuality is unlawfully changed or manipulated. The genetical treatment of early embryos is, so the moderate reading goes, directly comparable to any conventional treatment of fetuses and neonates who cannot give their consent to the procedures.

There are two intrinsic problems within the "moderate" view. First, defining "good health" is not an unambiguous matter, and one could well argue that human caprice always enters germ-line gene therapies through the particular definition employed. Second, the moderates of the Commission seem to assert that eugenic programming can be absolutely prohibited due to unnaturalness even though germ-line gene therapies cannot. This is a highly controversial view which presupposes that a tenable distinction can be drawn between the two practices.

The alleged difference between gene therapies and genetic improvement programs is that the former is aimed at eliminating hereditary diseases while the latter is intended to bring about or intensify some positive qualities in future individuals. This distinction is unclear, as it is obvious that illness may hinder the development of certain positive qualities and promote the development of others. A physically disabling disease, for instance, may prevent the individual from being strong and athletic—which are often regarded as "good qualities"—but it may indirectly promote the individual's willingness to learn cognitive and artistic skills—which are also often considered good.

D. A Critique of the Commission's View

The divergence of opinions within the Commission is an interesting detail, provided that the unnaturalness objection can be regarded as tenable. There are, however, several good reasons for thinking that this is not the case.

First, the argument from unnaturalness seems to apply to many practices which have been traditionally considered quite acceptable. If genetic engineering is to be condemned due to its power to change individuals by technical means, then most medical interventions should be condemned as well. Surgical operations, for instance, often alter people by transforming them from fatally ill patients into perfectly healthy citizens. And

changes of personal identity may be even more drastic in the case of radical psychiatric treatments.

Second, the Commission's argument presupposes theoretical elements which are by no means universally accepted. The report's entirely biological view concerning personal identity is a case in point. According to the view, human beings are who they are and what they are almost exclusively owing to the arrangement of their genes. Culture, education, and social environment cannot significantly change the individual's identity, only biotechnology can do that. Very few philosophers today believe that such a strict biological definition of personality and individuality could be credibly defended.

Another presupposition in the report which can be criticized is its underlying view of human freedom and independence. The argument requires that human beings can be free from each other's influence in the sense that people are not "manufactured" by other people. This is obviously true if the manufacturing of people is understood literally: human beings cannot at the moment be mechanically created by each other except in science fiction. But when it comes to less obtrusive types of interaction, it is also true that people simply cannot survive and function without the often restrictive and molding presence of other people. Human freedom without the individual's dependence on others is only an abstraction.

Third, the unnaturalness objection presented in the report rests on the assumption that genetic engineering would undermine the worth, humanity, and dignity of the individuals produced by using the technique. This assumption is not only dubious but it may be positively insulting toward those human beings who will be born in the future genetically altered or cloned, perhaps against prevailing laws. The depth of the actual insult depends upon the interpretation that one gives to the Commission's view.

One possibility is to state that, according to the report, genetically engineered individuals would in fact lack humanity, dignity, and personal freedom because their chromosomes have been tampered with. This line of argument would obviously be unreasonably unfair toward the individuals in question.

Another possibility would be to assume that the Commission does not discuss the objective worth of human life in the first place, but the individual's subjective sense of worth in her or his life. The argument would then be that genetic engineering is wrong because the knowledge of one's "artificial" and "unnatural" origin reduces one's sense of worth and dignity. And yet another possibility is to claim that other people's adverse attitudes will make genetically engineered individuals unhappy.

The remarks concerning attitudes can, no doubt, be

valid in predictable circumstances. But since people's attitudes toward themselves and toward others are subject to change, the argument in this form is conditional. If genetically altered human beings can be expected to have difficulties in coping with the question of their origins, then these difficulties may constitute a weak case against germ-line gene therapy, cloning, and eugenic programs. But this does not imply that these practices could be categorically rejected because of the attitudes that people happen to have.

III. DEONTOLOGICAL OBJECTIONS TO GENETIC ENGINEERING

A. The Role of Rules and Limits

The fundamental tenet of deontological ethics is that there are certain things that should not be done whatever the consequences of the omission. Ruth Chadwick has distinguished four different ways to intrepret this tenet. First, deontological ethicists can argue that there are duties which should not be overridden by consequentialist considerations. Second, they can maintain that there are rights which should not in any circumstances be violated. Third, it can be claimed that there are, irrespective of our duties and rights, boundaries which must never be crossed. And fourth, one possible line is to state that certain things count as "playing God" and should not therefore be attempted.

The first three lines of argument have not been fully developed with regard to the morality of genetic engineering, but brief descriptions of their possible characteristics follow in the next three sections. The playing-God argument, however, has been thoroughly analyzed by Ruth Chadwick, and it will be given a more detailed treatment in Section III.E.

B. Duties

What duties can we have with regard to genetic engineering? The fundamental duty postulated by Immanuel Kant, the foremost advocate of the intellectualistic version of deontological thought, is our obligation to treat the humanity in ourselves and in other persons always as an end and never as a mere means. This obligation, which we owe to our fellow humans but not to the members of other species, is based on our nature as rational agents. The way we ought to treat animals, plants, and other life-forms is determined by our duties toward ourselves and other persons, not by their nonexistent worth as ends in themselves.

The difficulties of applying Kant's views to the manipulation of nonhuman beings by recombinant DNA techniques include the fact that he did not clearly specify what our duties as regards animals, plants, and other nonhuman organisms are. He believed that violence and cruelty toward animals set a bad example to our treatment of other people, and that not even inanimate objects should be wantonly destroyed because that would prevent others from making use of them. But how these remarks should be interpreted in the context of genetic engineering in a purely nonconsequentialist analysis remains an unanswered question.

When it comes to the alteration of the human genome, Kant's principle is open to two readings. Many 20th century philosophers have thought that our humanity as persons can be best enhanced by respecting our preferences and freely made choices. If this is true, then somatic cell therapies and also germ-line gene therapies should presumably be condoned whenever the patients themselves decide that these treatments are beneficial to them. But Kant's own interpretation seems to be, rather, that our essential humanity should take priority over our contingent and irrational desires and decisions. Reason commands us to obey the moral law, usually against our own inclinations, and our true humanity can be respected only by protecting our rational nature against our externally generated impulses. This view bears a resemblance to the teleological doctrine, examined in Section II, which emphasizes the immorality of acting in unnatural ways, and it can most probably be criticized along the same lines.

C. Rights

Do individuals have specific rights with regard to their basic biological constitution? Is there, for instance, a right to genetic treatment which is equal to the right to be aided by other medical means? Or do future generations perhaps have a right to the genetic disorders of their ancestors—a right which would cancel the permission of present generations to undergo germ-line gene therapy? And if rights like these do exist, what is their foundation, and how can they be justified?

One answer to these questions can be found by studying the prevailing laws. In the *legal* sense, individuals are endowed with the rights and liberties that can be derived from national and international laws and statutes. But there are two problems with placing too much trust in this solution. First, the legislation concerning genetic engineering has not yet been completed, and it is impossible to foresee what rights individuals will acquire in the process. Second, even if the legislation had been completed, the prevailing laws could be morally condemnable. Lawgivers have been known to pass laws which, in further analysis, have turned out to be counterproductive or conducive to suffering and injustice.

Another possibility is to argue that individuals have *moral* rights which are not necessarily enforced by existing laws but which in the majority of instances should be. The justification for these moral rights can be found in either consequentialist or nonconsequentialist thinking. In the former case, the principles for postulating them are sketched in Sections I and IV. In the latter case, the answers must be mainly sought in natural rights theories, which range from the Catholic teaching of Thomas Aquinas to the libertarian doctrine originated by John Locke and recently revived by Robert Nozick. The Catholic view derives its definition of the concept of "natural" from the essential features of the ideal human being, and it is therefore open to the type of criticism presented in Section II.D. The libertarian view has been, and probably can be, applied only to the economic aspects of genetic engineering, such as the right of individuals to patent life-forms. According to this type of thinking, next to no restrictions should be set on the development and use of biotechnology.

D. Limits

Some people believe that there are limits which should never be crossed by human actions even if these actions cannot be identified as unprofitable, harmful, or unnatural, and even if they do not prevent agents from performing their duties or force them to violate the rights of others. One version of the view states that people are playing God if they cross certain boundaries—this notion is scrutinized in the next section. The only other version that theorists have come up with so far is emotionalistic, and it states that the morality of actions can be determined only by recourse to sentiments. According to emotion-based deontological thinking, if we feel strongly enough that genetic engineering is bad, evil, disgusting, or immoral, it ought to be banned.

The main difficulty with the emotion-oriented model is that it is exceedingly relativistic. In most cases there is no consensus concerning feelings, and many questions remain unanswered. Whose feelings should be respected? Should genetic engineering be banned only if *everybody* feels that it is bad? Or is it sufficient that the *majority* feel that way? Or perhaps prohibitions ought to be employed if a significant *minority* nurtures these feelings? Or should we say that if *anybody* feels this way, biotechnology ought to be rejected? There seem to be no good responses to these queries either at the general level or in the context of biotechnology.

E. Playing God

The argument of "playing God" in medicine has been thoroughly scrutinized by Ruth Chadwick. According to her analysis, the objection that an action is wrong because it is an instance of playing God has two different meanings in two different kinds of setting. In the context of sensitive medical decision making, the point of the objection is that human beings are in no position to decide legitimately about each other's fates on the basis of quality-of-life judgments. In the context of new medical technologies, again, the crux of the argument is that actions describable as playing God can lead to disastrous and unpredictable consequences. These two aspects are both present in certain forms of genetic engineering, such as germ-line gene therapy, and it is therefore useful to take a closer look at Chadwick's account.

1. Sensitive Medical Decision Making

With regard to the decision aspect of the playing-God objection, Chadwick distinguishes three major lines of argument, two of which she finds untenable.

First, the wrongness of playing God can be based on the idea that it is God's prerogative to give life and to take it away. Active euthanasia, for instance, has been attacked by referring to this notion. But the problem here is that no reasonable morality condemns doctors and nurses who do their best to save and prolong lives, although this work can, according to the interpretation, also be described as playing God.

Second, the point of the objection may be that in certain matters the natural course of events should be preferred to human interference. An example of such matters is the reallocation of health through medical decisions. To kill one patient in order to save two others would be the best thing to do in crude utilitarian terms, but it would also be a hideous instance of playing God. In situations like this, so the argument goes, doctors can act morally only by letting nature take its course. The obvious difficulty in this second interpretation is that whatever decisions doctors make, they cannot help playing God in the defined sense. Refusals to alter the "natural" course of events affect the patients and their lives as much as any positive action.

Third, the formulation that Chadwick finds plausible and morally relevant is founded on the equality and limited knowledge of human beings. In matters concerning life and death we may justifiably feel that no one else is qualified to judge whether our lives are worth living. This conviction stems from two factors. On the one hand, it can be argued that every human life has equal value, and that no person or group has the right to make decisions concerning the lives of others on assumptions of inequality. It is not, for instance, justifiable to allocate scarce life-saving medical treatments on the basis of quality-of-life measurements. On the other hand, even assuming that some human lives are more valuable than others, the judgments concerning them may require superhuman capacities. The traditional theological assumption is that while human beings are imperfect and their knowledge limited, God is omniscient. This implies that even if God, as an omniscient being, could pass valid judgments concerning human lives, the comparisons made by human beings would still be mere arrogant instances of playing God.

As Chadwick notes, the playing-God objection may in this third form have some moral relevance as a reminder of the limits of our knowledge. It may also serve as a warning against employing irrelevant criteria, like life quality, in the inescapable human decisions concerning life and death. But the objection is not by itself sufficiently strong to refute any actual practices, and it cannot be directly applied to genetic engineering, as the persons whose lives are compared, for instance, in decisions concerning germ-line gene therapy, do not exist at the time of the choice.

2. New Medical Technologies

With regard to the technology aspect of the playing-God objection, Chadwick argues that divine omnipotence rather than divine omniscience provides the key to this side of the issue. People who oppose activities like genetic engineering or artificial reproduction typically see these technologies as attempts to rival God's power by trying to create life or life-forms. When it comes to artificial insemination and *in vitro* fertilization, the counterargument can be made, as Chadwick in fact does, that reproductive technology only aims at rearranging materials, not at creating previously nonexistent entities. The same is not, however, quite true with regard to genetic engineering, which may, after all, create completely new life-forms. Admittedly, new life-forms have been created for centuries by animal and plant breeding. But these processes have been relatively slow, and humans have not been explicitly included in the program. The opponents of genetic engineering may wish to argue that there are certain limits beyond which human beings cannot go without unlawfully playing God.

If this idea of fixed moral limits is taken seriously, the next step is to find out where the lines have been drawn and by whom. Chadwick considers three possibilities.

First, playing God can be understood literally as a transgression of the invisible boundaries that separate immortal gods from mortal human beings. People who try to assume the role of gods are guilty of what the ancient Greeks used to call *hybris,* that is, excessive pride. In Greek mythology, overstepping the limits set by a divine will was generally punished in unusual and cruel ways. This literal interpretation of the playing-God objection is clear and intelligible, but its value as a moral guide is suspect. No critical morality can be based on the assumption that divine beings have set us limits which they continuously protect. Even if one believed in the existence of such divinities and in the sacredness of their will, it would be impossible to discover what the chosen deity would want us to do. In fact, one could well argue that the humans who pretend to be acquainted with the divine will are putting themselves in the divine role, and thereby themselves playing God.

Second, the playing-God objection in the context of medical technologies may also be meant to state that the natural environment as a whole sets certain limits to our actions. Humankind has during the last few decades acquired powers which could be used to destroy most of the biosphere. Many people seem to think that genetic engineering is one of these powers, and they fear that, for instance, the release of genetically altered organisms into the environment may have irreversible ecological consequences. Assuming that we are interested in the preservation of the biosphere, this objection against genetic engineering does indeed have some moral relevance. But the problem is that the appeal to consequences, which gives this argument its weight, also deprives it of its categorical disguise. It would, no doubt, be pragmatically unwise to destroy the only environment where we can live at present, but this does not amount to a categorical, or deontological, rejection of genetic engineering. The wrongness of the activity remains conditional upon the consequences.

Third, the limits of playing God can be set by human beings on the ground that certain actions, especially technology-related actions which have never been taken before, are liable to produce unforeseen, unpleasant, and unpredictable consequences. Despite the appeal to consequences, this approach may be genuinely categorical, since no weight is given to the nature of the feared outcome or to the probability or improbability of its occurrence. According to Chadwick, the logic of the playing-God objection here is that the unknown consequences of going beyond (present) human limits cause fear, anxiety, and uneasiness in many people. Some of these people believe that we will be faced with unimaginable disaster if new technologies are implemented. Others may have the feeling, unjustified perhaps but nonetheless painful, that divine retribution will follow the alleged human arrogance. Still others may be worried about the preservation of the current worldview, which may suffer from the breakdown of its customary limits.

None of these negative feelings amounts, by itself, to an independent refutation of new technologies, including genetic engineering. But as Chadwick points out, the appeal to unforeseen consequences may be taken as a counsel advising us to be very careful in assessing certain delicate decisions. If the pros and cons of a given new technology are otherwise equal, the scales can be tipped by the unpleasantness inflicted on people by the mere thought of the innovation.

IV. GENETIC ENGINEERING AND RISK

The concept of risk is one of the most important elements in negative utilitarian analyses of biotechnology. The term, or its linguistic equivalents, can be found in teleological and deontological arguments as well, but the role of risk in these is nullified by the absolute nature of certain perceived dangers.

A. The Risks Involved in Biotechnology

Risk can be defined as the possibility or probability of a loss, an injury, an unwanted outcome, or an undesired result. The main risks involved in genetic engineering are, as the positive utilitarian analysis in Section I partly indicates, the following.

The *release* of genetically altered organisms in the environment can increase human suffering when medical measures are concerned, decrease animal welfare in experiments or through the use of recombinant DNA techniques in breeding, and lead to ecological disasters. The *containment* of biotechnological material in laboratories and industrial plants involves two dangers: first is the possibility of an accidental release, and second is the increased probability with which uncontrolled releases can produce undesired results.

A risk that lies between these "scientifically controllable" dangers and the more indirect political hazards of biotechnology is the probability of inadequate containment and irresponsible releases, which can be prompted by the economic self-interest of research groups and industrial corporations. The difference between this type of risk and the more controlled hazard is that while in the ideal case of balanced decision making the risk lies between the act—of containment or release—and harm, in this case it lies between the agent and the act. What we primarily fear in this case is human weakness or immorality. What we primarily suspect in the case of scientific risk assessment is carelessness or negligence in the calculation of possible outcomes.

The purely social and political dangers of genetic engineering include the possibility of increased economic inequality and the possibility of large-scale eugenic programs and totalitarian control over human lives. The risk in these cases lies clearly between the agents and their actions. If multinational corporations choose to supersede the national products of Third World countries by their own biotechnological substances, millions of workers will in a few years' time be unemployed. And if governments decide to develop racial programs and surveillance systems based upon the achievements of genetic engineering, the undesired outcome is certain, not possible or probable.

In debates concerning the risks of biotechnology the social and political dangers are not discussed as often as the hazards of responsible and irresponsible containment and releases. A partial reason for this can be that economic inequality and totalitarian measures are not seen by all as unwanted, undesired, or evil. Another partial explanation could be that the probability of these outcomes is small, especially in the assessment of particular biotechnological innovations or products. It is difficult to see a connection between, say, a technological process designed to produce inexpensive pharmaceuticals on the one hand, and the emergence of an unjust, totalitarian political order on the other.

B. The Morality of Risk Taking

If risk can be defined as the probability of harm, then how should we define the concept of "acceptable risk," on which analyses of the morality of risk taking often center? Is a risk acceptable if the *probability* of harm is on a reasonable level, or should we require that the expected harm is also tolerable? The quick answer to this question is that the acceptability of a risk is the product of the acceptability of the expected harm and the acceptability of its probability. But acceptability to whom, and when, and on what criteria?

Industrial corporations have a tendency to treat risks as probable costs. This is not always commendable, because some of the harms inflicted by the production and marketing of goods cannot be easily compensated to those whom the harm befalls. When, for instance, the directors and engineers of an American automobile company noticed that they had produced a car which exploded in a rear crash if the speed was right and the left rear blinker was on, they went on to market the model on the ground that the overall economic loss incurred by the expected lawsuits would be lower than the price of repairing the cars. This decision cost many people their lives and caused others inordinate suffering, and although the statistics were correct, the company's policy was clearly immoral. At the very least, the buyers should have been given the chance to decide for themselves whether or not they wanted to take the risk, perhaps by purchasing the car at a lower price. Death and suffering caused by attempts to make an economic profit are not commensurable with the work and capital invested in the enterprise.

How, then, should we define the acceptability of the risks of genetic engineering? The assessment should probably in each case be left to those who can be harmed by the decision in question. Economic risks are acceptable, if they are condoned by the biotechnological corporations and governments who take them. The risks imposed on laboratory personnel by the containment of dangerous materials ought to be evaluated by the laboratory personnel themselves. All other risks involved in genetic engineering are more or less universal, and should therefore be assessed—and eventually accepted or rejected—as democratically as possible.

Scientists, industrialists, and autocratic political decision makers can argue against democratic risk assessment by claiming that their expertise enables them to predict with greater accuracy the consequences of policies. If the choices are left to democratic processes, they say, many good outcomes which would have been perfectly safe fail to come into existence, while many undesired results are brought about by the prevailing lack of knowledge.

What this objection overlooks is that the acceptability of a risk for a given group is not determined exclusively by the facts of the matter, but also by the way the members of the group perceive the facts, and by the way they evaluate them. People cannot fully commit themselves to decisions which are based on epistemic and moral values that they do not share. Thus if anything goes wrong with the predictions of the experts, people feel and are entitled to resent the consequences of the authoritarian choices. The risks taken by experts on behalf of others are therefore unacceptable. But if risk taking is based upon the considered choices of those who can be harmed by the consequences themselves,

the situation is different. Even if the undesired outcome is realized, the risk is acceptable, because it is embedded in their own system of ethical and epistemic values.

V. CONCLUSIONS

In sum, the actual consequences of genetic engineering are difficult to assess, and it would therefore be unwise to make decisions concerning the use of recombinant DNA techniques on purely positive utilitarian grounds. The same conclusion applies to teleological objections against genetic engineering, and to most deontological critiques which are based on duties, rights, or absolute limits. The playing-God objection has two conditional readings that can be useful as reminders of our lack of knowledge and our need to respect people's feelings. Probably the best way to evaluate the morality of genetic engineering is to analyze the risks involved in its implementation, but it should be kept in mind that mere scientific risk assessment is not sufficient. When the risk of harm is more or less universal, the acceptability of the risk must be decided as democratically as possible.

Bibliography

Chadwick, R. (1989). Playing God. *Cogito* **3,** 186–193.
Chadwick, R., Levitt, M., Häyry, H., Häyry, M., and Whitelegg, M. (Eds.) (1996). "Cultural and Social Objections to Biotechnology: Analysis of the Arguments, with Special Reference to the Views of Young People." Centre for Professional Ethics, Preston, UK.
Dyson, A., and Harris, J. (Eds.) (1994). "Biotechnology and Ethics." Routledge, London/New York.
Harris, J. (1992). "Wonderwoman and Superman: The Ethics of Human Biotechnology." Oxford Univ. Press, Oxford/New York.

Geneticization

ROGEER HOEDEMAEKERS

University of Nijmegen, The Netherlands

GLOSSARY

genetic determinism A tendency to view genes as causes of human health, disease, and behavior rather than as contributive influences.

genetic essentialism Genes are seen as containing the very essence of human existence, defining the fundamental characteristics of human existence and human identity.

heuristic tool An explanatory concept, idea, or hypothesis.

GENETICIZATION was introduced as a term by Abby Lippman in 1991, and she used it as a kind of umbrella concept to denote a number of moral concerns associated with the ever-growing influence of genetics in medicine and society. No precise definition was offered at the time and deliberately so. Lippman wanted this concept to be broad, loose, and fluid so that it could encompass new and unanticipated forms of interaction between genetics, medicine, and society. However, in order to be used as an explanatory tool for analysis and reflection on various processes associated with the growth of genetic knowledge and the application of genetic technology, a definition of geneticization is needed. A broad working definition that can bring to light and explore various societal and cultural processes associated with the increasing dominance of the molecular life sciences has been proposed by van Zwieten and ten Have (1998a). They define geneticization as a "process whereby more and more phenomena in human existence are brought within the sphere of influence of genetics" (398).

I. INTRODUCTION

In technology assessment the focus is usually on the more immediate impact of a new technology or product and on appropriate use. The wider implications of the introduction of a new technology as well as its transforming powers often remain unexplored. Following an analogy with the concept of medicalization, which helps explore and analyze the increasing influence and power of medical technology at various levels, geneticization can be seen as a concept that can uncover the intricate interaction between genetic technologies, medicine, and society. It can be employed to explore the wider and usually less visible moral implications of the developments in the molecular life sciences and their potential to transform society or societal segments.

One point of criticism regarding the geneticization concept is that it lacks precision. Geneticization appears to be a fluid, broad, and open concept. An answer to this critique it that it is meant to be so, as it makes it suitable for incorporating new developments. Another point is that societal effects are not ignored but are actually debated. This is only so in a limited way and the geneticization thesis covers much wider implications than the immediate effects for individuals. A third point is that there is a lack of empirical evidence for the geneticization thesis. But the thesis is not a sociological explanation of new phenomena. It should be taken as a more philosophical concept which enables us to change our perspective in moral debate. Here a working definition of geneticization is discussed which may facilitate further research into the geneticization thesis. Finally, the relevance of the change in perspective implied in the concept of geneticization for technology assessment is illustrated.

II. MORAL ASSESSMENT AND TECHNOLOGY ASSESSMENT

In technology assessment, and also in bioethics, there is a predilection for particular sorts of issues. There is usually a focus on moral questions flowing from the use of a specific technology and the aim is to calculate, weigh, and evaluate immediate effects, and to control, regulate, or eliminate problems with effectiveness, safety, costs, and potential harm. Bioethics and technology assessment also tend to be individual-oriented and focused on concrete and anticipated effects of a product on the end user. Their ultimate aim is ethically responsible or appropriate use of an end product.

This approach is in line with two dominant approaches in bioethics in Western culture, the consequentialist approach, with its focus on effects and a balancing of advantages and disadvantages, and the so-called "principle approach," where the implications of a new technology or product are assessed by applying four well-known principles—"respect for autonomy," "promotion of justice," "promotion of well-being," and "avoidance of harm." A disadvantage of this orientation toward individual issues is that broader societal effects tend to be neglected. And although bioethics and technology assessment also aim to evaluate the impact of new technologies on society, they are usually only concerned with the more immediate societal effects on individuals and groups. More indirect and less easily visible societal effects tend to remain out of sight.

Besides questions about appropriate use of a technology's end product, there are other ways to evaluate technologies. They concern viewing the technology itself as problematic. The focus is on medium- and long-term, more indirect effects on individuals, institutions, society, and culture, and their interactions. New genetic technologies can recast the way health problems are defined and experienced, and this may have wide repercussions.

For example, predictive or presymptomatic genetic testing can turn "healthy" persons into "potentially ill" persons if a test reveals that they are carriers of a gene associated with disease. Here a whole new group of patients is created, the "not yet ill," with a different kind of behavior. In cases where untested persons with a specific gene are not conscious of its existence and are not concerned about it, a positive test will bring awareness as well as concern, and will influence behavior. These persons will be more alert to symptoms of the predicted disease and probably therefore see a doctor more frequently. They might adapt their lifestyle and diet.

New technologies can also relocate or transform problems. Not being able to conceive has always been viewed as a serious problem, but new reproductive technologies (such as *in vitro* fertilization) tend to transform this condition into an infertility problem for which a technical solution can be offered. A consequence is that these new reproductive techniques, even if not very effective, have made the inability to conceive more of a problem and more unacceptable to many Western women. And if the problem persists it is not nature which is blamed, but the physician who has failed. The problem is perceived differently because of the availability of a promising solution. New technologies can also lead to a redefinition of important concepts. The best-known example probably is the way death has been redefined under the influence of the use of respirators and resuscitation techniques.

Novel technologies can also lead to narrowing a range of possible solutions toward a preferred solution. An example is genetic screening or testing for Down syndrome. More and more people see termination of pregnancy here as the preferred solution. And the more people that perceive this as an acceptable solution, the more difficult it will be to have a baby with Down syndrome—not only because this will be frowned upon, but also because fewer facilities to care for these babies may be available.

Changes in application criteria of new technologies also occur. As soon as a specific technology has been accepted it may be used for purposes other than those it was developed for, and the criteria for application will shift. In medical practice, for example, new technologies bring about new dividing lines between what is considered medical and nonmedical in nature. For example, genetic testing and screening was first used to detect future serious disease. Later new objectives became accepted, such as genetic testing for "reassurance," "enhancement of autonomy," or "enlarging the scope of action."

There will be shifts in application: groups of people for whom the technology is "indicated" will shift. For example, DNA testing during pregnancy does not only concern serious early onset diseases anymore. Parents also wish to have a genetic test to find out if their future child has a gene associated with a risk, e.g., for breast cancer. It is not serious disease itself, but the risk of serious (and late-onset) disease that becomes a criterion for genetic testing. Thus the technology itself can become a source of moral issues.

A difficulty is that long-term effects may remain hidden for quite some time and do not manifest themselves in different societies, cultures, and institutions in the same degree. In order to be able to focus on these less easily visible long-term processes—to analyze, explain, or warn against them—moral assessment needs to have a wider and more contextual perspective. It should not only explore the manner in which important societal and cultural values influence the building of knowledge and development of technology in a specific scientific domain, but should also assess the way in which a specific domain of science and technology influences and even transforms important moral and cultural values. For this, new concepts can be helpful.

III. MEDICALIZATION AND GENETICIZATION

One concept useful for uncovering less easily discernible patterns resulting from the increasing knowledge and use of technology in medical practice has come to be known as "medicalization." This term came into use in the 1970s and brought into focus the growing influence and power of medical knowledge and technology in society. Increasingly this concept came to be used to explain the complex interaction between new medical technologies and the sociocultural domain. It helped identify encompassing processes occurring at several societal levels. The first is the conceptual level, where medical vocabulary is used to define an already existing problem. The second is the institutional level, where medical professionals legitimize a problem. The third is the level of the doctor–patient relationship, where diagnosis and treatment of a problem occur. And at a fourth, the sociocultural level, the concept of medicalization is linked with the expansion of professional power over wider spheres of life and a dependency on professional and technological intervention.

In the same way as the concept of medicalization is useful for exploring the interaction between medical technology, medical practice, and society, the concept of geneticization can be seen as a concept that can uncover the intricate interactions between genetic technologies, medicine, and society. It can be employed to explore the societal and cultural impact of the molecular life sciences. Indeed, it is possible to see geneticization

as a more specific form of medicalization, as do van Zwieten and ten Have.

By analogy, processes of geneticization can also be studied on various levels. The first is the conceptual level. Genetic terminology such as "genetic health" or "genetic disease" is used to (re)define health problems. The second is the institutional level, where specific genetic expertise is required to deal with (health) problems (e.g., genetic diagnostic and predictive testing). The third is at the doctor–patient level, where genetic models of disease explanation are increasingly used, and where a shift from treatment to preventive measures can be perceived. At the fourth, the sociocultural level, genetic information and technology leads not only to changing individual and social attitudes toward reproduction and prevention and control of disease, but also to changes in personal or family relationships, ascription of individual responsibility, and changing views on human identity.

The analogy is not perfect. For example, at the level of the doctor–patient relationship, one of the presuppositions underlying the original concept of medicalization was that the patient was rather passive and willing to accept the health professional's views. An important change since the 1970s is that the patient has become more autonomous. In present-day medical practice, individual autonomous decision making and informed consent have become important aspects of medical practice, especially in Western nations. This is clearly visible in clinical genetics with its moral requirements of nondirectiveness and respect for individual autonomy. Yet, a question remains of whether processes of geneticization at other levels (institutional, societal) are not countering the tendency toward greater patient autonomy.

The concept of medicalization brought to light specific philosophical and ethical issues not clearly visible at the time in the medical–philosophical debates. Thus it could function as a heuristic tool, an explanatory concept, because it redirected and refocused moral discussion by creating different moral perspectives. The concept of geneticization can also be seen as an explanatory concept because it helps us change our perspective and draws our attention away from individually oriented moral issues to larger socioethical issues. Thus it poses a challenge to the dominant practical bioethics approach with its emphasis on individual autonomy and responsibility, informed consent, and morally appropriate use of genetic technology. It may also challenge the dominant approach in technology assessment, which aims to assess the direct impact and consequences of technological developments for humans, society, and the environment, and to develop proposals for appropriate use of new technological products.

IV. A FLUID AND BROAD CONCEPT

Various (slightly) different, but rather encompassing, definitions are in use to describe the complex interaction between genetics, medicine, and culture. Lippman described geneticization in 1991 as

> an ongoing process by which differences between individuals are reduced to their DNA codes, with most disorders, behaviors and psychological variations defined, at least in part, as genetic in origin. It refers as well to the process by which interventions employing genetic technologies are adopted to manage problems of health. (1991, 19)

Two years later she described geneticization as

> the ongoing process by which priority is given to differences between individuals based on their DNA codes, with most disorders, behaviors and physiological variations (including such things as schizophrenia and high blood pressure as well as perfect pitch and the ability of children to sit still to watch television) structured as, at least in part, hereditary. (1993, 178)

In a more recent text she describes geneticization as a term which captures the growing tendencies to distinguish people from one another on the basis of their genes and to define most disorders, behaviors, and physiological variations as wholly or in part genetic in origin. She regards geneticization both as a way of thinking and as a way of doing, with genetic technologies being used to diagnose, treat, and categorize conditions previously identified in other ways. For Lippman it is an ideology as well as as a set of practices—a combination that has great potential to divert attention from the structural changes necessary for true health promotion by reinforcing "healthism" and its assumptions of individual responsibility for the maintenance, if not the improvement, of a "disease-free" existence.

Lippman distinguishes several levels at which this process is visible. At the language level she objects to the inherent notion of determinism that is often present in genetic language (e.g., the gene for Alzheimer's, the gene for obesity, or the gene for breast cancer). This kind of language reorients in certain ways how problems will be defined, viewed, and managed. Labeling a disease as genetic tends to direct what sort of treatments or preventive measures are taken.

At the doctor–patient level geneticization introduces or reinforces a reductionist approach to health and disease, because focus increasingly shifts to genes as the causative factors of disease, even if for many disorders the interactions of multiple genetic and (internal and external) environmental factors must be assumed. Lippman sees geneticization as a contemporary expression of "lifestyle" medicine as practiced in North America where internal genetic factors can be managed by behavioral adaptations or lifestyle changes; disease prevention, not health, is the issue; and health becomes the responsibility of the individual. This genetic approach to dealing with health problems is incomplete. Not only biological determinants but also social and physical environments, economic conditions, gender, race, personal lifestyle, and available health services are determinants of health and disease and can be used to reduce or eliminate suffering. The current emphasis on genetic causes tends to remove these other causes from sight.

The concept therefore also has a political dimension, because it redefines what are taken to be significant differences between people. It also empowers people and institutions to make these new definitions. It is a kind of colonization process. Genetic technologies and approaches are applied to and take over areas not necessarily genetic in nature.

Finally, geneticization has a critical dimension. Lippman considers it a useful instrument for reorienting social and ethical analysis of genomic projects. It is insufficient merely to analyze the potential of genetic information and develop ways to avoid or control undesirable effects. The concept invites a thorough critique of the assumptions, rationales, and practices in (medical) genetics.

Hubbard and Wald do not really define the concept, but describe aspects of a process of geneticization. According to them our "current infatuation with genetics pushes genes into the foreground" (1993, 67). It "makes us focus on what is happening inside us and draws our attention away from other factors that we should be considering" (p. 2). They contend that, "considering the variety of social and economic risks all of us face, it seems a distraction from our obvious, daily problems to focus on the risks we may harbor in our genes. It brings about an unwanted individualization of responsibility for our own health and that of our children" (p. 66), and "erodes our social cohesiveness" (Hubbard 1995, 294).

The current preoccupation with genetics is also explored by Nelkin and Lindee. They have examined popular sources such as TV, radio talkshows, comic books, and science fiction, and according to them our modern society is pervaded with genetic imagery. The gene is perceived as a powerful image—deterministic and "central to an understanding of both everyday behavior and the 'secret of life'" (1995, 2). The gene has become a sacred entity and "a way to explore fundamental questions about human life, to define the essence of human existence" (p. 40). For these authors the gene has become a symbol, independent of biological definitions, a metaphor, and a way to define personhood, identity, and relationships. It is not only used to explain health and disease, but is also linked with discussions of responsibility, guilt, and power. For this tendency to use genetic imagery to define fundamental human characteristics, these authors use the term "genetic essentialism" instead of the concept of geneticization, which refers more to a process of increasing influence and dominance of the molecular life sciences in medicine and society.

Other recent publications also perceive the gene as an image or symbol with the potential to transform our human understanding and our existence. Koechlin and Amman, Katz Rothman, and Van Dijck tend to use even broader terms, imagery, or descriptions to convey the powers of transformation of the molecular technologies. For example, Katz Rothman writes,

> Genetics isn't just a science. It's becoming more than that. It is a way of thinking, an ideology. We're coming to see life through a "prism of heritability," a "discourse of gene action," a genetics frame. Genetics is the single best explanation, the most comprehensive theory since God. Whatever the question is, genetics is the answer. (1998, 13)

This sort of terminology illustrates that there is a different way of looking at the implications of genetic technology. The concept of geneticization goes far beyond the narrower perspective of individual and (short-term) social effects (discrimination, stigmatization). But it may itself also be seen as problematic as there are various, slightly different descriptions of this process. All definitions and descriptions describe mechanisms of interaction between genetic technology and society, medicine and genetics, and genetics and culture, but the descriptions are, on the whole, widely encompassing and not very precise.

V. CRITIQUES

The geneticization thesis is not universally supported. It has been rejected on various grounds. The first is that the concept is useless, because there actually is a public debate on the social consequences of genetics. It suggests that new phenomena and situations will flow from genetic technology, but psychological and societal effects of genetic technology have in fact been known for a long time, as pointed out by Niermeijer. This critique is not justified. It is true that in many Western countries a public debate is going on about social effects, but the debates are usually concerned with the more immediate impact of genetic information and technology on individuals, such as forms of genetic discrimination by insurance companies, discrimination and stigmatization of population groups in which a specific genetic mutation is frequent, or discrimination of existing patients with the condition or of handicapped people. The orientation on wider cultural and social implications of the new genetics is new.

The second critique, made by Hedgecoe and others, is that the geneticization thesis is not founded on empirical evidence. It is believed to be only a sweeping claim. There is no adequate proof that the public perception of genetics is deterministic. Nor is there good evidence that indicates that the use of genetic explanations is on the increase. Implicit in this critique is the assumption that good theories can only be based on empirical evidence. This assumption, however, does not appreciate the methodological difference between an empirical explanation and a more philosophical method, which tries to make a point plausible by means of a few well-chosen examples. The geneticization thesis is not a sociological explanation of new phenomena. It should be taken as a concept which enables us to change our perspective in a moral debate. It refocuses and redirects moral discussion, uncovers specific aspects, and brings in new perspectives in the debate on present-day genetics.

A third point of criticism, also made by Hedgecoe, is that the concept is unclear. There are different descriptions and it is difficult to distinguish the concept of geneticization from related concepts such as genetic essentialism and genetic determinism. Also, a precise definition is missing. Descriptions of the concept are comprehensive and wide-ranging. But according to Lippman this is meant to be the case so that it can include new social and cultural phenomena and evoke a more critical stance toward genetic technology itself. This, however, raises questions about its usefulness for more systematical research into the interactions of genetics, medicine, society, and culture.

VI. A DEFINITION

An attempt to create a more systematic and useful definition of geneticization was made by van Zwieten and ten Have. They take the concept of medicalization as a model for exploring and analyzing the various manifestations of geneticization. A process of medicalization has been described at different levels [the individual level, the doctor–patient relationship, the institutional (health care organization) level, and society] and from different perspectives (changing individual attitudes, changing power relations, changing language, and changing societal attitudes). In principle geneticization can also be described at different levels and from different theoretical perspectives, and this creates at least 16 different research areas (which may in part overlap). The following working definition is proposed, which contains all important elements: a (1) process whereby (2) more and more (3) phenomena in human existence (4) are brought (5) within the sphere of influence of (6) genetics.

1. Geneticization is not static, but a *process* of societal change found at different levels, e.g., the language level: people talk more and more about healthy asymptomatic carriers of a gene associated with late onset disease as if these people actually have the condition. At the individual level women begin to experience their pregnancy differently under the influence of prenatal diagnostic possibilities, which Katz Rothman describes as the "tentative pregnancy." Women feel pregnant only after they have been informed about the results of prenatal testing.

2. More and more can be interpreted quantitatively and qualitatively. There are more and more conditions where treatment changes under the influence of genetic knowledge. But there is also an increase of phenomena which until recently were not associated with genetics but now are, such as a closer association of genes with human identity.

3. Affected phenomena in human existence can occur at different levels. This has been disclosed in a small-scale study of the beta-thalassemia genetic screening programs on Cyprus, long presented by professionals as a successful genetic population screening program. In that community genetic education by health authorities and health professionals involved the creation of a web of social pressure to increase compliance with the screening. Besides health professionals, church authorities were also involved, demanding that people planning marriage show evidence that they had been screened for beta-thalassemia. Also, it was not only the burden of the condition on the patient and parents which constituted an argument for genetic screening (and the subsequent option of prenatal diagnosis and termination of pregnancy in the case of a positive test result)—the burden of this disease on the community was also an important reason for the population screening programs, and this was a new element. At the health care level the program had the consequence that selective abortion in the case of an affected unborn lost its exceptional character and became not only more acceptable for the population, but also standard medical practice. The diverse compliance strategies have also had consequences at the individual level: the web of social pressure compromises free choice and ascribes responsibility to individuals for the decisions taken and blame in the case of noncompliance.

4. As research into this process has hardly begun, it is difficult to explain at this moment how a process of "bringing under the influence of genetics" actually takes place at different sociocultural levels. Such research will probably benefit from insights in more general processes describing the interaction of science and society.

5. The sphere of influence can have different concrete forms in the description of geneticization. The form depends on the theoretical perspective taken (e.g., a feminist perspective, or power structures), and also on the level at which the process is studied (language, individual, institutional, sociocultural). All sorts of combinations are possible, which reveals that there are a variety of possible forms of geneticization.

6. The term "genetics" here refers to the whole domain of the molecular life sciences, biotechnology, the development of products and their application, and services involving genetic technology.

VII. GENETICIZATION AND TECHNOLOGY ASSESSMENT

It may be observed that the above working definition is broad and comprehensive, but it contains all relevant elements of a geneticization process, and can therefore structure and facilitate further research exploring less visible, slow, or more general—but probably more fundamental transformative—processes at individual, social, cultural, and institutional levels.

Technology assessment is usually concerned with the immediate impact of a specific product. If it is also concerned with moral implications, a concept such as geneticization, with its focus on the wider implications of genetics technology at various levels of society, seems useful for conveying the idea that it could make sense to explore a wide range of moral implications flowing from the introduction and use of a specific technology at various levels. It can in fact add an extra dimension to moral assessment by refocusing and redirecting moral attention, and could stimulate more moral research on collective and indirect consequences. The following examples in the domain of genetics and genetic technology may illustrate this.

1. In genetics research genes are identified for further use and product development—a process in which large amounts of money, public and private, are invested. The expectation is that the genetic information will lead to great changes in the prediction, prevention, diagnosis, and treatment of all major diseases. More and more genetic explanations for disease will be given. This may well have repercussions for the funding of research areas studying other disease or health-risk factors such as poor housing conditions, working conditions, environmental pollution, or violence.

2. In health care, the increasing use of predictive DNA testing may shift the emphasis from treatment to prevention. The positive effects of this shift are usually emphasized. For apparently healthy individuals, predictive testing means that if they know about the presence of a genetic risk factor associated with a specific

condition, they can change their lifestyle or diet and achieve a better quality of life. The immediate (and negative) emotional and psychological effects of possessing genetic information associated with serious conditions have been the subject of research, but wider implications, such as a growing tendency to ascribe responsibility for disease to the individual and to stigmatize individuals who refuse genetic testing, have received far less attention. Another, more indirect, community effect may be a shift in health allocation resources, giving preference to prevention of future disease at the expense of treatment, cure, and care of existing patients or the elderly.

3. When predictive testing or screening during pregnancy spreads, there may be more indirect implications such as changing attitudes toward reproduction. More and more emphasis could be laid on prevention of future sufferers of disease (through termination of pregnancy) and not prevention and treatment of future disease itself. Institutionalization of selective abortion as a means to prevent suffering—as in Cyprus—may lead to shifting standards of what is deemed acceptable (normal) in a newborn and what is not acceptable (abnormal). And many individual decisions to undergo genetic testing can have the collective effect of generating greater social pressure on others to also use this test. This may raise questions about autonomous decision making and free choice.

4. In the past people have been discriminated against or stigmatized on the basis of certain (visible) physical or personality traits, e.g., various forms of mental disability, intelligence, or homosexuality. More recently (some) societies have become more tolerant and accepted (some of) these traits as variations of normality. Research is currently conducted to identify genes associated with these personality traits. If this leads to the development of diagnostic tests that can identify these traits there may be an increasing demand to perform prenatal diagnosis and, in the case of a positive test result, a subsequent abortion. This possibility of screening out personality traits in the same way as gene-determined or gene-influenced diseases may counter societal tendencies toward greater tolerance for "abnormal" personality traits.

5. Genetic technologies have also brought new terms in medical practice such as "genetic health" and "genetic disease." Use of these concepts can be misleading since in the majority of cases genes are but one factor in a complex organic system in a complex environment. The terms may promote a reductive and one-sided approach to disease in medical practice. A concept such as "genetic risk" can also lead to misperceptions. Isolating a genetic risk factor from the biological and biographical system can lead to misunderstanding the overall risks to health and to neglect of other health risk factors. This may raise the question about the real value of genetic information.

6. A new developing field is pharmacogenetics. DNA screening devices are planned to be used to characterize genetic profiles for response to toxicity and side effects of drugs. This is said to have great individual benefits. The expectation is that better optimized drug dosages will lead to more effective individual treatments. A morally problematic and more collective side effect of such an approach is that the use of DNA tests for this purpose will divide patients up into groups for which a specific drug is effective or not. A more immediate effect may be discrimination, and an important question is how insurance companies will deal with these groups of patients. Wider implications are that the (usually larger) group of patients where the drug has no serious side or toxic effects will become the "normal" group and the much smaller group for which the drug does have serious effects will become the "abnormal" group. For private companies it will not be commercially advantageous to develop drugs especially for the much smaller group. The moral issue that comes into play is what should be done for this group and how this should be achieved.

7. Health effects are not the only subject of moral research. A process of geneticization can also generate new goals in medical practice. In principle there are currently three important objectives of genetic testing: for medical benefit, to enhance well-being (by offering "certainty" or "reassurance"), and to enhance autonomy (genetic testing aiming to facilitate making important life choices). Testing for reassurance is not completely new, but enhancement of autonomy is. And the moral issue is whether this is an acceptable goal of medical practice. This may be so if autonomy is instrumental in the prevention of suffering of the individual or family, but it is less clear if autonomy is used in the sense of giving people more control of their lives.

These few examples may reveal that besides moral concerns about immediate effects at the individual and societal level, a shift in moral perspective, as implied in the concept of geneticization, is possible and useful. Awareness of a process of geneticization introduces other moral concerns. If moral assessment is included in technology assessment, this means that analysis of the immediate implications of a technological product is not sufficient. Analysis of the various forms of interaction of a specific technology with society, or societal segments, is also required. It would involve exploration of the issues resulting from this interaction at various levels, and examination of the transforming and conditioning powers of a specific technology at various stages of a product's life cycle—from basic research, patenting, and research and development, to product development, promotion, and actual use at different societal levels.

Bibliography

Baird, P. A. (1990). Genetics and health care: A paradigm shift. *Perspectives in Biology and Medicine* **33**, 203–212.

Conrad, P., and Schneider, J. (1980). Looking at levels of medicalization: A comment on Strong's critique of the thesis of medical imperialism. *Social Science and Medicine* **14**, 75–79.

Crawford, R. (1980). Healthism and the medicalization of everyday life. *International Journal of the Health Services* **10**, 365–388.

Galjaard, H. (1994). *All People Are Unequal* (in Dutch). Amsterdam: Balans.

Hedgecoe, A. (1998). Geneticization, medicalization and polemics. *Medicine, Health Care and Philosophy* **1**(3), 235–243.

Hoedemaekers, R. (1998). Predictive genetic screening and the concept of risk. In *The Genetic Testing of Children* (A. J. Clarke, Ed.), pp. 245–264. Oxford: Bios Scientific.

Hoedemaekers, R., and ten Have, H. (1998). Geneticization: The Cyprus paradigm. *Journal of Medicine and Philosophy* **24**(4), 274–287.

Hoedemaekers, R., and ten Have, H. (1999). Genetic health and genetic disease. In *Genes and Morality: New Essays* (V. Launis, J. Pietarinen, and J. Räikä, Eds.), pp. 121–143. Atlanta: Rodopi.

Hubbard, R. (1995). Transparent women, visible genes and new conceptions of disease. *Cambridge Quarterly of Healthcare Ethics* **4**, 291–295.

Hubbard, R., and Wald, E. (1993). *Exploding the Gene Myth: How Genetic Information Is Produced and Manipulated by Scientists, Physicians, Employers, Insurance, Companies, Educators and Law Enforcers.* Boston: Beacon Press.

Illich, I. (1975). The medicalization of life. *Journal of Medical Ethics* **1**, 73–77.

Jonsen, A. R., *et al.* (1996). The advent of the "unpatients." *Nature Medicine* **2**, 622–624.

Katz Rothman, B. (1993). *The Tentative Pregnancy. How Amniocentesis Changes the Experience of Motherhood,* 2nd ed. New York: Norton.

Katz Rothman, B. (1998). *Genetic Maps and Human Imaginations: The Limits of Science in Understanding Who We Are.* New York: Norton.

Koechlin, J., and Amman, D. (1997). *Mythos Gen.* Rieden bei Baden: Utzinger/Stemmle Verlag.

Lewontin, R. C. (1991). *Biology as Ideology: The Doctrine of DNA.* Concord, Ontario: House of Anansi Press.

Lippman, A. (1991). Prenatal genetic testing and screening: Constructing needs and reinforcing inequities. *American Journal of Law and Medicine* **17**, 15–50.

Lippman, A. (1992). Led (astray) by genetic maps: The cartography of the human genome and health care. *Social Science and Medicine* **35**, 1469–1476.

Lippman, A. (1993). Prenatal genetic testing and geneticization: Mother matters for all. *Fetal Diagnosis and Therapy* **8** (suppl.), 175–188.

Lippman, A. (1998). The politics of health: Geneticization vs health promotion? In *The Politics of Women's Health. Exploring Agency and Autonomy* (S. Sherwin, Ed.), Chap. 4. Philadelphia: Temple Univ. Press.

Nelkin, D., and Lindee, S. (1995). *The DNA Mystique. The Gene as a Cultural Icon.* New York: Freeman.

Nelkin, D., and Tancredy, L. (1994). *Dangerous Diagnostics. The Social Power of Biological Information,* 2nd ed. with a preface. New York: Basic Books.

Niermeijer, M. F. (1998). Geneticization. Misleading ignorance by incorrect information (in Dutch). *Medisch Contact* **53**, 641–642.

Sykes, R. (1997). *The new genetics: A universal panacea or Pandora's box?* Royal College of Physicians Lumleian Lecture, 15 May.

ten Have, H. (1995a). Letters to Dr Frankenstein? Ethics and the new reproductive technologies (editorial). *Social Science and Medicine* **40**, 141–146.

ten Have, H. (1995b). Medical technology assessment and ethics: Ambivalent relations. *Hastings Center Report* Sept–Oct., 13–19.

van Dijck, J. (1998). *Imagination. Popular Images of Genetics.* London: Macmillan Press.

van Zwieten, M., and ten Have, H. (1998a). Geneticization, a new concept (in Dutch). *Medisch Contact* **53**, 398–400.

van Zwieten, M., and ten Have, H. (1998b). Geneticization. Postscript (in Dutch). *Medisch Contact* **53**, 642.

Zola, I. K. (1975). In the name of health and illness: On some sociopolitical consequences of medical influence. *Social Science and Medicine* **9**, 83–87.

Genetic Research

R. O. MASON* and G. E. TOMLINSON†

Southern Methodist University and †University of Texas Southwestern Medical Center

GLOSSARY

alleles The small sequence differences in DNA that form the genetic material at a particular locus on a specific chromosome and that may be inherited along with different versions of the same gene.

autosomes Any chromosome other than a sex chromosome. In humans there are 22 pairs of autosomes.

base pair A bond between two bases or nucleotides. Adenine bonds only with thymine (A-T or T-A) and guanine only with cytosine (G-C or C-G).

bases (nucleotides) Elementary units that occur in DNA molecules and determine their specific form. The four kinds of bases are adenine (A), cytosine (C), guanine (G), and thymine (T).

Bayesian analysis A form of probability calculus in which the probability of a hypotheses h being true given the evidence e is that of e, given p, multiplied by the independent probability of h, and divided by the independent probability of e: $\mathrm{Prob}(h\,|\,e) = \mathrm{Prob}(e\,|\,h) \times [\mathrm{Prob}(h)/\mathrm{Prob}(e)]$. In general the increase in probability which a hypothesis gains when its consequences are confirmed by means of a research project is proportional to the improbability that the research study would yield those consequences. Due to Reverend Thomas Bayes (1702–1761).

carrier An individual who has both the normal and an altered or mutilated form of a given gene, which generally causes that person to be susceptible or predisposed to a particular phenotype, trait, or disease.

cause/effect In its strictest usage, a relationship between two things, such as a genotype to a phenotype, in which the first is both necessary and sufficient for the occurrence of the second.

chromosome The threadlike linear strands of DNA and associated proteins in the nucleus of animal and plant cells that carry the genes and function to transmit the hereditary information. Each human being has 23 pairs of chromosomes, which are found in every cell of his or her body. One chromosome of each pair comes from each parent.

DNA Deoxyribonucleic acid is the primary genetic material of cells. It is composed of four molecules called bases or nucleotides which, like letters of the alphabet, spell out an organism's genetic information.

DNA sequence The specific nucleotide order of DNA.

dominant A dominant phenotype requires only a single copy of an altered gene from either parent to manifest its presence.

gene A hereditary unit that occupies a specific location on a chromosome, contains the instructions for a given protein, and determines a particular phenotype in an organism. Each gene has a unique DNA sequence and a specific location, called its locus, on a specific chromosome. Genes exist in a number of different forms and can undergo mutation.

genome The set of genes characteristic of each species revealed by the base set of chromosomes which is species specific. In humans, the genome consists of 23 pairs of chromosomes.

genotype The genetic constitution of an organism or a group of organisms. More precisely, a combination of alleles present at a locus, or a number of loci.

locus A place on a chromosome or on a chromosome pair occupied by a gene or by two alleles of the same gene.

mutation Local change in the genetic information carried by the DNA or a change in DNA sequence. This change may be responsible for causing a disease.

pedigree A diagram of a family history indicating family members and their relationship to a given subject; in practice a family tree.

penetrance The frequency, usually expressed as a percentage, with which a particular geneotype produces a specific phenotype in a group of organisms.

phenotype The observable physical or biochemical characteristics of an organism as determined by both genetic makeup and environmental influences. Examples include eye color, skin pigmentation, body build, sexual orientation, and observable diseases (for example, breast cancer).

polymorphism The occurrence of different forms, stages, or types in individual organisms or in organisms of the same species, independent of sexual variations. Also called normal variants.

producer/product A relationship between two things, such as a genotype to a phenotype, in which the first is necessary *but* not sufficient for the occurrence of the second.

protein A molecule composed of one or more chains of amino acids in a specific order that is determined by the sequence of nucleotides in the gene coding for that protein. Proteins are required for the structure, function, and regulation of the body's cells, tissues, and organs.

recessive A recessive phenotype, trait, or disorder requires two copies of an altered gene, one passed on from each parent, in order for a phenotype to manifest itself.

somatic cell A cell that is not a sex cell.

GENETIC RESEARCH is undertaken to increase society's understanding of the genetic composition and hereditary nature of organisms, especially human beings. It is a science of both discovery and manipulation (engineering), and is pursued with the long-term aim of securing improvement in the human condition.

A gene is a heredity unit and a unit of information. It is composed of a sequence of DNA (nucleic acid found in the nuclei of cells) and contains instructions for producing a given protein which, in turn, contributes to a particular characteristic, trait, or phenotype of an organism. Genes replicate and transfer from cell to cell via chromosomes. Each gene has a unique DNA sequence and a specific location on specific chromosomes (its locus). The genetic constitution of an organism or a group of organisms is its genotype. The traits that organisms exhibit are called phenotypes and include characteristics such as eye color, skin pigmentation, body build, and observable diseases, for example, cystic fibrosis, sickle cell anemia, and, in some instances, cancer. More questionably of purely causal genetic origin, but of considerable social concern, are phenotypes involving qualities of temperament and behavior that may contribute to, on the negative side, feeblemindedness, epilepsy, insanity, alcoholism, prostitution, criminality, and poverty, or, on the positive side, high IQs or superior athletic performance.

People (other than geneticists) do not care much about genes per se. They care about phenotypes. Phenotypes are both the subject and the object of many human values. Their presence or absence is consequently a contributor to many ethical issues. Phenotypes are determined by an organism's genotype in conjunction with its environment, although the proportional contribution varies in different situations. For example, current research indicates that cystic fibrosis is almost entirely genetic in origin and that 40% of adult-onset diabetes cases are perhaps due to genetic predisposition, while AIDS has only a minor underlying genetic component (although recent research indicates that as many as 1 in 100 people may have a mutation of the CCR5 gene that prevents the AIDS virus from docking on it).

Nevertheless, genetic research is socially important since nearly all human diseases, except physical injury, are affected by inherited changes in the structure or function of a person's DNA. Research has also shown that genes are not necessarily rigid pieces of information with predetermined outcomes. Changes in their biological, physical, and chemical environment—such as nutrition, pollutants, stress, education, and parenting—can turn genes on or off, or can change their phenotypic expression during an organism's development process, and this happens at various rates. (The frequency within a population that a genotype is manifested as a certain phenotype is referred to as its penetrance.) This has moral consequences: the agents who create an organism's biological, physical, or chemical environment may be as responsible for its phenotypes as are its genes.

Many ethical issues also stem from scientific findings about the relationship between a given genotype and a phenotype. If a given genotype is necessary and sufficient to bring about a phenotype—as appears to be true in just a very few instances such as cystic fibrosis, Tay-Sachs, disease and the retinoblastoma gene that leads to cancer—then the relationship is purely cause and effect (cause/effect). The moral concerns focus on two issues: how best to treat the genetic cause or how to help the subject live in serenity with the effects. On the other hand, if a genotype is necessary but *not* sufficient to bring about a phenotype—that is, if elements of the organism's environment are coproducers of the phenotype and are actively operative as well—then the relationship is one of producer and product (producer/product). This is the most common relationship. For example, most of the cancer-causing agents, such as those resulting in breast and lung cancer, are of this type. Under conditions of producer/product all of the moral issues associated with social, cultural, and economic environments—such as the availability of medical treatments, of education, of employment, of insurance, or of certain restraints on working conditions—play a crucial role because they also contribute to the production of the phenotype under investigation.

From the standpoint of social values, the fact that the environment plays a role in the production of phenotypes means, among other things, that the grand debate pitting beliefs in genetic determinism—the notion that genes have the power to determine social and personality phenotypes—against those of reactive environmental dogmatism—the environment is the overarching determinant—is misplaced. The ethical policies necessary to deal with genotype/phenotype as producer/product relationships must be based instead on a deeper and more complex understanding of the underlying science of genetics and its nuances.

Genetic information is inherently personal. Each individual's unique DNA sequence reveals considerable intimate information about that individual and about his or her family, including future biological makeups. Thus, genetic research produces information that may be used to harm as well as help people. Others can use it to discriminate against the individual. In the extreme, by means of gene therapy, it can be used to manipulate an individual's genetic composition.

Genetic research, in summary, is the application of the scientific method to the study of four related areas: (1) genes, genetic material, and their mutations in all of the many different forms they may take, that is, genotypes; (2) the phenotypes that are associated with or predisposed by specific genotypes; (3) the effects of different environments on a given genotype which ultimately produces phenotypes; and (4) kinships, the hereditary social groupings or families in which both genotypes and phenotypes are found.

I. A BRIEF HISTORY OF RESEARCH IN GENETICS

Charles Darwin's discovery of evolution laid a conceptual foundation. The modern science of genetics followed in about 1900, with rediscovery of Gregor Mendel's work (circa 1866) on inherited traits. In general, the Mendelian system holds that each inherited phenotype is determined by the combination of two genes, one from each of the parental reproductive cells. A gene may play one of two functional roles: dominant, for which only a single copy is needed from either parent to manifest its presence, and recessive, for which two copies, one from each parent, are required. A recessive gene stems from an allele that does not produce a characteristic phenotype or trait when a dominant allele is also present; that is, the recessive is dominated. Thus, a dominant gene masks the presence of a recessive gene which, in turn, lies "hidden" in the organism's DNA but still may be passed on to subsequent generations. The hidden gene may be inherited under conditions of dominant susceptibility in which a single copy of an altered gene that is inherited from either parent increases the subject's susceptibility to being affected by a given disease although other events are also required. The breast cancer genes BRCA1 and BRCA2 cause a dominantly inherited susceptibility. Many of the ethical issues created by genetics research are founded on distinctions made between dominant and recessive genes, who their carriers are, and what the underlying conditions of inheritance might be.

Some very emotional ethical issues depend also on the type of cell affected: germ-line or somatic. Germline cells, also called "gametes," include ovum and sperm cells (or one of their developmental precursors). More specifically, they contain the appropriate chromosomes to make them capable of fusing with similar cells of the opposite sex to produce a fertilized egg. Any genetic alteration that is present in the germ line manifests itself in every cell derived from the fertilized egg and thus is present in all body cells in the next and potentially subsequent generations. In this way germline cells directly effect inheritance. Since future generations—and potentially entire subsequent races of people—may be affected by research on germ-line cells, the ethical issues raised by this line of research are considerable and virtually unprecedented.

On the other hand, there are genetic events which occur at the cellular level in a formed body organ that do not directly affect inheritance. These cells are called somatic, or "body," cells. The ethical issues generated by research on somatic cells, while significant, are similar to those encountered in other areas of medical ethics and bioethics since in a given instance only one individual is primarily affected.

Genetic research into the presence of dominant or recessive genes in either somatic or germ-line cells brings together two streams of scientific inquiry. One is field-based research undertaken without the collection of biological material and in the absence of laboratory analysis. Field methods include research into inheritance and the construction and study of pedigrees consisting of databases and charts of information pertaining to an individual's ancestors. These data sources are used to determine patterns of Mendelian inheritance of identifiable phenotypes, especially of familial diseases.

The other line of inquiry is laboratory based. It includes analysis of several different types of genetic material and different levels, ranging from elemental bases to full genomes. One type of research focuses on bases or base pairs which reside in genes, and which may exhibit sequence differences. These differences are called alleles and are located, at the next level of complexity, in chromosomes, chromosomes bands, or fragments that are, in turn, contained in cells. When a sequence difference has an impact on the phenotype (usually taking the form of a disease) then it creates a mutation. Other sequence differences may not have an affect on a phenotype. In this case they are considered a normal variant or polymorphism. The genes contained in an organism when considered in their totality are called its genome. Information resulting from both field and laboratory studies are combined with information about an organism's environment. The outcome is used to help geneticists make inferences about relationships between genotypes and phenotypes.

For the most part, some form of Bayesian analysis is ultimately used to bring the results of these two lines of inquiry together. Evidence adduced from research studies is used to establish one of two probabilities or likelihoods. One is the probability that a subject will possess a specific genotype given that he or she is observed to express a specific phenotype—$\text{Prob}(G \mid P)$. For example, current research indicates that of the group of women in the general population who are diagnosed with breast cancer before the age of 35 years, the likelihood is approximately 0.10, or 10%, that they have the germ-line BRCA1 mutation on chromosome 17. The second probability is of subjects discovered to have a given genotype manifesting a specific phenotype—$\text{Prob}(P \mid G)$. For example, women who test positive for the BRCA1 mutation have a lifetime likelihood of contracting breast cancer that may be as high as 0.90, or 90%. These Bayesian likelihoods are frequently arrayed by age in morbidity or mortality tables. New research results, including those on the effects on different environmental conditions and different treatments or therapies, are used to "condition"—increase or de-

crease—these probabilities based on the improbability of obtaining those results. The widespread use of likelihood analysis emphasizes the fact that genetics research is rarely deterministic. An element of uncertainty or risk is almost always present. Consequently the moral judgments made about genetics research must be based on an estimate of risk as well as on notions of good or bad, right or wrong, or just or unjust outcomes.

Three major scientific discoveries have shaped modern genetics research. The first came in 1953 when James Watson and Francis Crick discovered the fundamental structure of DNA—that its strands, connected by hydrogen bonds between pairs of bases, are coiled in a double helix. This was followed during the 1970s with the discovery that strands of DNA can be spliced, recombined, and cloned by subjecting them to restriction enzymes. These enzymes recognize particular sequences in a piece of DNA and cut it at locations that are within or near those sequences. If the fragments are of appropriate size and character, they can be combined with the DNA of other organisms and the new host can be prompted into making additional copies of "recombinant" DNA. Once a gene is cloned it can be subjected to considerable further research and used to develop diagnostic, predictive, therapeutic, identification, and other techniques. The third innovation occurred during the mid-1980s when Kary Mullis invented a simple but revolutionary technique—polymerase chain reaction, or PCR—for reproducing billions of copies of a single piece of DNA. A fragment of DNA may be inserted into a bacterial or yeast DNA. As the gene replicates and divides it is copied in bacterium growth in the culture. In one of many applications, PCR has enabled genome researchers to merge short stretches of DNA to make genetic linkage maps and a variety of physical maps. PCR together with rapidly increasing knowledge of human DNA has opened up new opportunities for direct diagnosis at the DNA level, including genetic tests such as that for cystic fibrosis.

This kind of applied research relies on two general strategies. One is called "positional cloning" and begins with a phenotype (e.g., a disease), finds associations on a map of the genome, from these associations deduces the composition of the causal genes, and infers their function from the result. This approach has been used to identify the culprit genes in cystic fibrosis, Huntington's disease, and breast cancer (i.e., BRCA1 and BRCA2). The second is called "functional cloning." It also begins with a phenotype, but then goes on to determine the chemical functions that give rise to it, deduce the originating gene, and from that infer the gene's location on the genome map. Functional cloning has been used effectively in discovering the genetic basis of sickle cell anemia and thalassemia.

The result of genetic research to date may be summarized in terms of five overarching and significant scientific findings: One, genes are inherited. Consequently, knowledge about family pedigrees makes a difference. Two, identifiable mechanisms exist by which genes figure (either as cause/effect or as producer/product) in the determination of all life processes. They do this by directing the synthesis of cell proteins. Three, processes exist by which genetic mutations, alterations in either gene of chromosome structure, may contribute to the production of specific, identifiable phenotypes such as diseases. Four, by means of a technique know as cloning—using, for example, PCR technology—multiple copies of a fragment of DNA can be made. This makes it possible to reproduce, manipulate, and study large quantities of a particular genetic material. Five, in some cases genetic material can be inserted into the body of an organism, including human beings, and thereby used to change that organism's inherited genetic structure, leading also to a change in its phenotypic manifestations. These five findings result in moral and ethical issues that lie at the very heart of all those concerning human life, the human condition, future generations, and the nature of society.

Of particular contemporary interest is the human genome, a collection of about 3 billion base pairs. The entire genome contains about 100,000 genes. The average gene consists of from about 2000 to 200,000 base pairs. Only one base pair, however, needs to be altered, deleted, inserted, or substituted by another to cause a predisposition to a disease or another phenotypic trait. The human genome is being explored by the Human Genome Project, an international consortium organized to map, sequence, and decipher it in its entirety. It began in October of 1990 and is expected to complete its first phase, consisting of producing a genetical map, physical location map, and DNA sequence, in the first years of the 21st century at a total cost of about $3 billion. On May 8, 2000, the international consortium announced that it had completed a "rough draft" of the human genome (about 85%) and that chromosome 21 had been entirely sequenced. J. Craig Venter of Celera, however, promised that his private company would reach a 99.99% accurate sequence by the end of the year and ahead of the consortium. When the mapping is completed, scientists will have a full characterization of the human genetic complement that is analogous in its informative content and use to the familiar anatomical charts seen in textbooks or hanging on physicians' walls. Just as anatomical charts represent the structural components of the human body and characterize the ways in which human physiology carries out its bodily functions, the human genome map will have several similar uses: to diagnose genetic disease, to identify individuals who

are at risk of contracting a genetic disease, and to develop disease prevention therapies.

This line of genetic research is already helping society learn a great deal about the genetic makeup of individuals, especially their propensities to disease or to abnormalities of some kind. Indeed, it is believed that virtually all diseases have a contributory genetic component as well as an environmental component (although the proportions vary for different diseases). That is, diseases result, in part, from misspellings of base pairs. This research is also revealing a startling amount about its possible applications to various types of medical and health care decisions: diagnosis and prevention, drug therapy, and gene therapy or genetic engineering. Among the board questions of applied ethics genetics research raises are, will the genetic information be used for good purposes or bad? What kinds of manipulations, that is, gene therapy, if any, should be permitted? What are the implications for humankind's understanding of ourselves as human beings and of the human condition in general? In order to examine these questions it is instructive to see how they arise during the processes of conducting the research itself.

II. AN OVERVIEW OF THE GENETICS RESEARCH PROCESS

The conduct of genetics research proceeds in several closely related, sometimes overlapping or iterative, processes. In summary, they are as follows: (1) defining the research problem, determining its objectives, and allocating resources to it; (2) creating a model or framework within which the research problem can be defined; (3) developing from the model the hypotheses and explanations to be tested or explored; (4) specifying the data that will be required, the number of observations, and their form and precision; (5) identifying the subjects or other sources from whom the data will be collected; (6) specifying the protocols under which the data will be collected; (7) collecting the data according to the protocols; (8) transmitting the data to a central point; (9) analyzing the data; (10) producing the research results; (11) storing the original data and the results and transmitting them when needed; and (12) determining the use to which the results will be put. Crucial ethical issues can arise during all of these processes. Processes 5, 6, 10, 11, and 12, however, are replete with them.

There are several types of parties who participate in, or are implicated by, genetics research. Several important ethical relationships are formed between these participants, including scientific investigators to funders; investigators qua teachers to their students; investigators to their staff, investigators qua scientists to their subjects; investigators qua physicians to their subjects (patients); investigators and their peers (individually and as members of professional societies); and scientific investigators and society. Genetic research places considerable stress on each of these relationships and, in some cases, has caused them to be reformulated. Of special importance are the investigator qua physician and/or qua scientist relationships. At the outset of their research some scientists who are searching for genes assume that they have a trivial, if any (distant at most), relationship with their research subject. Canons of objectivity have reinforced this belief. As it becomes clearer, however, that a certain genetic sequence is associated with a particular disease phenotype among their subjects, it is morally more difficult for researchers and physicians to maintain this assumption of independence and separation. Ethical issues such as validation, consent, and disclosure come to the fore, putting the scientist's role values in conflict with those, say, of physician–patient or physician–family relationships.

Some of the fundamental ethical values that are relevant for guiding the conduct of genetics research and forming ethical relationships among its participants include: truthfulness, loyalty, trustworthiness and fidelity, stewardship, privacy, autonomy (self-determination), respect for others, validation, nonmaleficence (do no harm), beneficence, justice, and virtue.

- Truthfulness—speaking honestly and in accord with the reality of a situation—is essential for the scientists themselves and for the other participants so as to remain open to the ideas and conclusions presented by investigators and for approaching the scientific ideal of objectivity by overcoming bias and self-interest. Since the public bases much of its behavior on scientific results, the falsification of results can lead to great harm.
- Loyalty to one's profession and its code of conduct of conduct is essential for building trust.
- Trust and fidelity, in intentions, reliability, discharging duties and obligations, and honoring of colleagues, is essential for people to have confidence in research results.
- Stewardship—defined as the researcher's responsibility for safeguarding subjects, resources, institutions, and research data and results—is essential for the enterprise of science to succeed.
- Privacy—"the right to be let alone" and to keep one's personal information inaccessible to others—should be protected.
- Autonomy or self-determination of subjects, patients, and other participants—defined as the ability to be free and self-governing in decision making—leads to dignity, free inquiry, and creativity among participants and is essential for ensuring respect for the dignity of research participants and their uncoerced consent to participate. It is a libertarian principal that gives individuals a right to determine what will be done to their bodies and their genetic information.
- Respect for others—loving one's neighbors as one loves one's self—is essential for honoring others' autonomy and for giving them their due.
- Validation—the scientist's dual obligation to make sure that the findings are verified and reproducible prior to publishing and prior to being given to subjects or those who are responsible for their care—is essential for according others respect and for preventing the doing of harm.
- Nonmaleficence—avoiding doing harm and placing proper constraints on the power of knowledge—is essential to the respect for humanity and the dignity of all research participants.
- Beneficence—providing benefits, balancing benefits against risks, and using science to secure improvement in individuals and society—is an important goal of genetics research in particular and of research in general. It is also a major value of all medical ethics and bioethics.
- Justice—to see that benefits, costs, and risks are distributed among all participants fairly—is essential for the enterprise of science to fulfill its role in society, especially a democratic society. Human beings have little or no control (at this time) over their initial genetic lottery. Genes resulting in desirable and undesirable phenotypes are not distributed equitably, but are dispersed randomly by nature and are, therefore, undeserved. Philosopher John Rawls' "difference principle" implies that social justice calls for justifying the receipt of desirable inqualities by benefiting the least advantaged members of society—those whose lot it was to receive the undesirable genes. Under this principle, for example, wealthy people would be precluded from receiving genetic enhancements (say for attractiveness, intelligence, or athletic ability) unless it could be demonstrated that the condition of disadvantaged people would be improved.
- Virtue—aspiring to be a "good" scientist and characteristics of a moral agent such as courage, temperance, prudence, justice, honesty, etc.—is essential for ensuring that the scientist's motives and character are of the highest moral quality. For example, the most reliable protection for research subjects is provided by researchers who are informed, conscientious, compassionate, and responsible, that is, virtuous.

III. ETHICAL ISSUES ENCOUNTERED DURING EACH RESEARCH PROCESS

A. Problem Definition and Funding

A research agenda is a manifestation of human values. Major ethical questions generated during this stage in genetic research concern who sets the agenda and the form it takes.

In the early days indivdual scientists pursuing their own curiosity generally selected the research topics to be studied. As the amount of resources required to conduct both field and laboratory genetics research has increased, however, governments, private foundations, and more recently private business enterprises have played a more central role. In the United States, programs managed by the National Institutes of Health (NIH), the Department of Energy (DOE), and the Department of Defense (DOD) have become the primary sources of funds, and hence they are highly influential in setting the agenda. In one enlightened set of events, an examination of ethical issues has been a part of the United States' Human Genome Project, funded by the NIH and DOE, since its inception. Under the influence of its first director, James Watson, about 3% of its annual budget is devoted to the study of ethical, legal, and social implications under a program whose acronym is "ELSI." (ELSI currently receives about 5% of the budget.) This is one of the first instances in history of a program dedicated to studying the ethical implications and consequences of basic research prospectively and, hopefully, preemptively.

For some specific kinds of genetic diseases, private foundations, such as the American Heart Association, Muscular Dystrophy Association, and Susan G. Koman Foundation for Breast Cancer, constitute alternative sources of research funding. Each of these foundations fund research aimed at decreasing the mortality and morbidity of given diseases. Both governments and foundations, with input from scientists, however, generally set the overarching agenda and then evaluate individual scientist's project proposals according to the guidelines they establish. The guidelines are usually based on political considerations and personal preferences. Thus, during the process of program guideline development, important stakeholders' needs, such as patients with or without a given disease, could be overridden. Moreover, by responding to these programs and their guidelines in order to get support for their research, scientists often must give up research ideas in other areas which they think are more valuable.

For example, a group of concerned families whose members have a history of a disease called dysautono-mia found that the affliction was of little scientific interest to researchers. Dysautonomia is a degenerative neurological disease which occurs predominantly among Ashkenazi Jews and is currently incurable. It results from inheriting a recessive gene found in about 1 in 30 Ashkenazis. Thus, the chances both members of an Ashkenazi couple will carry the mutated allele are about 1 in 900 [$(1/30) \times (1/30)$]. Based on Mendel's formula, there is a 1 in 4 chance that each child born to the couple will inherit copies of the gene from both parents and, therefore, have the disease. This yields an incidence of about 1 in 3600 within the Ashkenazi population. People of Ashkenazi descent, however, make up a very small percentage of the U.S. population; so, what is a significant incidence of disease for them within their subgroup does not figure so prominently in the concerns for the nation as a whole. This may be one reason why very little research effort was devoted to the disorder. Nevertheless, these peoples' loved ones had been suffering from it, and, this led members of the concerned families to set up the Dysautonomia Foundation to support research into the disease. As of 1996 the markers for the gene have been identified and, if the necessary pedigree information is also available, a prenatal diagnostic test may reveal whether or not the fetus has dysautonomia. Some observers believe that these grassroots efforts make for the most effective and just overall allocation of research resources; others believe that they tend to dissipate valuable resources best used elsewhere.

The dysautonomia case is also an example of the problem of opportunity costs associated with the allocation of research funds and of finding the most effective organizational mechanisms for using them. These questions have been raised in genetics research since the early 1980s when national budget deficits placed political pressure on government allocations. The National Institutes of Health and the Department of Energy currently spends about $200 million annually on genetic research. Continual efforts to cut this budget have resulted, among other things, in an arrangement known as the cooperative research and development agreement (CRADA) between federal agencies and private enterprises. This is a joint agreement in which scientists, often university based, can participate without a legal conflict of interest. Some observes believe, however, that these arrangements divert research efforts away from basic research and redirect them toward producing more commercializable knowledge.

Genetic knowledge bestows power on its possessors. Because of their central role in all life, the results of research in genetics are a source of social and economic power raising all of the attendant concerns people have about the moral and ethical use of power, and, it is doing this at all societal levels. Nationally, concerns about "Big Brother" have reemerged. As early as 1971 Joseph Fletcher observed of genetic research, "Even though its medical aims were only to gain control over the basic 'stuff' of our human constitution it could no doubt also be turned into an instrument of power." (1971. Ethical Aspects of Genetic Controls: Designed Genetic Changes in Man, *New England Journal of Medicine,* Vol. 285, pp. 776–783). Insurance companies and employers potentially have the power to select or deselect their customers or employees based on their genetic makeup. Individuals potentially have the power to select the genetically "best" mate for themselves or for their relatives and to decide whether or not to carry a pregnancy to term.

Any research project that might serve to redistribute the social balance of power should be examined very carefully. Increased private sector involvement is bringing this concern to public attention. As the profit-making potential of genetic tests and therapies has become more evident, commercial diagnostic labs, pharmaceutical companies, and other business firms, both established and newly created, have become a major source of research funding and a major user of research results. By 1994 the pace of research in the biotechnology industry was generating about one new gene or so a day, one new company per week, and one new drug a year for potential use in therapy. In 1996 the biotechnology industry consisted of over 1300 companies, employed about 100,000 people, had sales of over $7 billion, and had reached a market capitalization of over $45 billion. Their total contribution to R&D was about $15 billion and exceeded $50 billion in 2000. This rapid pace of innovation has increased the output of research—but, it has also created puzzling conflicts of interest among researchers, universities, the scientific community, the public, and business.

In a harbinger event, the Biogen Company was set up in the early 1980s by Harvard professor Walter Gilbert, a Novel Laureate and developer of DNA sequencing technology, under the assumption that the commercialization of his research was an appropriate extension of his work as a scientist. This and many other similar events since have raised several important questions of ethics, including conflict of interest. Should the scientific agenda be shifted from seeking basic knowledge (the university's traditional role) to seeking commercializable, profitable knowledge (the private sector's role)? Should the public agenda for producing basic knowledge be augmented to include the development and production of usable genetic products which benefit the public? Since most basic research is funded by public tax dollars, the scientific establishment may have an ethical obligation to facilitate the return of the "fruits" of their re-

search to the public directly. Ultimately, this raises questions about the ethics of cooperative capitalism. Does competition among scientists working under grants from different companies jeopardize the open and free flow of scientific communication among the scientific community? What are the implications of the commercialization of scientists' research for their obligations and allegiance to their universities or other employing research organizations? Since universities may also profit from controlling the publication of research, what obligations do they have to ensure the timely dissemination of research results produced by their faculty and staff? And, given that a university's faculty works on industry-sponsored research grants, does this tend to compromise the education of their students and research fellows?

By late 1999 several private sector firms were in a race with the NIH-sponsored Human Genome Project to complete the sequencing. Celera Genomics, headed by J. Craig Venter, intended to complete the entire sequence by spring 2000. Incyte Pharmaceuticals planned to map only the most active 10% of the human genome by that time. There were discussions about the research rivals collaborating in order to complete the sequencing more quickly. Under an accord known as the Bermuda Agreement, the rivals submitted new DNA data every 24 hours to a public data bank. This was controversial. Most scientists believed that the human genome should be in the public domain; Celera, however, planned to file patent applications on any commercially valuable DNA sequences and favored a disclosure period longer than 24 hours and closer to 3 months. Celera, Incyte, and Hyseq charge substantial subscription fees for widespread access to their genetic library. Executives at these firms contend that the real value of genomic data lies in the analysis of it and that only under a for-profit business model will the information be effectively mined. Hyseq is marrying its genetic research with electronic commerce and selling and licensing genetic information–gene sequences, homology, and expression data—over the Web through a site called Genesolutions.com. By using a fee-per-item plan, Hyseq hopes to reach a larger customer base for its proprietary gene sequences and related non-published data.

Genetics research is raising many concerns about the appropriate roles of the public sector and the private sector in an information society. Those who favor commericalization of genetic research stress benefits such as providing sequence data more quickly, making biologically important work a research priority, freeing scientists to work on projects that apply newfound sequence data without restraints, creating a useful public–private partnership, and providing valuable competition to stimulate both private and public sector development. Others see significant disadvantages such as companies claiming ownership to parts of the genome by means of patents, profits flowing to the commercial companies while the costs remain in publicly funded centers, an emphasis on "quantity" of data produced rather than its "quality," a slowing of the distribution of new results from daily updates—as public centers currently do—to quarterly or periods advantageous to the company, and an erosion of international agreements on DNA patent rights. The patenting of DNA sequences raises deep questions about intellectual property rights. Which portions of human DNA have moral significance? DNA is produced originally by nature. Who owns it? Even if a scientist develops a DNA sequence in the laboratory that is other than a natural form, does it become the scientist's property and thus become patentable? Will restrictions on patents impede genetic research?

B. Creating or Adopting a Research Model

Knowledge is conditioned by historical and cultural forces. As Thomas Kuhn has pointed out for science in general, and Ludwik Fleck for medicine in particular, "paradigms" and all of the hidden, underlying assumptions in the theories or models used in research are shaped by these social forces. These forces are at work when scientists create the theories, models, or frameworks within which their research problems are to be defined. Since these theories reflect human values, they therefore have moral—and sometimes far-reaching—implications.

During the 1930s, for example, research in genetics was thriving in the former Soviet Union. Under the leadership of academician T. D. Lysenko, head of the All-Union Institute of Genetics and Selection in Odessa, the research was conducted under the Marxist assumption that human beings could be changed by altering society. Furthermore, these newly acquired phenotypes could be inherited by subsequent generations. This Stalin-inspired theory is sometimes called "reactive environmental dogmatism," or the doctrine of acquired characteristics. The theory, which implies that genes are either irrelevant or totally mutable by environmental forces, was applied in 1942 to Russian agriculture with disastrous effects. Following the theory's edicts, Lysenko instructed Siberian peasant farmers to plant winter wheat, which survives only in mild southern climates, in the harsh northern ground, sprinkling the seed among the short, stiff stalks that remained in fields after harvesting the previous season's spring wheat. (Which only grows during the warmer summer months.) The theory implied that the winter wheat would become acclima-

tized to its new environment and yields would soar. They did not, and, a reign of famine ensued.

C. Developing Hypotheses

Whatever theory or model is adopted, it is used to derive the hypotheses and explanations to be tested or explored. Canons of free and open scientific inquiry require that any hypotheses arrived at this way should be open to investigation. Again, however, social and political forces can intercede. Some hypotheses may be too "hot" or culturally incorrect to study. One such hypothesis stems from the theory of biodeterminism, which implies that human intelligence can be meaningfully abstracted and measured by a single number. This measurement can then be used to rank all people and all races on a linear, ratio scale of, as Stephen Jay Gould puts it, "intrinsic and unalterable mental worth." A recent controversial study in this tradition by Hernstein and Murray entitled *The Bell Curve* attempts to show that certain races are biologically, and hence genetically, inferior to others. The study has been attacked on moral as well as scientific grounds. Some people believe that it is simply unethical to explore hypotheses and explanations such as these. Nevertheless, when all genes have been mapped and when the gene frequencies present in each of various populations are known, people will be tempted to make judgments about "good" and "bad" genotypes as they are found in different racial groups. Is this ethically acceptable?

D. Data Specification

Unexpected ethical issues may arise during the process of specifying the kind of data and the number of observations that will be required and their levels of accuracy and precision. The results of research in genetics will be used for making medical and policy decisions that affect the lives of people. Thus, the geneticist is responsible, in the first instance, for ensuring that the form of the data is analytically appropriate for testing the hypotheses under investigation. This requirement may be difficult to satisfy since it may require access to pedigree information or classes of subjects that are unavailable. Next, a decision must be made as to the number of observations to collect. Statistical theory is used to project the minimal number required; but economics and availability always play a pivotal role. It is often the case that the cost of acquiring data increases exponentially with the number of observations, especially after some modest number is exceeded and local availability is exhausted. The judgments researchers make at this juncture may carry many future consequences. Peer review of scientific results is one safeguard against the most disastrous of these consequences; but, it is not infallible.

E. Selection of Subjects

Many human values relate to the subjects that are used for genetics research. The questions, "What research is permissible with certain subjects, especially human subjects?" and "How are subjects to be selected?" are paramount during this phase. Considerations may be divided into three parts: dead material, plants and animals, and human beings. With respect to dead material, many research studies benefit greatly from the use of surgical samples archived in hospital pathology labs. But, this practice violates some peoples' beliefs. In some cultures and among some peoples there is a strong belief that a being's spirit remains intact long after the body has ceased to function. Hence, harm can be done to the organism after it is pronounced dead, unless proper cultural precautions are taken and proper respect is accorded the remains. This concern extends especially to inheritable material such as DNA and is complicated by the fact that the DNA of persons or organisms long deceased can reveal a great deal about them and their descendants, including those currently living.

Similar concerns have also been voiced with respect to research using animals as subjects. For example, Peter Singer's *Animal Liberation,* published in 1975, argued for animal rights in research on the basis that they, too, show evidence of suffering, that they are aware of the suffering, and that they are not expendable. Recent applications of social contract theory also support this view. Since so much genetics research has been done on fruit flies, mice, nematode worms, and other beings that reproduce rapidly in captivity, these are not idle concerns for geneticists.

The selection of human subjects, however, presents the most compelling ethical issues. At this point the genetic researcher's role as a scientist comes into conflict with her role as a physician. As Leon Kass avers, "the physician must produce unswervingly the virtues of loyalty and fidelity to his patient." In contradistinction, the scientist is trying to determine the validity of a formally constructed hypothesis or answer a general scientific question. The ultimate goal of genetics research is to find (1) broad explanations about, and (2) therapeutic modalities to treat anyone who happens to have, certain conditions. Consequently, during a research project a "subject," having been converted from "someone in particular" to "anyone who happens to have this genotype or phenotype," becomes an anonymous "object"— essentially a nonperson. This change of status begins as soon as the subject agrees to participate in a research project. Two great moral traditions also come into conflict here. One is Immanuel Kant's view of human beings as bearers of respect and dignity, and therefore not to be treated as means only to a researcher's or society's ends. It argues against the "anyone" subject treatment in experiments. The other is utilitarianism in the tradition of Jeremy Bentham and J. S. Mill which seeks to find the greatest good for the greatest number and thereby allows the researcher to make defensible trade-offs among people in a pleasure (benefit) and pain (cost) calculus. Whichever tradition prevails, one minimal obligation overrides all others: there is a generally accepted moral imperative not to treat people *solely* as anonymous subjects. This moral principle has an important history.

The Nuremberg War Crimes Trials at the end of World War II revealed that prisoners of war and citizens were forced to participate in experiments that either maimed or killed many of them. At the Dachau concentration camp, for example, in 1943 the German military placed subjects in very low-pressure chambers and in tanks filled with iced water to determine the limits of their endurance and existence. In other concentration camps subjects were purposefully inflicted with different types of wounds and with malaria, typhus, yellow fever, and other diseases to learn more about the affliction and their possible treatments. The outrage at these revelations resulted in the Nuremberg Code of 1946, which established that the voluntary consent of a human subject to an experiment was absolutely essential. This was followed by the Helsinki declarations of 1964 and 1975 which set the following basic principle: Every biomedical research project involving human subjects should be preceded by careful assessment of predictable risks in comparison with foreseeable benefits to the subject or others. Concern for the subject must always prevail over the interests of science and society. Disclosure of the Tuskegee syphilis study and San Antonio contraception study among others in the United States served to reinforce the needs for these codes. In these cases, the researchers' duty of beneficence was obviously violated and the subjects were precluded from achieving any sense of autonomy. These codes require that special attention be given to the selection of children, prisoners, or other people in positions of lesser power as subjects, and have led to the development of protocols and principles of informed consent discussed in the next section. While these historical cases seem obviously unethical by modern research standards, similar, although perhaps more subtle, concerns can be raised about contemporary genetics research. For example, should genetic tests be administered to subjects to study diseases for which there is no known intervention?

F. Specifying Protocols for Data Collection

A typical research project involving human subjects and aimed at understanding a specific gene or gene marker should follow an acceptable protocol. It begins by mailing requests to participate to members of a population chosen for its relevance to the phenotypes or mutations under study. Those who ignore or refuse the request are dropped. Consent (i.e., informed consent) documents are sent to those who agree to participate and a baseline survey is administered to them. Upon receipt of the survey a genetic counseling session is held. Psychological screening at this point may result in some subjects declining to continue, being deferred, or dropped. Blood is then drawn from the subjects thus selected and its DNA is analyzed. In general, two outcomes are possible at this point: The tests reveal that the subject is either a carrier or a noncarrier. Based on the outcome, genetic, psychological, surgical, oncological, or other relevant forms of counseling are made available to both classes of subjects. If the experiment involves a therapy then subjects are assigned, in some cases randomly, to groups, some of which receive the therapy to be tested the other a placebo or perhaps an unsatisfactory treatment.

The research data generated by these kinds of projects should be collected with strict adherence to previously approved protocols, for both animal and human research. While each project presents its own individual problems, the following guidelines should provide some basic protections for the subjects: All subjects should receive genetic counseling before and after genetic testing, and they should receive psychiatric screening typically before testing and counseling by a family therapist afterward depending on the results of the tests. Subjects are encouraged to bring a nonkindred support person to the sessions. Strict confidentiality is maintained. DNA analysis is undertaken with constant attention to accuracy and validity by using a clinical diagnostic laboratory approved by an appropriate agency. At any time during the research the subject is free to withdraw (i.e., informed nonconsent). And, most importantly, the subject expressly gives his or her informed consent.

Informed consent involves more than just having subjects sign forms or sharing their information with researchers in decision-making processes that determine their participation in the research. It should be based on the concept of autonomous authorization by the subjects. Among the criteria for autonomous authorization are subjects' substantial understanding of the research and its possible harms, benefits, and risks; a lack of coercion or control by others (that is, no elements of force, fraud, deceit, duress, overreaching, or other ulterior form of constraint or coercion); and full compliance with social and legal rules of consent. In order to achieve this the researcher must ensure that each subject satisfies the following criteria: she is competent; she is disclosed and receives adequate information about the plan of the research; and the disclosure is understood clearly as to what was asserted, the nature of the procedures involved, the foreseeable consequences and possible outcomes that might follow, and the distinction between being a subject for research and being a patient for health care. So informed, the subject voluntarily agrees to participate according to the plan of the research; gives consent by making an overt decision to participate; and formally authorizes the plan of research. The America Society of Clinical Oncology, for example, recommends that prior to conducting genetic testing the following items should be discussed with each subject:

1. Information on the specific test being performed
2. Implications of a positive and negative result
3. Possibility that the test will not be informative
4. Options for risk estimation with genetic testing
5. Risk of passing a mutation to children
6. Technical accuracy of the test
7. Fees involved in testing and counseling
8. Risks of psychological distress
9. Risks of insurance or employer discrimination
10. Confidentially issues
11. Options and limitations of medical surveillance and screening following testing

Regardless of the particular protocol used, it should be evaluated by a group of peers from within the investigator's institution. Since July 1966 all parties receiving funding from NIH have been required to submit their protocols to a panel of peers known as an institutional review board (IRB). The proposed protocols and plan of research must be described to the IRB so that it can determine the risks to subjects, estimate the likely benefits to both the subjects involved and society at large, and determine whether or not the overall benefits outweigh the risks.

One especially challenging ethical issue for researchers and IRBs to resolve involves deception or intentional nondisclosure of facts to subjects. Many research design plans use techniques such as randomized trials with blind or double-blind treatments in which neither the researchers nor the subjects know which particular treatment or "control" group each subject is assigned to. This technique eliminates several major sources of bias and thereby improves the scientific validity of the results. But, since the subject is not fully informed, blind experiments may generate subject anxiety or even inflict harm on a subject, and also compromise the requirement of informed consent. In particular, the researcher qua physician's obligation to provide the very best treatment for his patient comes into conflict with the researcher qua scientist's obligation to produce valid information and make it available for the benefit of society. Only if the researcher is in a state of equipoise—that is, he believes that the severity and likelihood of harm and good are evenly balanced for either treatment—are randomization and blind trials ethically acceptable without dispute. Even if it is decided that it is ethically acceptable to begin a study, it may become ethically problematic to continue. This issue of ethical stopping rules is discussed in the next section.

The researchers and their staffs may also be at risk and they deserve as much attention as the subjects. Biological and physical safeguards should be used in all DNA research, especially if new organisms are involved. Biological barriers include bacterial hosts that will not survive in the material being tested and vectors able to grow only in specific hosts. Physical barriers include the use of gloves, goggles, hoods, clean rooms, filters, and other containment technologies. Both types of barriers should be deployed with a set of enforced organizational rules and procedures designed to protect the safety of all research workers. In general, the principles of informed consent indicate that all researchers and their staffs should be fully informed about the hazards involved before they agree to work on a project; they should be properly trained in safety and containment procedures; and they should be monitored throughout the experiment for exposure to dangerous substances and for their general health.

G. Collecting Data According to Protocols

The ethical planning of a research project must be followed by the ethical management of its execution. At a minimum this requires ensuring that the study is run and the data are collected according to the protocols. This means that all researchers and their staffs must have appropriate training and education in ethics as well as in the project's protocols. They must also develop a deep concern for the integrity of the information gained and for its potential impact on the subjects. Constant monitoring of the subjects and their well-being is essential and this raises the issue of ethical stopping rules. In some kinds of studies, such as those for cancer or AIDS, subjects may exhibit serious side effects and the researcher has to "unblind" the experiment to provide proper treatment. If blind or random assignment is stopped, the scientific value of the study may be lost or compromised and the participation of the other subjects wasted. Moreover, the scientist's qua physician responsibility to the subjects changes dynamically as the results of the experiment emerge. Once a researcher, for example, has formed a view about a new treatment or the

condition of a subject—good or bad—should the blindness and randomization be continued? The obligation of beneficence requires that as soon as a researcher believes that he can help the subject he should intercede and provide the help. Trials are also usually halted if the data show that there is only a small likelihood, as a function of statistical variation or the standard deviation—say 0.01 or 0.05—that the results are due to chance rather than a true relationship. At this point the efficacy or safety (or lack thereof) of the treatment or the statistical generalizations is established. If there are too few observations made to reach this point then the study will lack statistical significance and scientific validly. However, if more observations are made, say to improve the statistical significance, this will result in delays in making the knowledge available and securing its benefits.

H. Data Transmission

There is an important intermediate, and often overlooked, step between collecting and analyzing research data. Observations acquired from disparate points must be brought together and delivered to the central point at which they will be analyzed. Also, biological samples, such as DNA, are often shared by more than one investigator and may be examined in different research laboratories within the same or different institutions. This creates a point of vulnerability. Data security and confidentiality requirements are in danger of being breached. Consequently, values for accuracy and reliability are important; staff training is essential; and adequate resources must be allocated to the job.

I. Analyzing Data

There is a temptation for researchers to want to prove their hypothesis and to produce publishable results, perhaps even groundbreaking results. Among other things, receiving substantial project grants and academic promotions depend on it. Consequently the researcher's values for scientific integrity and truth telling are put to the test. In some noteworthy instances the researchers have succumbed to the temptation to "fudge the results." In 1981 in a highly publicized case, John Darsee of the Harvard Medical School admitted to falsifying the results in one of his research papers, presumably to continue to get grants and to keep his research scientist job. Subsequent investigations uncovered a troubling fact. Not only had Darsee been dishonest, but his coauthors and peers had failed to detect the flaws before the work was published. They, too, bore a responsibility for scientific integrity. Their failure to discharge it seriously compromised the system of guarantorship upon which integrity is based.

These concerns were magnified in a case *The New York Times* questioned might be "A Scientific Watergate?" Thereza Imanishi-Kari of Tufts University published a paper in a 1986 issue of *Cell* concerning genetic influences on the immune system which claimed to show that inserting a gene from a foreign mouse into a certain, different strain of mice caused changes in the host mouse's stock of antibodies. Noted geneticist and Nobel laureate David Baltimore was a coauthor. Margot O'Toole, a postdoctoral researcher in molecular biology at Tufts, doubted the results and openly questioned the validity of some of the analysis at the time. For this she was dismissed from her job. The case was brought to the NIH and reached the floors of the U.S. Congress. Laboratory notebooks of the experiments were subpoenaed by the Secret Service, whose forensic studies of the notebook pages, inks, and counter tapes from assays of radio-labeled reagents tended to support the conclusion that some entries had been wrongly dated and possibly fabricated. The report of a NIH committee in 1990 found Imanishi-Kari guilty of "serious scientific misconduct," stating that she had repeatedly presented false and misleading information. O'Toole was praised for her dedication to the truth, her courage, and for blowing the whistle. But, these findings were appealed, and in 1996 a panel of the Department of Health and Human Services ruled that the evidence did not support the government's scientific misconduct case. Nevertheless, it found that the paper as a whole was "rife" with errors of all sorts and that Baltimore, although never accused of fabrication, as a coauthor must share some of the responsibility.

Vigilance must be constant. Even the most highly regarded researchers can be victimized. Because a junior colleague had fabricated data, Francis Collins, who heads the Human Genome Project, decided to retract five research papers on the role of a defective gene in producing leukemia. An anonymous peer reviewer's observation that the data "suggested intentional deception" caused Collins to investigate and discover the source of the fabrication.

Collins concludes that "there is no fail-safe way to prevent this kind of occurrence if a capable, bright, motivated trainee is determined to fabricate data in a deceptive and intentional way, short of setting up a police state in your laboratory" (quoted from *The New York Times* Nov. 6, 1996). Collins' timely and full disclosure by means of a broadly distributed "Dear Colleague" letter served to reduce the negative impacts of the falsification.

Two ethical issues are put into relief by these cases: the temptation for scientific misconduct and the respon-

sibility of peers. The guarantor of the results of most scientific inquiry is the agreement of the community of scientists. It is apparent in the Imanishi-Kari case that both her coauthors and the review panel at *Cell* failed to discharge this responsibility, whereas in the Collins case they did, albeit belatedly. This concern has resulted in the NIH and Alcohol, Drug Abuse and Mental Health Administration (ADAMHA) publishing a set of principles of scientific integrity to be given to staff and students at institutions receiving grants, and encouraging the delivery of training programs to all researchers. Education in ethics for scientists and staff, whether informal or formal, is essential. Topics to be covered include responsible authorship, responsible reviewership, the recording and retention of data, conflict of interest, issues of human and animal experimentation, and professional standards and codes of conduct. The American Association for the Advancement of Science has also promoted the need for professional standards and codes of conduct.

J. Producing and Publishing the Research Results

The norms of science call for open and free inquiry and the free exchange of information and materials among scientists. This includes publishing results promptly and making them available publicly, usually via academic journals. Individual scientists, however, sometimes have incentives to either delay publication in order to cash in on the findings beforehand or to patent the results. These problems will be exacerbated in the near future. It is likely that genetics research in general, and the Human Genome Project in particular, will be the source of increasing amounts of basic information that can be used for targeting new drugs and genetic therapies. This will become the mainstay of the pharmaceutical industry. Investments and commercial application of the discoveries often require protection of the intellectual property rights involved by means of patents. This raises many questions. Who owns the data and the research results—scientists, business, universities, the public, or others? Can they be patented? By sequencing any fragments of DNA and relating the results to Human Genome maps it is possible to discover new genes and determine their functions, even if they are arbitrary fragments of genes with no known use. Should an individual or corporation be able to lay claim to any such sequence? In 1991 the NIH filed for four patents for fragments of gene sequences that had been obtained from copies of DNA using rather standard techniques. The applications were rejected by the U.S. Patent Office in part because they failed to meet the main criteria for patentability—innovation or novelty, commercial utility, and being nonobvious. But, as more uses of genetic research become evident, it can be expected that the pressure to patent will continue. The issues of publication and ownership also relate directly to some of the ethical issues that emerge during the storage and retrieval of research information.

K. Storage and Retrieval of Research Information

Several crucial issues arise with respect to the storage and retrieval of genetic materials and research results. Since genetic materials such as DNA splices or sperm can be used to alter life, the highest regard must be given to their security and safety. People will be tempted to use them for purposes other then those intended when they were collected. For example, the University of California, Irvine, was sued because allegedly some donors' sperm was substituted for others. DNA tests on the children conceived *in vitro* supported the allegation. It was contended that the University failed to install adequate security measures.

Respecting subjects' privacy is another issue. Much of the genetically relevant information collected about subjects and the results of studies conducted on them reveal considerable intimacies about the subjects and their families–information most people generally want to keep private. Safeguarding against the unwanted disclosure of private information has always been a problem, even prior to the development of molecular biology. Lay and professional people have always drawn anecdotal conclusions about a person's genotype from the phenotype he or she manifests and from the traits of relatives. The results of the new genetic tests, however, provide far more comprehensive and reliable information about a person than this, and they can be used for genetic discrimination or stigmatization. Protection against this requires norms of confidentially on the part of researchers. The scientific investigator–subject relationship is akin to the physician–patient relationship. It must be based on trust and a due respect for the autonomy of the subject not only as a moral obligation, but also because the continuance of the scientific enterprise and the willing participation of subjects require it. This means that, in general, genetic information should be kept materially secure from an unwanted physical intrusion and institutionally secure from unwarranted acquisition by means of procedures to safeguard confidentiality.

In practice this may be difficult to achieve. Much of the information is placed in computerized databases which may be subject to unauthorized access. Insurers and employers have strong motivations to acquire the data, as do prospective adoptive parents. Even acceding to seemingly appropriate requests can lead to unantici-

pated disclosures. In one case an attending physician requested genetic information about his patient from members of a genetics research project he knew the patient and her family had participated in. The research team released the information under strictures of confidentiality because they believed that the information would be used to help the patient. The physician accepted the information in good faith and based his diagnosis, in part, on it. However, not fully understanding the consequences of his actions, he recorded the genetic test results in the patient's bedside chart. This made it semipublic information and soon family members, nurses, various medical personnel, insurers, and others were privileged to information about the woman that was intended to be restricted only to the physician's and the researchers' eyes.

A similar issue arises when researchers publish their results. Usually group statistics are presented covering a large enough number of subjects to protect their anonymity. But sometimes small samples are used or unique individual cases are discussed using pseudonyms. In these cases, although the researchers attempt to maintain anonymity, it is sometimes possible, with the use of little additional knowledge, to unveil the subjects' true identities.

Subsequent research and use of research results in genome databases depend on the accuracy and validity of the data. Recent studies have shown that some databases such as GenBank, the DNA Database of Japan, EDJ, and the Protein Data bank are accumulating errors. At a minimum these errors will cause confusion and inefficiencies in research. In the extreme, erroneous results could be produced that will lead to harmful application to people.

A new scientific discipline called "bioinformatics" has been created to meet the technical challenges of storing and sorting genetic information and materials. This discipline, too, must find its ethical base. The fundamental moral quandary facing bioinformaticians and genetic researchers with respect to the storage and retrieval of genetic information and material is challenging. It pits the researchers' scientific needs to participate fully in the democratization of genetic science and to create a collegial partnership in the production of knowledge—one that may also serve his or her own self-interests for acclaim or promotion—against the subjects' rights to privacy and autonomy.

L. Using Genetic Research Results

The value in genetics research, and hence the source of many of the ethical issues it engenders, lies in the uses to which it can be put—good and bad, just or unjust. The possible applications of the research results are quite numerous and growing rapidly as the pace of research increases. They can be classified into five general categories of use: diagnosis, prediction, identification, therapy, and enhancement. Significant ethical issues arise for each of these uses.

1. Diagnosis

Genetic research results are used to develop genetic tests. The diagnostic use of genetic testing occurs either (a) when a particular phenotype is observed, usually a disease or an abnormality, and the investigator wants to determine its genetic origins, i.e., its genotype, in the hopes of providing useful information to the patient, or (b) when a particular individual is asymptomatic but wants to know what diseases he or she may be susceptible to. Fragments of a subject's DNA are isolated and used to determine how the sequence of base pairs he or she carries compares with normal and mutant alleles at the loci from which the fragments were extracted. This information is used to determine the disease's genetic causes or, when used in conjunction with other diagnostic techniques, to reduce the ambiguity in diagnosis. Diseases such as Tay-Sachs, cystic fibrosis, and sickle cell anemia can be diagnosed in this way. In all of these, clinical symptoms are apparent early in childhood. Diagnostic results may thus serve as the basis for a therapy. While this is a powerful tool the ethical issues it engenders are similar to those addressed by medical ethics in general. A difficult moral dilemma is raised when a patient is diagnosed with a serious genetic disease. Where does the physician's or researcher's foremost ethical duty lie? Should the patient's right to confidentiality be honored? Or, do possibly affected third parties, such as blood-line members or a spouse, have a right to know the test results?

2. Prediction

Prediction involves determining the probability that a particular phenotype will occur in the future given that the subject has been diagnosed as having a specific genotype. Predictive uses of genetic information include protecting individuals from getting illnesses to which they are genetically predisposed and preventing the transmission of genetic predisposition to the next generation. On the surface these are laudable goals, but reaching for them also generates several types of difficult ethical issues. First, since the occurrence of a phenotype is the result of the workings of the genotype and the environment as they interact over time, calculating probabilities and producing morbidity tables is an error-prone activity, potentially leading to harmful, inaccurate conclusions. The likelihood of producing false positives or false negatives is increased. Second, and most significantly, more morally questionable actions stem from

prediction than from diagnosis. Genetic predictions play a prominent role in the ethical issues of the use and abuse of genetic testing in the workplace, by insurance companies, for other types of carrier testing, and in making abortion decisions. For example, concern has been expressed that employers will seek genetic information about current or prospective employees and use it to deny them jobs. Similar concerns are expressed that insurance companies will use genetic information to deny people health care or life insurance. In these cases an individual's rights are placed in opposition to those of the employing organization or insurance company.

3. Identification

Polymerase chain reaction makes it possible to multiply small quantities of DNA into large quantities relatively easily and inexpensively. This means that DNA samples found in trace materials such as hair, blood, or semen can be reproduced and compared with DNA samples deposited in a master databank and used to identify individuals by means of genetic "fingerprinting." The forensic use of the technique, using items picked up at a crime scene, is relatively well established. In fact, judges are being educated in its scientific underpinnings. It is fairly well established that DNA tests can be used effectively to exclude an individual from being the person whose DNA was tested. However, an error-free positive identification is more difficult to obtain since it involves many complex probability calculations and the assumption that a combination of specific repetitive DNA sequences from the person tested is very unlikely to match the DNA of anyone else. The ethical issues created by forensic use are similar to those found with the use of other criminal evidence-generating techniques. The possible use of genetic fingerprinting for other forms of identification raises important ethical issues of informed consent and invasion of privacy. Proposals have been made to collect DNA information on convicted prisoners, parolees, probationers, and suspected criminals. To date, data from DNA companies have been used to exclude accused suspects as well as to add to the probability that they were the perpetrators. The possibility of abuse of DNA information is one reason many people are leery of contributing to DNA banks and why security and confidentiality are such important factors.

4. Therapy

Genetic research informs and enables two types of therapy: drug and genetic. Drug therapy is primarily used to cure a disease although some drugs are also used as a preventive measure. Drugs are essentially foreign substances which when introduced into the body have a biological effect. New drug therapies are developed by figuring out what genotypes are at the root of a particular phenotype. A drug then is developed that targets the relevant genes, or the proteins for which they are the blueprint, and alters or suppresses the genes' activity. These developments have led further to the new research field of pharmacogenomics in which functional genomics is translated into drug therapeutics. One's genetic constitution can affect the therapeutic effect and toxicity of a drug. For example, research reveals that the potential consequences of administering the same dose of a drug to individuals with different drug metabolism genotypes or different drug receptor genotypes can have substantially different results. Marketing these new drug therapies is a major business opportunity for pharmaceutical firms. This is one of the reasons the firms want to set the research agenda for genetic research and own or control its results. The ethical issues raised by drug therapy, however, are similar to those raised by other forms of pharmacology.

Gene therapy, in contradistinction, introduces normal genes into a subject's cell nuclei in order to repair, replace, or compensate for a defective or mutilated allele. Gene therapy research has addressed rare genetic diseases and complex multifactorial afflictions such as cancer, AIDS, and heart disease. It has the potential to relieve the social burden of these diseases. There are two basic types of cells leading to two types of therapy: germ-line (sex) cells and somatic (nonsex) cells. Germline interventions replace an allele in all sex cells, including eggs and sperm. These new cells are inherited by the individual's descendants. The ethical arguments in favor of germ-line therapy are manyfold: one can spare descendants from possibly receiving and transmitting disease-related genes; it can keep descendants from having to undergo somatic cell gene therapy or other forms of therapy; it is more economical than repeating somatic gene therapy for successive generations; it replaces abortion or selective discarding as personal options; it protects individuals who have disabilities; and by preventing disease and promoting health it supports science's duty to remove sources of potential harm to society. Ethical arguments against germ-line gene therapy include its possible use for eugenics (producing "superior" beings or robot-like slaves); it is contrary to religious beliefs against exceeding human limits or the natural law; it reduces diversity in the human genome and thereby interferes with human evolution; it breaches the child's right to receive an untampered genome; it favors wealthy and powerful individuals; and it may result in unanticipated negative effects. The ethical issues raised by somatic cell therapy are similar to those related to other medical interventions. While some ethicists argue that somatic cell gene therapy should be used only after alternative therapies have proven unsuccessful, others contend that if the procedure is safe it should be employed early in the disease process so that the chances of preventing deterioration in the patient's condition are improved. At this time, however, the risks of unforeseen damage are too great to encourage most somatic interventions. Gene therapy can be used to cure a disease, but it also creates a temptation to enhance a person's own capacities or those of his or her children.

5. Enhancement

The broad term used to describe the goal of enhancing the human condition by means of genetic manipulation is "eugenics." Positive eugenics uses the results of genetic research to produce people with "superior" phenotypes; negative eugenics seeks to improve the human condition by eliminating biologically "inferior" people or unwanted clusters of phenotypes from the population. Genetic engineering is based on several evolving science-based techniques by which these goals can be achieved. Both of these forms of enhancement, in Benedikt Hårlin's words, must be based on crucial scientific and value judgments about what are "normal and abnormal, acceptable and unacceptable, viable and non-viable forms of the genetic make-up of individual human beings before and after birth." In a democratic and market-oriented society many eugenic decisions are made by individuals themselves, resulting in a kind of "homemade" eugenics—one often favoring the wealthy or powerful. In a totalitarian or authoritarian society, despots—a Hitler or Stalin—make the decisions and this can result in genocide as well as attempts to create a super race.

The risks in doing this have been captured well by Monette Vaquin (quoted in Bodmer and McKie, 1994, p. 173):

> Today, astounding paradox, the generation following Nazism is giving the world the tools of eugenics beyond the wildest Hitlerian dreams. It is as if the unthinkable of the generation of the fathers haunted the discoveries of the sons. Scientists of tomorrow will have a power that exceeds all the powers known to mankind; that of manipulating the genome. Who can say for sure that it will be used only for the avoidance of hereditary illnesses?

IV. CHALLENGES TO OUR UNDERSTANDING OF HUMAN BEINGS AND SOCIETY

Who are we as human beings? What is the meaning of life? What is a "good" or "superior" human being? What is a society or a community? What are its members' responsibilities to it? These questions have motivated much of scientific and ethical inquiry from the beginning. Genetics research brings them even more prominently into the forefront and promises to take much of the mystery, and, perhaps, the variety, out of life. New views of nature are also evolving. Part of the mystery and, hence, motivation of the human condition is a striving for self-discovery, personal expression, and actualization. We do not know what we might become and many of us have faith that we can shape the process of becoming who we want to be. Genetic research challenges these beliefs. Genetic predispositions—our genotypes—greatly condition our physical makeup as expressed throughout our lives and, to some extent, our traits and characteristics—our phenotypes. All control is not in our own hands. But, as the discussion above of the producer/product relationship between genotype/phenotype describes, all is not out of our hands either. The issues of applied ethics concerning the creation and maintenance of a nurturing and supporting environment are still with us. The challenges presented by genetics research to applied ethics are twofold: to accept with serenity what is beyond our ability to change, and to have the wisdom and courage to change those things we can to improve the human condition. We must use the research to create a healthier society and not to intercede when improvement is not likely to ensue or harm will result.

As the discussion of eugenics suggests, there is a great temptation for members of society to use the results of genetics research to abort or otherwise eliminate human beings or other organisms whose genetic makeup leads to undesirable phenotypes; to discriminate against and stigmatize them; to use the new genetic technologies to attenuate undesirable effects by means of drug therapy; or to reconstitute a person's or other organism's genome by means of gene therapy or by means of genetic engineering. Some of these actions, such as eliminating fatal inherited diseases, may be morally justifiable. Others, such as using genetic therapy for cosmetic purposes, are likely not. Many such possible actions will remain morally ambiguous. Bioethics must continue to address them.

One crucial element in the moral evaluation of genetics research and its uses is our concept of community. In *Earthwalk* (1974), Philip Slader draws a key distinction between a network and a community. At one end of a continuum a network is a group of homogeneous individuals who take pride in their commonalties. At the other end is a community which is composed of a rather heterogeneous group of people—young and old, rich and poor, large and small, sick and healthy, the wise old man, the village idiot, the butcher, the baker, the candlestick maker, and the like. Communities celebrate their differences. And, they draw strength and a sense of individuality from them. The temptation to use genetics research to alter our fates serves to move society from its

historical and natural foundations in community toward those of a network. This not only has implications for contemporary moral values, but also has implications for the long-term survival of the human species itself, perhaps the greatest value of all. It reduces variety in the gene pool and thereby may eliminate a vital, yet currently unrecognized, genetic response to some new environmental force unfolded by Darwinian evolution.

Genetics research can lead to a change in an individual's sense of his or her role in society and responsibility to it. Too strong a belief in genetic determinism will cause people to eschew personal responsibility: "My genes made me do it!" The abiding concerns of applied ethics—virtue and good motives for one's actions, compliance with one's duties, and choosing acts that result in the overall greatest social happiness—require that every moral agent assume responsibility for his or her actions. The findings of genetics research in general and of the studies and tests performed on an individual in particular should not be used to rob human beings of that most fundamental of human characteristics, responsibility.

Bibliography

For an in-depth treatment of the systems approach to scientific inquiry and of information handling as used in this article see:

Churchman, C. W. (1971). *The Design of Inquiring-Systems: Basic Concepts of Systems and Organization.* New York: Basic Books.

Mason, R. O., Mason, F. M., and Culnan, M. J. (1995). *The Ethics of Information Management.* Thousand Oaks, CA: Sage.

Some relevant readings in genetics research include:

Bodmer, W., and McKie, R. (1994). *The Book of Man: The Human Genome Project and the Quest to Discover Our Genetic Heritage.* New York: Scribner.

Cook-Deegan, R. (1994). *The Gene Wars: Science, Politics, and the Human Genome Project.* New York: Norton.

Kitcher, P. (1996). *The Lives to Come: The Genetic Revolution and Human Possibilities.* New York: Simon & Schuster.

Lyon, J., and Gorner, P. (1995). *Altered Fates: Gene Therapy and the Retooling of Human Life.* New York: Norton.

Murphy, T. F., and Lappe, M. A. (1994). *Justice and the Human Genome Project.* Berkeley: Univ. of California Press.

Rosenberg, S. A. (1992). *The Transformed Cell: Unlocking the Mysteries of Cancer.* New York: Avon.

Watson, J. D. (1968). *The Double Helix.* New York: Penguin.

Related readings in medical ethics and bioethics may be found in:

Beauchamp, T. L., and Childress, J. F. (1994). *Principles of Biomedical Ethics,* 4th ed. New York: Oxford Univ. Press.

Bulger, R. E., Ed. (1993). *The Ethical Dimensions of the Biological Sciences.* Cambridge: Cambridge Univ. Press.

Engelhardt, H. T., (1986). *The Foundations of Bioethics.* New York: Oxford Univ. Press.

Faden, R. R., and Beauchamp, T. L. (1986). *A History and Theory of Informed Consent.* New York: Oxford Univ. Press.

Fletcher, J. (1971). Ethical aspects of genetic controls: Designed genetic changes in man. *New England Journal of Medicine* **285,** 776–783.

Kass, L. R. (1985). *Toward a More Natural Science: Biology and Human Affairs.* New York: The Free Press.

Genetics and Behavior

GARLAND E. ALLEN
Washington University

GLOSSARY

genetic marker Any detectable element at the genetic level (chromosome, segment of DNA) that can be followed in successive generations within a breeding group (family lines). Chromosomal markers include particular band patterns, extra segments, knobs, or other distinguishable features of chromosome morphology that can be observed (microscopically); molecular markers can be identified indirectly by extracting and separating components of an organism's DNA (using such separatory techniques as gel electrophoresis). The appearance of markers is correlated with the appearance of phenotypes in given breeding lines to suggest the possible existence or location of genes for the trait in question. Markers do not, however, mean that genes for the trait necessarily exist or are located at the position in the genome occupied by the marker itself.

genome A general term for the totality of genetic information contained within an organism or a species. Thus, geneticists speak of the genome of a particular individual, meaning the genes carried by that individ-

ual, or of a whole population, such as the human genome.

genotype The basic genetic elements (genes, DNA segments) inherited by an individual from its parents and capable of being passed on to its offspring. The genotype is distinguished from the phenotype, or outward appearance of the organism as it develops.

phenotype The appearance of an organism for a given trait (anatomical, physiological, behavioral) produced by an interaction between the organism's genotype and the environment in which the organism developed. Phenotype, as opposed to genotype, cannot be directly passed on to an individual's descendants.

polymorphism The occurrence of varying forms of a trait, gene, or marker within a population. Most complex traits in animal, plant, and human populations are polymorphic (in humans, for example, hair color, eye color, blood groups, etc.).

BEHAVIOR GENETICS is a field of the biological sciences that deals with elucidating the genetic, or inherited, components of animal behavior. The field at present encompasses a large group of researchers whose focus is behavior in nonhuman animals, ranging from one-celled protozoa to nonhuman primates. A somewhat smaller group of researchers focuses primarily on human behavior where, for practical and ethical reasons, genetic data cannot be obtained by the usual methods (planned breeding and rigorous control of environmental conditions under which offspring develop).

I. INTRODUCTION

Most workers in the field of behavior genetics, regardless of the organism(s) with which they work, harbor some hopes that their work will throw light on general aspects of human behavior and its origins. However, many are highly skeptical of naive attempts to reason by analogy from other animals to human beings, or to directly apply findings in nonhuman species to humans. This schism has plagued the field of behavior genetics for a century or more, and manifested itself overtly in a split within the Behavior Genetics Society (an international professional organization) in 1995. Partly as a result of growing distrust of exaggerated claims for a genetic basis of many human personality and behavioral traits, and partly in protest against remarks made by the president of the society (claiming that behavior genetics studies might well show that African-Americans were genetically inferior to Caucasians in intelligence), a number of researchers left the society to form a new organization that would focus on studies of nonhuman animals. Despite these controversies, however, the field of behavior genetics holds much potential for elucidating the evolution and adaptive significance of behavior in many animal species.

Behavior genetics gains much of its interest and controversy when applied to human behavior for a number of reasons. First, unequivocal data are difficult to come by, and thus require considerable interpretation and qualification. Second, past history has shown that claims about the genetic causes of specific behaviors (usually what were considered undesirable behaviors) have often been put to misuse, as in the justification of genocide, or restricting the reproduction of individuals or families claimed to harbor genetic defects. Third, claims about the genetic basis of behavior today have obvious and unavoidable implications for social policy. Researchers do not study the possible genetic basis of such conditions as alcoholism or manic depression, for example, for purely abstract purposes; the ultimate aim is eventually to reduce the amount of overt alcoholism or depression in society as well as to help individual alcoholics or depressives.

However, genetic claims about behavioral (as well as all other) traits have almost always been accompanied by the underlying assumption that what is genetic cannot be changed: "Genes are destiny," as this view is sometimes characterized. Thus, behavior genetics claims raise a host of ethical and social issues, all of which revolve around the question of how society should react to individuals diagnosed with any one of a number of behavioral or personality traits that are claimed to have a significant genetic basis. For example, what social or legislative policies should or should not be adopted with regard to the perpetuation of such traits? In the past, it has been claimed that so-called genetically defective individuals should be (1) sterilized, (2) hospitalized, (3) forbidden to marry, (4) refused immigration status, (5) treated with drugs or other external "therapies," and so on. An even deeper question arises as to the extent to which social, moral, and ethical decisions should be based on genetic (or any scientific) claims. Science may inform by presenting information, but it does not necessarily dictate policy. The complex moral and ethical issues thus raised by behavior genetics—along with its often questionable, or at least highly controversial, data—have raised the question of whether research on such issues (the genetic basis of schizophrenia, criminality, compulsiveness, homosexuality, etc.) should be funded at all. There is no agreement on this issue among researchers in the field or among ethicists and specialists in science policy. However, the issue is significant enough that calls for halting such research have been heard in recent years in the United States, England, and Germany.

II. HISTORICAL BACKGROUND

The question of why animals behave as they do—the broad similarities within species as well as the differences between species—has perplexed naturalists and philosophers for centuries. Hippocrates (460?—377 B.C.E.) and Aristotle (384—322 B.C.E.) both wrote on animal behavior, developing the first western notions of instinct as built-in or natural (as opposed to learned) behavior. In the Greek sense, instinct was seen as necessary for the survival of lower animals, but ascending the *scala naturae* (ladder of nature or "Great Chain of Beings," which ordered organisms from simplest to most complex) showed that as complexity increased, reliance on purely instinctual behavior decreased, with learning becoming a more dominant aspect of behavior. For the ancients, humans showed only the most general kinds of instinctual behavior: self-protection and survival, maternal care, and sociality. In Greek terms, humans were distinguished from other animals by their self-awareness and capacity to learn and engage in abstract thought. Conversely, the lower an animal ranged on the "Chain of Beings," the more it relied on instinctual, or built-in, behavior.

Some of the first and most thorough treatments of instinct theory can be found in Darwin's *On the Origin of Species* (1859) and (more explicitly) in his *Descent of Man* (1871) and *Expression of Emotions in Animals and Man* (1872). Darwin emphasized that instincts had to be studied as behaviors arising from a (at the time unknown, but assumed) biological cause which was shared among members of a species. Darwin made the further assumption that such behaviors were largely inherited, a view that was necessary for his theory of evolution by natural selection. According to Darwin, new instincts evolved through the action of natural selection on variations in previously existing behaviors. In Darwin's sense, behaviors that were learned could not be affected by natural selection, and thus were not instinctual, properly speaking. He also recognized that in higher animals instinctual behavior was often overlain by learned behavior. The adaptive value of instinctual behavior was that it provided an organism with a repertoire of immediate, ready-made behaviors that enhanced its survival without having to go through the sometimes long and haphazard process of learning. At the same time, instinctual behaviors were not easily modified to fit new or complex circumstances; the ability to learn, and therefore modify, behavior to meet new circumstances could thus ultimately be seen as a more highly evolved adaptation. Knowing nothing of genetics in our modern sense, Darwin thought that some learned (acquired) behaviors could be passed on to an animal's offspring and thus ultimately become part of the species' instinctual repertoire.

Darwin's ideas stimulated much work on the nature of instinct in the late 19th and especially 20th centuries. With the rediscovery of Mendelian genetics in 1900, the study of what has come to be regarded as "behavior genetics" may be said to have begun. Mendel's work provided a predictive and experimentally based system by which to determine the pattern of inheritance of many kinds of traits. Although behavioral traits were not among the first to be examined by the new school of Mendelian genetics, after about 1910 there was a considerable attempt made at applying Mendel's work to a wide variety of human behaviors. Numerous studies of the genetics of alcoholism, prostitution and sexual immorality, manic depression, criminality, feeblemindedness, muscial ability, genius, sea-faringness, and the like abounded in the early decades of the century. Most were based on naive applications of Mendelian principles to family pedigree analyses, but they stimulated a whole movement known as eugenics, or the attempt to determine the hereditary basis of human social and behavioral traits with an eye to eliminating those that were considered "undersirable" and promoting those that were deemed "desirable."

The rise of behaviorism in psychology during the first decades of the 20th century deflected interest from the study of instinct by focusing instead on learned or conditioned behavior. Behaviorists such as John B. Watson (1878–1958) and B. F. Skinner (1904–1990) assumed that organisms (at least higher vertebrates, including humans) began life with a minimum of inherited behav-

ior, with most of their specific responses being learned by repetitive conditioning. For example, while the general behavior associated with foraging, or searching for food, might be called instinctual (animals do not have to be taught to want food and to look for it), the processes of finding food on a daily basis, in specific localities, using specific techniques, or associated with specific activities (pushing a food bar in a cage), were all learned. Thus, until the 1930s the study of instinctual behavior and its possible genetic components was pursued by relatively few biologists of psychologists.

One of the important developments stimulating work in the genetics of behavior was the growth of a field of biology known as ethology. Ethology, pioneered by such investigators as Konrad Lorenz in Germany, Niko Tindbergen in Denmark, and Daniel Lehrman in the United States, was the attempt to study animal behavior from an observational and comparative point of view, with the aim of understanding the evolution of particular behaviors. Ethologists made detailed studies of animal mating, feeding, defensive, aggressive, and parental care behaviors and attempted to see how these behaviors evolved from simpler ancestral forms. What ethologists brought to the fore was the issue of behavior as an inherited, adaptive trait that was capable of modification by natural selection. If this were true, then it was imperative to find out more about the nature of the inheritance of complex behaviors, about which very little was actually known at the time; by making this gap in knowledge so obvious, ethologists stimulated other biologists to search for the genetic basis of specific animal behaviors.

Early attempts to study the genetics of behavior in larger animals such as dogs by American researchers John Fuller and John Paul Scott in the 1930s–1950s were problematical: the behaviors were complex and variable even when environmental conditions seemed stable, breeding was slow, and results were inconclusive. Among the earliest successful attempts to work out the genetic basis for a specific animal behavior focused on the social insects. Experiments in the 1950s showed that among honeybees (*Apis mellifera*) there were two different behavioral responses to the death of larva within a cell of the honeycomb (where the larvae normally develop). In one strain (called "hygienic") the workers removed the dead larvae and discarded them outside the hive; in the other strain (called "unhygienic") workers left the larvae in the cell to decompose. Crossing purebred hygienic bees (symbolized *hh*) with purebred unhygienic ones (symbolized *HH*) produced hybrids that were all unhygienic (symbolized *Hh*) but which, when crossed with each other (that is *Hh* × *Hh*), produced an expected Mendelian ratio of three unhygienic (either *HH* or *Hh*) to one hygienic (*hh*) offspring. According to Mendel's principles, unhygienic behavior was said to be "dominant" over hygienic, while hygienic was said to be "recessive" to unhygienic. Further experiments showed that this behavior could be broken down into two different components: uncapping the cell (taking off the seal from the top) and removing the larva. Each of these behaviors appeared to be controlled by a different set of genes: uncapping/non-uncapping and removal/nonremoval. Work such as this showed clearly that at least some basic animal behaviors could be inherited according to simple Mendelian principles.

In the time since such work was begun behavior geneticists have uncovered the genetic basis for many instinctual behaviors in organisms as widely diverse as fruit flies (*Drosophila*), spiders, roundworms, mice, and chickens. With the advent of molecular genetics, the chemical basis of actual genes has been localized to specific segments of DNA in the organism's genome. Behavior genetics has thus become a major field within modern biological science.

III. BEHAVIOR GENETICS TODAY

The aim of behavior genetics was set forth clearly in the first issue of the journal *Behavior Genetics* in 1970: ". . . behavior genetics is simply the intersection between genetics and the behavioral sciences." Although from the beginning, human behavior was considered an important component of the field, many investigators focused their attention on less complex animals, especially the small fruit fly *D. melanogaster,* since so much was known about its genetics. Pioneers in that work included American psychologists-geneticists such as A. Manning and Jerry Hirsch, both of whom carried out breeding and selection experiments on *Drosophila* to study the mode of inheritance of behavior and its modification through selection. For example, in 1961 Manning carried out two selection experiments for mating speed in *D. melanogaster,* producing a slow and a rapid strain that were easily distinguishable. Hirsch selected for various behavioral traits over hundreds of generations, producing strains with widely divergent behavior patterns. Such experiments showed clearly that specific behaviors in fruit flies have a distinct genetic component and that the traits can be altered by natural selection. Manning, Hirsch, and others also showed that mating behavior in *Drosophila,* like response of worker bees to dead larvae, consists of a number of separable components.

In the 1980s and 1990s Ralph Greenspan and his associates at New York University identified these components and traced out their neurological and genetic bases. Gene mutations have been observed to affect such features of courtship as the "mating song," produced as sound pulses when the male flaps his wings in certain rhythms. Males with a mating song gene mutation called "period" flap their wings at different intervals than normal males. The result is that mutant males have less success copulating with females than normal males do. The "song" is just one aspect of a complex courtship ritual that involves specific male and femal motions, wing positions, extension of the proboscis (a long feeding device extruded from the mouth), licking of the genitals, and copulation itself. That these individual components can be isolated and studied at the genetic and neurological level demonstrates the power of behavior genetic analysis when carried out under rigorous laboratory conditions.

More recent work has focused on interspecific comparisons in *Drosophila,* again using mating behavior. Behavioral hybrids produce courtship responses that are often intermediate between the two parental species, thereby reducing the number of successful copulations. Such experiments have thrown much light on Darwin's hypothesized mechanism of sexual selection. A variation of natural selection, "sexual selection," was introduced by Darwin to explain the almost-universal sexual dimorphism (distinctly different) in both physical traits (male–female differences in coloration or plumage in birds, or hair distribution in mammals) and behaviors (maternal versus territorial behavior in female and male baboons, respectively). Such persistent differences of forms *within* a species seemed difficult to account for in terms of traditional natural selection. Darwin concluded that females choose among competing males for a partner, selecting the male with, for example, the brightest plumage or the most distinct courtship behaviors. Thus, from an original population of undistinguished males in the population, through sexual selection males of many species evolved distinct male-associated characteristics (comb of the cock, mane in male lions, etc.) that seemed to serve no other function than as an attraction to the opposite sex. By way of such arguments, behavior genetics, like its ancestor ethology, has always had a close association with the study of evolutionary processes.

Modern-day behavioral geneticists work with a much larger variety of animals than fruit flies and roundworms. Many studies have been carried out with mice, especially those laboratory strains whose genetics is thoroughly understood. And despite the difficulty in working with them, dogs continue to be a popular object of behavioral studies. Primates such as chimpanzees or macaques have also been used, though for obvious reasons (like humans they have small numbers of offspring and their gestation periods are long) genetic data are more difficult to come by.

In the past decade behavior genetic research on humans has increased at a great rate, despite the contentiousness and public sensitivity to the issue. Behavior geneticists such as Robert Plomin at Pennsylvania State University and Joel Gelernter at Yale University argue that the field has been misrepresented. They point out that many human behavioral genes, or at least chromosomal or molecular markers thought to be associated with specific genes, have been correlated with specific behavioral types such as Tourette's syndrome (leading to uncontrollable movements and speaking), schizophrenia, manic depression, alcoholism, attention deficit hyperactivity disorder (ADHD), and homosexuality, to name just a few. These correlations suggest strongly that there might be a significant genetic component to these behaviors. Human behavior genetic researchers emphasize that they do not discount the role of environment, nor the additive effect of many genes impinging on any given behavior. In fact, they make a point of emphasizing that the outcome of any behavioral development in humans (or any other organism) is of necessity the product of genes interacting with environment. They therefore argue that the old nature–nurture dispute is meaningless. All traits, including behavioral ones, are a product of the combined effects of heredity and environment.

Since 1998 one of the areas of human behavior genetics to make the most significant progress in trying to pin down genetic influences is research in schizophrenia. Researchers Robert Friedman, Sherry Leonard, and their colleagues at the University of Colorado Medical Center in Denver have studied a number of what are known as nicotinic receptors in the surface of cells in the brain and central nervous system. These receptors are involved in mediating a wide variety of behavioral responses, since nicotinic receptors are among the most common in the central nervous system. There are over a dozen different types of nicotinic receptors, but the researchers have identified a mutant repeat region on chromosome 15 that affects the portion of the gene that enables it to be active (the promoter region). Nonschizophrenics have two copies of this duplicated region while schizophrenics have only one. The studies show that people who have the promoter mutation have a 50% chance of getting schizophrenia while those who lack it have only a 3% chance. Both figures suggest that having the mutation does not ensure that schizophrenia will develop, nor does lacking the mutation mean that the patient will never experience schizophrenia. As in all such gene–behavior relationships, many factors interact to produce a given outcome. Researchers acknowledge, for example, that even with the gene mutation present some traumatic event (physical or psychological) is usually necessary for schizophrenia, or at least some of its symptoms, to develop. In this case, at least, a gene, rather than merely a marker, has been

localized that is known to code for a protein that appears to have a distinct behavioral effect.

Basic methods of research in human behavior genetics usually begin with a definition of the behavioral trait in question—alcoholism, schizophrenia, violent/aggressive behavior, homosexuality—followed up by determining criteria for diagnosis (i.e., guidelines for identifying who does and does not display the condition). For example, in genetic studies of crime and alcoholism, psychiatrist C. Robert Cloninger at Washington University Medical School used police and temperance board records from Sweden (where good public health records have been kept for over a century) to classify individual subjects as either "criminals" or "alcoholics" or both (he used the existence of three or more citations in the public record to establish that an individual was an "alcoholic" or "criminal"). Many psychiatrists in the United States and abroad use the American Psychiatric Association's *Diagnostic and Statistical Manual IV* (now in the second version of its fourth edition) as the criterion for diagnosing individuals with one or more mental illnesses. Behavioral geneticists emphasize that it is necessary to establish unambiguous definitions of traits before setting out to investigate their inheritance patterns.

A second step is to trace the occurrence of the trait in a given family or group of relatives. The traditional method for recording such data is by constructing a family pedigree chart for the trait through as many generations as possible; in more recent times researchers carry out genetic analysis by identifying chromosome or DNA markers. In genetics a "marker" is some detectable region of a chromosome that appears more frequently in individuals who possess a particular trait than in individuals who do not possess the trait. Cytological markers are visible under the microscope (a physical protrusion from the chromosome or special banding pattern observed in chromosome preparations). Molecular markers are segments of DNA detected by special molecular probes. In either case, the presence or absence of the marker is correlated with the presence or absence of the trait in the individual's behavior (the Xq28 marker on the human X-chromosome was correlated with homosexual behavior in 33 out of 40 pairs of gay brothers in a study by Hamer and colleagues in 1993). It is important to note that markers are not equivalent to genes affecting the trait; they only provide some sort of clude about a region of the chromosome where the gene or genes for the trait might be located.

A third component of the method is to analyze statistically the frequency of correlation between marker and visible trait in the family in question, in collateral family lines, and in the population at large. A useful tool for this purpose is known as the analysis of variance, and an associated calculation is the heritability of a trait.

While some behavior geneticists today have abandoned heritability, it has been a staple of human behavior genetics for over fifty years. Heritability is a technique that attempts to partition that part of a trait that is affected by heredity from that which is affected by environment, so that a heritability value of 0.8 (= 80%) can be interpreted to say that 80% of a given trait *might* be ascribable to hereditary effects. It does not say that 80% of the trait is genetically determined. The term "heritability" in its technical sense has led to considerable confusion in the literature, both scientific and popular, since it is often interpreted to mean "inherited," as in "trait X is 80% due to genetics and 20% due to environment." Heritability calculations only state that, all other things being equal, a given heritability (say, 0.8) means that 80% of the trait *might be ascribed* to genetics. It is, of course, the "all other things being equal" part (i.e., knowing the genetic relationships between organisms in the sample, and knowing specific and relevant features of the environment in which the individuals have developed) that is the catch. For human behavior, the latter set of conditions are particularly difficult to assess with any accuracy. (Do two children brought up in the same household have the same environment? Do children brought up in different households have significantly different environments? What counts as significant components of the environment in terms of effects on adult traits?) The statistical analysis part of human behavior genetics has always raised sticky methodological issues.

Finally, once the data are analyzed the behavior geneticist is faced with trying to draw some conclusions about the degree to which a given trait might be affected by some genetic component. This is where even the most staunch behavior geneticist admits there are great pitfalls. The most common is the tendency to overinterpret the data. Are there any genetic effects to be discerned at all? If so, do they appear to be single-gene effects (very few complex traits in humans or any other animals appear to be attributable to single genes), are they additive effects (the presence of two genes yields roughly twice the effect of one), or are they nonadditive (two genes yield three times the effect of one, and so on)? Are the relative comparisons (control groups) available for judging the possible genetic effects in the observed group? For example, if a particular chromosome marker is found in a group of people who show a particular behavioral trait, it would be necessary to know the prevalence of the marker in the general population to draw any conclusion about the possible genetic effects in the observed group.

Critics of human behavior genetics such as Jonathan Beckwith of Harvard Medical School, Peter Breggin, Director of the Center for the Study of Psychiatry

(Bethesda, MD), and Steven Rose at the Open University in Great Britain disagree with the idea that the findings of behavior geneticists are conclusive in any way. They point to a number of methodological problems that have undermined virtually all studies purporting to have found a genetic determiner for any specific human behavior. The flaws generally fall into the four methodological areas dicussed above.

First is the problem of defining human behavioral traits in such a way that they can be diagnosed by any well-trained observer. This becomes difficult, the critics point out, for traits such as criminality, aggressiveness, alcoholism, manic depression, schizophrenia, homosexuality, and many others. These are very complex behaviors and judging whether an individual really fits into the trait category can be quite subjective—was Robin Hood a criminal or a hero when he stole from the rich to give to the poor? Psychiatrists are debating today whether schizophrenia and manic depression are really different diseases or varying manifestations of the same disease. Furthermore, the case where some real genetic differences may exist between individuals—for example, some people can metabolize alcohol more readily than others, and thus exhibit a much greater tolerance—may not be the same as that of social behavior or trait (in the example, "alcoholism"). With such wide areas of possible disagreement about what constitutes a particular trait, critics point out that it is no wonder behavior geneticists have a difficult time even replicating each other's work, much less carrying out a clear genetic analysis.

At the second step—gathering data on families, siblings, adoptees, etc.—critics argue that many human behavior genetic studies are faulty in a number of different ways. Some have too small a sample size, a special problem for those studies using monozygotic (identical) twins raised apart (Cyril Burt, on whose famous twin studies so much research on the inheritance of IQ has been based, managed after 40 years to accumulate only 53 pairs of twins). Others do not institute proper controls. A 1994 study by Dean Hamer and colleagues at the National Cancer Institute found a genetic marker on the X-chromosome in 33 out of 40 pairs of gay brothers, suggesting to the team that there might be genes located on the chromosome near that marker predisposing those individuals to homosexual behavior. The study did not provide information on whether other brothers (non-gay) in the same families did or did not have the marker. If other sons in the families had the marker, the correlation with homosexuality would be meaningless. Other problems with this second step include non-standardized methods of determining which individuals have the trait in question (different diagnostic procedures or assessments made under different conditions), and bias in selecting subjects for the study (many studies get their subject by asking publicly for volunteers, which can bias the results toward individuals who are outgoing personality types or might exaggerate claims about themselves to remain in the study).

Critics also argue that in the third step of human behavior genetics research—introduction of various statistical procedures—researchers often misuse or misapply particular techniques. One of the most commonly misused and misunderstood statistical procedures is that of heritability. As pointed out above, heritability does not mean "inherited," though many behavior geneticists do not make the distinction clear, especially when talking to reporters or giving popular talks. A more common problem, however, is the failure of those using heritability to take into account two underlying constraints on the method. First, any heritability estimate is limited to the given population, in a given environment. Thus, the heritability estimate for a trait in population A cannot be applied to population B, since there is no guarantee that either the genetic or the environmental components of the two populations are comparable. Critics point out that Berkeley psychologist Arthur Jensen committed this error in his famous paper of 1969, in which he applied heritability estimates of IQ based on British Caucasian students (population A) to American Caucasian and African-American students (populations B_1 and B_2). Such a comparison is invalid by the rules of heritability analysis. A second assumption of heritability is that all members within a single population (for example, population A) share the same or nearly similar environments. Since heritability as a technique was introduced in the 1930s primarily as an aid to animal and plant breeders, the assumption, under most breeding conditions, was reasonably safe. However, applied to humans—for example, to Caucasian and African-American populations in any large American city—the assumption is clearly untenable. In addition to these more esoteric statistical procedures, many human behavior genetic studies have been based on extremely small sample sizes, a feature that makes all statistical analysis meaningless. One study of homosexuality consisted of a total of five twin pairs—three male and two female. Even a study of 40 siblings, as in the example of Hamer's research on homosexuality, is tiny by comparison to the number of organisms used in any behavior genetic study of nonhuman animals.

It is in the fourth step or procedure in human behavior genetics—drawing conclusions from the data—that critics find some of the most flagrant violations of sound scientific procedure. On the one hand the conclusions can sound impressive but be enormously trivial. As critics have put it, to claim that a particular study shows that genes and environment interact in producing a par-

ticular behavioral trait is such a truism as to say nothing of any interest. *Every* trait—physical, chemical, or biological—in every organism is the result of *some* interaction between genetic and environmental components. The important question in any given case is to show how and under what conditions (genetic and environmental) a particular behavioral outcome will result. Few if any human behavioral genetic studies have been able to make such a relationship clear or precise. A second problem is in a sense the flip side of the same coin: overinterpretation of the results of a given study. Both Jensen's study of racial differences in IQ in the 1960s and Hamer's study of the hereditary component of male homosexuality in the 1990s were guilty of overinterpretation. Whether in the verbal form of "the gene for . . ." or "the genes for . . . ," overinterpretation gives the impression, especially among lay readers, that the biological evidence is much stronger than it is. According to most critics of human behavior genetics research, given all the problems with carrying out research on this topic, virtually all strong claims are guilty of overinterpretation.

One aspect of human behavior genetics work that has also bothered critics, as well as other members of the scientific community at large, is the tendency of researchers to talk freely, and often in exaggerated terms, to the press. While announcing new scientific findings in press conferences has become more prevalent in all areas of science than it used to be, it is particularly disturbing in areas with considerable political implications. Critics point to the 1990 example when the discovery of a putative gene for "alcoholism" (the D_2, or dopamine receptor locus, on chromosome 11) was announced with great fanfare as front-page news in many newspapers and as cover stories for a number of magazines. Failure of other researchers to replicate the study has resulted in the ultimate discrediting of the work; announcement of this failure got no publicity at all. Critics argue that those who do work in the area of human behavior genetics should use more than the usual amount of caution in announcing any purported discoveries in the press or in other arenas.

A particularly egregious aspect of publicizing results in the press that has been troublesome to behavioral biologists occurred in early 2000 with a flurry of prepublication attention given to a book claiming that there is an evolutionary basis for rape. In *A Natural History of Rape: Biological Bases of Sexual Coercion,* Randy Thornhill, who studies insect mating behaviors, Craig Palmer, an anthropologist, and Margo Wilson, argue that rape is evolutionarily advantageous for males and has been selected over the last million or more years as a way of maximizing the transmission of a male's genes to the next generation. Coming with a full set of instructions that included admonitions to women that they should not dress "provocatively" if they want to avoid being raped, the book was strongly criticized by geneticists and evolutionary biologists as representing the worst of excesses in human behavioral genetics. With little empirical evidence to back up such claims, evolutionary models have repeatedly been fabricated from conjecture, another version of what evolutionists referred to in the 1970s as "just-so-stories" (after Rudyard Kiplings's famous book of the same title in which, among other essays, was a famous discourse on "How the Elephant Got His Trunk"). Evolutionary biologists have criticized "just-so-stories" for being imaginative scenarios supported by no evidence other than that they sound possible. In the case of *A Natural History of Rape,* ethicists argue that widespread belief in such ideas, especially by young males, could lead to an increase in the incidence of rape on the grounds that young men were only "doing what comes naturally." The further prospect that the biological argument might be used in court to exonerate or gain a lighter penalty for perpetrators of rape is also seen as raising legal as well as moral and ethical questions.

IV. ETHICS AND HUMAN BEHAVIOR GENETICS

As with virtually all areas of scientific research, the nature of claims made and conclusions drawn can have profound effects for matters of ethics. In human behavior genetics the effects can be particularly far-reaching when they affect attitudes toward health care (including mental health), education, official public policy (such as sterilizing or incarcerating persons claimed to be genetically defective), and, ultimately, the issue of research involving human subjects. The last is the most extreme and probably involves the least amount of contention. Especially given the widespread use of human subjects, including identical twins, by the Nazis for genetic studies, the western scientific community has adopted explicit and rigorous constraints on the use of humans as subjects for scientific research. For human behavior genetics, this means that the avenues open to animal behavior geneticists—planned breeding and raising of offspring under highly controlled environmental conditions—are not an option. There seems to be little doubt about this issue among either researchers or laypersons.

More problematical are ethical issues raised by overinterpretation of results and the ensuing effect such conclusions—especially when widely disseminated in the press—can have on public attitudes and policy. Today, a claim that a trait is *genetically determined* (or at least largely so) is still misunderstood as saying the trait cannot be altered, suggesting that only a limited number

of options exist for treating individuals who harbor undesirable traits. Despite considerable exaggeration by some molecular geneticists, gene therapy as a significant remedy for even well-identified genetic traits is a distant hope. Thus, claims by human behavior geneticists that alcoholism, schizophrenia or manic depression, disposition to criminal activity, and attention deficit hyperactivity disorder (ADHD) are genetic is tantamount to saying that the root cause cannot be changed, only the external manifestations "managed," usually in today's terms by drug therapy, and if that does not work, institutionalization and/or sterilization.

These approaches raise a host of ethical issues. First, the very claim that a given behavioral condition is genetic, in the face of so much questioning and controversy within the scientific community, raises questions about whom to believe. If the conclusions are so ambiguous, is it ethical to be treating children diagnosed as having ADHD with Ritalin (a drug that decreases activity) instead of looking to boring or overcrowded school conditions, or problems within the family, as the source of the behavior? Proponents of the genetic point of view, such as Dr. David Comings of the City of Hope Medical Center in California, argue that traditional psychological counseling and other supposedly curative approaches have not worked, and that drug therapy makes the life of the child, his teacher, and his parents much less stressful, and actually increases school performance noticeably. Critics point out, however, that the gene–drug approach ignores problems in the school and home, and thus allows the conditions generating ADHD to persist—like allowing a person to continue eating a poison while continuing to give them a curative. The problem still remains.

In addition, critics argue, drugs like Ritalin or Prozac (an antidepressant) are too new for their long-range effects to really be known, especially when administered over a lifetime. According to this critique, researchers are, today, using people, specifically children, as guinea pigs in large-scale experiment of social control. Some critics even liken the use of Ritalin to counteract behaviors like ADHD to the widespread use of "soma" in Aldous Huxley's *Brave New World;* for Huxley's characters, soma, available at water-fountain-like dispensaries, relieved the stress, or clouded people's perceptions, engendered by a totalitarian society.

Human behavior genetics raises the difficult ethical issues of where to draw the line between using science to help individuals improve their lives and using science for social control. Even if all the claims of human behavior genetics were taken at face value, the implications of these findings for social policy are not clear. People who are truly genetically defective, it can be argued, deserve more of society's support and resources, not less. But, as critics of the human behavior genetic program point out, at the present time with insurance companies refusing to cover "prior conditions" in the clients, genetic arguments provide an easy way to exclude a large number of people from their policy rolls. The problem is even more acute when the scientific evidence itself is highly debatable.

According to some of the most extreme critics, the whole focus on genetic explanations is economically driven. At a time when both private corporations and government are cutting back, providing a medical-genetic explanation for social-behavioral problems provides a cheap solution. Drugs such as Ritalin are expensive, but not nearly so much as one-on-one psychological counseling or therapy. Drug therapy is even cheaper than paying people higher wages, increasing (or at least not cutting back) benefits such as health care, or reducing the stress in the workplace by hiring more workers. According to this scenario the funds provided for research into, and popularization of, human behavior genetic work only serve as a smokescreen to hide the true causes of widespread mental and personality disorders.

From a biological point of view, there is nothing that says that even if a trait is significantly affected by genes, it cannot be changed to some degree. People have genetically determined disease like diabetes, yet over much of a lifetime it can be managed by controlling diet and taking artificial insulin. People with genetically defective eyesight wear glasses. People with genetically determined brown hair convert it to blonde by hair dyes, and so on. But these are only corrective measures. In addition, biologists now know that all genes are influenced by various environmental factors during embryonic development and to varying degrees throughout the rest of their lifetime. Thus, even if a child inherits a tendency to produce more dopamine receptors than normal, a particular type of home environment can alter the expression of that gene. Genes do not unfold automatically into adult traits, but rather have a "norm of reaction" over which their expression is altered by varying environmental factors. Geneticists know very little about the norm of reaction for any genes, especially putative genes for behavioral traits. Thus, most biologists would warn heavily about relying on genetic claims regarding human social and personality traits as a basis for formulating medical or social policy.

Bibliography

Allen, G. E. (1996). The double-edged sword of genetic determination: Social and political agendas in genetic studies of homosexuality, 1940–1994. In *Science and Homosexualities* (V. A. Rosario, Ed.), pp. 242–270. New York: Routledge.

Barinaga, M. (1994). From fruit flies, rats, mice: Evidence of genetic influence. *Science* **264**, 1690–1693.

Greenspan, R. (1995). Understanding the genetic construction of behavior. *Scientific American* **272**, 72–78.

Mann, C. C. (1994). Behavior genetics in transition. *Science* **264**, 1686–1689.

McDonald, K. A. (1994). Biology and behavior. Social scientists and evolutionary biologists discuss and debate new findings. *Chronicle of Higher Education* **14**, A10–11, A21.

Plomin, R., Owen, M. J., and Mcguffin, P. (1994). The genetic basis of complex human behaviors. *Science* **264**, 1733–1739.

Genetic Screening

RUTH CHADWICK

Lancaster University

GLOSSARY

gene chip An array of several thousand different complementary DNA probes arranged on a tile using silicon chip technology. Each of these probes will only bind to a stretch of DNA carrying its complementary code. Thus if DNA is swept over the chip and binding takes place, it is possible to determine the sequence of the attaching part.

genetic screening The determination of the prevalence of a gene in an asymptomatic population or population group where for any given individual there is no reason to believe he or she has the gene in question. Normally contrasted with the genetic *testing* of an individual for whom there may be some reason to think he or she is at risk, e.g., because of family history.

multifactorial In contrast to a single-gene disorder (where a single gene accounts for the presence of disease), a condition under which a number of genes, environmental causes, or both are implicated.

multiplex testing The use of tests to identify genetic status for multiple disorders or mutations at the same time.

GENETIC SCREENING may be defined using slightly different emphases. Thus the Danish Council of Ethics has defined it as the study of the occurrence of a specific gene or chromosome complement in a population or population group (1993, p. 56). The Council of Europe, in its 1992 Recommendation, added that there should be no previous *suspicion* that the individuals have the condition (pp. 9–10)—the Nuffield Council that there should be no *evidence* that the individual does, thus replacing a subjective test with an objective one (1993, par. 1.9).

I. INTRODUCTION: GENETIC SCREENING

In light of developments in human genome analysis, the prospects not only for individual genetic testing but also for population-wide genetic screening programs are increasingly an issue. To some extent testing and screening raise analogous issues, such as the potential for both harm and benefit arising in relation to the management of results, and "testing" issues will be mentioned where relevant. In other respects they are different, e.g., in relation to the justification of the intervention. This makes any assessment of costs and benefits (understood in a wide sense) different also. Whereas testing is typically carried out when a patient seeks help, a population screening program may be initiated as a public health measure. What is called "cascade screening" refers to the active seeking out of relatives of an identified case and offering testing.

There are of course different types of screening, each raising somewhat different issues. The categories of genetic screening may be divided by population group according to the stage of life. For example, screening may be carried out on fetuses, on neonates, on children, and on adults. The development of techniques of preimplantation diagnosis has opened up the possibilities of testing and/or screening embryos. Types of genetic screening may also be classified according to the type of condition being screened for: a late-onset disorder such as Huntington's disease; a predisposition to a multifactorial condition such as one of the cancers; or carrier status. Within these categories the nature of the specific condition and the range of possible interventions following a positive result will have implications for certain ethical issues. The possibility of screening for predisposition to behavioral and mental differences is arguably more controversial than screening for susceptibility to what are regarded as physical disorders (Nuffield Council on Bioethics, 1998). The diagnosis of mental disorders has always been particularly contested, and when a predictive diagnosis is concerned in a case where a particular genotype is associated with a specific form of mental disorder, this will be even more so. In predictive testing generally, there is a risk of confusing "genetically predisposed" with "genetically predetermined." In the case of mental disorders, there may be additional factors associated with the self-fulfilling prophecy of a predisposition to, e.g., depression.

The possibilities of "multiplex" testing and screening further complicate the picture since technological advances such as the gene chip make it possible to test for a number of conditions at the same time.

II. CRITERIA FOR THE INTRODUCTION OF GENETIC SCREENING

What is the justification for introducing a population genetic screening program? Guidelines have been developed by a number of policymaking bodies around the world. The Wilson–Jungner principles on screening, which were not specific to genetic screening, emphasized the relevance of the condition sought, stating that it should be an "important problem." What counts as important is, however, a matter of debate. Important to whom? What counts as an important problem may be influenced by factors such as social class, ethnicity, and gender.

Some, though not all (see Health Council of the Netherlands, 1994), later guidelines have stated that the condition sought should be "serious." The Nuffield Council on Bioethics (1993), for example, referred to the Clothier report's recommendation that the first candidates for gene therapy should be for conditions which are life-threatening or seriously handicapping, and for which treatment is either not available or unsatisfactory (Clothier, 1992, para. 8.6). The Council acknowledged, however, that for screening the criteria are likely to be wider than that, but that it is difficult to define them precisely. In light of this it took a negative approach of exclusion rather than inclusion. Screening should not be offered for conditions which, while having a genetic component, are not diseases. The so-called "gay gene" was offered as an example here (para. 10.3).

In attempting to establish what is regarded as serious disease, the sociocultural context cannot be ignored. Suppose, however, that what counts as sufficiently serious *has* been agreed upon. The question then arises as to what benefits should be expected of the program. Wilson and Jungner ask whether treatment is available. The Nuffield Council lists the availability of therapy as one of six criteria identified as important at the end of its report, while acknowledging that if it is not, it does not mean that screening is not worthwhile (1993, para. 10.21)—other benefits can result from screening.

The Danish Council of Ethics recognizes this in saying that evaluation of a screening program depends on the available "scope for action" (p. 50). This could encompass a broader interpretation than simply the availability of therapy, such as the making of reproductive choices—if screening is carried out, say, prenatally or before conception—and alterations of lifestyle, for example, in the case of screening which identifies persons with a high risk of developing a condition, where the risk can be reduced by change of diet. Even where some form of therapy is available, however, the case for screening may be disputed, as has been the case with hemochromatosis, a common disorder in the Caucasian population which results in excessive accumulation of iron. There are other variables to consider, including costs—financial, social, and psychological.

It has been argued by John Bell that a major driving force toward the implementation of screening may be the development of drugs along genetic guidelines. The interest here would be in screening and testing not only for genetic susceptibility to side effects to drugs, but also for the potential for better response rates. This could lead to the stratification of the patient population in ways to be taken into account by physician prescribers. The interpretation of the principles concerning the nature of the condition sought and the scope for action would in this context need reinterpretation.

they are directly of clinical importance (Sect. III, para 11).

III. IMPLEMENTATION AND QUALITY CONTROL

The Wilson–Jungner principles mention that the test should be acceptable to the population. This is not emphasized in the Nuffield or Danish reports. Perhaps it is thought that DNA testing can be done in a noninvasive way (e.g., by a mouthwash test) so that it is not necessary to have this requirement. There are still matters of inconvenience to be considered, however.

The accuracy and predictive power of a given test are important, especially when taking into account whether the benefits of taking the test are likely to make the possible inconvenience worthwhile. False positives and false negatives are sources of harm in themselves. Predictive power will be especially significant when considering screening for *predispositions* to multifactorial diseases—for example, in a disease which counts as an important problem, and which has a genetic component, but where there is a low concordance rate between identical twins, such as rheumatoid arthritis. If we know that the presence of a particular gene is a factor, but is without a high predictive value, the most we might be able to say is that someone *without* the gene is very unlikely to suffer from the condition in question. To those people the result of a test might bring considerable relief, but would this be worth doing? Would the criterion of "scope for action" be met? This is not clear. There is a further problem about the extent to which those being screened will understand the weakness in predictive power of particular tests.

Genetic counseling is commonly cited as an important stage in the implementation of screening programs and may be considered an aspect of quality control. The Nuffield Council says that counseling *should* be available (1993, para. 10.5); the Danish Council says it *must* be available (p. 66). The Danish report says that it must be nondirective (p. 66). The Nuffield report, however, recognizes that counseling is unlikely to remain completely neutral and that there are disadvantages in counseling that seems cold and unhelpful (para. 4.21).

The Danish report says that screening should be so organized as to examine only those factors on which the screening is focused (p. 48), but since it is not possible to exclude secondary findings, they recommend follow-up research to establish what importance such findings have acquired (p. 67). The Council of Europe provides that unexpected findings may be communicated only when

IV. RIGHTS TO KNOW AND NOT TO KNOW

It has already been mentioned that population screening may be part of a public health agenda, e.g., to reduce the prevalence of a disorder. The purpose of providing information via genetic screening, however, may be conceived differently in terms of facilitating choice. The principle of autonomy plays a major role in most if not all discussions of the issues, whether explicitly or implicitly. In the Danish report the duty to help is itself interpreted in terms of autonomy (p. 63) rather than, for example, promoting the health of the individual or the genetic health of the population. In medical practice generally, it is said there is a duty to help where possible—in effecting a cure, for example— but the duty to help in genetics is characterized differently, because the aim is not to effect a cure but to provide information which may be important in facilitating autonomy. Here autonomy appears to be interpreted in a wide sense—in other words, what is at issue is not self-determination in specific situations of choice, but the empowerment of people to think for themselves and take charge of their lives. However, there may come a point at which so much information is forthcoming that it undermines rather than promotes autonomy:

> It is perfectly conceivable that the greatest problem will be the volume of accessible information on the individual's personal sphere, in which case respect for the personal sphere will outweigh the motive for helping people achieve greater autonomy. In such situations, the implementation of genetic screening will not be ethically defensible. (p. 63)

There are also hints in the Nuffield report of implicit adherence to autonomy in the wide sense. It speaks of "the value to those being screened of the knowledge gained" (1993, para. 10.21). In paragraph 8.20 it is said that "the benefits [of screening] should be seen as enabling individuals to take account of the information for their own lives and empowering prospective parents to make informed choices about having children."

Elsewhere and in other reports there appears to be support for a narrower interpretation in terms of self-determination with regard to specific choices; this underlies the concern for informed consent therein. This is precisely, however, what is going to become increasingly problematic as human genome analysis makes possible, e.g., in the United States, screening for multiple disorders. The Committee on Assessing Genetic Risks has suggested that to minimize this problem, multiplex tests should be grouped into categories raising analogous issues.

The Danish report is relatively unusual in mentioning the duty to help in specific terms, though arguably this is an important factor in considering the ethical aspects of screening. If screening is withheld, individuals who could benefit from it are denied those benefits. There are well-known, though contested, arguments to suggest that the duty to help is less stringent or urgent than the duty to avoid harm, and screening has the potential to harm, by raising anxiety, changing people's self-image, and paving the way for genetic discrimination. The harm caused by failure to screen and that caused by implementing the screening program are both relevant in evaluating a proposed program. The increasing realization that genetic information may not be simply autonomy-enhancing but also a burden has led to claims of a right not to know genetic information, rather than a right to access it.

Suppose, for example, that a predictive test were found for severe mental disorder x, which was considered to undermine an individual's capacity for autonomous decision making. In the event that there was an effective preventive strategy, the individual could then, it might be argued, make an autonomous decision to have the therapy before the onset of symptoms, thus avoiding any later worries about informed consent. Even setting aside queries about predictive power, however, this possibility only gives rise to other ethical issues—should mental disorder "x" be eliminated, even before we know how, when, or even if it might manifest itself? The argument for a right not to know in this sort of situation might be supported by considerations of identity and integrity of the person, in addition to protection from psychological and social costs.

In turn this has been challenged by the argument that there is a duty to provide (and be informed of) information relevant to making decisions, as well as by an argument, based on grounds of solidarity, for a responsibility to share genetic information that can have potentially adverse consequences for other people, and hence a responsibility to be provided it.

V. INFORMATION: ACCESS AND CONTROL

The interests of parties other than the persons undergoing screening and testing include, first, family members. Whereas in bioethics generally most questions have typically been framed in terms of the individual, genetics is concerned with *relatedness*. The Nuffield report says, "Thus the status of genetic information raises ethical questions that differ significantly from the normal rules and standards applied to the handling of personal medical records" (1993, para. 1.10). The implications for family members include both informed consent and confidentiality of medical information. When an individual gives informed consent in this context he or she may be consenting to something that affects relatives. Health care professionals may be faced with difficult issues concerning whether or not to disclose genetic information to relatives of an affected individual. The important question for ethics is *how* the family enters the debate. Is it, for example, an independent value, or is it an extension of the ethics of individualism? The first of these can quickly be disposed of. It is not the integrity of the family itself that is at issue, as it has been, for example, in debates about the well-being of the nuclear family. Within an individualist ethic, genetic information concerning the person screened may have implications for the life choices of his or her spouse or genetic relatives, so their interests are at stake as well. In this situation of conflicting interests, an emphasis on the traditional confidential nature of the professional–patient relationship is likely to lead to a presumption against disclosure of information without consent.

The situation becomes more complicated when third-party interests other than family members are introduced: the interests of social institutions such as insurance companies and employers. Employers may well find it useful to screen potential employees for genetic predispositions such as susceptibility to toxins in the workplace, or indeed for certain behavioral traits. Insurers have an interest in information about the risk status of the insured.

The difficult questions in this area include not just a balancing of the interests of different parties but also the conceptual issue of the extent to which genetic information is different in kind from other medical or employment-relevant information. There are also wider questions of social policy and social justice.

VI. EUGENICS, DISCRIMINATION, AND STIGMATIZATION

Another focus of ethical concern arises out of perceptions of the public health agenda in genetic screening as the prevention of avoidable genetic disorders or a reduction in their incidence. From this perspective the emphasis on facilitating autonomy and choice is not held to be convincing. Either overtly or through the cumulative effect of the choices of large numbers of individuals, policies that are regarded as eugenic may be feared. In another context an argument against allowing people to choose the sex of their child is that it will lead to bad consequences for society, e.g., imbalance in the sex ratio, or that it will have undesirable implications for the position of women in society. Groups representing people with disabilities make similar points about genetic screening. The suggestion is that it is, at least in the reproductive context, implicitly eugenic and will lead both to less respect for the rights of disabled people and to fewer facilities in terms of support for them.

While there may be strong logical arguments to the effect that genetic screening does not necessarily discriminate against people with disabilities, nevertheless as long as the point is perceived in this way, the potential for adverse consequences is considerable. One of the most significant implications of genetic screening is the potential effect on our very understanding of concepts of normality and disability. The conceptual and political contest over our understanding of deafness, for example, becomes intensified. This may be the case even in relatively noncontroversial contexts: "Normality is a relative concept, and any health-related examination therefore includes a risk of the examinees feeling or being felt to be thought of as abnormal or just plain ill" (Danish Council, p. 60).

Some people object to the language of normality and abnormality, suggesting that it is not ethically neutral. There is, then, an issue of whether it is possible to be concerned for the genetic health of the population without adopting discredited eugenic policies—it might be thought to imply intolerance of what is regarded as imperfection. Arguably, however, if health of the population is a good that public health medicine has a duty to pursue, then why is not *genetic* health one as well? Reasons for holding that there is a moral difference are unclear, but the prior question is one of conceptual unclarity. The unclarity perhaps relates to the concept of "health," however, not the "genetic" element specifically. In one interpretation the content of the purported duty would be to reduce the incidence of genetic disorders, analogously to reducing the incidence of, say, infectious disease. The question concerns the extent the goal could be achieved while safeguarding the interests of individuals. The tension is reflected at deeper level by the theoretical ethical question of the opposition between autonomy and community.

The issue of abnormality is related to that of stigmatization. The Nuffield report is fairly optimistic about the dangers of stigmatization. They say that "such evidence as exists suggests that current genetic screening programmes need not result in any significant stigmatisation" (1993, para. 8.14). It may be right to say that previous disturbing precedents arose largely because of inadequate or poorly understood information. A question must be asked, however, about the significance of such factors as the social context, target population, and nature of the disorder in question. It may be argued, for example, that there is a danger of racism in discus-sions of genetics in cases where some minority ethnic groups have a higher frequency of a particular gene.

In this particular area the tension between the interests of individuals regarding control of their genetic information and the interests of others in having access to it becomes acute. The potential benefits of greater sharing of genetic information, in terms of the interests of genetic relatives, future people, and society as a whole, may be great but the question is whether they are sufficiently compelling to justify the imposition of individual responsibility to share it. The most promising way forward seems to be an attempt to mediate between the interests of the individual and community via the *encouragement* of sharing of information rather than by its *requirement*.

VII. CONCLUSION

As the amount of knowledge about the human genome increases, the possibilities for genetic testing and screening multiply. How we should deal with the volume of information forthcoming, however, remains a contested area with regard to justifications for acquiring the information by screening programs and how to control access to the information so acquired. Consideration of the applications and implications of genetic information gives rise to extremes of both optimism and pessimism, and in light of the ways in which this information has the potential to challenge the boundaries of our concepts, e.g., of health, disease, and normality, the applicability of ethical frameworks is also tested. Hence we see reconsideration of autonomy, the right to know, informed consent, and confidentiality in this context.

Bibliography

Bell, J. (1998). The new genetics in clinical practice. *British Medical Journal* **316**, 618–620.

Chadwick, R. (1999). Criteria for genetic screening: The impact of pharmaceutical research. *Monash Bioethics Review* **18**(1), 22–26.

Chadwick, R., *et al.*, Eds. (1997). *The Right to Know and the Right Not to Know.* Aldershot: Avebury.

Chadwick, R., *et al.*, Eds. (1999). *The Ethics of Genetic Screening.* Dordrecht: Kluwer.

Chadwick, R., and Levitt, M. (1999). Genetic technology: A threat to deafness. *Journal of Medicine, Healthcare and Philosophy* **1**, 1–7.

Clothier, C. M. (Chairman) (1992). *Report of the Committee on the Ethics of Gene Therapy.* London: HMSO.

Committee on Assessing Genetic Risks (1994). *Assessing Genetic Risks: Implications for Health Policy.* Washington, DC: National Academy Press.

Council of Europe (1992/1994). Recommendation R(92)3 on Genetic Testing and Screening for Health Care Purposes (reprint). *Bulletin of Medical Ethics* Feb. 9–11.

Danish Council of Ethics (1993). *Ethics and Mapping of the Human Genome.* Copenhagen: Danish Council of Ethics.

Health Council of the Netherlands (1994). *Genetic Screening.* The Hague: HCN.

Holtzman, N. A., and Watson, M. S., Eds. (1997). *Promoting Safe and Effective Genetic Testing in the United States: Final Report of the Task Force on Genetic Testing.* Baltimore: Johns Hopkins Univ. Press.

McGleenan, T., *et al.,* Eds. (1999). *Genetics and Insurance.* Oxford: Bios.

Nuffield Council on Bioethics (1993). *Genetic Screening: Ethical Issues.* London: Nuffield Council on Bioethics.

Nuffield Council on Bioethics (1998). *Mental Disorders and Genetics: The Ethical Context.* London: Nuffield Council on Bioethics.

Rhodes, R. (1998). Genetic links, family ties, and social bonds: Rights and responsibilities in the face of genetic knowledge. *Journal of Medicine and Philosophy* **23,** 10–30.

Shickle, D., and Chadwick, R. (1994). The ethics of screening: Is "screening-itis" an incurable disease? *Journal of Medical Ethics* **20,** 12–18.

Thompson, A. K., and Chadwick, R. F., Eds. (1999). *Genetic Information: Acquisition, Access and Control.* New York: Kluwer/ Plenum.

Wilson, J. M. G., and Jungner, G. (1968). *Public Health Papers,* Vol. 34: *The Principles and Practice of Screening for Disease.* Geneva: World Health Organization.

Genetic Technology, Legal Regulation of

TONY McGLEENAN

The Queen's University of Belfast

GLOSSARY

DNA Deoxyribonucleic acid is the chemical substance which makes up a gene and which contains an individual's genetic code.

gene The basic unit of heredity consisting of a sequence of DNA which occupies a particular location on the genome.

genome The total genetic complement of an individual or of a species.

germ line gene therapy A scientific procedure which alters the reproductive cells of an individual, thereby altering the genome permanently for future generations.

somatic cell gene therapy A scientific procedure which alters the genetic structure of the ordinary cells of an individual.

GENETIC TECHNOLOGY promises to deliver tremendous health and social benefits to the human race. It may be possible to utilize genetic technology to eradicate some of the major genetic diseases that currently truncate the human life span. Advances in genetic screening and testing will enable us to determine our susceptibility to certain diseases, including some of the major multifactorial diseases such as cancers and heart disease. It may then be possible to make suitable lifestyle changes in order to minimize the risks from these illnesses. However, alongside these potential benefits, genetic technology also carries significant potential for harm. Gene therapy techniques could be used to interfere with an individual's reproductive cells and so alter the genetic composition of generations of offspring. Genetically modified organisms may be accidentally or deliberately released, causing untold environmental damage. Unscrupulous regimes may attempt to link genetic technology with political ideology to create stronger, taller, or more intelligent human beings. Genetic screening techniques could be employed to provide intimate information about individuals which could be used by insurers, employers, or governments to deny basic benefits such as health and life insurance, housing, and employment to those seen as high risk. Prenatal genetic testing could be utilized in order to facilitate the termination of pregnancies where the fetus is found to have less than optimum genetic health. Because of the potential for genetic technology to both benefit and harm the human race, discussions about means of regulating the use of these techniques have begun in many countries. This article examines the different modes of regulation adopted at both international and national levels, and comments on the trends in genetic regulation.

I. INTERNATIONAL REGULATION

A. International Declarations

1. The Declaration of Iyanuma

At the 1991 Council for International Organisations of Medical Sciences (CIOMS) conference on Genetics, Ethics, and Human Values, the Declaration of Inuyama was formally adopted. Many of the provisions of the Declaration have gone on to inform other international documents addressing the issues raised by gene therapy. Section 1 of the Declaration affirms the need for gene therapy to conform to the ethical standards of research and for any knowledge acquired by genetic techniques to be used appropriately. It states that somatic cell gene therapy should be approached like any other form of therapy but that any use of genetic technology which is intended to improve or delete aesthetic, behavioral, or cognitive characteristics is prohibited except where such action is associated with the treatment of a disease. Section 6 of the Declaration states that germ line gene therapy should only be acceptable when all risks have been eradicated.

2. The Valencia Declaration

At the Second International Cooperation Workshop of the Human Genome Project the Valencia Declaration was adopted. This stated that somatic cell gene therapy was an acceptable form of treatment for certain diseases. The use of germ line gene therapy was opposed on the basis of the technical difficulties which confronted it and the ethical controversy which surrounded the technique.

3. European Convention on Bioethics and Biomedicine

The European Convention on Human Rights and Biomedicine was finally agreed upon by the Committee of Ministers of the Council of Europe in November 1996. The Council of Europe was established in 1949 to promote political, legal, and cultural cooperation among member states. It is entirely distinct from the European Union, a fact of some significance in relation to the enforceability of the Council's norms. The Parliamentary Assembly of the Council of Europe began drafting a Bioethics Convention in 1991. The Committee of Ministers issued a directive to the Committee on Bioethics to "study the set of problems posed for law, ethics and human rights by progress in the biomedical sciences . . . with a view to harmonising the policies of the member states as far as possible."

After a protracted and controversial discussion period the Convention was endorsed by all but 3 of the 39 participating nations in 1996 and was signed by 5 participating nations in 1997. The Convention marks a significant attempt to address the diverse dilemmas of bioethics through the use of a human rights framework. The Convention consciously follows the model of the 1950 Convention for the Protection of Human Rights and Fundamental Freedoms (ECHR), at times even borrowing language and phrases form the ECHR.

The rationale for the adoption of a new Convention rather than simply amending other Council resolutions and recommendations was set out by Palacios, rapporteur to the General Assembly. First, it was argued that advances in biomedicine were moving at such a pace that the laws in the various member states were not able to keep pace with the developments. Second, concerns were emerging that given the rapid pace of development, and the fragmentation of approach among member states, "havens" could emerge for research where scientists could exploit lack of regulation in order to evade the legal restrictions in force in their own countries. Autonomy and self-determination are the core principles underpinning the Convention. Many of the safeguards built into the Convention are dependent on the informed consent of the individual. This has been cogently criticized by Hennau-Hublet, who argues that this represents

> the consecration of an Anglo-American concept that in practice, if not also in theory, tends to reduce the person to an act of volition of which the human body is only an object. This is all the more serious because we have no reason to believe that expressions of will alone cover the real interest of the person or affirm the primacy of the human being. (1996, p. 25)

The autonomy model has been adopted in numerous state jurisdictions as the means of protecting individuals from the social consequences of genetic information. As has been noted, the use of individual autonomy is not necessarily an effective safeguard against these undesirable social consequences. Where the possibility exists for powerful social actors, such as employers and insurers, to seek genetic information for their own ends, the autonomy of individuals is unlikely to provide adequate protection.

A specific section of the Bioethics Convention is devoted to bioethical problems related to the human genome. The issue of genetic privacy is addressed in the terms of the Convention, albeit in rather opaque language. Article 12 addresses informational concerns in relation to genetic screening:

> Tests which are predictive of genetic diseases or which serve either to identify the subject as a carrier of a gene responsible for a disease or to detect a genetic predisposition or susceptibility to disease may be performed only for health purposes or for scientific research purposes, and subject to appropriate genetic counselling.

As is frequently the case with international human rights texts, the language of this article is open in nature. The restriction of genetic diagnostics to "health purposes" and "scientific research purposes" clearly has implications for those who would seek to use genetic screening in the workplace or in the classification of life and health insurance premiums.

The precise meaning of Article 12 can be further illuminated by reading it in conjunction with the provisions of Article 11, which states that "any form of discrimination against a person on grounds of his or her genetic heritage is prohibited." This provision represents a clear attempt by the Council of Europe to address the concerns arising from the potential use of predictive molecular analysis in employment and insurance. Concerns that the provisions of the Convention will have a significant impact on such matters are evident from the response of the Association of British Insurers, which publicly denounced parts of the Convention, claiming, in particular, that Article 11 (prohibiting any discrimination on genetic grounds) was unenforceable. Articles 11 and 12 appear to suggest that genetic testing carried out in the workplace may only be acceptable where it has been performed for the health benefit of the employee. Consequently, the use of such diagnostic techniques to exclude from the workplace those who have a predisposition to ill health would appear to be prohibited. The wording of the provision raises a presumption against the use of genetic testing in the workplace for employment purposes, yet uncertainties remain in relation to testing which may have a health and employment purpose, as Lawton argues:

> An employer may very well have a mixed motive for genetic testing. But the employer's asserted motive must be the health of the employee. Allowing the employer, and not the employee, the right to decide whether to accept a job that exposes the employee to adverse health consequences because the employee has a genetic susceptibility to certain environmental agents creates improper incentives for employers. (1997, p. 365)

This would accord with the approach adopted within a large number of the member states who contributed to the drafting of the Convention. Similarly, Articles 11 and 12 appear to suggest that genetic testing for the purposes of risk classification for insurance would also be unacceptable. One key issue which is not addressed in the Convention is whether individuals who have taken a genetic test for "health purposes" will be required to disclose that information to an insurance company.

Article 11 of the Convention contains a broad prohibition on discrimination based on genetic heritage. This raises questions about which characteristics can be considered part of an individual's "genetic heritage." If gender is part of the genetic heritage then this would seem to prohibit the common insurance practice of offering differential rates to male and female proposers based on the actuarial differences in the mortality tables. Indeed the terms of Article 11 seem to threaten the entire insurance industry practice of differentiating between different insurance risks, since underwriting practice is based on setting different premiums for individuals based on their potential health risk, something which is inextricably linked with their genetic heritage.

4. UNESCO Declaration on the Human Genome and Human Rights

The United Nations Educational, Scientific, and Cultural Organisation (UNESCO) formally adopted a Universal Declaration on the Human Genome and Human Rights in 1996. The International Bioethics Committee of the United Nations had been mandated in 1993, by 185 member states, to consider the possibility of establishing an international legal framework for the protection of the human genome. The response is in the form of a declaration rather than a legally binding treaty, because of the apparent political need for flexibility. Lenoir argues that the Declaration has a twofold purpose:

> It protects the rights and liberties of individuals and also enshrines the role of science and knowledge in helping civilisation to progress. The declaration is also designed to remind the international community of its duty of solidarity towards poorer countries from the benefits of biomedical progress. (1997, p. 33)

This attempt to balance the somewhat divergent claims of social solidarity with the protection of individual human rights permeates the entire Declaration. Article 1 states that "the human genome, inasmuch as it underlines the fundamental unity of all members of the human family and the dignity with which each is endorsed, is a common heritage of humanity."

The issue of genetic privacy is addressed in Articles 8 and 9 of the Declaration. Article 8 is of relevance to the privacy issue because it refers to the concept of genetic discrimination, which relates to the question of whether insurance companies are entitled to have access to the results of a genetic test. It states that "no one may be subjected to discrimination based on genetic characteristics that is intended to diminish, or has the effect of diminishing, human dignity or impairing the right to be treated equally." This formulation is of interest in relation to the genetics and insurance debate. If genetic information is used in the formation of an insurance contract then it is highly likely that this practice will have the "effect of . . . impairing the right to be treated equally."

Article 9 of the Declaration explicitly refers to the issue of genetic privacy: "Genetic data associated with a named person and/or stored or processed for the purposes of research or any other purpose must be held confidential and protected against disclosure to third parties." This would also seem to stand against the possibility of those with a commercial interest in private genetic information, such as employers and insurers, being able to demand access to that information. The Declaration has an avowed determination to uphold both individual human rights and broader notions of social solidarity. The issue of solidarity is addressed in a separate section of the document under the heading, "Duty of Solidarity." Article 15 declares that "states must guarantee the effectiveness of the duty of solidarity towards individuals, families and population groups that are particularly vulnerable to disease or disability linked to anomalies of genetic character." This is a strong aspiration but nowhere does the document address the issue of how social solidarity can be fostered in a regulatory climate where individuals have the right not to disclose genetic information "for any purpose" to third parties. Nor is it apparent how the terms of Article 17, which mandates states to "foster the international dissemination of scientific knowledge," can be fulfilled when individuals have been granted an explicit right which frustrates this very purpose.

Like the Council of Europe Convention, the legal status of the UNESCO Declaration can be ascertained by comparison with another long-standing international human rights document, the 1948 Universal Declaration of Human Rights. Lenoir argues that the Human Genome document is analagous to the 1948 instrument which, while not strictly binding, is referred to in many jurisdictions as a source of legal inspiration and which has, in fact, been integrated into the modern constitutions of both Spain and Portugal.

B. European Union Regulations

There are many hundreds of regulations, directives, and decisions in European Union law which are directed at the field of genetic technology. However, most of these measures are derivative of a number of key directives which have been adopted in order to ensure a coordinated approach throughout the member states of the European Union to regulation. The most significant European laws to date have involved the development of biosafety directive relating to contained use and deliberate release of genetically modified organisms. These provisions are considered directly applicable in the member states and consequently the national governments must implement these norms although they retain some discretion as to the detail of the legislation. The following directives have been implemented into

the national laws of most member states in the European Union.

1. Contained Use Directive 90/219 EC

This directive seeks to prevent environmental damage from accidental release of genetically modified organisms (GMOs) by ensuring harmonization between the laws of the member states which relate to the use of GMOs "with a view to protecting human health and the environment." The directive obliges member states to ensure that all appropriate measures are taken to avoid adverse effects on human health and the environment which might arise from the contained use of GMOs. The directive imposes a significant administrative burden on member states in the interests of biosafety, including reporting requirements and the introduction of guidelines for the safe handling of GMOs. At the end of each year member states are required to inform the European Commission of all instances of the contained use of GMOs.

2. Deliberate Release Directives 90/220 EC

The objective of this directive is

to approximate the laws, regulations and administrative provisions of the Member States and to protect human health and the environment: when carrying out the release of genetically modified organisms into the environment, when placing on the market products containing, or consisting of, genetically modified organisms intended for subsequent deliberate release into the environment.

The directive contains 23 detailed articles which impose a series of conditions upon any researcher who plans to release genetically modified organisms into the environment.

Most member states have implemented the contained use or deliberate release directives in some form. One area of controversy in relation to the deliberate release directive is whether it will apply to human beings who have undergone some form of somatic cell gene therapy. Such an interpretation would require the introduction of a significant number of procedural safeguards to the process of somatic cell gene therapy trials. The directive is not explicit on the point of human gene therapy. The definition of a genetically modified organism simply states that it refers to "an organism in which the genetic material has been altered in a way that does not occur naturally by mating and/or natural recombination." The legislation contains appendexes which state that the process of genetic modification is not to be taken to include *in vitro* fertilization. It remains silent, however, on the question of human somatic or germ line gene therapy, which could lead to the interpretation that these treatments could come within the bounds of the legislation. The matter has been clarified somewhat by the 1993 marketing authorization directive 2903/93 EC, which states that "Art 11 to 18 of Directive 90/220 EC shall not apply to medicinal products for human use containing or consisting of genetically modified organisms." It would appear therefore that the other articles of the directive do in fact apply to human gene therapy trials.

3. Marketing Authorization Directive 2309/93 EC

In 1993 the Council of the European Communities laid down procedures for the establishment of a European Agency for the Evaluation of Medicinal Products. The purpose of this legislation was to introduce a standardized system of regulation of medicinal products. Before any such product can be placed on the market in any member state of the European Union it must receive market authorization from the European Agency. In particular, the directive states that where a medicinal product involves a genetically modified organism as defined in 90/220/EC, the application for market authorization must be accompanied by a copy of any written consent from the competent authorities to the deliberate release of the GMOs into the environment; a technical dossier; an environmental risk assessment; and the results of any investigations performed for the purposes of reasearch or development. Article 6(4) states that the European Agency has a duty to consult with the biosafety bodies established in the member states and should draw up guidance for the use of GMOs in humans in concert with such bodies and with other "interested parties." The directive introduces a number of new requirements in the name of "pharmacovigilance." Thus adverse reactions to any medicinal product authorized for use in the EU must be reported to the Agency.

Various other organs of the European Union have issued official statements or policy documents which relate to the question of genetic technology. In 1989 the European Parliament published a resolution on the ethical and legal problems of genetic engineering which addressed both gene therapy and genetic screening. The Council issued a decision in December 1994 adopting a specific program of research and technological development in biomedicine and health (Decision 94/913/EC). Part of this program relates to research on the human genome, and the decision contains an apparent prohibition on the use of germ line gene therapy. It states that

> no research modifying of seeking to modify, the genetic constitution of human beings by alteration of germ cells or of any stage of embryo development which may make these alterations necessary, will be carried out under this programme.

Similarly, the Group of Advisers on the Ethical Implications of Biotechnology who report to the European Commission issued an opinion in 1994 on the ethical implications of gene therapy. This report described the criteria under which somatic cell gene therapy could be undertaken in the United Kingdom and then went on to state that germ line gene therapy was not "at the present time" ethically acceptable.

II. NATIONAL REGULATION

A. Australia

The regulation of medical research in Australia adopts the two-tier model of regulation. Overall regulation is coordinated by the National Health and Medical Research Council (NHMRC), a body which was placed under a new statutory framework by the National Health and Medical Research Council Act of 1992. This body supervises all experimentation on human subjects and requires the establishment of Institutional Ethics Committees (IEC's) as a precondition for research funding. IECs are required to monitor the progress of the project and to ensure that appropriate procedures for obtaining consent are followed. Unusually for a system of local ethical review, the IECs monitor not only those projects funded by the NHMRC but also projects which are funded internally by the particular academic institution and those which are funded externally by other bodies. For this reason they have been described as the "linchpin" of the Australian system by Chalmers.

The National Health and Medical Research Council Act of 1992 established a new regulatory framework and an Australian Health Ethics Committee (AHEC) to which the IECs must report. In the field of gene therapy IECs are guided by NHMRC Guidance Note 7, introduced in 1987 and modified in 1992, which states that while somatic cell gene therapy is acceptable, germ line gene therapy is not. Note 7 regards somatic cell gene therapy as experimental treatment and stipulates that the IEC should satisfy itself that the research is ethical and that the investigators are competent. In 1994 the NHMRC established a centralized Gene Therapy Committee (GTC) closely following the model of the Gene Therapy Advisory Committee (GTAC) in the United Kingdom and adopting to a large extent the protocols developed by the Recombinant DNA Advisory Committee (RAC) in the United States. The GTC does not hold public meetings. This two-tier ethical review system is supplemented in the area of safety by the Genetic Manipulation Advisory Committee (GMAC), which supervises the use of novel genetic manipulation techniques and is primarily concerned with the containment of genetically modified organisms in order to prevent escape of such entities into the environment. Most human gene therapy protocols are exempted from review by GMAC.

Genetic technology is also regulated under the umbrella of assisted conception laws. Australia has a federal and state system of government and the individual states have opted for slightly different forms of regulation for the new reproductive technologies. Three states, Victoria, South Australia, and Western Australia, opted for statutory frameworks. In Victoria the Infertility (Medical Procedures) Act of 1984 established a standing committee which operates as a licensing authority. South Australia, under the Reproductive Technology Act of 1988, and Western Australia, with the Human Reproductive Technology Act of 1991, opted for a more explicit licensing system. In each state an annual license is required which stipulates the conditions which must be met in the use of reproductive technology. Three states, Queensland, Tasmania, and New South Wales, have opted for a system of self-regulation rather than a statutory framework.

B. Austria

The Austrian government enacted in July 1994 the Gene Technology Law *Genetechnik* Federal Law BGB 510/1994. This is a broadly based piece of legislation which regulates all work with genetically modified organisms and the use of genetic testing and gene therapy in human beings. Section 1 states that the aims of the legislation are first to protect the health of man and his descendants from damage which may be caused by manipulation of the human genome, by genetic analysis, or by the impact of genetically modified organisms. Second, the legislation is intended to support the application of genetic technology for the sake of human well-being by creating a legal framework for research, development, and application of genetic technology. Section 4 of the Act applies to the use of gene analysis and gene therapy techniques in human beings. Section 64 contains a direct prohibition of intervention in human reproductive cell lines—thus germ line gene therapy is clearly prohibited.

Section 65 of the Gene Technology Law deals in some detail with the issues arising from genetic screening and testing technology. The law states that genetic testing or screening can only be carried out at the request of a doctor specializing in medical genetics. The analysis cannot be performed without the written consent of the patient. Prenatal genetic screening must only be carried out where it is medically necessary to do so. Under Section 70, the doctor carrying out the analysis is required to advise the person whose DNA is being ana-

lyzed to inform relatives of the test where the testing of relatives would verify the diagnosis, or where it is believed that there will be a risk for the relatives of the individual. The legislation also contains substantial provisions on the storage and protection of data derived from genetic screening.

C. Belgium

Although, like many other nations, Belgium does not have a specific genetics law, it was one of the first nations to attempt to address the problems which might arise from genetic technology in that it established a Higher Council on Human Genetics in 1973. More recently the Crown Order of 14 December 1987 established explicit standards for the operation of centers for human genetics. In the Law of June 25, 1992, the Belgian government precluded insurance companies from requesting or using genetic information in their determinations of life insurance contracts.

D. Canada

The use of gene therapy in Canada is largely supervised by the Canadian Medical Research Council (MRC). The use or development of germ line gene therapy is completely prohibited. The MRC does permit the use of somatic cell gene therapy provided that:

- The disease is attributable to a single-gene disorder
- The genetic anomaly must give rise to a seriously debilitating disease or premature death
- The disease cannot be successfully treated by any other method

These guidelines were drawn up by the MRC in 1987 and were accompanied by a recommendation that a National Committee for the Examination of Gene Therapy should be established. In 1993 the Royal Commission on New Reproductive Technologies also recommended that all gene therapy protocols should be evaluated by a national assessment committee.

A new legislative initiative to regulate genetic technology in Canada was introduced in the House of Commons in June 1996. This followed the recommendations of the Final Report of the Royal Commission on New Reproductive Technologies. The Human Reproductive and Genetic Technologies Act created a regulatory framework which includes a national agency, operating at the federal level, which establishes national standards for the use of genetic materials, issues licenses, and ensures compliance with the legislation. Section 3 of the Act states that the objectives of the legislation are to

protect the health and safety of Canadians in the use of human reproductive materials for assisted reproduction, other medical procedures and medical research; to ensure the appropriate treatment of human reproductive materials outside the body in recognition of their potential to form human life; and to protect the dignity of all persons, in particular children and women, in relation to uses of human reproductive materials.

The Act itself prohibits 13 practices which are set out in Section 4 of the proposed legislation. Among the prohibited practices are sex selection for nonmedical purposes; commerical sale of eggs, sperm, or embroys; germ line genetic manipulation; cloning of human embryos; commercial surrogacy; research on human embryos, older than 14 days after conception; and the creation of embryos only for the purpose of research.

E. Denmark

Danish Law 353 of 3 June 1987 established the Danish Council of Ethics, which has produced several significant reports on the area of genetic technology. In 1989 the Danish Council produced a report on the "Protection of Human Gametes, Fertilized Ova, Embryos and Fetuses" which argued for a prohibition on germ line gene therapy. The report was more accommodating toward somatic cell gene therapy although the Danish government stated that any such intervention would only be tolerated if it could be shown that there was no danger that somatic cell therapy could alter the germ cells.

F. Finland

In 1995 Finland passed the Gene Technology Act (337/1995). The aims and scope of this legislation are established in Section 1, which states,

The aim of this Act is:

1. To promote the safe use and development of gene technology in a way that is ethically acceptable; and
2. To prevent and avert any harm to human health, animals, property or the environment that may be caused by the use of genetically modified organisms.

The legislation is supervised by the Finnish Ministry for Social Affairs and Health insofar as it relates to general matters potentially affecting human health. It is jointly controlled by the Ministry of the Environment, which has jurisdiction over those aspects of genetic technology and genetically modified organisms which may have some impact on the environment.

In addition, the legislation establishes a broader regulatory framework through the introduction of a Board for Gene Technology. The details and functions of this board are more fully established in the Gene Technology Decree (821/1995) of 24 May 1995, which provides that, *inter alia,* the Board shall act as a registration

authority, that it should review on a case by case basis the initial use of any new genetically modified organisms, that it should provide guidelines for institutions engaged in genetic technology research or production, that it has the power to impose sanctions on institutions using genetically modified organisms under Section 22 of the 1995 Act, and that it should have responsibility for the contained use and deliberate release of genetically modified organisms.

In addition to the Gene Technology Board, Section 6 of the Act also makes provision for the utilization of such "expert authorities and institutions" as may be specified by decree. This is developed in the Gene Technology Decree, which lists five expert authorities and four expert institutions.

G. France

France has developed one of the most comprehensive systems of regulation of genetic technology. This reflects the fact that the issues relating to genetic technology have provoked a considerable degree of public interest and debate within France. A number of influential reports on the issue of genetic technology have emanated from France, including the Lenoir Report, the Serusclat Report, the Bouliac Report, and the Matie Report. The *Comite consultif national d'ethique pour les sciences de la vie et de la sante* (CCNE) has twice reported on the questions raised by genetic technology. In 1990 the CCNE formally prohibited all forms of germ line gene therapy and limited the use of somatic cell gene therapy to monogenic heriditary disease with a grave prognosis. The permissible scope of somatic cell gene therapy was extended by a second opinion of the CCNE in 1993 which stipulated five requirements which must be met before any form of somatic cell gene therapy could go ahead:

1. Any procedure designed to change the physical or mental characteristics of an individual, except in the case of serious disease, is prohibited
2. Somatic cell gene therapy trials must be preceded by adequate prior examination of animals to determine the effectiveness and safety of the techniques
3. Somatic cell gene therapy may only be used in individuals suffering from a condition for which there is no available effective treatment and the prognosis for which is sufficiently serious to warrant the potential risks
4. The results of all research must be monitored for scientific and technical validity and by the CCNE for ethical standards
5. Any information about the potential for the research must be released with objectivity, restraint, and realism to avoid unduly heightened expectations

In addition to the CCNE two other bodies are concerned with the regulation of genetic technology in France, the *Commission de gene genetique* (CGG) and the *Commission de genie biomoleculaire* (CGBM). These two bodies are charged with ensuring compliance with the EC directives on controlled use and deliberate release. They in turn report to the *Agence du medicament,* which has two commissions with competence in this area, the Commission on Viral Safety and the Commission on Clinical Trials. These commissions are directly connected and report to the French Minister for Health. In common with the two-tier system operated in many countries, the French also have local ethical review which takes place under the auspices of the *Comite consultif de protection des personnes se pretant a des recherches biologiques.*

H. Germany

There are at least three areas of law which regulate genetic technology in Germany. First, the use of genetically modified material is regulated by the Arzneimittelgesetz (AMG—The Law of Medical Drugs). Sections 40–42 of the AMG regulate the safety of the patient who is undergoing any form of medical treatment. The AMG requires that a local ethical committee give approval to any clinical trial. The law requires that the patient consent to the intervention, having been informed of the nature and effect of the treatment. The patient must be insured and has the right to be compensated for any damage which occurs as a consequence of the treatment.

Of perhaps greater importance is the more controversial *Embryonenschutzgesetz* (ESG—Embryo Protection Law). As with the United Kingdom's Human Fertilisation and Embryology Act, the ESG states in Section 5 that the alteration of human germ line cells is a criminal offense. The ESG differs from similar legislation in other states insofar as it introduces an element of intent. Thus where the germ line alteration is unintentional, i.e., as a consequence of radiation treatment or chemotherapy, the action does not fall within the prohibition in Section 5. While this may appear to be a sensible rule, it has given rise to controversy as it would appear to permit unintentional germ line gene therapy which occurred during the development of somatic cell gene therapy techniques.

The use of genetic technology in Germany is also circumscribed by the relatively rigid rules of the German medical profession. Most significant among the rules which have been developed by the German medical profession is the principle that a physician need not seek full legal and ethical approval for the "single attempt to heal." Thus an exception is made to the normally very rigid codes of practice where the treatment in question is an experimental treatment carried out as a matter of last resort.

Also of importance is the *Gentechnikgesetz* (GTG—The Law on Genetic Technology). This law does not refer directly to the use of genetic technology on human beings but is, rather, a biosafety measure designed to comply with the requirements of European law. Before any clinical trial involving genetically modified material can begin, the safety of the proposed procedure must be assessed by the Central Commission for Biological Safety.

I. Netherlands

Genetic screening has been regulated in the Netherlands since 1992 when the *Staatsblad van het Koninkrijk der Nederlanden* (Population Screening Act) was enacted. This law requires that the minister approve any genetic screening program before it is implemented and requires that he be advised on the matter by the Health Council. The Act defines population screening in broad terms as

> a medical examination which is carried out in response to an offer made to the entire population or to a section thereof and which is designed to detect diseases of a certain kind or certain risk indicators either wholly or partly for the benefit of the persons to be examined.

A license for the screening program will be refused if it is scientifically unsound, if it conflicts with statutory regulations, or if it involves risks that outweigh the likely benefits. Following the implementation of this law the Health Council of the Netherlands published a report from the Committee on Genetic Screening. This report states that screening should not necessarily be restricted to certain categories of serious genetic disease, but rather takes the view that the provision of comprehensible information is essential to maintain the autonomy of the individual.

Gene therapy legislation was introduced in 1993 to implement the EC directives on genetically modified organisms into Dutch national law. The Dutch Health Council reported on the issue of gene therapy in 1989 and found that somatic cell gene therapy was an acceptable experimental procedure which could be undertaken provided the following safeguards had been met:

- The protocols must be approved by the Central Committee on the Ethics of Medical Research
- The procedures must be monitored to ensure that they comply with the terms of the submitted protocol

- The protocols will be the subject of specific requirements in relation to the scientific value and safety of the project

Germ line gene therapy was also considered by the Health Council and they recommended that a moratorium on germ line research be introduced because of the unduly high risks posed by the technique.

The terms of the Medical Examinations Act of 1998 prevent insurance companies in the Netherlands from seeking disclosure of the results of any genetic test where the amount being sought is less than NLG 300,000. The legislative scheme explicitly prohibits certain types of medical examination. An individual must not be asked whether he or she will suffer from any hereditary, untreatable, or serious disease unless they are already expressing symptoms of the disease. In addition, the legislation attempts to prohibit the acquisition of genetic information through the results of genetic tests taken by relatives. Insurers are prohibited from asking whether any blood relative has any hereditary, serious, untreatable disease, even if the individual is exhibiting symptoms of that disease. For insurance policies with a value in excess of the threshold of NLG 300,000 for life insurance and NLG 60,000 for disability policies, the insurance companies are at liberty to seek disclosure of existing genetic information, but not to require an individual to take a genetic test. The approach adopted in the Netherlands is one which is, in effect, a relatively common practice in the insurance industry. A system which triggers requests for disclosure or examination is already well established for certain insurance products.

J. Norway

In August 1994, the Storting passed one of the more restrictive laws relating to the regulation of genetic technology. The premble to the Act Relating to the Application of Biotechnology in Medicine (1994) states that

the purpose of this Act is to ensure that the application of biotechnology is utilised in the best interests of human beings in a society where everyone plays a role and is fully valued. This shall take place in accordance with the principles of respect for human dignity, human rights and personal integrity and without discrimination on the basis of genetic background based on ethical norms relating to our western cultural heritage.

Section 7 of the Act deals directly with the issue of gene therapy and states that the human genome can only be altered for the purpose of treating serious disease or preventing serious disease from occurring. In common with many other regulations, it states that germ line gene therapy is prohibited. In addition to the terms of the legislation the Act also requires that no treatment involving gene therapy can go ahead without first ob-

taining permission from the Ministry of Health which has constituted a Biotechnology Advisory board. The Norwegian statute also refers to the issue of genetic testing in Section 6. Genetic testing is permitted, although Section 6(2) states that it may only be carried out when there is a clear diagnostic or therapeutic objective. Perhaps more importantly, given the dangers of genetic discrimination, Section 6(7) of the Act states that "it is prohibited to request, receive, possess or make use of genetic information concerning an individual that results from a genetic test." Similarly it is illegal to inquire as to whether or not a genetic test has actually been carried out.

K. Sweden

In Sweden genetically modified material, which would include the type of products to be used in gene therapy, fall into the category of drugs, and are therefore regulated under the Swedish Medical Products Agency. Under Law 114 of March 1991 authorization from *Svensk forfattningssamling* is required before investigations or testing of DNA can be carried out. Under Law 115 of March 1991, this body must give its approval to any somatic cell gene intervention. This law states that experimentation on fertilized oocytes must not have the objective of developing methods of "causing heritable genetic effects." Approval is also required before any genetic testing is carried out for diagnostic purposes. In addition to the national body, Sweden has also developed a system of regional advisory ethics committees which assess the scientific validity of research protocols.

L. Switzerland

Switzerland, perhaps uniquely, has actually amended the Federal Constitution by the insertion of a new Section 24 which states that "man and the environment are protected against the abuse of genetic engineering and reproduction technology." The amendment states that interventions which affect the genetic heritage of human gametes are not permissible; that the germ lines of non-human species should not be transferred into human beings; that human germ line cells should not be the subject of commercial transactions; and that an individual's genetic composition may only be analyzed, recorded, and disclosed with his or her consent.

In addition to this constitutional regulation, individual cantons have also enacted laws which impact the field of genetic technology. A directive issued in the canton of Basel-Land in 1987 relating to the use of *in vitro* fertilization techniques provides that interventions on the genetic material of human cells are to be prohibited. In April 1994 the canton of Geneva passed regulations on clinical research involving interventions in the field of human genetic manipulation. This legislation requires that research in the field of genetic manipulation is to be undertaken solely under the auspices of Geneva University or a body of similar academic standing, and stipulates conditions which must be met by research protocols contemplating the use of genetic manipulation techniques. Since 1993 the Regulation for Intercantonal Control of Medicinal Products has been in force. Currently drug and clinical trials are regulated at the cantonal level but the new regulation harmonizes the regulatory procedure and defines the tasks of the local ethical committees. This regulation closely affects the development of gene therapy trials in Switzerland.

The issues of contained use and deliberate release of genetically modified organisms into the environment are addressed in the Law on Environmental Protection, which in conjunction with the Law on Epidemics has led to the establishment of an Expert Commission for Biosafety. The use of genetic technology in Switzerland is also constrained by the guidelines of two national bodies—the Swiss Academy of Medical Sciences (SAMW) and the Swiss Commission for Biological Safety (SKBS). The SAMW guidelines require that any research project involving human beings must gain the approval of an ethics committee before proceeding. The SAMW is not a state body and this system of regulation is a voluntary one except in those cantons where the guidelines have been incorporated into legislation. The SAMW has determined that genetic technology poses such significant problems that a special subcommittee on gene therapy has been established.

While ethical oversight is mainly the preserve of the SAMW, issues of safety are considered by the SKBS. This organization registers the interests of researchers and in 1992 issued Guidelines for Work with Genetically Modified Organisms to researchers planning to submit a protocol involving genetic technology. These guidelines are broadly based on the Points to Consider document developed by the U.S. National Institutes of Health. The SKBS is organized along the lines of the RAC in the United States. The current procedure for the submission of a gene therapy protocol in Switzerland is that first of all the protocol submitted to the SKBS should conform with the NIH Points to Consider document. Within three months an expert subcommittee of the SKBS should review the protocol. The applicant should then inform the cantonal authorities while the project is submitted to a local ethics committee. The SKBS informs both the federal authorities and the public of its findings. A duty is imposed upon the researcher to report any adverse findings to the SKBS along with a duty to submit a final report.

M. United Kingdom

In the United Kingdom regulatory mechanisms, both legislative and procedural, have been established in a number of areas which relate directly and indirectly to the use of genetic technology. In each of these areas the regulatory framework has been set up following the publication of a government report.

In 1984 the Warnock Committee was established to consider the ethical issues arising from the development of *in vitro* fertilization (IVF) techniques. The Committee reported in 1985 and the findings of this report were finally embodied in the Human Fertilisation and Embryology Act of 1990, which established a regulatory body, the Human Fertilisation and Embryology Authority, to operate a licensing scheme for IVF treatment centers. The Act is a significant piece of legislation for many reasons, but in the area of genetics it explicitly addresses the question of germ line gene therapy. Section 3.3.d. of the Act states that "a licence cannot authorise . . . (d) replacing the nucleus of a cell or embryo with a nucleus taken from a cell of any person, embryo or subsequent development of an embryo." However, despite this apparently rigid prohibition, the licensing authority still retains a degree of discretion as to what type of procedures it is prepared to authorize.

The issue of gene therapy was explicitly addressed by a House of Commons committee chaired by Sir Cecil Clothier. In common with similar studies elsewhere, the Clothier Committee concerned itself primarily with the dichotomy between germ line gene therapy and somatic cell gene therapy, concluding that germ line gene therapy was unacceptable. It recommended that somatic cell gene therapy be permitted as it posed no new ethical problems, but included the caveat that it should be considered an experimental procedure and thus be subject to higher standards of scrutiny than an ordinary medical treatment. The major practical recommendation of the report was to set up the Gene Therapy Advisory Committee, a nonstatutory committee charged with reviewing the ethical acceptibility of proposals for gene therapy. Despite the terms of reference of the GTAC, it would appear that, like many similar bodies, it concerned itself primarily with the scientific merit of gene therapy protocols than with the ethical acceptability of the proposals.

The entire field of genetic technology was surveyed by the House of Commons Select Committee on Science and Technology in a report published in July 1995. (HMSO. July 1995). This report examined the ethical issues arising from genetic technology and made a number of criticisms of existing procedures and recommended the setting up of a Human Genetics Commission to regulate the advance of genetic technology.

Following a review of biotechnology regulation in 1999, a new Human Genetics Commission was established in the United Kingdom in February 2000. This body takes over the supervision of all aspects of genetic research involving human beings in the United Kingdom.

N. United States

Many of the components of the U.S. approach to the regulation of genetic technology have been emulated and implemented elsewhere. Indeed, this has happened to the extent that there are calls for the American model to become the standard for a harmonized global system of regulation.

Gene therapy protocols are the subject of regulation by two national bodies and one local one. At the national level are the Food and Drug Administration (FDA) and the Recombinant DNA Advisory Committee (RAC) of the National Institutes of Health (NIH). The FDA is charged with reviewing all new and investigational gene therapy protocols which take place in the United States. This work is done at the Center for Biologics Evaluation and Research (CBER). The RAC has a somewhat different ambit in that it is charged with reviewing all gene therapy protocols which use NIH funds. The RAC is a multidisciplinary body made up of scientists, physicians, lawyers, and ethicists which holds its meetings in public. At the local level all uses of gene therapy must gain the approval of an Institutional Review Board (IRB).

The process has recently been streamlined with the introduction of a consolidated review system. Thus a project which is publicly funded is first assessed to determine whether it is a novel protocol. If it is a new technique then the application must go to both the RAC and the FDA. If, on the other hand, the technique is not new then the review process is deferred by the RAC and oversight is carried out by just the FDA. Currently, most of the U.S. gene therapy protocols follow this procedure.

If the application is novel and approval is obtained from the FDA and RAC, then the investigators also need FDA approval of an Investigational New Drug (IND) in order to be exempt from the FDA premarket licensing requirements. The IND approval is obtained after review by a team of experts who either give permission to proceed or put the application "on hold." The team is made up of a primary product reviewer, a pharmocology and toxicology reviewer, a clinical reviewer, and any other consultant who are considered to be necessary. The investigators are informed of the concerns which led to the imposition of the "hold," and if these concerns are addressed then the study may proceed. Typical concerns which would arrest a trial would be

that there is an unreasonable or significant risk, or that the reviewers have not been presented with sufficient information to assess the risk. Once the protocol is given permission to proceed, any adverse events which occur must be reported to the FDA. In addition, there is a duty to submit an annual report on the progress of the research.

Clinical trials in the United States are separated into three phases. Phase I trials involve the initial introduction of a product into humans in order to determine safety and toxicity with increasing dose. Phase II trials seek to determine short-term side effects and risk, as well as evaluating the efficacy of the product in a more controlled study. Phase III trials are expanded trials which are designed to gather additional information on efficacy, safety, and toxicity so that an overall risk to benefit analysis can be made.

The RAC system of review has proved to be popular because of the transparency and consistency of its processes. However, it is not without defects. It is limited to reviewing proposals which are funded by the NIH, and cannot therefore exercise jurisdiction over externally funded research. In addition, the RAC is itself a subcommittee of the NIH and there is therefore a possible conflict of interest in that the NIH is both the instigator, funder, and regulator of research.

III. DISCUSSION

The international community of legislators have responded with commendable speed to the perceived threats of a technology which is still in its infancy. The timely introduction of a panoply of different forms of legislation illustrates that interference with the human genome is a concept which provokes deep concerns. Given the substantial jurisprudential and cultural differences which exist between national legal systems, there would seem to be a remarkable degree of homogeneity among the legislative responses. However, this confluence of opinion cannot in itself be taken to imply that the legislative responses which have appeared to prevail are in fact the best means of minimizing the dangers of genetic technology, nor that these responses are philosophically sound. Legislation which introduces procedural safeguards without some substantive basis or ethical underpinning may prove only to be of value in the short term. Procedural safeguards, and "soft" regulatory systems generally, are dependent on those who supervise the licensing procedures. The bodies which oversee these procedural safeguards are often largely made up of individuals sympathetic to the practices in question—in this case scientists. There is then a possibility that the regulatory system will not be applied as rigorously as might have been anticipated by those who designed the system.

Having catalogued the regulatory responses of some countries at the leading edge of genetic technology, it is possible to discern a number of common trends. First, in relation to gene therapy, many of the countries have opted for the two-tier method of ethical oversight. This typically involves a local research ethics committee reviewing the proposed protocol as if it were any other clinical trial. The role of these local bodies differs slightly in the various countries, but generally they are tasked with assessing the ethical and scientific merit of the protocols. The argument for doing this at a local level is that there may be localized cultural difficulties with carrying out such research in a particular area and that community representatives should have some form of input into the process.

The second tier involves review at national level, and in the United States, United Kingdom, and Australia, this is carried out by a national committee of experts who meet to discuss approval for specific genetic manipulation projects. These committees tend to be scientifically dominated, despite the fact that their remit is usually confined to the assessment of ethical issues related to the project and that the assessment of the scientific merit of the procedure tends to form a separate part of the review process. The tendency to overlap and to overemphasize the need for scientific appraisal of gene therapy proposals, often at the expense of more rigorous ethical review, is one of the shortcomings of many of the two-tier systems of ethical review.

Another recurring feature of the regulatory landscape is that many of the countries have opted, often at an early stage in the regulatory debate, to utilize assisted conception legislation as a means of regulating genetic technology. This is true of Australia, Germany, the United Kingdom, and Sweden. One weakness of this approach is that assisted conception laws were drafted as a specific response to the advent of *in vitro* fertilization techniques in the late 1970s and early 1980s. While such legislation often refers to the need to avoid interference with the reproductive cell lines of human beings, the provisions relating to genetic technology tended to be introduced as amendments in response to particular controversies. As such many of these measures may be of limited efficacy in relation to regulating the reality of germ line or enhancement gene therapy techniques.

The legislative responses to gene therapy are almost unanimous in their adoption of the somatic cell–germ line dichtomy. This split has been in evidence from the earliest reports and inquiries into the issue of gene therapy such as the Declaration of Inuyama and the Clothier Report. In most instances it has been argued that somatic cell gene therapy poses no new ethical problems, but should be regulated as if it were an experimental treatment. Germ line gene therapy is seen to pose too many risks both in terms of biosafety and in terms of potential damage to future generations. There are a few exceptions—the Netherlands, for example, has rejected this rigid approach to germ line gene therapy and appears to recognize that there may in fact be good reasons for permitting limited use of such techniques.

But however attractive the distinction between germ line gene therapy and somatic cell gene therapy might appear to regulators and legislators, the distinction between the two may not be as scientifically, or indeed ethically, sustainable as was once thought. The two-tier method of review and the FDA–RAC model developed in the United States and much copied elsewhere is based on a case by case review system. There are inherent dangers built into such a system of operating. While dangers of a slippery slope which will take us into the realm of supermen and wonderwomen are probably greatly exaggerated, it is possible that such a system of ethical review is more likely to move the regulatory system in a liberal direction as fine distinctions and comparisons are made and as a body of regulatory "precedent" develops over time. This is not to argue that moving in a liberal direction is necessarily a bad outcome, but rather that without a coherent agreement on the underlying ethical principles which govern gene therapy research, it will be difficult to prevent this form of regulatory drift.

Despite the considerable degree of legislative effort and energy which has been expended on this topic there is still only a very small amount of legislation which has been drafted specifically to tackle the problems of genetic technology. Perhaps genetic technology does not pose problems of such a difficult nature that they require a specific legislative response. Alternately there may not exist the will or the energy within the various legislatures to tackle the complex moral issues which genetics can raise. Much of the regulation of genetic technology which does exist tends to come under the umbrella of broader measures intended to tackle issues such as assisted conception or reproductive technology. One consequence of this lack of topic-specific legislation is the degree of overlap which tends to occur between the regulation of genetic technology as it is applied to human beings and the broader regulation of biotechnology and the use of genetic technology in agriculture and industry. This is perhaps most pronounced in Europe where there are EC directives relating to the use of "genetically modified organisms" alongside existing national legislation on biosafety or reproductive technology. The overall effect can be to confuse rather than clarify the situation. A further weakness of the current approaches to the regulation of gene therapy is that there is little evidence of the regulatory mechanisms pursuing a proactive research agenda. The scientific developments are constantly progressing and consequently an onus lies upon the regulatory systems to at least keep pace with, and preferrably to anticipate, future developments.

The examination of the national responses to the problems posed by genetic technology also appears to reveal a disproportionate weighting of legislation in the area of gene therapy and a relative dearth of regulation in the area of genetic screening. While the possibility of manipulating genetic material for therapeutic purposes tends to capture the headlines and the imaginations of legislators, it is the arguably less newsworthy issue of genetic analysis by screening and testing which poses the more immediate dangers. Gene therapy techniques may pose dangers for the future, but genetic screening and analysis poses immediate problems of discrimination and invasion of privacy. Surprisingly, there has been little direct regulatory response to this issue. Many countries have enacted disability discrimination legislation; others have introduced measures which seek to safeguard the methods of storage of genetic test samples and results. The Austrian legislation, for example, contains detailed provisions on the storage of genetic information. The Ethical, Legal, and Social Implications (ELSI) research group of the Human Genome Project has drafted a model Genetic Privacy Act which directly addresses some of these problems. These regulatory responses share a common basis in that they follow the procedural model of regulation by introducing safeguards to minimize access to information.

Bibliography

Bilbao Declaration (1994). *International Digest of Health Legislation* **45**(2), 234–237.

Billings, P., and Beckwith, J. (1992). Genetic testing in the workplace: A view from the USA. *Trends in Genetics* **8**.

Chalmers, D. (1994). Institutional ethics committees and the management of medical research and experimentation. *Australian Health Law* **3**, 37.

Cohen-Haguenauer, O. (1995). Overview of regulation of gene therapy in Europe: A current statement. *Human Gene Therapy* **6**, 773–785.

Council for International Organisations of Medical Sciences (1991). The Inuyama Declaration. In *Genetics, Ethics and Human Values: Human Genome Mapping, Genetic Screening and Gene Therapy* (Proceedings of the XXIVth CIOMS Round Table Conference) (Z. Bankowski and A. M. Capron, Eds.), pp. 1–3. Geneva: CIOMS.

Fletcher, J. C., and Anderson, W. F. (1992). Germ line gene therapy: A new state of debate. *Law, Medicine and Health Care* **20**, 26–39.

Gene Therapy Advisory Committee (March 1995). *First Annual Report.* Department of Health, London.

Gene Therapy Advisory Committee (April 1996). *Second Annual Report.* Department of Health, London.

Hennau-Hublet, C. (1996). Le Projet de Convention de Bioéthique du Conseil de l'Europe: L'espoir d'une protection élevée des droits de l'homme. *Revue du Droit de Santé*, 25.

Kielstein, R., and Sass, H. (1992). Right not to know or duty to know? Prenatal screening for polycystic renal disease. *Journal of Medicine and Philosophy* **17**, 395.

Lappe, M. (1991). Ethical issues in manipulating the human germ line. *Journal of Medicine and Philosophy* **16**, 621.

Lawton, A. (1997). Regulating genetic destiny: A comparative study of legal constraints in Europe and the United States. *Emory International Law Journal* **11**, 365.

Lenoir, N. (1997). UNESCO, genetics and human rights. *Kennedy Institute of Ethics Journal* **7**, 31.

McGleenan, T. (1995). Human gene therapy and the slippery slope arguments. *Journal of Medical Ethics* **21**, 350.

Matthewman, W. D. (1984). Title VII and genetic testing: Can your genes screen you out of a job? *Howard Law Journal* **27**, 1185.

Medical Research Council of Canada (1990). *Guidelines for Research on Somatic Cell Gene Therapy in Humans.* Medical Research Council, Ottawa Canada.

Rothstein, M. A. (1993). Discrimination based on genetic information. *Jurimetrics Journal* **33**, 13.

Science and Technology Committee. (July 1995). *Human Genetics and Its Consequences.* HMSO. London.

Subcommittee on Human Gene Therapy, Recombinant DNA Advisory Committee, National Institutes of Health (1990). Points to consider in the design and submission of protocols for the transfer of recombinant DNA into the genome of human subjects. *Human Gene Therapy* **1**, 93.

UNESCO (1997/1998). Universal Declaration on the Human Genome and Human Rights. Document 27v/45 adopted by the 31st General Assembly of UNESCO, Paris, Nov. 11, 1997. Reprinted in *Journal of Medicine and Philosophy* **23**, 334.

Valencia Declaration on Ethics and the Human Genome Project (1991). *International Digest of Health Legislation* **42**(2), 338–339.

Walter, L. (1991). Human gene therapy: Ethics and public policy. *Human Gene Therapy* **2**, 115.

Zimmerman, B. K. (1991). Human germ line gene therapy: The case for its development and use. *Journal of Medicine and Philosophy* **16**, 593.

Genome Analysis

ADAM M. HEDGECOE
University College London

GLOSSARY

big science A popular term for large-scale, very expensive projects that usually involve large installations or equipment, or more rarely, a large number of smaller, less expensive research centers.

DNA Deoxyribonucleic acid, the material contained in all living things that makes up a gene and that contains an individual's genetic code.

gene The basic unit of heredity, consisting of a stretch of DNA that carries instructions telling the body how to make a particular protein.

geneticization The process by which diseases (especially in humans) and behaviors are described and understood in genetic terms.

genome The sum total of all genes carried by any individual organism or species.

Human Genome Diversity Project A contemporary research effort aimed at describing the genetic differences and similarities of various human populations or ethnic groups.

Human Genome Project The attempt to analyze and identify all the genes, and sequence all the DNA present in human beings.

mapping A generic term used to describe the activity of finding out the position of various genes in the genome.

reductionism The increasingly fine-tuned analysis of scientific processes, "reducing" the explanation of processes to smaller scale.

sequencing The act of determining the exact molecular sequence of genetic material.

GENOME ANALYSIS is the investigation into the genetic information carried by individual living things. This information is of great interest to scientists, doctors, and other experts, and the ethical aspects of its application are huge. There are also ethical issues involved in simply investigating the genome, irrespective of the use to which the information produced is put.

Large amounts of money are currently being spent on the Human Genome Project, the international effort to discover everything about the human genome, and it is expected that the final results, with a complete description of the human genome, will be available by the early years of the next century. Thus, there is urgent need to address the ethical issues raised by genome analysis.

I. INTRODUCTION

The aim of this article is to present the ethical issues associated with the investigation, analysis, and discovery of information present in the human genome, that is, the sum total of genetic material carried by any one person. The concentration will be on human genetics, because although some of the issues are related to analysis of the genomes of other species, these problems are most acute when seen in the context of humankind.

The first difficulty with this aim is the fact that much of the ethical writing on genetics is associated with the *use* of genetic information discovered in the analysis of the genome (e.g., genetic testing and health insurance, gene therapy, etc.) rather than the simple fact of the discovery of that information in and of itself. Quite often, genomic analysis is not felt to be ethically difficult, only the consequences of its information. For example, the Declaration of Inuyama of the Council for International Organizations of Medical Sciences, states that "efforts to map the human genome present no inherent ethical problems." But others have rejected this viewpoint, making clear that "Gene Mapping is in itself problematic and of concern in ways that ongoing research on the social, legal, and ethical consequences of the information deriving from genome projects cannot address." (A. Lippman. (1992). Led (astray) by genetic maps: The cartography of the human genome and health care. *Social Science and Medicine*, 35 (12), 1469–1476.) This article will outline those instances when human genome analysis *is* ethically significant in itself. This article will concentrate on the huge, multinational program known as the Human Genome Project (HGP) as the best-known example of genome analysis, but as will hopefully become clear, many of the ethical issues associated with genome analysis are not limited to large-scale projects like the HGP, but also to the idea of genome analysis itself.

II. TECHNIQUES OF ANALYSIS

The human genome is often described as a "blueprint" or a "book" that lists the secret of humanity, but the analogy most used in the actual science of molecular biology is that of "mapping." The four ways of analyzing a genome are genetic mapping, physical mapping, cDNA mapping, and DNA sequencing, and these can be seen as four separate scales or levels of resolution with which to look at a genome.

A. Genetic Mapping

Genetic mapping (also known as genetic linkage mapping) shows the relevant locations of specific DNA markers along a chromosome. These markers are either active DNA regions (i.e., genes, or "exons") or areas for which no particular use has been discovered ("junk"

DNA, or "introns"). These markers are used to trace the familial variations of various observable ("phenotypic") characteristics, whether they be diseases or physical characteristics such as eye color. When genes are passed from parents to children, the process involved may split up any two markers on the same chromosome; the closer the markers are together the less likely they are to be split up, and so the rate at which various markers occur can be used as a means to estimate the distance between markers on a chromosome. Thus, even if the actual gene for a particular disorder is not known, it is possible to identify those markers that are normally associated with it (and that do not occur in individuals without that disorder), and thus a test for genetic diseases can be produced.

B. Physical Mapping

The lowest resolution physical map is the chromosomal map, which assigns genes and other identifiable DNA fragments to their various positions on the chromosome, with the distance between fragments measured in base-pairs, which are the combinations of the four chemicals that make up the "rungs" of DNA's double helix structure. Beyond these lower resolution maps, there are high resolution methods that can allow the identification of DNA pieces in stretches of genome measuring only 10,000 base-pairs long.

C. cDNA Mapping

The complementary (cDNA) map includes only the expressed regions of DNA; that is, it includes only the genes, without the interspersing junk DNA. Because these maps represent only that DNA that codes information, they are felt to be useful in the identification of those genes responsible for disease causation.

D. Sequencing All the DNA

The ultimate map of a genome involves "sequencing" it; determining the exact sequence of base pairs along each chromosome. This involves techniques of purification and separation of DNA sequences and in the future will use cloning techniques and PCR (polymerase chain reaction, a means of "amplifying" quantities of a particular DNA stretch).

III. THE HUMAN GENOME PROJECT

All these techniques are currently being used in the HGP. This has variously been described as "a biological moonshot," producing the "Rosetta stone for studying human biology" and reading the "book of man." Once past all the hyperbole, the HGP is still an extremely impressive undertaking. In 1988 estimates were that it would cost $3 billion in total. In 1995 this was estimated at only $200 million and a rough "working draft" is expected in the spring of 2000. Originally conceived of in the mid to late 1980s, the HGP only really began producing results in the 1990s, but the developments of sequencing technology and knowledge about human genetics has developed at a far faster rate than many people expected, resulting in downward revisions of the end date of the project. The HGP is overseen by the Human Genome Organization (HUGO), which has the responsibility for disseminating new information and for coordinating the research efforts of teams in the various countries taking part.

Although the HGP is "only" genome analysis on a grand scale, there are ethical issues that relate to it, and not to genome analysis itself. These are to a large extent a result of the sheer size of the HGP and the criticism that such "big science" arouses.

A. Big Science

Although there seems to be no cut-off point in terms of cost between big science and the rest of science, most commentators would agree on examples of the former if presented with them. Often they involve large, expensive installations or equipment. Examples of this sort might include the CERN accelerator near Geneva used for investigating subatomic particles, or the Hubble telescope launched by NASA. The HGP is a different sort of big science, because it is not about one large installation, carrying out all the genome analysis work, but a network of mainly preexisting labs and centers, scattered around the world, supported by funding from their own countries. Thus it is the overall cost of manpower, new computer banks to store the information, and personnel costs that raise criticisms. The arguments against big science suggest that large sums of money are spent, largely inefficiently (due to the large bureaucracies generated), on scientific topics that may turn out to be just a "fad"; this money could be redistributed to other, smaller-scale projects with greater effect. Those in support of the HGP cite it as the only means by which such a massive undertaking could be completed; the human genome is so immense that only an equally huge project could analyze it properly. In many ways, the HGP is only "big" in comparison with the size of studies that are normally carried out in the biological sciences. For example, in 1991, the U.S. government spent $136 million on the HGP; but in comparison with the average U.S. $300 million (on average) spent on the development of a single pharmaceutical (bearing in mind the

potential benefit for human health that the HGP embodies) it seems rather small.

In fact, the case could be made that objections to the HGP on the grounds that it is too big are not really ethical at all. It is not that most critics are suggesting that the money be spent on an activity such as feeding the Third World, or increasing welfare benefits. They would prefer that the money be spent on other areas of biology or science. Thus, the concerns raised are mainly ones of efficiency and effectiveness, rather than inherently ethical.

B. Access to Information

The problems that still remain concern access to, and the ownership of, the information that results from the analysis being carried out. Although public funds are going into the HGP, so is money from private firms, especially pharmaceutical companies. The issue then arises of who should be allowed access to which data. At the beginning of the project, James Watson, then head of the U.S. NIH/DoE genome project, threatened to withhold access to genomic information from the Japanese, unless they in turn allowed others access to their own research. While there was some sympathy for this position, criticism was voiced by those who felt that the human genome is a world resource, and not the property of any one person or organization to control. At a more practical level, doubts have been raised about the point of an international collaborative project, if the results are subject to government imposed scrutiny. There is a history of political use of scientific information, such as the U.S. National Library of Medicine denying Soviet users access to the Genbank and PIR databases during the Soviet invasion of Afghanistan. The initial result of the U.S. threats was wider interest in funding the HGP, so perhaps it could be said that they were successful; in the long term, such a tactic could be counterproductive leading to tit-for-tat denials by national governments.

At the level of personal ethics, individual researchers may have an interest in delaying publication of information for a period of time, to enable their own research to progress, while denying others the chance to catch up. This would seem to breach the generally accepted ethical requirements for researchers to share data with others, especially in those cases where public funding is involved.

More recently, the issue of access has arisen, not in terms of national efforts but with the increased commercialization of the HGP. The best-known example is that of The Institute for Genomic Research (TIGR), which has a large database of expressed sequence tags (EST). These are short stretches of DNA associated with particular genes, which can be used to identify the postion of genes on any stretch of DNA far quicker than with other methods. TIGR originally tried to patent these ESTs, but was told by the National Institutes of Health that because they were not useful in and of themselves, they were unpatentable. TIGR then agreed to allow other researchers to access the database and its 160,000 ESTs, but on the one condition that if any marketable product was produced as a result of the use of the ESTs, then TIGR, or rather its commercial arm, Human Genome Sciences (HGS), and its partner and financial backer, SmithKline Beecham, would be allowed first refusal on marketing that product. This idea produced alarm in the scientific community, which had assumed that the information was to be made available with no conditions attached. It has been stated that because researchers need access to other DNA sequences in order to compare the accuracy of their own research, the denial of access to such a large database will impede the overall progress of the HGP. TIGR's reply is that they have to maintain some sort of control over their work in order to justify the huge investments made by SmithKline Beecham, and that by not publishing their database, they are allowing people who do discover something useful via one of the ESTs to patent it, because such information would not be in the public domain and thus not infringe the "unobvious" requirement of patentability. Recently, the scientific mind behind TIGR, Craig Venter, has stimulated more controversy by his claim that he will privately sequence the human genome faster than the publicly funded project. This in turn will have an effect on researchers' personal ethics, as they try to balance a "scientific duty" to publish their work, and the commercial need for trade secrecy. As well as patenting issues, the involvement of companies in the storage of genetic information can be controversial. In Iceland, there is concern over whether a national DNA database (which has recently been approved) should be managed by the commercial company deCODE. Fears have been expressed about the security of the information held and the use to which it might be put (R. Chadwick, 1999. The Icelandic database—do modern times need modern sagas. *BMJ* **319**, 441–444).

IV. UNESCO DECLARATION OF THE PROTECTION OF THE HUMAN GENOME

As a result of general concern about developments in human genetics, UNESCO set up (in 1993) an International Bioethics Committee, which has published several papers. Its main work involved drafting the Declaration on the Protection of the Human Genome and Human Rights, which aims at guidelines that should be borne in mind by those who are carrying out research into the Human Genome Project, although its overall remit is much broader.

The document itself contains 23 articles, starting with the proclamation that "the human genome is the common heritage of humanity." This is an important step, because this is a very public expression of what is considered ethical in the realm of genetic investigations. The declaration then goes on to outline acceptable research behavior in the areas that relate to genetic therapy, screening, and genetic privacy. Criticism has been voiced about the thinking behind such a declaration and the wording used to express these thoughts.

One question arises with the reference to the genome as "common heritage of humanity." The explanatory notes to early versions of the declaration make explicit reference to international treaties that have been drawn up, comparing the genome to such *limited* resources as the seabed. But the appropriateness of this comparison has been questioned; the human genome (either in its material form, or in terms of the information it contains) cannot be used up in the same way as more "traditional" resources such as fossil fuels or mineral deposits, so the exact reasoning behind this link remains unclear. Similarly, confusion rests on the assumption that the human genome is in *need* of protection; from whom or what is not made clear. While the originators of the declaration may have felt that its content was relatively straight forward, it can be argued that it rests upon certain assumptions (such as "the human genome is a finite resource") that are not at all certain. Certainly, if the human genome *is* a resource that belongs to the whole of humanity, this raises questions about whether it is right that private companies should be involved in the HGP, or, separately, whether they should be allowed to profit from the work they do. It also leads us to ask, who should fund the HGP, one answer being all those countries that can afford to, although we could also insist that this does not exclude those who cannot afford it, from benefiting from the results of the research. But whatever queries there may be over the actual semantics of the Declaration, that such a document has been drawn up highlights the fact that investigations into the human genome raise ethical issues, independent of the use to which genetic information is put.

V. THE HUMAN GENOME DIVERSITY PROJECT

Initially proposed by several American academics in 1991, this project had not started by 1996 (despite being supported by HUGO) and has now foundered as a single coordinated research proposal. This delay was largely due to ethical concerns. The Human Genome Diversity

Project (HGDP) aims to help us "find out who we are as a species" by preserving cells from various ethnic groups and aboriginal populations around the world, to give students of human genetics access to various exotic gene pools. This work is also of interest to anthropologists, epidemiologists, and even linguists. It has produced ethical questions of its own and has led to a great deal of controversy.

A. Rationale for the HGDP

The proposers of the diversity project claim that however useful the HGP is in refining genetic analysis techniques and information storage systems, its one Achilles heel is that it is explicitly mapping a single genome (not the genome of any one individual, but a combination of genomes). Thus it ignores one of the most interesting aspects of human genetics, the variation between populations. Each of our genomes is a unique product of cultural, environmental, and historical factors, many of which we share with people who have the same history that we do. Thus there are genetic variations between different ethnic groups, or populations. As well as the more academic interest in genetic variation, there is the practical fact that by finding out why some groups are genetically different from others, we will discover more about diseases prevalent among particular groups and thus help forward the HGP's stated aim of improving knowledge about human health.

B. Methods

The HGDP requires that between 400 and 500 ethnic groups, or "populations" should be identified to represent humanity. Preferential treatment would be given to those populations that conformed to one or more of six criteria, ranging from those with the best knowledge of processes that impact on their genetic composition, through those populations that might be most useful in identifying specific genetic diseases, to those that are in danger of losing their genetic, cultural, or linguistic uniqueness. This last criteria means that a great number of indigenous populations are suitable for sampling in the HGDP, and this has raised a number of ethical problems (see below). There are three types of sampling that will be carried out on the populations identified. A large number of genetic markers (150) from single individuals or a smaller number from between 20 and 25 individuals. In both cases they will denote blood that will then be transformed into "immortal" cell lines that will preserve the biological material for indefinite periods of time. A larger number from each population (100–200) may donate some blood and other material (such as hair tissue) that will be used for studies using

larger numbers, but with finite material. The total estimate was at first put at between $25 and 30 million for 5 years; this is about 1% of the projected total cost of the HGP as a whole, and it is claimed that bearing in mind the usefulness of the information produced, this should not be regarded as excessive by any means.

C. Benefits

There are four keys areas of research that will benefit from the HGDP:

1. The Origin of Modern Humans

HGDP information will help researchers trying to determine whether humans evolved in Africa and then migrated (the "Out of Africa" thesis), or developed simultaneously across the globe. The information gathered would also shed light on more recent migrations (such as the spread to the Americas) and support archeological evidence that is, at the moment, indeterminate.

2. Social Structure

Anthropologists will gain important clues about mating patterns (such as whether males or females leave the tribe to mate) from DNA samples. Because Y chromosomes are only found in males, it is hoped that the "route" taken by particular DNA can be traced.

3. Adaptation and Disease

Perhaps the most obviously "useful" reason for the HGDP is that the information gained will show whether variations in disease patterns between various populations are due to adaptation to local conditions or merely random changes in genetic make up.

4. Forensic Anthropology

This will add to our sum of knowledge about simply knowing which population is where, and how populations differ from each other.

D. Criticism of the HGDP

1. Racism

Critics of the HGDP claim that it is applying modern technology to an outmoded concept, that of "race." Anthropologists have questioned the central assumptions of the project, that isolated populations are genetically discrete, that human groups are defined solely in terms of genetics, and that genetic differences between groups are greater than those between individuals. There are worries that by focusing on race, however scientifically defined, the HGDP may inadvertently foster racist attitudes. A UNESCO subcommittee set up to examine this issue found that it is not necessarily the case that the supporters of the HGDP are racists (although this claim *has* been made by some indigenous peoples) but that they are naive in not accepting the dangers that their research may produce. This is denied by the HGDP researchers themselves, who claim they are aware of the risks, and that their work is in fact more likely to *reduce* racism, by exposing the scientific fallacy of racist doctrine, than to encourage it. This case has been put very strongly by Luca Cavalli-Sforza, one of the proposers of the HGDP, who states "there is no documented biological superiority of race, however defined. Nowhere is there purity of races ... No damage is caused to humans by racial mixture ... In fact, the concept of race can hardly be given a scientific, careful definition." (Cavalli-Sforza L. (1994, September 12). *The Human Genome Diversity Project: Address to a Special Meeting of UNESCO. Paris*). The question still remains, though, whether the information resulting from the HGDP could be used by those with less concern for scientific facts, and more dangerous motives.

2. Informed Consent

The leaders of the HGDP intend to enforce the requirement of informed consent when collecting the genetic material, but this has run into objections from indigenous peoples' groups, on both conceptual and practical grounds. Individuals who work with native populations have questioned whether proper *informed* consent can be obtained in the short time envisaged by the HGDP teams. It can take anywhere from 6 months to 5 years for even a minority of individuals to achieve levels of understanding that would normally be considered acceptable. The conceptual difficulty rests on the question of who is justified in giving consent? Although individuals are donating material, the information sought is exclusive to the ethnic group concerned (that is the raison d'être of the HGDP after all). Who then should give permission? The individuals, the whole ethnic group, a majority or a representative (such as a chief, or tribal council). Native rights groups have attacked the HGDP on the grounds that it has failed to address these issues adequately. (E. T. Juengst (1998). Groups as gatekeepers to genomic research: Conceptually confusing, morally hazardous, and practically useless. *Kennedy Institute of Ethics Journal* **8**, 183–200.)

3. Patenting, Profiting, and Ownership

The ownership of genetic material and information is highly controversial anyway, but it has added over-

tones in the realm of human diversity. A great many objections have been raised to the suggestion that scientists from the developed world will take genetic material from indigenous people, isolate important genes from it, and then make a profit out of any products that are developed. Discussions about this took place in March 1995 as the United States issued a patent on genetic material taken from a member of the Hagahai people from Papua New Guinea; the material is useful in detecting HTLV-1-related retroviruses. Although this research was not connected to the HGDP and the Hagahai are not on the provisional list of candidate populations for it, opponents of the HGDP, now dubbed the "Vampire project," suggested that the HGDP would merely continue the same practices, except on a larger and far more organized scale.

This has been explicitly rejected by the leaders of the HGDP, who claim that no commercial use can be made of indigenous genetic material without the express permission of regional committees that represent all interested parties in the project (including the sampled populations). Moreover, any money gained from these products will be shared with the indigenous groups, either as direct royalties, or as funds distributed by a neutral external body. With the demise of the HGDP as an organized project, this sampling is now being carried out piecemeal by academics and pharmaceutical companies; there is little or no formal regulation of this sampling.

4. Genetic Imperialism

The above complaints combine to form the objection that the HGDP is just the latest in a long line of various forms of colonization, exploitation, and imperialism, carried out by the Western world upon developing countries. In previous times the object of colonization was the land where the indigenous peoples lived. Today, it is their genetic material. This position is eloquently outlined by Aroha Te Pareake Mead, of the Indigenous Peoples' Biodiversity Network, who states "For Maori, and many others, the human gene is geneology ... It is difficult to articulate the degree to which indigenous and Western scientific philosophies differ on such a fundamental point ... but ... it is the difference in understanding of the origin of humanity, the responsibility of individuals and the safety of future generations which sits so firmly at the core of indigenous opposition to the HGDP ... This type of research proposes to interfere in a highly sacred domain of indigenous history, survival and commitment to future generations" (Mead, A. (1995), Correspondence: Response to Draft UNESCO Bioethics Paper). Such objections would appear to be at a ethical level far removed from the practical difficulties of ensuring that proper informed consent is achieved for all participants in the project, and it is not clear how to respond to such criticism (M. Dodson and R. Williamson (1999). Indigenous peoples and the morality of the Human Genome Diversity Project. *Journal of Medical Ethics* **25,** 204–208).

VI. GENETICIZATION

Criticism of genome analysis is not limited to large-scale, international projects like the HGP and HGDP. The very act of sequencing a genome, it is claimed, whatever the information it contains is used for reinforces certain beliefs about human genetics and encourages "geneticization." This "refers to an ongoing process by which differences between individuals are reduced to their DNA codes, with most disorders, behaviors, and physiological variations defined, at least in part, as genetic in origin ... Through this process, human biology is incorrectly equated with human genetics, implying that the latter acts alone to make us each the organism he or she is." (Lippman, A. (1991) Prenatal genetic testing and screening: Constructing needs and reinforcing inequalities. *American Journal for Law and Medicine,* **17**(1&2), 15–50). Genticization is a form of scientific reductionism, in that it describes events at a higher level (that of physiology or psychology) in terms of a lower one, molecular biology. Much of force of this reductionism is a result of the information to which genetic material might or will be put to use when it becomes available (such as genetic tests by health insurers), but much of it can also be traced back to the language used to describe the Human Genome Project in particular, and genome analysis in general.

A. Mapping

Genome analysis aims to produce maps, of varying scales. The image of the map is a particularly powerful one, especially as we no longer expect to see "here be dragons" but rather "X marks the spot." Modern maps are held to be definitive indicators of the absence or presence of an object, whether it be a road, a stream, or a gene "for" a disease. The mapping metaphor currently at use in molecular genetics suggests a degree of certainty and clarity about genetic information that is not reflected in terms of actual results. Abby Lippman has criticized the use of such metaphor (and its attendant correlate, the "blueprint"); "Mapmaking, whether of the body or of the earth is as much political and cultural as it is 'scientific'. It is a social activity ... because it is an expression of and influence on social values ... geneticization gives mapmakers ... tremendous power ... for defining how we think of ourselves and others and for determining who will manage us as individuals

and as a society" (Lippman, 1992). She proposes an alternative metaphor, taken from management theory, that of the organogram, the bureaucratic design of an organization. This would at least have the advantage of obviously displaying its limitations, which cannot be said for maps.

B. Genetic Stories

Mapping is just one of a range of stories that are told by scientists, doctors, and politicians about genetics. These stories are also told by members of the public; the gene is no longer just a scientific entity. As Dorothy Nelkin and Susan Lindee have documented, the image of the gene and DNA's double helix permeate popular culture with a variety of images; "Geneticists also refer to the genome as the Bible, the Holy Grail and the Book of Man. Explicit religious metaphors suggest that the genome ... will be a powerful guide to moral order." They warn that "the images and narratives of the gene in popular culture reflect and convey a message that we will call genetic essentialism [geneticization]" (Nelkin D., & Lindee, M. S. (1995) *The DNA Mystique: The Gene as a Cultural Icon.* New York: Freeman). The main objection to such use of language is not that there is an explicit plan to promote a genetic view of the world. Rather, the position is that the exciting and groundbreaking work currently being carried out in the area of human genetics reinforces geneticization as a side effect of the production of genetic information, regardless of the actual use to which that information is put. Proponents of this position do not object to genome analysis as it *might* be carried out, but in the way it is currently carried out. There might be ways of investigating the human genome and publicizing this information that would not enforce an overly genetic view of the world. Unfortunately, it is not at all clear that the metaphors criticized are interpreted in this way. Empirical work (C. Condit (1997). Public discourse about genetic blueprints: Audience responses are not necessarily deterministic or discriminatory, *AAS Symposium 17/2/97*) suggests that the blueprint metaphor does not have to be seen as deterministic. Other research casts doubt on the idea that media discussions of genetics are currently more deterministic than in the past (C. Condit, N. Ofulue, and K. Sheedy (1997). Determinism and mass media portrayals of genetics. *American Journal of Human Genetics* **62,** 979–984). The truth is probably more complicated than originally suggested by the critics of geneticization.

VII. CONCLUSION

Genome analysis, especially in the form of the Human Genome Project, holds great potential for our understanding of ourselves of disease and of life itself, but the concept of genome analysis is an extremely potent one. The language used and the images presented carry with them the potential to reinforce an extremely limited view of what human beings are, and with it the possibility of harmful application of genetic information. It is not that the ethical problems associated with genome analysis *have* to be so. With awareness of how powerful genes are in people's minds and in our culture, their actual power and use can be controlled.

Bibliography

Hubbard, R., & Wald, E. (1993) *Exploding the Gene Myth.* Boston: Beacon Press.

Macer, D. (1991). Whose genome project?. *Bioethics,* **5**(3), 183–211.

Marteau, T., & Richards, M. (1996). *The troubled helix: Social and psychological implications of the new human genetics.* Cambridge: Cambridge University Press.

Mauron, A. (1995). HGP: The Holy Genome Project? *Eubios Journal of Asian and International Bioethics,* **5,** 117–119.

UNESCO IBC (1995). Preliminary draft of a universal declaration on the human genome and human rights.

UNESCO IBC (1995). Report on bioethics and population genetics research.

Hazardous and Toxic Substances

KRISTIN SHRADER-FRECHETTE

University of South Florida

GLOSSARY

compensating wage differential The increases in individuals' wages that are paid to them in order to compensate them for higher workplace risks that they bear as a consequence of their particular jobs.

discount rate The annual percentage by which economists reduce the sum of money representing the value (the costs or benefits) associated with some event, act, or process that occurs in the future. It is the opposite of an interest rate, the annual percentage by which a sum of money increases.

economies of scale Financial savings that are possible, for example, to a company, because of the great size or volume of the operations.

externality Social costs (such as pollution) or benefits (such as clean air) that are not traded on a market and that are not part of benefit–cost calculations.

genetic damage Damage to the genes and chromosomes that is passed on through heredity to one's descendants.

liability The state of being legally or financially responsible for something or someone.

maximin A rule for decision making, often societal decision making, according to which one acts so as to avoid the worst outcome.

pesticide Chemical used to kill insects or weeds.

risk assessment A set of mathematical and scientific techniques for identifying, estimating, and evaluating the probability and the consequences associated with various public health and environmental threats, such as those from hazardous waste facilities.

statistical casualties Premature deaths calculated to occur on the basis of a particular exposure to a hazard or to a toxic chemical, even though the people killed by a particular threat sometimes are not easily identifiable because their deaths (for example, through cancer) often are not immediate and because the people usually are not under medical examination.

statutory Pertaining to a law, rule, or statute, usually of a government.

toxic tort A civil wrong for which a party injured by toxic chemicals is entitled legally to compensation.

tragedy of the commons The damage that occurs to everyone as a result of some individuals trying to benefit themselves personally by using common resources. For example, everyone is harmed by air pollution because each person attempts to save money by being a polluter.

HAZARDOUS AND TOXIC SUBSTANCES are products or by-products of manufacturing, scientific, medical, and agricultural processes, and they have at least one of four characteristics: ignitability, corrosivity, reactivity, or toxicity (T. P. Wagner, 1990. *Hazardous Waste Identification and*

Classification Manual. Van Nostrand–Reinhold, New York). Hazardous and toxic substances become wastes only when they have outlived their economic life. They include solvents, electroplating substances, pesticides such as dioxin, and radioactive wastes. Toxic substances, a subset of hazardous substances, have the characteristic of toxicity: the ability to cause serious injury, illness, or death.

Many people first became aware of the threat of hazardous and toxic substances from Pullitzer Prize nominee Michael Brown's *Laying Waste: The Poisoning of America by Toxic Chemicals* (1979), a frightening account of the cancers, deformities, and deaths caused by hazardous waste in the Love Canal neighborhood in New York. Rachel Carson also helped to educate people to the dangers from hazardous and toxic substances when she wrote *Silent Spring,* her 1962 classic analysis of the dangers of pesticides. U.S. Congressman James Florio said in 1981 that management of hazardous wastes is the single most serious environmental problem (M. Greenberg and R. Anderson, 1984. *Hazardous Waste Sites: The Credibility Gap.* Rutgers, New Brunswick, NJ). In the United States alone, approximately 80% of hazardous and toxic wastes has been dumped into hundreds of landfills, ponds, and pits. It has polluted air, wells, surface water, and groundwater. It has destroyed species, habitats, and ecosystems. It also has caused fires, explosions, direct contact poisoning, cancer, genetic defects, and birth defects.

I. INTRODUCTION

Hazardous and toxic substances are at the center of conflicts between the chemical industry and the public. When the U.S. Environmental Protection Agency in 1995 released its landmark study of the dangers of dioxin, a by-product of manufacturing chlorine, the Chlorine Chemistry Council immediately tried to downplay the dangers. Dioxin causes immune system damage, infertility, cancer, and death. Dioxin is produced whenever industry generates, uses, or burns chlorine or chlorine-derived products. Although PVC (polyvinyl chlorine) plastic is the largest use of chlorine, the pulp and paper industry uses much chlorine in bleaching. As a consequence, incinerating paper, plastic, and other chlorine-containing materials creates dioxin, a dangerous product of many incinerators. In most nations of the world, including the United States, there are few restrictions on the use of chlorine, and it is widely used in paper and pulp bleaching.

Although regulation of hazardous and toxic substances always lags behind the social, ethical, and political need for such regulation, there are a number of laws aimed at reducing the threat from hazardous and toxic substances. For example, to protect workers and the public from the dangers associated with hazardous substances, the United States has passed laws such as the Hazardous Materials Transportation Act, the Atomic Energy Act, the Clean Water Act, the Clean Air Act, the Resource Conservation and Recovery Act (RCRA), the Toxic Substances Control Act (TSCA), and the Comprehensive Environmental Response, Compensation, and Liability Act (CERCLA or Superfund). Laws such as these have many provisions, including those requiring behavior such as monitoring pollutants, reporting spills, preparation of manifests describing particular wastes, and special packaging for transporting specific types of hazardous materials. Smelter emissions, for instance, are regulated under the Clean Air Act, and mining-caused water pollution is regulated under the Clean Water Act (J. Young, 1992. *Mining the Earth:* Worldwatch Institute, Washington, DC). RCRA was passed to fill a statutory void left by the Clean Air Act and the Clean Water Act; they require removal of hazardous materials from air and water but leave the question of the ultimate deposition of hazardous waste unanswered. RCRA addresses the question of the handling of hazardous waste at current and future facilities, but it does not deal with closed or abandoned sites. CERCLA focuses on hazardous waste contamination when sites or spills have been abandoned; through penalties and taxes on hazardous substances, CERCLA provides for cleaning up abandoned sites.

Despite the various provisions of U.S. laws governing hazardous substances, use of toxins and management of hazardous wastes raise a number of ethical issues that have not been adequately addressed by existing laws and regulations. Most of these ethical questions are related either to the equity of risk distribution or to the assessment and regulation of societal risks associated with hazardous substances. The equity issues include siting facilities that produce or use hazardous materials; rights of future generations to protection against hazardous and toxic substances; worker rights to such protection; public and worker rights to free and informed consent to exposure to hazardous and toxic substances; rights to compensation for risk or damage caused by hazardous or toxic substances; and rights to due process in the event that one is harmed by toxic or hazardous substances through the fault of other people. Questions about risk assessment and regulation include appropriate ethical behavior under conditions of uncertainty about the hazards or toxins; where to place the burden of proof in alleged harms involving hazardous substances; and the right to know of workers and the public who may be exposed to hazardous and toxic substances.

II. NIMBY AND EQUITY ISSUES

In 1996 a scientific team directed by famous lead researcher Herbert Needleman showed that inner-city children with the highest concentrations of lead in their bones were those most likely to exhibit abusive, violent, and delinquent behavior. The lead poisoning—arising from auto exhausts, gasoline fumes, paint, chemicals, and pipes—lowered their intelligence and contributed to impulsive behavior, hyperactivity, and attention-deficit disorder, all of which help cause delinquency. Even after the researchers controlled for factors such as single-parent homes, parents' IQ, and child-rearing practices, they found that lead levels in the bones of young children are accurate predictors of later delinquency and crime as adults. The lead example illustrates clearly that those who can afford to avoid hazardous and toxic substances typically do so. Virtually no people want toxic substances or hazardous wastes used or stored near them. Hence they cry, "not in my back-yard"—NIMBY. The NIMBY syndrome also arises because of the fact that any hazardous technology is unavoidably dependent upon fragile and short-lived human institutions and human capabilities. It was not faulty technology, after all, that caused Three Mile Island, Bhopal, Love Canal, or Chernobyl. It was human error. Likewise it could well be human error that is the insoluble problem with using toxics and managing hazardous wastes.

Because of the biological, chemical, or radiological dangers associated with hazardous and toxic substances, and because of the potential human errors associated with their use and disposal, most people try to avoid them. Those who cannot easily do so are usually poor or otherwise disadvantaged, like the inner-city victims of lead poisoning from auto exhausts and gasoline fumes.

Questions about the equity of risk distribution are central to the issue of managing hazardous substances and toxins, because poor, uneducated, or politically powerless people typically bear the gravest threats from such substances. Many people in developing nations—such as thousands of people in Bhopal, India, for example—have already died as a consequence of exposure to hazardous substances. Such deaths occur because economic comparisons of alternative technologies and different sites for hazardous facilities typically ignore ethical considerations. Instead technological and siting decisions occur because of attempts to minimize market costs. Economists ignore the externalities (or social costs), such as the inequitable distribution of health hazards, associated with toxic chemicals. They also ignore the risk–benefit asymmetries associated with using toxic substances or managing hazardous wastes. Because geographical and intergenerational inequities

are typically "external" to the benefit–cost evaluation scheme used as the basis for public policy, such externalities are almost always ignored (K. Shrader-Frechette, 1985. *Science Policy, Ethics, and Economic Methodology.* Kluwer, Boston).

Because rich and poor people are not equally able to avoid threats from hazardous and toxic substances, public and workplace exposure to such hazards raises a number of questions of equity. Many of these questions concern geographical equity, intergenerational equity, or occupational equity. Intergenerational equity problems deal with imposing the risks and costs of hazardous wastes and toxic substances on future generations. The geographical equity issues have to do with where and how to site waste dumps or facilities using toxic substances. The occupational equity problems focus on whether to maximize the safety of the public or that of persons who work with toxic substances or hazardous wastes, since both cannot be accomplished at once (see R. Kasperson, Ed., 1983. *Equity Issues in Radioactive Waste Management.* Oelgelschlager, Gunn, and Hain, Cambridge, MA; K. Shrader-Frechette, 1993. *Burying Uncertainty: Risk and the Case against Geological Disposal of Nuclear Waste.* Univ. of California, Berkeley).

The key ethical issue raised by concerns about intergenerational equity is whether society ought to mortgage the future by imposing its debts of buried hazardous wastes on subsequent generations. In virtually every nuclear nation of the world, current plans for future storage of high-level radioactive waste, for example, require the steel canisters to resist corrosion for as little as 300 years. Nevertheless, the U.S. Department of Energy and the U.S. National Academy of Sciences admit that the waste will remain dangerous for longer than 10,000 years. Government experts agree that, at best, they can merely limit the radioactivity that reaches the environment, and that there is no doubt that the repository will leak over the course of the next 10,000 years (Shrader-Frechette, 1993). If society saddles its descendants with the medical and financial debts of such waste, much of which is extremely long lived, such actions are questionable: this generation has received most of the benefits from the use of industrial and agricultural processes that create hazardous wastes, whereas future persons will bear most of the risks and costs. This risk/cost–benefit asymmetry suggests that, without good reasons or compensating benefits, future generations ought not be saddled with their ancestors' debts (Shrader-Frechette, 1993).

Moreover, any alleged economies associated with long-term storage of hazardous and toxic wastes are in large part questionable because of their dependence on economists' discounting future costs at some rate of x percent per year. For example, at a discount rate of 10%, effects on people's welfare 20 years from now count only for one-tenth of what effects on people's welfare count for now. Or, more graphically, with a discount rate of 5%, a billion deaths in 400 years—caused by leaking toxic wastes—counts the same as 1 death next year. A number of moral philosophers, such as Parfit, have argued that use of a discount rate is unethical, because the moral importance of future events, like the death of a person, does not decline at some x percent per year.

Another issue related to intergenerational equity is what sort of criteria might justify environmentally irreversible damage to the environment, like that caused by deep-well storage of high-level radwaste. Nuclear waste management schemes which are irreversible theoretically impose fewer management burdens on later generations, but they also preempt future choices about how to deal with the waste. On the other hand, schemes which are reversible allow for greater choices for future generations, but they also impose greater management burdens. If society cannot do both, is it ethically desirable to maximize future freedom or to minimize future burdens?

In the absence of knowledge that society can successfully store hazardous waste for centuries, the technical problems associated with it are forcing people to take a great gamble with the freedom and the security of future persons. This is a gamble that today's descendants will not breach the repositories through war, terrorism, or drilling for minerals; that groundwater will not leach out toxins; and that subsequent ice sheets and geological folding will not uncover the wastes.

In addition to temporal inequities associated with different generations, using and storing hazardous and toxic substances also raises questions of spatial or geographical inequity. One such issue is whether it is fair to impose a higher risk (of being harmed by leachate from a hazardous waste dump, for example) on a person just because she lives in a certain part of the community. Likewise, is it ethical for one geographical subset of persons to receive the benefits of products created by using toxic substances, while a much smaller set of persons bears the health risks associated with living near a hazardous waste dump or near an industry employing toxic materials? How does one site (or transport) hazards equitably?

The most serious problems of geographical equity in the distribution of the risks associated with hazardous wastes and toxic substances arise because of developed nations' shipping their toxic substances and hazardous wastes to developing countries. One-third of U.S. pesticide exports, for example, are products that are banned for use in the United States, and many of these exports are responsible for the 40,000 annual deaths caused by pesticides, mainly in developing nations (K. Shrader-Frechette, 1991. *Risk and Rationality.* Univ. of California, Berkeley). The UN estimates that as much as 20% of hazardous wastes is sent to other countries, particularly other nations where health and safety standards are virtually nonexistent. The Organization of African Unity has pleaded with member states to stop such traffic, but corruption and crime have prevented the waste transport from stopping (B. Moyers, 1990. *Global Dumping Ground.* Seven Locks, Washington, DC).

Exporting toxic substances and hazardous wastes may be the current version of infant formula. During the last three decades, U.S. and multinational corporations have profited by exporting infant formula to developing nations. They were able to do so only by coercive sales tactics and by misleading other nations about the relative merits and dangers of the exports. As a consequence, many children in developing nations have died because multinational corporations encouraged their mothers to buy infant formula (even when they had no clean water or sterilization equipment) and to abandon breast feeding.

Some of the greatest risks associated with toxic substances and hazardous wastes, whether in developed or in developing nations, are borne by workers. One of the main questions of occupational equity is whether it is just to impose higher health burdens on workers in exchange for wages. Is it fair to allow persons to trade their health and safety for money? This question is particularly troublesome in the United States, which has a double standard for occupational and public exposures to hazardous and toxic substances. In other nations, such as the Scandinavian countries, Germany, and the former USSR, standards for exposure to public and occupational risks from toxins is approximately the same. Because the United States follows the alleged "compensating wage differential" (CWD) of Adam Smith, U.S. regulators argue that, in exchange for facing higher occupational (than public) risks from toxic substances, workers receive higher wages that compensate them for their burden. Other countries do not accept the theory underlying the CWD and argue for equal health standards (Shrader-Frechette, 1991).

III. CONSENT ISSUES

One reason that the theory underlying the CWD is questionable is that it presupposes that workers exposed to hazards have given free, informed consent to the risks. Siting hazardous facilities and employing persons to work with toxins typically require the consent of those put at risk. Yet, from an ethical point of view, those most able to give free, informed consent—those who are well educated with many job opportunities—are usually unwilling to do so. Those least able to validly consent to a risky workplace or neighborhood—because of their lack of education or information and their financial constraints—are often willing to give alleged consent.

It is true that the 1986 U.S. Right-to-Know Act requires owners or operators of hazardous facilities to notify the Emergency Response Commission in their state that toxins are present at the facility. To some extent, this act contributes to information about hazards, but at least two factors suggest that the requirement is insufficient to provide full conditions for the free, informed consent of persons likely to be harmed by some hazardous substance. One problem is that owners/operators (rather than some more neutral source) provide the information about the hazard. Often those responsible for toxic substances and hazardous wastes do not inform workers and the public of the risks that they face, even after company physicians have documented serious health problems. Moreover, if employers in the chemical industry, for example, expend funds to assess worker health, typically they engage in genetic screening so as to exclude susceptible persons from the workplace, rather than in monitoring their health on the job so as to protect them (Draper, 1991). Another difficulty with ensuring free, informed consent, even under existing right-to-know laws, is that the existence, location, and operational procedures of the dangerous facility itself are likely things to which citizens and workers did not give free, informed consent in the first place, although they may be informed after the fact.

Sociological data reveal that, as education and income rise, persons are less willing to take risky jobs, and those who do so are primarily those who are poorly educated or financially strapped. The data also show that the alleged compensating wage differential does not operate for poor, unskilled, minority, or nonunionized workers. Yet these are precisely the persons most likely to work at risky jobs like storing nuclear waste. This means that the very set of persons *least* able to give free, informed consent to workplace risks from hazardous and toxic substances are precisely those who most *often* work in risky jobs (Shrader-Frechette, 1993).

Likewise, at the international level, the persons and nations least able to give free, informed consent to the location of facilities using toxins are typically those who bear the risks associated with such sites. Hazardous wastes shipped abroad, for example, are usually sent to countries that will take them at the cheapest rate, and these tend to be developing nations that are often ill informed about the risks involved. Recently the United Nations passed a resolution requiring any country receiving hazardous waste to give consent before it is sent. Because socioeconomic conditions and corruption often

militate against the exercise of genuine free, informed consent, however, just as they do in the workplace, it is questionable whether the UN resolution will have much effect (Shrader-Frechette, 1991).

If it is difficult to ensure that the risks posed by toxic substances and hazardous wastes are ones to which persons actually give free, informed consent, then medical experimentation may have something to teach society about the ethical constraints on dealing with hazardous waste and toxic substances. Scholars know that the promise of early release for a prisoner who consents to risky medical experimentation provides a highly coercive context which could jeopardize his legitimate consent. So also do high wages for a desperate worker who consents to take a risky job provide a highly coercive context which could jeopardize his legitimate consent. Likewise, providing financial benefits for an economically depressed community provides a coercive context in which the requirements for free, informed consent are unlikely to be met. This means that society must either admit that its classical ethical theory of free, informed consent is wrong or question whether current laws and regulations provide an ethical framework in which those closely affected by hazardous substances genuinely give informed consent to the risk.

One consent-related area of needed improvement in laws regulating hazardous substances, for example, would be to include mining among industries required to report their toxic emissions to state and federal regulators. Utah's Bingham Canyon Copper Mine, owned by Kennecott Copper, ranks fourth in the United States in total toxic releases, yet it and other mining companies are not yet required to report their releases (Young, 1992).

Another aspect of consent problems with hazardous substances is political. Liberty and grassroots self-determination require local control of whether a hazardous facility is sited in a particular area. Yet, equality of consideration for people in all locales, and minimizing overall risk, requires federal control. Should the local community be able to veto a given site, even though that site may be the best in the country and may provide for the most equal protection of all people? Or does one say that the national government can impose a hazardous facility on a local community even though the imposition is at odds with free and self-determined choice?

On the one hand, national supremacy is likely to protect the environment, to avoid the tragedy of the commons, to gain national economies of scale, and to avoid regional disparities in representing all sides of a controversy. National supremacy also is likely to provide for compensation of the victims of one region for spillovers from another locale and to facilitate the politics of sacrifice by imposing equal burdens on all. On the other hand, local autonomy is likely to promote diversity, to offer a more flexible vehicle for experimenting with waste regulations, and to enhance citizens' autonomy and liberty. It also is likely to encourage community coherence through participation in decision making; to avoid inequitable federal policies; and to avoid violations of rights.

IV. LIABILITY ISSUES

One of the greatest ethical problems with regulating hazardous and toxic substances is that current laws do not typically provide for full exercise of due process rights by those who may have been harmed by toxins or by hazardous wastes. One reason is that many of the companies that handle toxic substances do not have either full insurance for their pollution risk or adequate funds themselves to cover damages. In the United States, RCRA and CERCLA, however, require such companies both to show that they are capable of paying at least some of the damages resulting from their activities and to clean up their sites.

Part of the problem is that enforcement of liability and coverage provisions of laws such as RCRA and CERCLA is difficult, and many components of the hazardous waste industries operate outside the law. Another difficulty is that most insurers have withdrawn from the pollution market, claiming that providing such insurance would leave them exposed to enormous payments for claims that would bankrupt them.

But if insurers fear the large liability claims in cases involving hazardous substances, so do members of the public. In Yucca Mountain, Nevada, for example, chosen as the likely site for the world's first, permanent, high-level nuclear waste disposal facility, local residents and the state have asked for unlimited strict liability for any nuclear waste accident or incident. The U.S. Department of Energy position, solidified by the Price–Anderson Act, is on the side of limited liability. The U.S. nuclear program, including radioactive waste management, has a liability limit that is less than 3% of the government-calculated costs of the Chernobyl accident, and Chernobyl was not a worst-case accident (see Shrader-Frechette, 1993).

The ethical judgment that government ought to limit liability for hazardous waste and toxic substances incidents is questionable because liability is a well-known incentive for appropriate, safe behavior. Also, if hazardous and radioactive waste sites are as safe as the government proclaims, then there is nothing to lose from full and strict liability. Third, it is questionable whether government officials should have the right to limit due-process rights under law. Such a limitation may mean

that, in the case of an accident at a hazardous waste facility, the main financial burdens will be borne by accident victims rather than by the perpetrators of the hazard. But such a risk–benefit asymmetry appears to be inequitable. Moreover, since much less is known about the dangers from hazardous wastes and toxic substances than about more ordinary risks, it seems reasonable to guarantee full liability. Finally, the safety record of hazardous facilities, in the past, has not been good. Every state and every nation in the world has extensive, long-term pollution from toxins. Even in the United States, the government has been one of the worst offenders. A recent Congressional report argued that cleaning up the hazardous and radioactive wastes at government weapons' facilities is now an impossible task because it would cost more than $300 billion (Shrader-Frechette, 1993). Such revelations suggest that full compensation for potential victims of hazardous substances is ethically desirable (Shrader-Frechette, 1993).

V. ETHICAL RULES FOR BEHAVIOR UNDER UNCERTAINTY

Ethical difficulties such as inadequate compensation for victims of hazardous and toxic substances, and inequitable distribution of the risks associated with hazardous wastes—as well as the uncertainties and potential harm associated with such substances—provide a powerful argument for reducing or eliminating exposure to them. Society needs to move ''beyond dumping'' and, to some degree, provide market incentives for reducing the volume of toxic substances and hazardous wastes (B. Piasecki, Ed., 1984. *Beyond Dumping: New Strategies for Controlling Toxic Contamination*. Quorum, London.

In 1996, the U.S. National Academy of Sciences began a long-term study of xenoestrogens—hazardous chemicals that look nothing like the female sex hormone estrogen but have similar effects; these chemicals include PCBs, DDT, and other pesticides and industrial chemicals. The organochlorine chemicals that are xenoestrogens essentially behave like female hormones. Even at the lowest levels, parts per trillion, they have caused massive feminization and eventual extinction of amphibian and bird populations and massive decreases in sperm counts. Because of the groundbreaking work of scientists like Dr. Theodora Colborn, people now know that male sperm counts have been steadily decreasing, across all species, in developed nations at least since 1940. Human sperm counts have decreased by 42% since 1940, and female breast cancer has been increasing by at least 1% per year since 1950. The culprit in both cases appears to be extremely low levels of organochlorine chemicals.

Because it is difficult to perform adequate tests for health effects of very low-level chemical exposures—thousands of persons must be tested in order to detect a low-level effect—and because so many toxic substances produce health effects synergistically, there are many uncertainties about actual exposure to hazardous substances and about the effects of such exposures (N. Ashford and C. Miller, 1991. *Chemical Exposures: Low Levels and High Stakes*. Van Nostrand–Reinhold, New York). An additional equity and uncertainty problem is that children are much more susceptible to the effects of hazardous and toxic substances. In 1993, a landmark U.S. National Academy of Sciences Committee (National Research Council, 1993. *Pesticides in the Diets of Infants and Children*. NRC, Washington, DC) argued that existing standards for pesticides in the diets of infants and children were not adequate to protect their health. Differences in phenotypical characteristics among individuals often vary by a factor of 200, also causing extreme differences in responses to toxins and contributing to the uncertainty problem. All of these factors cause great uncertainties regarding exposure to, and effects of, hazardous substances.

Moreover, the industries that produce toxic substances and hazardous wastes—and that profit from them—perform almost all of the required tests to determine toxicity and health effects. Pesticide registration decisions in the West, for example, are tied to a risk–benefit standard according to which scientific and economic evidence are considered together. With industry doing most or all of the testing, and with environmental and health groups being forced to argue that the dangers outweigh the economic benefits of a particular pesticide, there is much uncertainty about the real hazards actually faced by workers and consumers.

Human error and crime are also large contributors to uncertainty about the possibility and the effects of exposure to toxins. This uncertainty raises additional questions about appropriate ethical behavior under conditions of uncertainty. Whenever one is dealing with a potentially catastrophic risk, uncertainty about the likelihood of harm arises in part because of human factors. According to risk assessors, for example, 60 to 80% of industrial accidents are due to human error or corruption (Shrader-Frechette, 1993). At the nation's largest incinerator for hazardous wastes, run by Chemical Waste Management, Inc. (CMW), in Chicago, a 1990 grand jury found evidence of criminal conduct, including deliberate mislabeling of many barrels of hazardous waste and deliberate disconnection of pollution-monitoring devices. Moreover, in the United States, corruption in the waste disposal industry has been rampant

since the 1940s, when the Mafia won control of the carting business through Local 813 of the International Brotherhood of Teamsters. Today, three Mafia families still dominate hazardous waste disposal and illegal dumping: the Gambino, the Lucchese, and Genovese/Tiere crime groups (see A. Szasz, 1986. *Criminology* **24**, 1–27). Given the potential for human error and corruption, there is great uncertainty regarding whether hazardous wastes and toxic substances will be handled safely, with little threat to workers or to the public.

Because of these unknowns, several moral philosophers have argued that, in potentially catastrophic situations involving hazardous wastes and toxic substances, scientific and probabilistic uncertainty requires ethically conservative behavior (C. Cranor, 1993. *Regulating Toxic Substances*. Oxford Univ. Press, New York; Shrader-Frechette, 1991; Ashford and Miller, 1991). It often requires one to choose a maximin decision rule—to avoid situations with the greatest potential for harm—as John Rawls argued. Such conservatism and maximin decision making in the face of the threat of toxins might require tougher health and safety standards. It might also require greater assurances of free, informed consent and full compensation for victims of toxics. Likewise, it might be important, from an ethical point of view, to ensure that the burden of proof be placed on the users of hazardous substances, rather than on their potential victims. This, in turn, means that society may need many reforms in its law governing so-called "toxic torts" (Cranor, 1993). More generally, it is arguable that in situations of uncertainty, ethical conservatism dictates that one probably ought to minimize type II statistical errors (minimize false negatives, false assertions of no harm from some substance) rather than type I errors (minimize false positives, false assertions of harm from some substance). Most risk assessors, however, in evaluating potentially catastrophic risks from hazardous substances, argue that one ought to minimize type I errors, even though this policy does not contribute to the greatest protection of public health and safety (Shrader-Frechette, 1991).

VI. CONCLUSION

If ethics requires reforms so as to protect future generations, workers, and disenfranchised members of the public, then indeed most law, policy, and regulation regarding hazardous wastes and toxic substances may need revision. Formerly "safe," low-level doses of chemicals such as xenoestrogens or organochlorines are feminizing and then killing off many species, and similar effects on humans have been clear for 50 years. Society requires greater ethical attention to the issues of equity, consent, compensation, and uncertainty that surround its use and disposal of hazardous and toxic substances.

Bibliography

Carson, R. (1962). "Silent Spring." Houghton Mifflin, Boston.

Colborn, T., and Clement, C. (Eds.) (1992). "Chemically Induced Alterations in Sexual and Functional Development." Princeton Scientific, Princeton, NJ.

La Dou, J. (1992). First world exports to the third world. *Western J. Med.* **156,** 553–554.

National Research Council (NRC) (1993). "Pesticides in the Diets of Infants and Children." NRC, Washington, DC.

Nordquist, J. (1988). "Toxic Waste," Bibliography. Reference and Research Services, Santa Cruz, CA.

Piasecki, B., and Davis, G. (1987). "America's Future in Toxic Waste Management: Lessons from Europe." Quorum, New York.

Postel, S. (1987). "Defusing the Toxics Threat." Worldwatch Institute, Washington, DC.

Samuels, S. (1986). "The Environment of the Workplace and Human Values." Liss, New York.

Wynne, B. (Ed.) (1987). "Risk Management and Hazardous Waste." Springer-Verlag, New York.

Health and Disease, Concepts of

ROBERT WACHBROIT

University of Maryland

I. Concepts of Health and Their Impact
II. Articulating a Family of Concepts: Constraints and Proposals
III. The Issue of Normativism

GLOSSARY

biological function The characterization of a biological phenomenon in terms of its achieving a particular goal or effect, rather than in terms of its etiology or cause.

medicalization The process by which a phenomenon or issue not usually seen as a medical problem comes to be regarded as susceptible to medical analysis and response.

physiology The study of how biological systems and processes normally function.

THE CONCEPT OF HEALTH is crucial to medicine because it informs medicine's goals, scope, and criteria of success. To state it roughly and simply, ignoring many subtleties and qualifications, the goal of medicine is to maintain or promote health; the scope of medicine's concern is with problems of health; and when medical practice brings someone back to a state of health, (medical) success is rightly declared. And yet, there is considerable controversy not only over particular definitions of health but also over the general conditions that any proposed definition should satisfy. The presence of this controversy should not be surprising. Many ethical and public policy issues concerning medi-

cine are, in part, disputes over the definition or analysis of the concept of health and, with that, the concept of disease.

Analyses of the concept have variously described health as complete physical, mental, and social well-being; as what people in a culture value or desire; as whatever the medical profession decides to call "health"; or as nothing more than a theoretical concept in physiological theory. The purpose of this article is not to assess the merits of specific proposals but rather to clarify the general themes and concerns that have shaped the philosophical and ethical literature on the concept of health.

I. CONCEPTS OF HEALTH AND THEIR IMPACT

The significance of the concepts of health and disease can be illustrated in several ways. One of the clearest and most dramatic ways is by looking at the phenomenon of "medicalization."

It is common to come across cases where a problem previously understood in terms of socialization, individual personalities, and the like is now framed in the language of medicine, invoking the terms "health" and "disease." Alcoholics and persistent gamblers, rather than being seen as individuals with various psychological weaknesses or moral failings, may now be regarded as victims of a disease. The restless, rebellious child or the poor learner, rather than merely exhibiting youthful energy, poor upbringing, or mediocre intelligence, may be diagnosed as someone with a disease—for example, attention deficit disorder. Even street violence, a problem many might regard as paradigmatically a social problem, has sometimes been called a "public health problem" or a problem arising from various disorders concerning impulsivity control and aggressive behavior.

In many cases, the warrant for using the language of disease is not some empirical discovery about human physiology but instead a general argument arising from a particular conception of health or disease. For example, if a disease is understood to be an undesirable condition over which the individual has no direct control but that seems to be treatable with certain medications, then alcoholism, on this understanding, is a disease. Once a condition is classified as a disease, several conclusions are often immediately drawn: The condition needs to be *cured;* finding and administering the cure is the responsibility of health professionals (e.g., physicians); the cure may require surgery, medication, or, if the condition is particularly serious and contagious, isolation and quarantine. Plainly, classifying something as a disease is a serious matter.

The enormous impact of the concept of health also appears in disputes over the sort of things to which the concept applies. Everyone agrees that the concept applies to people—it surely makes sense to say that an individual is healthy or has a disease. Common practice would also suggest that the concept applies to animals and plants as well as to parts of individuals—for example, organs and tissues. Controversies arise, however, when we ask whether the concept applies to units larger than an individual or to parts smaller than tissues. Does it make sense to ask whether a particular population or national group is healthy? Does it make sense to talk about the health of a species or a race? Does it make sense to talk about healthy or diseased genes or to invoke the idea of "genetic health?"

The legitimacy of extending the concept in these ways is closely tied up with the issue of eugenics. The aim of eugenics, as it has often been understood in this century, is to improve the health of the species. In this extended sense, health is indicated not simply by the health of current individuals but also by the health of their potential offspring. That is to say, even if individuals show no signs or symptoms of disease, the likelihood of their giving birth to children born with some disease indicates that those individuals have a condition that constitutes a health problem. Some eugenicists argued that such individuals were not "genetically healthy" and could transmit their disease to the population through reproduction. The cure: ensure that these individuals do not in fact reproduce. The result—the American sterilization laws and the Nazi eugenic practices—was surely one of the darkest events in the history of medicine.

For another, perhaps more contemporary illustration of the significance different concepts of health can have,

consider the controversy over human genetic engineering. Many critics believe that there is an important moral difference between manipulating people's genes in order to cure them of some disease—so-called "gene therapy"—and manipulating their genes in order to enhance their physical or intellectual capacities. The former is seen to be on a par with any other therapeutic intervention; the latter, however, is seen to be morally suspect, perhaps a return to discredited eugenics. For example, manipulating a child's genes in order to cure him of dwarfism would not be more morally suspect than any other medical intervention, but manipulating a normal child's genes so that he grows to an above-average height would be seen as morally problematic. The line between therapy and enhancement is nothing other than the line between health and disease.

Finally, the importance of the concepts of health and disease in informing the character of medicine shows itself in this: when we have difficulty identifying a condition as healthy or diseased, it becomes quite unclear what, if anything, medicine should do about it. Consider the case of genetic susceptibilities. These are genetic abnormalities that lead to diseases only when certain environmental (including, in some cases, certain cellular) conditions are present. For example, mutations of the BRCA1 gene can lead to a form of breast cancer, but, it appears, only when certain other conditions, such as a family history of that disease, are present. Is the mere presence of a genetic susceptibility itself a disease? If we were to regard genetic susceptibilities as diseases, then the goal of medicine would become the cure, control, or elimination not only of diseases but also of genetic susceptibilities to disease. This view, however, comes uncomfortably close to some of the older, discredited forms of eugenics, where the goal was not simply health but also a kind of genetic purity. On the other hand, it is difficult to regard genetic susceptibilities as healthy conditions, because a person with a genetic susceptibility is at greater risk of contracting the disease. Presumably, the goal of maintaining health includes reducing the risks to health. The difficulty in construing genetic susceptibilities as either healthy or diseased conditions complicates our efforts to spell out the goals of medicine with respect to these (quite common) susceptibilities.

II. ARTICULATING A FAMILY OF CONCEPTS: CONSTRAINTS AND PROPOSALS

Health and disease are just two members of a family of concepts that includes illness, disorder, normality, abnormality, malady, function, malfunction, dysfunction, trauma, injury, and so forth. Some writers have argued that the relationship between different pairings is often more complicated than either synonymy or antonymy.

For example, is health nothing more than the absence of disease? At first it might seem so, but two considerations have led some to deny this. First, there are many conditions that are not usually called "diseases"—for example, broken bones, motion sickness, wounds, and so on—but whose absence is usually thought to be part of being healthy. Some have therefore tried to identify a generic term to cover all these cases—for example, malady—while others have responded by explicitly stating that the term "disease" is being used in their discussion in an especially broad way somewhat at odds with ordinary usage. A different consideration suggesting that health and disease are not opposites arises from noting certain comparative judgments regarding health. Even if two individuals are both free of disease (even in the broad sense of disease), one might be judged to be healthier than the other because, for example, the first is a star athlete while the other is sedentary. One response to such considerations is to examine first basic or minimal health in the hope that more positive conceptions can be seen as extensions of the basic conception. Furthermore, because incorporating positive conceptions of health into the goals of medicine introduces new controversial issues, establishing the basic conception also has the advantage of beginning the discussion with the simpler case.

Here is another conceptual problem: Are disease and illness the same thing? At first it might seem so. But while a disease can have asymptomatic phases, illness cannot: it would be (linguistically) odd to say of someone that he was physically ill but nevertheless felt fine. Some have claimed that illness is a certain class of disease—a disease that the individual experiences as an undesirable condition. Others have suggested that illnesses are not a subclass of diseases, because one can in fact be ill without having a disease—as, for instance, when one has a hangover.

The substantive issue underlying these apparently linguistic matters is whether the investigation of health or disease should be directed at the family of ordinary or scientific concepts. That is to say, should the aim be to analyze the concepts of health, disease, and so on, as reflected in everyday usage, or should the aim be to analyze these concepts as they are reflected in biological theories? This question must be addressed prior to any proposed analysis because the answer determines how the resulting analysis should be assessed. The concept of health is one of many concepts—others are those of space, time, heat, force, work, solidity, being a fish, and so on—that have an ordinary and a scientific understanding. The relationship between the ordinary and the scientific concept is a matter of some controversy, and so we cannot assume that the same analysis would apply to both. For example, we would expect an analysis of the ordinary concept of space as reflected in everyday use to be quite different from an analysis of the scientific concept of space as reflected in physicists' theories of the structure of spacetime.

Some commentators have denied that there are scientific concepts of health and disease, insisting that these concepts have no role in biological theories and explanations. But this claim probably reflects the influence of early logical positivism and its suspicion of the use of the concept of function in biology. Few would now deny that at least some members of the family of concepts play a significant role in biological theories. The impact of this acknowledgment on the assessment of proposed analyses should be clear. If the aim is an analysis of the ordinary concepts, then the goal is to construct a definition that captures how these concepts are typically used and understood; a failure to capture *any* ordinary use of the concepts would constitute a prima facie objection to the analysis. On the other hand, if the aim is to analyze the scientific concept, then the goal is to analyze the role these concepts play in the appropriate biological theories and explanations, notably those in physiology; any effort to capture usage outside of these scientific contexts would be beside the point.

Not surprisingly, deciding which investigation is the correct one to pursue is controversial because that decision is tied to the question of the relationship between medicine and biology. Putting the issue in its most extreme form: Are medical concepts simply biological concepts suitably framed and applied? Or is medicine an independent discipline whose use of biology is purely opportunistic? We will return to this dispute over which set of concepts is the appropriate one to analyze.

III. THE ISSUE OF NORMATIVISM

Most of the literature on the concepts of health, disease, and so on has been concerned not with the assessment of particular definitions but with the question of whether the concepts are value-laden. This has been called the issue of "normativism." In order to explain what is at issue, we should begin with the observation that ordinary judgments of health and disease have what we might call a normative "force" or "effect." That is to say, most people regard being healthy as a prima facie good thing and having a disease as a prima facie bad thing. The dispute is over how to explain these conclusions.

According to the normativists, value claims are part of the *meaning* of health and disease. For example, one version of normativism holds that having a disease is having a biological condition that society values negatively. Epilepsy is judged in our society to be a disease because, in part, it is judged to be a bad or undesirable condition. But in another society, where the epileptic's seizure is taken to indicate a valued encounter with the sacred, epilepsy is not judged to be a disease at all. Thus, the normative conclusion is a straightforward inference based solely on the meanings of the terms.

The nonnormativists, on the other hand, hold that the meanings of health and disease can be specified in a value-neutral way. According to one version of nonnormativism, disease is a deviation from normal biological functioning, where what the biological functions are and what counts as normal functioning is specified by biological theory, specifically physiology. (It is usually assumed by both sides that scientific *theory* is value-free. The tenability of scientific *practice* being value-free is a different and more controversial matter.) Epilepsy is a disease because it is a deviation from normal functioning; if another society does not regard it as a disease, then its members are simply wrong about human biology. The normative conclusion does not therefore follow directly from the meanings of "health" and "disease" but requires further premises of an explicitly moral nature. For example, some people have argued that assumptions about a theory of justice are required to explain the normative effect of (some) judgments of health.

It would seem therefore that the difference between the normativists and the nonnormativists turns on who can provide a better semantic theory of medical language. After all, it hardly seems plausible that simple (or sophisticated) introspective observation can determine whether the normative conclusion is based solely on the meanings of the terms or requires additional moral premises. But for the most part, the disputants have not engaged in a discussion of the merits of various semantic theories. Instead, the literature has attended more to constructing and responding to various counterexamples to the specific claims about whether the concepts of health are or are not value-free.

For example, a typical argument nonnormativists use against normativism is to identify cases where having a disease might be thought to be a good thing—for example, an individual might be happy to have flat feet (he wants to avoid the military draft), and people who contract the cowpox disease may count themselves fortunate (because contracting cowpox confers an immunity to the more serious smallpox disease). The thought behind these objections is that if being a bad or undesirable condition were part of the meaning of "disease," then examples such as these would be self-contradictions, which they are not. But this type of argument does not

work in the absence of an explicit semantics. Consider a concept that everyone would regard as value-laden—for example, the concept of beauty. We would not conclude that the concept is not value-laden after all because we can imagine cases where someone doesn't want to be beautiful or thinks being plain-looking is a good thing.

On the other hand, the common type of objection against nonnormativism is to point out counterexamples to Boorse's account, which is usually taken to be the paradigm nonnormativist position. Briefly put, Boorse characterizes disease as a departure from normal functioning, where "normal functioning" is understood in terms of what is statistically typical for the species. Thus, if an individual's thyroid—whose function is to secrete hormones important for metabolism—secretes an excessive amount of hormones, compared to what is statistically typical, then the individual has an abnormal— that is, a diseased—thyroid. While normativists may point to various exceptions to Boorse's account, perhaps the clearest counterexample is the one that Boorse himself raises. If malfunctioning is understood as deviation from species typicality, then the account cannot recognize widespread or universal diseases—for example, dental cavities or functional declines associated with old age. Boorse's response is to propose ad hoc additions to his account of disease or to regard such examples as in fact not counterexamples at all but rather themselves "anomalies deserving continued analysis." That is to say, the nonnormativist account is not being proposed as simply descriptive of medical usage; if certain examples cannot fit the account, so much the worse for these examples—we should reconsider calling them diseases. Such a response suggests that the semantics of the concepts of health and disease is not really the issue.

One suggestion for what might be at issue—in contrast to the fairly technical one of semantics—is the objectivity of judgments of health and disease. When a society claims that masturbation is a disease, does that claim rest on a biological fact, which either supports or refutes it, or is the claim nothing more than a society's values dressed up in medical language? If one society claims that homosexuality is a disease while another does not, are the different societies disagreeing about some objective, biological fact or is there no real disagreement, only a difference of culture?

Recasting the dispute in this way clearly assumes that values, and so normative conceptions of health and disease, are not objective. But there are several problems with interpreting the normativism dispute in this way. The move from a debate over whether the semantics of health and disease contain value terms to a metaphysical debate over whether the condition of being healthy or diseased is a fact or a value is a move that many philosophers are hesitant to make: the semantic "is–ought" distinction is far less obscure than the metaphysical fact–value distinction. Consequently, there is considerable controversy among philosophers over what "objectivity" means, so that few would say that the concept of objectivity is clearer or less controversial than the concepts of health and disease. Moreover, and in part because of this controversy, some philosophers would hold that values *are* objective, including some who argue for the normativist position.

Nevertheless, thinking of the dispute as one over the objectivity of judgments of health and disease does seem to point in the right direction. Perhaps a clearer way of identifying the problem would be in terms of the issues we mentioned earlier regarding the contrast between ordinary and scientific concepts of health and disease: Are the concepts of health and disease that characterize the aim, scope, and criteria of success of medicine scientific or ordinary concepts?

By and large, normativists focus on the ordinary concepts and judgments, because their aim seems to be to capture ordinary usage. For the most part they do not seem to be arguing that standard biological theories are (or must be) value-laden. Indeed, they do not seem to be drawing any conclusion regarding biological theories. On the other hand, nonnormativists appear to give pride of place to the scientific concepts and judgments of health and disease. They appear willing to revise or correct ordinary usage in order to smooth the transition to the scientific concepts. In their view, the apparently normative judgments about disease that the normativist identifies should really be understood as judgments about illness, a quite different concept. Of course, this suggestion is not intended to entail that a normativist could not challenge the value-free claims of biological theories, although the task of arguing the broader claim would dwarf the normativist's thesis regarding health. Similarly, it is conceivable that a nonnormativist would argue that ordinary judgments of health and disease are value-free as they stand, although, given the number and variety of apparent counterexamples that have been proposed, such a position would face considerable difficulties without an appropriate semantic theory of these ordinary concepts.

Once the dispute is cast in terms of a contrast between the ordinary and the scientific, the normativist and nonnormativist positions no longer seem diametrically opposed: The nonnormativist is right in that the concepts of health and disease, considered as scientific concepts in biology, are value-free. The normativist is right in

that the concepts of health and disease, considered as ordinary concepts that are shaped by our experiences of illness and disability, are value-laden. The question becomes, however, which set of concepts is appropriate for medicine, for these two kinds of concepts suggest two very different conceptions of medicine. According to one, medicine is the application of biological knowledge to diagnose and correct biological abnormalities. According to the other, medicine is a discipline that uses biological discoveries to advance and maintain certain values.

The second (normative) conception might be objectionable in that it seems open to abuse: the image of health care practitioners advancing a particular agenda of values calls forth all the problems raised by medical paternalism. This might make the apparently more modest first (nonnormative) conception more appealing until we realize that it depicts practitioners as, in effect, morally neutral technicians, which can itself lead to a different kind of abuse. The second conception does make clear that medicine and those who work within it have a distinctive moral responsibility. Presented in this way, the dispute is hardly an academic one.

Bibliography

Boorse, C. (1975). On the distinction between health and disease. *Philosophy & Public Affairs, 5,* 49–68.

Boorse, C. (1977). Health as a theoretical concept. *Philosophy of Science, 44,* 542–573.

Caplan, A. L. (1996). The Concepts of health and disease. In R. M. Veatch (Ed.), *Medical ethics,* pp. 49–62. Boston: Jones and Bartlett.

Caplan, A. L., Engelhardt, H. T., & McCartney, J. J., (Eds.). (1981). *Concepts of health and disease: Interdisciplinary perspectives,* Reading, MA: Addison-Wesley Publishing Co.

Culver, G. M., & Gert, B. (1982). *Philosophy of medicine: Conceptual and ethical issues in medicine and psychiatry,* New York: Oxford University Press.

Daniels, N. (1985). *Just health care.* Cambridge: Cambridge University Press.

Engelhardt, H. T. (1975). The Concepts of Health and Disease. In H. T. Engelhardt & S. Spicker (Eds.), *Evaluation and explanation in the biomedical sciences,* pp. 125–141. Boston: Reidel.

Fulford, K. W. M. (1989). *Moral theory and medical practice.* Cambridge: Cambridge University Press.

King, L. (1954). What is a disease? *Philosophy of Science, 12,* 193–203.

Margolis, J. (1976). The concept of disease. *Journal of Medicine and Philosophy, 1,* 238–255.

Reznek, L. (1987). *The nature of disease.* London: Routledge & Kegan Paul.

Wachbroit, R. (1994a). Distinguishing genetic disease and genetic susceptibility. *American Journal of Medical Genetics, 53,* 236–240.

Wachbroit, R. (1994b). Normality as a biological concept. *Philosophy of Science, 61,* 579–591.

Health Technology Assessment

RICHARD E. ASHCROFT

Imperial College School of Medicine

HEALTH TECHNOLOGY ASSESSMENT (HTA) is the assessment of health technologies (drugs, operations, techniques, modes of delivery, support systems, organizations) for efficacy, effectiveness, efficiency, cost-effectiveness, and social and ethical acceptability. HTA programs are now being set up all over the world, and have had widespread impact in producing national assessments of coronary care, "keyhole" surgery, and cancer screening programs, to name but three examples. Much of HTA is primary research—a phase III clinical trial is a technology assessment. Some HTA is secondary research, involving meta-analysis of existing research or audit of current practice. HTA shapes policy and practice through summary and analysis of research and audit of existing practice. The ethics of HTA is in part research ethics and in part the ethics of health policy, particularly the ethics of allocation of scarce health care resources (money, time, space, drugs, skills, etc.). The ethics of HTA research concerns how HTA is done, and the ethics of HTA as a policymaking tool concerns what is done with HTA findings and consequently which questions are posed in framing assessments.

I. BACKGROUND

A. Ethics in Health Care

Traditional medical ethics concentrated on the relationships between a doctor and each of his individual patients, and between a doctor and his fellow medical professionals. However, this has for many years been regarded as insufficiently narrow. No adequate health care ethics can think only of the "sacred dyad" of "one doctor, one patient." Every doctor has many patients, and most doctors practice within a health care system or institution. Doctors are by no means the only professionals responsible for patients, or involved in their care; these diverse relationships and responsibilities involve their own ethical issues.

An adequate ethics for health care must also take into account the "population perspective" of public health medicine and the impact of collective funding of health care (through insurance or state-financed health care). Also, health care is now practiced under a new constraint: once what could be done for a patient was constrained by what was known, where now it is often the case that what we can afford is more of a constraint than what we know. Technological change in medicine is occurring at a very rapid rate, and there is a seemingly constant growth in the number of options for treatment and care in all areas of medicine. These issues have made justice in health care a primary moral problem.

A final contrast, much remarked upon in the general medical ethics literature in the developed world, is the way the balance between beneficence and autonomy has shifted away from the primacy of beneficence (so-called "paternalism") toward the primacy of autonomy. This trend has been linked by several commentators to two social trends in capitalist societies: a putative moral and cultural fragmentation of society, such that we have become "moral strangers" to each other, unable to identify each other's best interests, and the rise of consumer society. It could be argued that informed consent began by being understood as a negative right to refuse certain things being done to you, but has become a positive right to demand that certain things be done for you.

In consequence of these factors, medical ethics seems now to require that technical choices are framed not only by the need to be beneficent efficiently and effectively, but also by the need to optimize social welfare (which may restrict some options) and to satisfy aggregate private welfare (arguably not the same thing, and arguably in tension with social welfare). Thus, the meaning of "health technology assessment" varies with the degree of emphasis placed on social solidarity or individual liberty in the society where HTA is done.

B. Ethics and Technology

Until relatively recently, the "ethics of technology" was very little discussed by philosophers in the analytic tradition which dominates philosophy in the English-speaking world. Marxist, feminist, and phenomenological approaches to the philosophy of technology exist, but until recently these operated at a level of abstraction quite divorced from the nuanced historical and sociological analyses of particular technologies produced within the same intellectual traditions. In essence this chasm could only be bridged by a methodology for an *ethical* assessment of technology which marries the general, normative, critical stances implied in Marx, Adorno, Heidegger, and others to the careful descriptive work produced by sociologists and historians. This methodology is beginning to be constructed; until it is, the broadly "positivist" approach to technology assessment generated within HTA remains the best available approach. This approach builds on the idea that some issues are matters of fact (prices, survival rates, and numbers of surgeons, for example) while others are matters of value (acceptability, priorities). Facts and values are logically unconnected; HTA can summarize the facts in the most usable form, but the choices to be made are based on values lying outside HTA's scope. So HTA, in this positivist account, produces the information which society and its members can then use to make choices, without predetermining those choices. Implicit in this account is the idea that the particular technologies under assessment are strictly morally neutral, and ethics enters only

in consideration of the uses to which they are put, the priorities set in their purchase, and access to them. This account may be rather persuasive when one considers the assessment of some technologies (e.g., different approaches to the management of stroke), more problematic when considering comparisons between different technologies (pharmaceutical and psychotherapeutic approaches to reactive depression), and perhaps beside the point altogether when considering some radical innovations (reproductive cloning).

The first part of this article considers HTA on its own terms, in the positivist framework here sketched. This account of the ethics of HTA I label the "thin account." The second part of this article considers the ethics of HTA from a more critical standpoint, which I label the "thick account." At present, the debate between the two versions of the ethics of HTA is very new, and no authorial preference should be read into the order or labeling of the two accounts. The thin account is, however, currently much better established in health care management and practice.

II. ETHICS AND HEALTH TECHNOLOGY ASSESSMENT: THE "THIN" ACCOUNT

On its own terms, HTA is an ethical program. It addresses beneficence and nonmaleficence directly, by optimizing the knowledge base and by direct promotion of best practice through its focus on effectiveness and efficiency. Indirectly, it addresses justice by trying to reduce waste. It is, perhaps, ambiguous with respect to autonomy. On the one hand, it promotes informed choice, but on the other, it restricts choice to those options which are measurable and pass its strict and technocratic conceptions of efficiency and effectiveness. It faces some problems in applying its method to individual care. And it can be insensitive to questions of value, both personal and social. What is best, technically, may not be what is "best for me." These problems are particularly acute when HTA becomes concerned with criteria of optimality related to cost-effectiveness, or with added-value in terms of quality of life.

There are three ways in which ethics can be related to HTA: ethics in HTA (ethical assessment as a component of health technology assessment); ethics applied to HTA; and HTA applied to ethics.

A. Ethics in Health Technology Assessment

Most HTA methodologists argue that ethical and/or social impact assessment should be a vital part of any HTA review. But in fact this is rarely done. In part, this may be because two different things—social impact assessment and ethical assessment—are conflated. The former is researchable by a number of quantitative, empirical methods, including modeling, while the latter is much harder to define. In any case, they are not the same, simply because what is socially acceptable need not be ethically right (bribery, capital punishment, slavery). More sophisticated is the notion that choice of a policy that is ethically right (fair, optimal) is a matter of selecting a policy which optimally satisfies preferences over the population of interest. This is a quasi-utilitarian strategy, which may work reasonably well—at least conceptually—for tradable economic goods and services. The contribution that economics has made to health care analysis should not be underestimated, nor should one assume that all economists are ethical utilitarians (or all ethical utilitarians are economists *manqué*). But two major problems with this approach are well known: first, satisfaction of preferences need not be ethical. Second, a simple optimizing strategy is not "person-regarding," and this can lead to a variety of problems.

So how could ethics be built into HTA?

First, most systematic reviews exclude studies which do not meet certain ethical standards, especially the standard of informed consent. Partly this is because this sanction is meant to discourage unethical experimentation (there is no point in doing a study which will not be published, or will not be cited or included in reviews). And partly it is meant to encourage a certain methodological homogeneity—consent does make a difference to patient behavior in studies, and so we need to compare studies that are methodologically equivalent.

Second, social impact assessment, while distinct from ethical assessment, is relevant to it, and is sometimes done, particularly when HTA turns its attention to questions of dissemination and implementation of HTA review recommendations.

Ethical assessment is rarely done, and this is for a number of reasons. It is hard to know what it would be, methodologically speaking. It is impossible to separate out a point of view from the review, and the question of "who judges" can never be entirely settled. Finally, HTA studies must be general in their applicability. Can ethics reviews be generalizable? In one sense they should be, if ethical standards are universally valid. But in another sense they could not be: assessment is so dependent on thick descriptions of situations that studies are inevitably local, and this is obviously so when one considers that "acceptability" is always relative to a local population. And from the implementation point of view, acceptability is a crucial determinant.

HTA's strengths are in technical assessments. If ethical impact assessment is so hard to do, perhaps it would be better not to do it, and to regard HTA as providing a menu of options for the public and policy makers and doctors to choose from in some reflective and transparent process of debate. There is one way HTA methods could contribute to this, which is through adoption of methods of focus groups and consensus conferences. But the "evidence" these generate is usually regarded as low quality and "noisy." Even "technical" assessments are local, in that "techniques" are always delivered in an institutional context, although the effect or significance of this "locality" for the implications of the assessment can usually be abstraced or minimized.

B. Ethics Applied to HTA

Does ethics have something to say about the process of HTA itself, and about the role HTA plays in health care policy and delivery? Yes—in two areas, that of fairness and that of the social context of HTA.

1. Fairness

Does HTA produce fair outcomes? It targets high-demand areas, and is thus subject to the "squeaky wheel" problem—treating areas in which public attention is strong. Lower demand areas or rare diseases tend not to get attention, unless some interest group gets involved (either patients afflicted or scientists who find the area interesting). Also, HTA focuses on uses rather than needs. For example, if a variation in use is detected, this is seen as a problem, usually in isolation from any consideration of underlying variations in morbidity (yes, we do more coronary artery bypass grafts here, but then we have more coronary heart disease here). Third, HTA tends to be intradisciplinary, if only because the studies it reviews are (trials of one surgical procedure against another, not of surgery against social care). Finally, HTAs generate a "collapse" problem: they select a unique best option, rather than a family or set of options. "Which is the best?" rarely has a unique answer, because answers must be purpose-relative. This is the same problem faced with Quality Adjusted Life Years (QALYs), which tend to be one-dimensional judgments of scalar quality, rather than multidimensional measures. But of course, multidimensional measures are impossible to rank in a single ordered chain, unless we impose some ordering rule (e.g., a "lexicographic" ordering as used by Rawls), which must in turn be justified.

2. The Social Context of HTA

The social context of HTA is also important. HTAs produce technical solutions to problems which are a mix of technical and value problems. Some ethical problems can be overcome by gathering more data, or by agreeing upon a given technical standard to use for calibrating techniques against. But most are simply not like that. And technical solutions have a way of closing off some problems at the cost of opening up new ones. Most important, technical solutions to ethical or political problems can be invoked to block political debate, when in fact they have been set up in such a way as to entail a given political outcome. "Neutral" technical solutions are rarely as "neutral" as they first appear.

Commentators from Eastern Europe and the Third World are very concerned about conflicts of interest in health policy making, which HTA can sometimes prevent (prostate screening) but sometimes promote (technology T is the best, but not currently available in country C, due to its expense—so does its being best mean that it ought to be purchased and sacrifices made elsewhere? Who decides? Under whose influence?).

As well as conflicts of interest between politics and health care and industry, there are also conflicts between professions in health care (doctors, surgeons, nurses, radiologists, managers). Technology T may be the best overall, but favoring it favors profession P, at the expense of profession Q, who are experts at providing the now disfavored technology S; funding T and not S reduces the power and job security of Q and raises the power and job security of P. This conflict goes all the way down to methodology: nurses often favor qualitative research because it concentrates on the care relationship, for example, and so may favor the nursing approach to health care delivery. Other nurses favor quantitative research because it is perceived to be more "scientific" and consequently it may erase some of the cognitive and status boundaries between medicine and nursing. HTAs remain unconscious of this, but in favoring randomized controlled trials (RCTs) as the gold standard of evidence they take a stand on this particular intra- and interprofessional conflict.

HTA frames a focus on efficiency issues in the application and development of technologies of health care. This can have ethical relevance in two ways: by improving allocative efficiency, we can be assured that an optimal "bundle" of health care goods and services is purchased, which in theory should mean that more people can be treated than under a less efficient allocation of resources. On the other hand, a focus on efficiency may mean that some individuals or groups receive less treatment than they might feel fair because the treatment they require may be regarded as not cost effective, relative to other treatments for other groups, and since this impacts on their health status, which is not normally considered an economic good, perhaps the "economic rationality" of distribution of health care goods and services does not reliably shadow the "ethical rationality" of distribution of health status. This may be doubly unjust in many cases, given that the distribution of ill health is generally known to be correlated with the distribution of poverty.

Nonetheless, the economic component of HTA *does* help in priority setting through identification of needs, and it *does* help achieve local maxima by targeting resources away from ineffective treatments toward effective treatments locally: say from ineffective to effective (or cost-ineffective to cost-effective) treatments of storke. HTA may help medium-scale resource allocation problems too, by deciding on which package of health promotion, primary and secondary prevention, and drug or operative procedures to buy, given a certain gross budget for stroke care. But HTA may have nothing at all to say about large-scale allocation problems (how much to spend on stroke relative to spending on psychiatric in-patient care, for example). Arguably HTA can provide some data here, as can public health, but most of the decisions are value-based: which things will our society make into priorities?

C. HTA Methods Applied to Ethics

HTA methods can be applied to ethics in three ways.

1. Reviews of Opinion and Guidelines

HTA methods can be used to generate systematic reviews of ethical opinion or guidelines. This is a very limited field at present, and perhaps in principle. Not much work has been done for reviews to summarize, and it is not clear that opinion studies tell us much about ethics, although they do tell us something about socially expressed values. Thus, opinion surveys will not tell us whether capital punishment is morally right, although they may tell us how many people think it is morally right (or wrong). Most reviews summarize data that can be aggregated and generalized. Opinion data are rarely generalizable across time and between countries. Also, we can measure preferences or beliefs, as noted, but these may have very little weight in determining the ethical principles or rules to be applied. So the utility of reviews—and (to a lesser extent) primary studies—may be small.

2. Empirical Data for Ethical Reflection

Nonetheless, many ethical questions are partly empirical. Some ethical questions do turn on matters of fact (some of the issues in debates about abortion may turn on matters of fact about development of the fetal nervous system, for example). Some others can be illuminated by matters of fact. If we want to know how to allocate resources between coronary artery bypass grafts (CABGs) and *in vitro* fertilization (IVF) services, then part of what we need to know is cost data, survival rates, and rates of successfully completed implantation and pregnancy—but only part of what we need to know is factual in this sense.

3. Evidence-Based Consequentialism

Some of the methods of ethics are more evidence-based than others. The importance of this depends on strictly philosophical questions about the proper foundation for ethical knowledge and practical judgment. In ethics, being evidence-based is not an end in itself, nor is it a test of validity or reliability. But if one believes—as many philosophers and professionals do—that the test of what makes an action good is whether its consequences are good (the best available), then one needs to know a lot of factual material about what the consequences of each action in the set of options are. And so "consequentialist" reasoning in ethics and policy making should rely on good HTA assessments as a primary source of information about the consequences of using particular techniques. In fact much consequentialist reasoning is of a peculiarly "armchair" sort, and seems to have little to do with evidence! By their nature, economists tend to be better than philosophers, at ensuring their claims are supported by empirical evidence and doctors occupy positions all along the scale.

4. Evidence-Based Methods for Implementing Ethical Requirements

Finally, HTA can produce assessments of some of the methods of ethics itself. For instance, we almost all agree (and are in some sense committed to agreeing by the UN Declaration of Human Rights, the World Medical Association Declaration of Helsinki, and the Council of Europe's Convention on Bioethics) that free, informed consent is a critical test of what makes medical care and medical interventions ethical. But little consistency is evident in what methods are used in obtaining such consent, while there are many studies of different methods and their effectiveness and acceptability to patients and health care workers. HTA review can assist here in pulling together the evidence.

HTA can review methodologies of ethics research (e.g., in the use of focus groups); methodologies of ethical assurance (e.g., of techniques in consent); measures of acceptance of process (e.g., public understanding and tolerance of clinical trials); and assessment of outcomes (political and social consequences of new technologies).

D. Patient's Informed Choice and HTA

At present, the main role of HTA is in forming informed choice by major purchasers of health care (states and private insurers). Consent is the paradigm of patient informed choice, because it is the point at which most patients decide on what treatment they wish to receive, albeit one constrained by what they are actually offered, and by what they (or the third-party payer—their family, insurer, or government) can afford. We could widen the scope of consent to include not only, "Do you want to undergo *this* or not (and this is the alternative)?" but also, "Here are some options, which would you prefer?" To the latter question, there is only one sensible answer—I want the best. So the content of the consent information must contain the information about each option, which allows the patient to determine which treatment is the best. Historically this was thought absurd—the only people capable of judging the best treatment in a given situation were trained doctors. Nowadays—as often in the past—patients have their own ideas about what is best for them. And rightly so. Because while we can often assume that "best" is best in a technical sense, even the meaning of that is context sensitive. It can mean "best as judged by the outcome of an HTA," but where no complete or reliable HTA is available, best might mean the new treatment currently being trailed (of course, this might in fact be the worst!)—and many patients do think new = state of the art = best, and demand accordingly. Alternatively, best might mean "best for me—given my condition, my body, and my values." And this "bestness" may have little to do with best in the HTA sense, particularly given that the outcome measures in HTA can be efficacy, effectiveness, survival, cost-effectiveness, and so on.

The role of informed choice can be inspected from two different points of view: can we fix on what patients want by looking at patterns of demand? HTA and audits have historically looked at patterns of use by doctors; public health and epidemiology look at patterns of need; and perhaps economists can study patterns of demand. The three patterns need by no means be isomorphic, but becuase of the way health care need in individuals is determined by the same people as those who provide the services that meet that need, one would expect the demand and use patterns to be more similar than the demand and need patterns and the need and use patterns. The difference between demand and need patterns could perhaps be reduced by improving the amount and quality of information—the market in health care is imperfect in many ways, but one key imperfection is the imperfection of information.

Alternatively, is satisfaction of economic demand (i.e., of patient or intermediate suppliers preferences) the same thing as making the ethically right allocation of goods? If we set funding levels in response to patients' informed choices, are we actually buying a fair portfolio and distributing from it fairly? At present patients have poor information and choose on the basis of it; if they had better information they would make better informed choices. But there is no guarantee that either pattern of choices is formally rational, and there is no guarantee that even a formally rational pattern of choices is ethical. Again, this is an area where empirical and analytic research is possible. The part HTA can play in this is twofold: it is a source of information for choice, and it can evaluate different methods of eliciting choice. What it cannot do is determine which are the ethically right choices, although it may be able to determine the formally rational choices.

E. Summary of the "Thin" Account

The thin account of HTA which I have given in this section is an attempt to describe HTA on its own terms. This description shows that HTA can describe on its own terms many of its own strengths (naturally), but also many of its weaknesses. HTA is one of the most important techniques for improving health care that we have, and long may it continue. But it has its limits and weaknesses. HTA is good at technical questions, and for that very reason we should be suspicious of its ability to resolve ethical, evaluative, policy, or political questions. Even within the realm of technique its methods are limited to determining local maxima. The public has the right to the best information it can afford—even data does not come for nothing—and HTA contributes to satisfying this right, and to satisfying the right to effective health care. Its contribution to just health care is much more ambiguous. We might agree that, on its own terms, HTA provides grounds for determining which technology or policy is most just *substantially*—even allowing that HTA cannot fully determine that judgment. Yet there are still grounds for querying whether its approach to such problems is *procedurally* just. In practice, the distinctions between theory and practice, fact and value, and technique and politics are rather vague in HTA. Hence, the HTA methodology has an in-built bias toward expert judgment and away from participation and debate. We turn now to the thick account of the ethics of HTA, which tries to address some of these issues.

III. ETHICS AND HEALTH TECHNOLOGY ASSESSMENT: THE "THICK" ACCOUNT

The thick account begins by summarizing some features of contemporary HTA which adherents of the thin account would probably concede. HTA is strongest in evaluating effects of technologies which can be quantified and measured. It assumes the following: to speak of "health technology" is meaningful; technology is neutral in itself; quantification is evaluatively neutral in itself; quantification provides a basis for unprejudiced comparison between quantifiable factors inherent in the technology; what cannot be quantified may be important, but lies outside the domain of HTA methodology; and where a technology is to be purchased, it should be assessed by HTA methods and compared with other technologies on that basis. Taken together, these propositions arguably imply that HTA gives us an evaluative structure for assessing any health technology. When comparing two technologies, we feed in background values which frame the assessment—for example, five-year survival rates in testicular cancer treatment, which represent a value judgment that survival for five years post-diagnosis is important (not all that controversial)—and we "run" the assessment. The results enable us to assign positions in the evaluative structure to the two treatments. If we decide that treatment A is better than treatment B on the basis of the HTA, but that we will purchase treatment B more than (or instead of) treatment A anyway, then, on the thin account, we are applying a different evaluative standard. In principle, we hope to elucidate what that standard is, use it in future similar HTAs, and justify our "odd" preference for B over A. If we cannot so justify our preference, then we are being formally irrational. On the HTA account, formal irrationality is the worst sort, as it is the one sort of inconsistency mathematical and logical systems cannot tolerate.

To see how an alternative account might be given of HTA, consider the following example. A certain fraction of the population of heterosexual couples in the developed world are unable to have children biologically related to them owing to infertility in the female partner, while the male is fertile. Allowing that this is regarded by a large subset of these couples as a problem, and allowing that this is a medical problem, there are a number of medical solutions available to this couple, which include egg donation and "straight" surrogacy (where the surrogate mother is inseminated by the commissioning father's sperm, generally without sexual intercourse). Imagine an HTA evaluation of these two solutions to female infertility. Egg donation is relatively uncommon; poses nontrivial medical and surgical risks to the donor; requires IVF treatment, which has a low success rate (measured by implantation or by pregnancy successfully brought to term); and requires a high degree of medical intervention throughout. "Straight" surrogacy is relatively cheap (even allowing for the surrogate mother's "expenses"); requires very little intervention; has a relatively high success rate (measured in terms of pregnancy brought to term); and its pitfalls (especially the difficulty of ensuring "handover" of the child, and long-term psychological morbidity to the child and to the parties of the agreement) are difficult to quantify. If we cannot quantify them, we can leave consideration of that to the consent process. On the basis of this informal comparison, we can see how to get the data we need, and I would bet that "straight" surrogacy would come out of the HTA process better than egg donation plus IVF.

Many people would think this was a reasonable conclusion, but equally many would not, and this disagreement cannot readily be explained in terms of formal rationality or irrationality. The sort of objection that would be raised to publicly supported or sanctioned surrogacy as the policy of choice for resolving female infertility would normally advert to some concept of exploitation, and traditional (non-Marxian) economic evaluation tends not to be able to cope with that concept. At the very least, this example may point to the insufficiency of HTA. It may show more than this: that HTA evaluations actually lead us seriously astray, either unwittingly or perhaps inherently. HTA evaluations might be inherently distorting in some contexts because they are not value-neutral themselves, as claimed.

In the rest of this section I will pursue these thoughts, drawing on some of the literature that makes up the thick account of the ethics of HTA.

A. Technology and Values

Writers within the main Continental philosophical traditions, together with most contemporary historians and sociologists of science and medicine, dispute the claim the technology is value-neutral. Heidegger offers a critique of technology in which his argument is that technologies such as hydroelectric power represent a dangerous transformation by man of nature into a mere "standing reserve" of resources for our use, the consequences of this transformation being the destruction of nature, and treatment of each other as merely resources for our self-interested projects. The Frankfurt School (especially Marcuse, Horkheimer, and Adorno) offer a similar diagnosis, but rather than treating this as a tendency born from philosophical metaphysics, they describe technology in a Marxist framework which concentrates on the alienation of labor and the atomization of society. Phenomenologists hold that technology is value-laden in virtue of its meanings for us in use, while pointing out (paradoxically) that "technical thinking" aims at ignoring or destroying the personal, situated meanings of objects and environments in order that techniques and technologies should be effective in all contexts.

Much of this work is stimulating on its own terms, but can easily be criticized as subjective in its evaluation, lacking explanatory power, being highly abstract (in its focus on an essence of "Technology"), and open to challenge at every turn. For instance, a Heideggerian might very well claim that IVF and reproductive medicine in general abstracts reproduction from sexuality (just as contraception abstracts sexuality from reproduction) and treats persons as sources of reproductive material, women as reproductive containers, and children as mere reproductive products—commodities, in Marxist parlance. Most writers in the traditions I have grouped together in this section would argue that there is something sinister in the idea of health as an object of "technology"; in a Marxist vein, one might argue that to think of health as a technological object only makes sense in the context of optimizing the performance of workers under capitalism. Yet this is merely tendentious redescription for most doctors, analytic philosophers, and childless couples: a redescription of rhetorical power but no analytic significance.

While this debate descends into name-calling, it is clear that the Frankfurt School and the various kinds of phenomenology are trying to make long-range assessments of technologies as historical phenomena which bind together historically contingent contexts of meaning, use, construction, and regulation, along with the contingencies of biology and medicine which are taken as definitive by the thin account of HTA. Currently the most politically and intellectually relevant appropriation of some of these intellectual trends is in feminist critique of technology, practitioners of which might well recognize themselves in the "Heideggerian" critique of IVF that was sketched above.

Arguably the long-term assessment of the significance of technology seems to be a project at odds with HTA evaluation, which tends to assume a static context of health, illness, and medical care, and which different health technologies modify very little in essence. While there are more sophisticated variants of HTA which try to sketch the impact of such technologies as their implementation ramifies throughout health care systems and society at large, the very difficulty of such "systems" or "scenario planning" approaches, and their dubious reliability and scientificity, makes such assessment within the HTA paradigm uncommon. Traditional Marxist economics did offer the prospect of a truly scientific modeling of the impact of technical change on society, but this approach has many problems of its own. The mainstream philosophical traditions outside the analytic tradition tend not to be "problem-solving" oriented, and tend not to make common cause with the analytical human sciences (psychology, economics, social policy analysis). It is unlikely that a "radical" appropriation of HTA will be forthcoming in anything like the form suggested by Marxist economics.

More fruitful (at least in terms of academic output) has been the growing field of science and technology studies, which is methodologically eclectic, drawing widely on methods from history, sociology, and Continental philosophy. Under the banners of "social construction of technology" (SCOT) and "actor network theory," rich and interesting accounts have been written of the way technologies involve social values and priorities, rather than "merely technical" choices.

If one takes these innovative approaches seriously, then the idea of a neutral HTA is difficult to make sense of; comparison of two technologies is never comparison only of two different solutions to the same problem. It is also comparison of two sets of values, and arguably of a third set (the set which defined *this* rather than *that* as the problem). For instance, a celebrated example of technology assessment (still ongoing at the time of writing) is the assessment of cochlear implants as a "cure" for deafness. In this assessment, the narrow problem is, "Do they work, and are they worth the money?" But the "real" problem is this: is deafness an illness to be cured or a disability to be corrected? Or is it merely a difference between individuals, similar to race, height, or language community? Traditional HTA cannot easily address the "real" problem. It can, however, indicate how effective cochlear implants are in achieving a certain level of hearing in deaf people who choose to undergo implantation. The philosophical debates here are much discussed, but note in particular the debate between a liberal account of technology, where the issue is basically consent, choice, and technical efficacy narrowly construed, and a disability-rights account of technology, where the issue is the promulgation of a technology whose nature is *essentially* to evaluate deafness as problematic and to be destroyed. A disability rights view of cochlera implants might compare this with an evaluation of a genetic therepy for curing (putatively) genetically disposed homosexuality or "race."

B. Technology and Democracy

Three accounts of HTA's political and social significance are of particular interest here. They have been chosen because of their intrinsic interest, but also because their influence on empirical medical and social research has been considerable in terms of method and analytic stance. Other approaches have also been fruitful (notably feminist scholarship), but methodologically these have great similarities to the three examined here, and are omitted for reasons of space. The Marxist account of technology sees technological change as increasingly routinizing and atomizing work processes and the roles of those employed in them. The Foucauldian

account of technology sees it as a diverse and ever more finely tuned set of mechanims for regulating and monitoring the social behavior, psychological attitudes, and choices of members of society, both repressive and productive of different forms of social and individual life. The second-generation critical social theory of Jürgen Habermas sees technology, in its political form of technocracy, as a predominant threat to democratic deliberation and choice.

1. Marxism

On the Marxist view, as mentioned above, to treat health as an object of technology is in part to consider health as a resource, or capacity, for productive functioning of labor in the capitalist process. Rationalization of productive processes, through identifying the technically most "efficient" means of production, serves (directly or indirectly) the principal aim of capitalist economic order—the generation and expropriation of surplus value. The humanist Marxism of authors such as Harry Braverman in the 1970s finds expression in several contemporary sociologists of technology, who write about the "rationalization of medical work" (identifying the ways in which technology assessment makes medicine, considered as a form of labor, more efficient and more "manageable") and the implications this has for the care of patients both in the hospital and in society at large. For example, it is not hard to see such trends as out-patient surgery, the shift to primary care as the site of health care "gatekeeping," and the recognition by doctors and nurses that they are employees as well as (instead of?) professionals as signs of the industrialization of medicine. Moreover, consideration of efficiency and effectiveness of such microprocesses of health care as obtaining patient consent has as much to do with optimizing the use of doctors' time as with ethical responses to patient well-being.

2. Foucault

Developing these themes in a rather different framework, Foucauldian scholars have identified the way particular technologies are used in a process of standardization and normalization of human individual and social diversity (for instance, through the therapeutic "production" of normal sexuality—consider the medical response to hermaphroditism and transsexualism).

While some of this work has similar characteristics to the pessimistic ideology of Heidegger and others, certain trends in health care, including HTA as a technique itself, have an undoubtedly Foucauldian ring to them—"clinical governance" being only the most obvious. The latter is the trend to quantification, tabulation, and comparison, which is the essence of "governance" (originally a term in business management, now applied to health services), and the reshaping of health care activities such that they can be measured and evaluated in ways that fit the governance paradigm. This movement in part responds to, and in part transforms, patients' demands for medical and health services accountability.

3. Habermas

While much Foucauldian work is sociologically and historically challenging and insightful, its ethical and political significance is vague, as Jürgen Habermas has argued. While Habermas's own work often suffers from the obverse fault (sociological emptiness and excess abstraction), a Habermassian response to the trends identified by Foucault has great ethical and political interest.

First, Habermas identifies the principal difficulty in post-World War II politics as the tension between democratic deliberation and expert technical rationality. A good example of this would be the state's response to genetically modified (GM) food in the United Kingdom. GM food represents a diffuse set of benefits and risks to the agrichemical and biotechnology industries, farmers, consumers, and the environment. Two broad responses to this topic are possible in Habermas's model: to seek democratic deliberation on the topic, recognizing this as a political debate over the interests of the different stakeholders, or to treat this topic as a technical problem in risk assessment and risk management, best carried out by experts qualified in the various fields. Within the latter paradigm of decision making, there is a role for consideration of ethical and other issues of public concern, but these are treated as "data" about "public attitudes and understanding" rather than as equal contributions to a political debate. Rather than a procedurally just approach to a multisided dialogue, the technical solution rules some interests out as irrelevant to the specification of the problem, some interests in as informed, and others out as uninformed, and seeks to combine all the "in" views into a consensus based on a technical calculus (of which cost–benefit analysis is the best known variant).

This is a somewhat unfair characterization of the complex and noisy process of government policymaking, but where Habermas wrote originally concerning nuclear power, defense, and industrial policy in West Germany in the 1960s, his model has a lot of plausibility regarding today's science politics. In the field of biomedical policymaking, his account underscores some of the more problematic aspects of "ethics committees" such as the National Bioethics Advisory Commission (in the United States) and the series of ethics committees set up in the wake of the Warnock Committee on Human Fertilisation and Embryology in the United Kingdom.

C. Summary of the "Thick" Account

This section began with a consideration of the ways the thin account of the ethics of HTA may be regarded as insufficient or misleading in the assessment of new technologies which involve particular substantive ethical issues (such as reproductive rights), and this section has offered a variety of different approaches to the ethics of health technology which could be regarded as alternatives to HTA.

Of the approaches surveyed, the Marxist approach (but arguably none of the others) has an alternative to "liberal" HTA which could function in similar terms by substituting Marxian economics for classical economics in economic evaluation. In most of the accounts discussed here (probably with the exception of Habermas), the issue is not to construct an alternative HTA program, but to question the very idea. If one takes seriously the ethical challenge laid down by HTA itself in Section II above, this is hardly satisfactory. Moreover, several of the approaches surveyed in this section (notably Marxist and Heideggerian-inspired approaches) have problems with the notion of ethics as well, or regard it with a sort of irony and nostalgia (Foucault). The social-constructionist approaches, while able to describe powerfully the ways in which values are built into technologies, have tended to be much less forthcoming when pressed for a normative assessment of those values and those technologies, although I would argue that there are similarities between the critical approach of many sociologists of technology and the sophisticated postmodern philosophy of Jean-François Lyotard.

There are, notwithstanding the above, attempts within the HTA paradigm to incorporate some of these critical approaches. So-called "fourth-generation" evaluation attempts to incorporate users' perspectives into technology assessment, and many HTA projects incorporate elements of qualitative research with patients and other interested parties at various stages of the assessment, sometimes including the design of the assessment itself. A good example of this approach is the ongoing Dutch HTA of cochlear implants, which has incorporated a very interesting initiative whereby both pro-implant and anti-implant members of the deaf community are encouraged to take part in the debate over the merits of the implant, in particular to help familiarize each other with each group's perceptions of their interests and motives. The theory underlying this initiative is that without this dialogue, the legitimacy of the evaluation will never be established, and the status of the implant will remain politically deadlocked. The design of this program was explicitly framed with reference to Habermas's discourse ethics, and this remains the "thick" approach most likely to contribute to the development and, perhaps, transformation of HTA in the next decade.

IV. CONCLUSION

HTA is a fascinating discipline, technically, ethically, philosophically, and politically. The range of ethical points of view one can take on the enterprise itself, and on particular HTA methods and topics of assessment, is very diverse. There is a major divergence in points of view regarding HTA, which I have indicated in my division of this article into two parts, labeled as the thin and thick accounts. Superficially the difference is one of political belief (liberals versus the rest), but matters are hardly so simple, as the dispute involves issues in the philosophy of science, the concept of rational choice, and arguments about the theory and practice of justice, both in society and in health care. I have tried to make both accounts plausible; the reader must decide between them, or find a way of integrating them or replacing them, herself.

Acknowledgment

An earlier version of Section II of this paper was published as R. E. Ashcroft (1999), "Ethics and Health Technology Assessment," *Monash Bioethics Review* **18**(2), 15–24. This article is adapted from this paper and parts of it reproduced by kind permission of the Editor of the *Monash Bioethics Review*.

Bibliography

Banta, H. D., and Luce, B. R. (1993). *Health Care Technology and Its Assessment: An International Perspective*. Oxford: Oxford Univ. Press.

Berg, M. (1997). *Rationalizing Medical Work: Decision-Support Technologies and Medical Practices*. Cambridge, MA: MIT Press.

Bijker, W. E., and Law, J., Eds. (1992). *Shaping Technology/Building Society: Studies in Sociotechnical Change*. Cambridge, MA: MIT Press.

Black, N., Brazier, J., Fitzpatrick, R., and Reeves, B., Eds. (1998). *Health Services Research Methods: A Guide to Best Practice*. London: BMJ Publications.

Elston, M. A., Ed. (1997). *The Sociology of Medical Science and Technology*. Oxford: Basil Blackwell.

Feenberg, A. (1991). *Critical Theory of Technology*. Oxford: Oxford Univ. Press.

Guba, E. G., and Lincoln, Y. S. (1989). *Fourth Generation Evaluation*. Newbury Park, CA: Sage.

Irwin, A. (1995), *Citizen Science: A Study of People, Expertise and Sustainable Development*. London: Routledge.

Ten Have, H.A.M.J. (1995). Medical technology assessment and ethics: Ambivalent relations. *Hastings Center Report* **25**(5), 13–19.

Wajcman, J. (1991). *Feminism Confronts Technology*. Cambridge: Polity Press.

Human Nature, Views of

STRACHAN DONNELLEY
The Hastings Center

I. The Philosophic and Practical Problem
II. Contemporary Applied Ethics and Views of
Human Nature
III. Views of Human Nature: Future Tasks

GLOSSARY

determinism The philosophic doctrine that all phenomena and events, natural or human, are determined
by antecedent, efficient causes, physical (material)
or other.
dualism Philosophic conceptions that posit fundamentally different and independent realms of reality, for
example, mind and body (matter) or God and the
world.
essentialism The philosophic doctrine that different
forms of existence, say a human being or a horse,
have a fundamental and unchanging essence (form)
or character.
idealism The philosophic doctrine that all phenomena
are essentially mental in character, that reality consists of minds and their experiences.
materialism The scientific or philosophic doctrine that
all phenomena or reality are essentially material
(physical) in character, that reality involves the interactions and causal relations among physical, material forces.
philosophic naturalism Philosophic arguments that find
no radical disjunctions between nature and human
life (individuals and communities).
worldview An overarching conception or imaginative

picture of the origin, nature, and significance of the
world and human life.

VIEWS OF HUMAN NATURE, as the histories of religions,
philosophy, and ideas tells us, are probably as old as human
communities themselves. The views are characteristically a
part of a larger imaginative picture that humans fashion
about themselves and their world: how the world and they
came into being; what the world and they centrally are like;
what is the final significance, meaning, or importance of
the world and themselves. Views of human nature and
accompanying worldviews are central to human cultures at
all times and places. There is no escaping them. These views
may be thoughtlessly inherited from the past or critically
faced in the present. Nevertheless, they importantly rule
our human lives. They are how we gain our moral, religious,
and cultural orientations. We have the Old Testament Book
of Genesis and the latest efforts to come to grips with
Darwinian evolutionary biology. Whatever the differences,
the human impulse and need are the same: to understand
and to orient ourselves in the world. The formal disciplines
related to this inquiry are philosophic cosmogony, cosmol-
ogy, anthropology, and psychology, but in some sense or
other we all do it. It is endemic to human and humanly
moral life. Thus, views of human nature are and must be
central to applied ethics.

I. THE PHILOSOPHIC AND
PRACTICAL PROBLEM

Our histories also tell us that worldviews and conceptions of human nature have been innumerably many

and remain a fundamental controversial issue. We cannot seem finally to agree upon who we are as human beings. Perhaps this indecision is inescapable, despite the importance of the issue. Views of human nature are strange conceptual beasts, neither fish nor fowl. On one hand, they are meant to describe who we are, to give us the fundamental facts about our nature or mode of being. They are meant to be descriptive and realistic. On the other hand, they tell us humans how we ought or ought not to be, what our particular significance is in the scheme of things, and what we ought to do or become. They are normative and prescriptive.

If views of human nature are inevitably both descriptive and normative, and if in addition humans must start their explorations of themselves from the prevailing ideas and knowledge of their time and place—ideas and knowledge that may not be fully adequate to the task of capturing who we are—then we may better appreciate why these views remain incomplete and controversial. If, moreover, humans in part define or determine who they are—if they are not fully determined by some given and fixed nature—then this also reinforces why each age and human community must decide its "descriptive-normative" view of itself. We are no exception.

No human community starts *de nuovo* with its views of human nature, but exists within historical and cultural traditions that it inherits and that importantly set the stage for contemporary intellectual explorations and speculations. Our Western and increasingly global heritage is exceedingly complex and fateful in determining how we view ourselves and our significance at the end of the twentieth century. Here I can only sketch out a few major issues, bones of contention, and crucial ideas inherited from the past with which we must come to grips in "re-viewing" our human nature. I will then briefly consider the relevance of these ideas for particular issues in contemporary applied ethics.

Traditionally views of human nature have conceived human beings as a species, or as a human community or a "people," or as individual selves. In the spirit of our times, I will focus on our individual selves. What are the major issues and historically the central bones of contention? The fundamental question, of course, is the nature of the human self: its fundamental and abiding character. What determines this character? Is it biology (physical nature), divinity, culture, or society, or the self itself (its reason or will)? Is the self singly and essentially determined by one of these "causes" or is it multiply determined by these several factors together, if not more? Correspondingly, with respect to the human self's status in the scheme of things, what is its relation to nature (its body and physical matter), God, human others, society and community, history, and eternity? Is the relation essential and constitutive of the self or merely accidental, something that does not touch its inner core or nature?

A. The Historical Drama of Ideas

These are philosophic issues that have been fought over for 2000 years or longer in the West. The drama of ideas has both Greek and Judeo-Christian, if not older origins, and I want to fasten on only a few particularly fateful quarrels, for they have crucially influenced rival views of human nature. The first quarrel concerns the human self's relation to the world and the beyond. The Pythagoreans and Plato speculated that the true or higher self had its origins elsewhere, outside the world that we experience in everyday life. Similarly, early Christians and Jews held that we have been created in the image of a God that transcends the natural universe.

The Presocratic Heraclitus retorted that the Pythagoreans spoke nonsense. We, along with everything else, are woven out of the fabric of the natural universe (the "Everliving Fire"). Aristotle analogously played down transcendent origins and emphasized our life within the world: our coming to know the natural universe (the Cosmos) and our active life within the polis or political community.

This fundamental split in worldviews has had a long consequent history. For our purposes, we note certain rival tendencies. First, the Pythagorean/Platonic/Judeo-Christian tradition has fostered various forms of philosophic or theological *dualism* that have tended to set human beings and their cultural communities over and against the natural world. In these dualistic perspectives, the meaning and significance of human life, our particular role in the scheme of things, has little or nothing to do with nature, which tends rather to drag us down and away from our best, most human selves (reason, free will, the capacity for caring love, preparation for an afterlife, etc.).

Closely coupled with the history of dualism is another tendency, again present at the beginnings: *essentialism*. Human nature, if not also the fundamental character of the natural world, is eternally fixed. We all are fundamentally the same. There is a universal human essence. Any differences between us are merely accidental and not finally important. The reason for this abiding human essence is that we are created according to a fixed human form (a "Platonic idea") or in the image of an unchanging God, even if this means that our unchanging nature is the exercise of a free moral will.

The rival tradition of Heraclitus and Aristotle (which later includes David Hume and his moral sentiments) has tended towards a *philosophic naturalism*. The natural world is not seen as an alien or essentially unfriendly place. Rather, it is conceived as a locus for significant and meaningful human activity, and there need not be any radical separation of nature and human culture (human communities). Human life and selfhood, as with Heraclitus, are seen to fit within the natural, worldly scheme of things. It is this tendency toward philosophic naturalism, which includes humans within the wider natural world, that has been more willing to abandon essentialism and look rather for complex and multiple worldly origins of the human self.

The philosophic struggles be can usefully seen in our ongoing disquiet over the issue of *freedom* and *determinism*, which pierces to the quick of our modern notions of ourselves and our human nature. Are we free human agents or are our lives determined by natural (physical, material) causes, cultural or societal forces, or some cosmic or historical fate? Are we swamped by external or internal forces and only have the illusion of individual freedom? The question seems crucial, for if we have no freedom—if we have no real capacity for choosing between viable alternatives of action and for affecting things for better or worse—then we cannot be held morally responsible for what we do. If we consider moral responsibility for ourselves and our world to be importantly constitutive of who we are and to be integral to our human nature, then we are in real trouble. We have lost our traditional philosophic, moral, and cultural moorings. How did we get into this conceptual and existential pickle, which so radically challenges our everyday self-understanding and conceptions of ourselves?

B. The Modern Period and the Legacy of Descartes

Arguably, we can credit this present human or spiritual crisis to a historical and unholy alliance and interweaving of dualism, essentialism, and naturalism. The fateful modern turn came in the seventeenth century with Descartes' particular brand of dualism, his bifurcation of ourselves and our world into mind and body, human beings and nature. Mind is the realm of language, thought, and freedom (free will), and only humans have minds (originated from a transcendent God). Nature is purely a realm of mechanical, material, blind, and valueless forces, ruled strictly by causal necessity (efficient causes). Perhaps this conception of nature well served the newly emerging Galilean-Cartesian-Newtonian physical sciences, but it has philosophically thrown us off-balance ever since. First and foremost it reinforced a radical distinction between human life (human individuals and their communities) and nature. Philosophically, it left no intelligible relation or connection between ourselves and nature, including our own organic bodies. (Why do we have them?) It thus thwarted any sensible or nuanced naturalism. We and nature were

dealt a further blow by Cartesian dualism aligning itself with essentialism. The human self is essentially and eternally the same: reason and will. Nature is essentially and eternally the same: matter or energy ruled by fixed causal laws and principles.

This may be a feast for the mathematically rational mind, as it was for Spinoza, but it has arguably been a disaster for philosophic interpretations of ourselves and our world, both human and natural. It has tended to undermine, skew, or obliterate an understanding of community or togetherness: our understanding of our relations to others, again both human and natural. In the original Cartesian vision there is a radical disjunction of *freedom* (the reason and will of individual humans) and *determinism* (causal necessity). We have "atom-individuals," self-sufficient mental substances essentially unrelated to one another, over and against the universal causal nexus of nature—humans essentially isolated from one another, if not their God, and alien to the "totally other" natural universe. (Note here the origin of later existentialist worldviews and moods of alienation both from nature and other human beings.)

For everyday life, this conceptual picture is untenable, but philosophy and the sciences have been more or less plagued by it ever since. It is hard to live with such a dualism, but if one wishes to escape the split and to side exclusively with the mind, the ploy of philosophic *idealism*, we are left with essentially unrelated mental subjects curiously in touch with mental others while enjoying an experience of a material world that is just that, mere experience or phenomena. (Today we would ask, "Where's the beef?") If this is too much for us to take, we can side with the material body and become philosophic materialists. But this allegiance characteristically has led to a naturalism plagued by essentialism and causal determinism. The human self becomes engulfed and obliterated in the causal mechanisms of nature. Certain forms of *behaviorism* notwithstanding, which espouse this position, this is an equally untenable interpretation of the human self and view of human nature.

In modern forms of *dualism*, *idealism*, and *materialism*, we are more often than not left with individuals understood as autonomous rational agents, laws unto themselves, or "epiphenomenal" selves, lawfully determined and unfree, caught in the universal causal web of nature—that is, no real human selves or agents at all. Often these inconsistent ideas are melded together. We can find the uneasy tension between these rival conceptions of the self expressed within major modern interpretations of human nature, including Darwin, Marx, and Freud. Biological species(understood as populations of constituent individuals), humans included, actively exploit an environment that "mechanistically" or causally determines their genetic and ultimately phenotypical transformations (natural selection). (The relation of organic freedom evidenced at least in the more complex animals and causal necessity implied by evolutionary biology remains a genuine puzzle for most of us.) Political revolutionaries individually and energetically promote a cause that is historically inevitable (the triumph of the proletariat and the communist state under the spur of economic determinism). Patients and therapists freely combine their wits to uncover the underlying causal determinants of behavior or phobias conceived under a theory of psychological determinism. We still witness the ongoing disease in contemporary *sociobiology*, *biomedical genetic research*, and *Skinnerian behaviorism* on the one hand and culturalists, including *deconstructionists* on the other—the modern heirs of materialism and idealism respectively. The one side easily tends to or endorses genetic and environmental (material) determinism; the other often recognize no natural or causal bounds whatsoever (the triumph of the protean mind). Yet we also witness recurrent attempts to overcome the philosophic hangovers of dualism and the ensuing alternatives of materialism and idealism by thinkers in search of a judicious middle way more adequate to our everyday experiences of ourselves and the world (Isaiah Berlin, Mary Midgley, and Hans Jonas, among others).

There are promising attempts to recover the tradition of the philosophic naturalism of a Heraclitus or Aristotle, minus essentialist and deterministic undercurrents or ideologies. There are genuine endeavors to take the worldly life of both humans and nature seriously and to give both human and natural history their due in the ongoing constitution of our human communities and our individual selves. We are still and will long remain on this search to understand ourselves and our nature, but applied ethics cannot await the outcome. On practical matters, we must decide and act now.

II. CONTEMPORARY APPLIED ETHICS AND VIEWS OF HUMAN NATURE

Views of human nature and their accompanying worldviews provide a moral framework or landscape within which to situate ourselves and reflect upon everyday moral quandaries. Presently applied ethics—biomedical, animal, and environmental ethics, among others—are characterized by a variety of moral frameworks and accompanying views of human nature, sometimes complementing one another, other times at loggerheads.

Given our prior historical discussions, I want briefly to sketch out three moral frameworks at work in con-

temporary ethical debates that (arguably) have their historical roots in idealism, materialism, and philosophic naturalism. We will then consider how well these frameworks elucidate pressing issues in practical ethics. There are, of course, other relevant moral frameworks deriving from various alternate philosophic, religious, or cultural traditions.

The first moral framework derives ultimately from Kant and his particular brand of idealism. According to Kant, the human self is characterized by a practical reason and a free will at least in principle untouched by nature or history. This reason and will and their self-legislating activities are the ground of the individual's moral worth and constitute humans as "ends-in-themselves," worthy of ultimate moral respect and concern. Here is an ethics and moral framework that dominantly emphasizes respect for individual persons, especially their capacities to make autonomous rational and moral decisions. It is also an ethics emphasizing justice: the extension of moral respect and concern to all "ends-in-themselves," human and perhaps other. This *"deontological" ethics* is known for rights and duties owed primarily to individuals.

The second moral framework is *"consequentialist"* and is chiefly concerned with the consequences of our actions as they affect the well-being or welfare of individuals or "moral subjects." This welfare is defined in terms of experiential pleasure, pain, and happiness; the capacity to pursue one's own life plans or subjective preferences; or whatever is deemed the moral good to be pursued and the harm to be avoided. Characteristically, this ethics is not rationalistic in the Kantian sense and seems comfortably compatible with, if not derived from materialism and determinism. The welfare of individuals—the pleasures, pains, happinesses, and subjective preferences—may all be caused and beyond the individual's ultimate control. However, if the welfare of individuals—the greatest good for the greatest number—is promoted, that is what morally counts. This too is an ethics of justice and equality—each moral subject is to count as one and only one—but with its emphasis on the summation of overall welfare, it can be less protective of the rights of individuals than deontological ethics. The consequentialist ethics that we know best is *utilitarianism*, derived from Bentham and J. S. Mill in the nineteenth century.

Deontological and utilitarian (consequentialist) ethics are undergirded and dominated by rival conceptions of human nature and the human self—the one more akin to the philosophic and dualistic traditions of Plato and Descartes (the rational self with divine origins); the other more earthbound, if not overtly materialist. Yet both share an emphasis on the individual moral subject, with perhaps an inadequate attention to worldly time, history, and relations to others. It is here that the tradition of philosophic naturalism reenters the picture, under the guise of what we might term contingent or historical naturalism to avoid any deterministic connotations.

Let me briefly characterize this contemporary philosophic naturalism and its conception of human nature. Whatever relations we might have to a divinity or eternity (the beyond), we are to the core worldly selves, fundamentally related to the body and wider nature, to other individuals and human communities, and to historical and cultural contexts. We are genuine human agents or actors, with a circumscribed and context-dependent freedom to choose among alternatives, to act in the world, to become our individual selves. (Nature and freedom are not radically opposed.) We have a genuine, if limited capacity to affect others and the world and thus to influence the future for better or worse. This everyday, commonsense view, which has long been held among moral and political philosophers (for example, Berlin, Midgley, and Jonas), is also supported by recent philosophic interpretations of *evolutionary biology*, which offer a radical critique of both causal determinism and essentialism as applied to organic life and recognize the role of chance and historical contingencies at work in particular natural and cultural contexts (Ernst Mayr).

In sum, this philosophic naturalism proposes that we are "embedded" human selves, with substantive moral responsibilities to ourselves, human others, and the natural world. What we do matters to how we, others, and the world become. This active implication in the everyday world in part constitutes and engenders our human meaning, significance, and particular, if modest role in the scheme of things.

A. Biomedical Ethics

1. Rehabilitation Medicine

How do such moral frameworks or landscapes contribute to the consideration of particular cases typically confronted in applied ethics? Consider *rehabilitation medicine*, which aims to serve patients suffering from severe and traumatic accidents or chronic and incurable afflictions. Imagine a young athlete, say a soccer player, who has permanently lost the use of his or her leg; or an accomplished cellist who ravaged by MS (Multiple Sclerosis) can no longer play the cello and perform before audiences; or a mother who through mental afflictions can no longer care for her children; or a philosopher who has suffered a major debilitating stroke. What are the typical biomedical ethical issues?

Standard bioethics and the combined moral frameworks of consequentialism and deontology tell us that a first duty is to do no harm, to attend to the individual's welfare and suffering, and above all to respect persons and get a truly *informed consent* for medical procedures or interventions. However, this counsel may be too simplistic, especially if we are imaginatively ruled by a simplified edition of human nature, for example, the idea that from a moral perspective we are essentially minds and rational decision-makers and not full-fledged *human organisms* ("mind-bodies"). From the perspective of philosophic naturalism, how could the patient readily give his or her informed consent? How would he or she know how to decide, given that the particular worldly, historical, and bodily self has been severely challenged or undermined, a core sense of self perhaps temporarily, if not permanently lost? What is the young athlete, the cellist, the mother, or the philosopher to do now? How do they replace the old meaning and significance of their lives? How should they reorient themselves? Should not a primary obligation of health care providers be to help the patient engender a "newly" active self, with new worldly life goals commensurate with new levels of capacity?

Moreover, we not only deal with individual patients, but with their families and friends, all those involved in intricate webs of intimately interconnected lives, each in his or her own way challenged or devastated by the patient's affliction. The whole web, with the patient at the center, needs care and moral attention. This attention may include consideration of particular cultural traditions and even where the rehabilitation ought to take place, especially if the patient and family are deeply rooted in their community and regional home and become disoriented in unfamiliar places. Such is the embeddedness of our individual selves in the everyday world. How well does either deontology or consequentialism, with their essentially "individual" selves, deal with these moral dimensions that arise from essential "worldly" relatedness?

This is not to speak of other patients in rehabilitation settings, the cultural and moral traditions of the health care professionals (which may themselves be diverse and conflicting), and the complex moral and societal context of the health care system in which the rehabilitation medicine is situated. Each of these elements of the moral landscape may or may not have a moral pull in particular decisions concerning the patient's care and rehabilitation. The recognition of this moral complexity is the price we must pay for understanding that we are fundamentally worldly selves, inextricably caught up with one another and the rest of the world.

Recognition of the complexity of our human nature and moral lives may help us understand better other issues in medicine and health care that presently vex us, for example, physician-assisted suicide and organ transplantation.

2. Physician-Assisted Suicide

The central moral imperatives of a bioethics dominantly centered on individuals—consequentialist and deontological injunctions to relieve suffering and to respect persons and their autonomy in decision-making—can powerfully combine to argue the case for *physician-assisted suicide*, which moreover carries its own intuitive appeal. Who would not want to end intractable and unrelievable suffering? Who would want morally to infantilize an individual by taking away moral agency over such an ultimately significant issue as one's own life and death? Yet things are not ethically so simple. We human individuals live with one another in communities that have their own complex habits historically fashioned, upheld, and dynamically transformed over extensive periods of time, stretching from the distant past into the indefinite future. One of our primary human responsibilities is to uphold the cultural and moral fabrics of our communities. Will physician-assisted suicide, primarily meant for individuals in extreme distress, undermine important community habits? Will it lessen our respect for human life and our sense of responsibility for the weak, vulnerable, and infirm, if not lead outright to the taking of lives unwanted by us, but not by the individuals themselves? Will we ordinary citizens come to ape, in the name of compassion, suicide-assisting physicians? These are real questions that haunt the moral landscape of physician-assisted suicide.

Moreover, how will physician-assisted suicide affect the moral fabric and habits of the health care professions, especially if individual professionals remain divided within themselves or among one another over the issue? Could the professions tolerate a universal right to physician-assisted suicide and still effectively perform their community functions and obligations? Should legitimate moral concerns for individuals trump legitimate moral concerns for the ongoing well-being of professions and communities? (This is a question deontological ethics must seriously ask itself). Does the moral decision moreover depend significantly on the different communities, cultures, and individuals involved? For example, what might prove to be morally tolerable or permissible in the Netherlands might be morally intolerable or dangerous in other countries or societies, given different moral and cultural climates and habits. If we are truly worldly selves, all relevant things must be carefully considered. Ethical concerns for individuals must be matched and balanced by ethical responsibilities to communities and societies.

3. Organ Transplantation

We find a parallel situation with *organ transplantation*, which on the surface seems such a straightforward issue. There is a significant and unmet need of human organs for transplantation into critically ill individuals, irrespective of the quality of their posttransplant lives (a question often overlooked). Human individuals do not need their organs after death and therefore should surrender them upon dying for the sake of the welfare and well-being of others. Within a consequentialist framework, what could be simpler? It seems morally incomprehensible that there remains a chronic shortage of organs for transplantation—unless one considers relevant and differing views of human nature.

Ethical arguments for organ donation work particularly well within dualistic, materialist, and utilitarian views of human nature. If the organic body is a mere mechanism or machine that has nothing essentially to do with our human selves, then it and its parts are of no direct moral concern. The parts of particular bodies, given certain biological limitations, are "fungible" or substitutable for one another. For the welfare of others, we should share these parts when no longer needed and get on with it.

However, if we are essentially worldly, bodily, and historical selves, things are not so straightforward. The human body typically has enormous meaning and significance for us. Think of the role of the human body in art, music, athletics, sexual relations, and family life. Think how important our individual bodies are in our own personal lives. These examples only touch the tip of the iceberg. For us, human bodies are anything but meaningless, insignificant, and valueless. It is not us that are odd, but a science or philosophy that assigns our bodies to a valueless moral limbo.

For morally weighty and altruistic reasons, I might want to give my organs to benefit others in extreme need. But I should not expect my wife and children upon my death easily to surrender my body for the organ harvest. They have long lived with the bodily me, and my body has not been morally and humanly insignificant to them. This goes beyond any erotic intimacy to palpable human realities of warmth and security, the exuberance of physical play, and a religious or philosophic awe before the very fact of intimate, bodily existence, us humanly organic ones mattering so much to one another. Nor should we be surprised at the moral burnout of the medical teams that harvest human organs, especially the nurses. The recently dead individual, just parted from its intricate web of personal lives, is soon a bag of bones and useless parts. This rapid and dramatic transformation seems morally difficult to digest, despite the very real benefits to others, precisely because our bodily nature lies within and not outside of traditional moral and religious landscapes that have a long human past and an abiding hold over us. We are morally and spiritually troubled by doing a good deed and perhaps rightly so.

4. Animal Donors and *Xenografts*

The moral complexity of organ transplantation does not vanish if we turn our attention away from humans to animals as the source of organs. For example, chimpanzees have been considered as a source for "bridge" hearts until a human donor becomes available. (Interestingly, both deontology and consequentialism have wavered on the morality of this proposal, depending on whether or not their respective moral concerns are extended beyond humans to other ("higher") animals.) But chimpanzees are our close evolutionary cousins, remarkably capacitated in their own right, and, unlike ourselves, are threatened with extinction. Baboons have experimentally contributed hearts and livers to human patients, but though not threatened with extinction, we worry about baboons transmitting unknown and lethal viruses into the human population, not to mention the moral obligations we might owe the baboons themselves.

The latest animal organ donor candidate is a *transgenic pig* so biotechnologically fashioned with human genetic material as to suppress immunological rejection of the foreign organ, for example, a heart. Beyond questions of scientific and biological feasibility, this again may seem morally unproblematic. If we raise pigs for food and other useful products, why not use them for spare organ parts, certainly a more compelling moral justification on consequentialist, if not also deontological grounds? But again this line of argument may put the issue in an overly simplified or reduced ethical landscape incommensurate with our human nature.

As worldly, historical, and bodily beings, we have a long entwined, if checkered history with nature and animals. Nature and animals have multiple values and significance for us, from the biological and economic to the scientific, aesthetic, moral, and religious. In ways that we are only beginning better to understand, nature and animals are complexly and intimately woven into our cultural capacities, habits, and achievements and thus into our very selves. Nature and animals, both positively and negatively, importantly matter to us. Moreover, the future fate of humans, animals, and nature are intimately conjoined.

What will be the moral fallout if transgenic pig hearts become a viable resource for human transplantation? Will we accept the porcine hearts into ourselves without a thought or with a new gratitude for pigs and animal

life? Or will we become profoundly disturbed and morally disoriented? Given the central symbolic and iconic significance that our hearts have for us—think of literature (*Heart of Darkness*), ordinary metaphors ("my heart goes out to you;" "our heartland"), and popular songs ("My heart dies for you")—can we readily incorporate the various long-standing cultural and moral meanings of pigs or other animals into the very core of our human bodily being? What will it humanly feel like to be so utterly dependent on another animal's organ? What will it do to our sense of self and moral, religious, and cultural orientations? Perhaps we cannot effectively answer these questions before the event, but this might not be such an easy and minor affair as boosters of organ transplantations (especially, the consequentialists) would like to think. Our human nature, as we find with human organ donation, is not so easily pliable.

B. Environmental Ethics

1. Wild Animals and Nature

Pigs have long been domesticated animals and thus assimilated into our human communities and ordinary cultural routines. For the moral better or worse, we do not give them much thought, and perhaps after all their hearts would slip into our bodies without fanfare. But most of animal life and nature is not domesticated. It has different meanings and significance for us. We have innumerable other relations to animals and nature that have become deeply rooted in our human lives and that require their own moral considerations. In North America we have had a long love–hate relationship with wild nature and animals. This relationship has become part of our collective soul and culture and still strongly animates us. Wild trout, salmon (Pacific and Atlantic), and rivers are a case in point. In various regions of the United States and Canada, there are energetic conservation efforts to save native trout, salmon, rivers, and surrounding watersheds, often met by equally energetic efforts to exploit these and other natural resources.

The conservation and exploitation of nature in North America raise deep moral and human passions, leading on occasion to acts of terrorism for the sakes of forests, wild and domesticated animals, or human communities. These deep-running passions make no sense on human-centered worldviews, deontological, consequentialist, or other. If we are essentially minds unrelated to the body and wider nature—or if we are essentially pleasure seekers and pain avoiders who can get our satisfactions chiefly, if not solely within the confines of human communities—why should we care so much about nature and animal life? How do we explain ongoing and raging controversies over human–nature relations and interactions, which are not considered to be merely pragmatic issues, but characteristically to have a central moral, if not religious significance?

If we adopt a philosophic naturalism and moral framework that presupposes that we are worldly, historical, and bodily beings, then these contemporary moral and political struggles over humans and nature become more readily intelligible and hopefully tractable. We better see what multiple moral needs and interests that we need to take into account. Considering human beings as outside of and radically over against nature blinds us to both what in fact is happening in the world and our moral duties to the future. We need somehow better to think humans and nature together and understand them in the final analysis as involved in a single and dynamic biospheric or worldly whole.

With mention of conserving trout, salmon, and rivers, we of course have entered upon applied environmental and conservation ethics, perhaps the outstanding challenge of our time and of the foreseeable future. The challenge is both practical (moral and political) and theoretical. It underscores that ideas, including ideas of human nature, no matter how speculative, really matter. How we conceive ourselves importantly determines who we are, how we feel, how we act, what we become. The human–nature crisis, impending and already here, presses us systematically to reconsider ourselves and thoughtfully to begin again, which is the original and ongoing task of philosophy.

III. VIEWS OF HUMAN NATURE: FUTURE TASKS

In this "descriptive-normative" task, we should, I think, embark along certain directions, decidedly beyond inherited frameworks of deontological and consequentialist thought. Again, from a moral, religious, and cultural point of view, we must consider nature and humans' ongoing interactions with nature seriously. This means taking the philosophic measure of contemporary Darwinian evolutionary biology and ecology, as coupled with other dominant cultural orientations towards nature. This enterprise inherently must be both speculative and critical. Certain major themes already seem evident. As against certain atemporal, essentialistic, and deterministic conceptions of humans and nature, the natural world conceived by most evolutionary biologists and ecologists is deeply historical, dynamic, multicaused, contingent (no grand cosmic plan), opportunistic, and context-dependent. Lives are lived and species evolve only in particular bioregional contexts, notwithstanding the complex interactions of bioregions within an overall biosphere.

Similarly, humans have evolved naturally and culturally in historical, dynamic, and bioregional contexts in response to natural, as well as human challenges and opportunities. There is an underlying and deep affinity between the processes of human communities and nature. Indeed this affinity may help to explain why there is so much nature in our human cultures and selves and so much evidence of ourselves in nature, interventions intended or unintended. Human life, immersed and embedded in natural reality, involves the creative modification of nature, in fact and in symbol. This mutual implication or immanence of human individuals, communities, and nature is the backbone of philosophic naturalism and its view of human nature.

This is not a politically innocent or uncontroversial view. No view of human nature, no matter how substantive or minimal (abstract), is uncontentious or above the fray. Philosophic naturalism has definite moral implications. For example, it suggests that our ethics of human–nature interactions should focus on long-term responsibilities to historical and dynamic processes (community and individual) and be regional. Human communities should attend to their regional ecosystems, landscapes, flora, and fauna. We need to understand our regional human and natural past (natural history and human–nature interactions) and opportunities for a vital human and natural future. We need to understand how our regional homes dynamically fit into wider bioregional and global systems (human and natural), influenced and influencing, and to recognize both local and global moral responsibilities. We need to fashion some form of cosmopolitan regionalism if we are practically and morally to save ourselves and nature from the looming pressures of human populations and exploitation of natural resources.

We will not meet this fundamental moral challenge if we do not get our heads straight about our own human nature and our place in the natural scheme of things. Such philosophic exploration is not our only task, but it is indispensable. The human world is ruled by ideas, views of human nature (clashing or convergent) in particular. In our everyday lives, nature and ideas are indissolvable.

Given the history of human life and thought, we should expect and hope that this philosophic naturalism will be fleshed out and corrected, if not superseded by more philosophically and ethically adequate views. Whatever, we must squarely and earnestly face the original and fateful Cartesian disjoining of mind and body, thought and feeling, freedom and determinism, and our human selves and the rest of nature. The subsequent history of moral and political theory and practice has had an ironic, if intelligible course. We have seen an ardent championing of individual freedom that has condoned an irresponsibility for human communities and wider nature. We have had an equally ardent championing of connections to human communities or nature that has meant the denial of individual freedom, self-hood, and thus moral responsibility altogether. Both of these theory-driven tendencies to disjoin individuals, communities, and nature are philosophically and morally bankrupt. We need more judicious and adequate middle way between human individuals, human communities, and the wider nature world. Our ongoing challenge is to conceive together humans and nature, freedom and causal influence. We need ethically to rediscover genuine human moral actors embedded in wider human and natural communities. There is no other sensible way to go. We are still very much amidst this unfinished philosophic and ethical business. In short, we have yet adequately to conceive and appreciate our human nature.

Bibliography

Arnhart, L. (1995). The new Darwinian naturalism in political theory. *American Political Science Review*, **89**:2, 389–400.

Berlin, I. (1979). *Four essays on liberty.* Oxford: Oxford University Press.

Donnelley, S. (1988). Human selves, chronic illness, and the ethics of medicine. *Hastings Center Report*, **18**:1, 5–8.

Donnelley, S. (1989). Hans Jonas, the philosophy of nature, and the ethics of responsibility. *Social Research*, **56**:3, 635–657.

Donnelley, S. (1995). The art of moral ecology. *Ecosystem Health*, **1**:3, 170–176.

Donnelley, S. (1995). Bioethical troubles: Animal individuals and human organisms. *Hastings Center Report*, **25**:7, 21–29.

Jonas, H. (1985). *The imperative of responsibility.* Chicago: The University of Chicago Press.

Mayr, E. (1991) *One long argument.* Cambridge, MA: Harvard University Press.

Midgley, M. (1995). *Beast and man.* London: Routledge.

Human Research Subjects, Selection of

RICHARD ASHCROFT

University of Bristol

I. Scientific Issues
II. Ethical and Social Issues

GLOSSARY

bias A quantifiable mismeasure of some determinate quantity due to an unrepresentative method of selecting subjects.
harm An actual physical or psychological injury to self or body.
risk A determinable or indeterminable chance of a harm.
subject A participant in research, whose body and/or psychology are the site of scientific investigations into their workings, possibly under the influence of innovative chemical or biological agents, or of some other innovative social or material practice.

THE SELECTION OF HUMAN RESEARCH SUBJECTS in medical and other scientific studies was until recently an issue in scientific methodology alone. Could we validly derive generally true beliefs about human physiology or the efficacy of some treatment from experiments using this subject or sample of subjects? Since the Nuremberg war crimes trials, revelations about scientific "misuse" of human subjects have forced a recasting of the selection issue. The contemporary form of the selection issue concentrates on the ethics and justice of enrolling or excluding particular individuals or populations. The watchwords are autonomy,

informed consent, and the interests of each subject. More recently the clarity of this model has been clouded by the recognition that the ethical consideration of any selection method and any experiment must consider the wider purpose of the experiment. This consideration must take into account the interests of members of society not selected into the experiment, possibly including future members. This is particularly relevant in cases where the investigation could proceed through a series of linked but limited experiments, or instead through a single (or smaller number of) large-scale inclusive trials which might involve more subjects here and now, but determine answers to more questions and sooner. Recently there has been a shift of emphasis back to issues of research methodology and reliability, as a consequence of the recognition that these issues have an ethical dimension.

I. SCIENTIFIC ISSUES

Scientific soundness is a necessary condition for ethical selection of subjects. In this section I will discuss why this is so, and indicate areas where the distinction between the scientific and the ethical may be confusing.

A. Researcher Obligations Related to Scientific Validity

The fundamental principle of research ethics is that bad science is bad ethics. The soundness of the scientific research proposal and the research design is a necessary, but not a sufficient, condition for a scientific experiment involving human subjects to be considered ethical. The principle holds good for a number of reasons, and is valid in all sciences where human subjects are used. In the first place, any scientific experiment involving human subjects will involve subjects giving up some of their time on the understanding that they are assisting scientific research. If that research is incompetently framed or carried out so that no reasonable scientific assessor would expect anything useful and novel could be learned in an epistemologically reliable way, then the subject's participation has been gained under false pretenses. This is perhaps no more than a breach of etiquette; but it may be more than that. A special case is the case of experiments carried out for the educational benefit of the student researcher. Here the subject has no right to expect competence from the student, and the subject's enrollment is conditional instead upon the educational utility of the experiment.

I have assumed so far that the experiment offers no significant possibilities of benefit or harm to the subject. I have also assumed that the subject is enrolling for no reason other than altruism, curiosity, or an interest in playing a part in scientific progress. In much research involving human subjects, researchers recognize that the inconvenience of taking part may be in excess of the interest of sufficiently many subjects for the validity of the experiment. Some financial or other inducement may be offered, in order to stimulate enough participants to come forward. This financial or other benefit to participants raises justice issues, which I will discuss in more detail later. But it should remind us here that scientific experiments cost money, and in the competitive environments of private and state scientific funding, it is almost always the case that where one project is funded, several others could not be. The incompetently designed and managed experiment therefore deprives society of the possible benefits of research which was not funded. This represents a waste of scarce resources, which might be regarded as unethical in many cases.

In scientific experiments which involve the use of possibly harmful agents or procedures, the unethical character of the badly designed experiment is more clear-cut. In the first place, the experiment may increase the already existent measure of risk involved in use of this agent or procedure. The researcher is under an obligation to minimize the risks which the subject must undergo, consistent with the aims of the experiment and the informed and reflective agreement of the subject to undergo those risks. Failure to obey this obligation represents negligence, and may do so even where the subject has given her explicit consent. Secondly, the incompetent and dangerous experiment may yield inconclusive results which are due to the performance of the experiment and which may necessitate repetition

of the experiment (after redesign) and hence involve exposure of further subjects to the experimental risks, delaying still further the licensing of a beneficial treatment (in medicine) or the proscription of some dangerous substance.

B. Subject Obligations Related to Scientific Validity

It should be noted that while all of these obligations and expectations rest upon the researcher's shoulders, they may also be incurred by the subjects themselves. There is little literature on the obligations incurred by subjects in experiments, largely because, as we shall see, most experimentation ethics is orientated toward a protection standard. Most work in this area is concerned with the safety and dignity of subjects actually enrolled in the study. However, it is clear that a subject who enters a study with a prior intention of breaking the protocol for malicious or self-serving reasons, or who develops such an intention while participating in the study, is acting unethically. This is because the subject's actions may cause the experiment to become invalid, and he is taking the place of some other would-be participant who would have acted according to the spirit of the agreement. It may, in exceptional circumstances, become arguable that an experimental protocol is unethical in its design, and some other principle may override the obligation to honor the agreement made with the researcher on joining the experiment. This is analogous to the arguments which may be made concerning civil disobedience or revolt.

C. Compliance

A second reason may be adduced why the issue of subjects' obligations has not received much attention in the literature. In medical research, the concept of noncompliance is used, which covers all forms of patient nonadherence to treatment instructions. Simply because patients may not comply for many reasons, which may include the side effects of the treatment, the practicalities of taking the treatment, or the inconvenience involved in the treatment, the concept of noncompliance is not a moral one. It is arguable that the medical concept is often applied in a moralistic way, as a mechanism of social control. But we may isolate the intention as the key here, and distinguish deliberate, premeditated noncompliance from other forms. In the same way, we have used the concept of competent design and performance of an experiment to distinguish experiments which are vague, ill thought out, or dangerous from experiments which are sound in conception, but which yield no conclusive or negative findings.

The importance of distinguishing moral from non-moral uses of the compliance concept is as follows. In any experiment which is designed to test the efficacy and utility of a medical treatment or social policy, it is important to know not only whether it is efficacious in ideal cases, but whether it is "workable." Patients or other subjects must be enrolled in the experiment in a way which will enable the workability of the drug or procedure to be tested. Compliance is a relational property: it is only in extreme cases that a patient may be a "noncomplier" by preference and habit. In most cases, most patients will find some treatments hard to comply with, and vice versa. In some circumstances the researcher may decide that some subject is so unlikely to comply with the experimental protocol that they are unsuitable for enrollment. But, particularly in medical cases, where the researcher is a doctor under an obligation to do the patient some good where possible, this judgment will never be made lightly.

D. Competence, Importance, and Reliability

To define what will count as a competently designed research protocol we need the following. An experiment must be designed such that it can give a reliable answer to a well-posed scientific question which is of importance to the theory or applications of the science in question. The "science" can be construed quite broadly here: any predictive or policy science involving human beings and their behavior, psychology, physiology, biology, or medicine falls within the domain of interest.

Defining "importance" here is difficult to do analytically. As already noted, many experiments involving human subjects are pedagogical exercises rather than being intended to add something new to human knowledge. I will concentrate on experiments which are intended to play a part in research proper rather than in training. The same principles apply in each case. In the training experiment, more is known about the substance or procedure under investigation, so the principles can be applied more easily (for instance, more is known about the risks, harms, and benefits involved, so they can be explained more exactly and, usually, controlled more precisely). Supervision by a senior researcher is required to ensure that the procedures of the experiment are carried out to the normal standards of safety and competence. Thirdly, and finally, the subject is being asked to participate in a training exercise rather than innovation, and this will have a bearing on how likely each particular person will be to consent and on what his or her expectations are.

The importance of an experiment is contextual: simple experiments are often more important than complex ones; some measurement experiments are of great rele-

vance, while some hypothesis testing may have few practical consequences or theoretical ramifications; and some attempts to replicate earlier data are more significant than the original "discovery." It is possible for a test of some drug's effectiveness as a treatment for most instances of a condition to show that while it is efficacious as an antibiotic, it may be less useful than another treatment for some reason. So giving a definition of "importance" is probably futile. It is a task that is performed sociologically by the grant-application process (albeit in a somewhat unsatisfactory way, if most disappointed researchers are to be believed!).

Reliability in method is easier to determine in most sciences, particularly those of a statistical nature, such as empirical psychology or clinical and social medicine. Mathematical methods exist for determining how many subjects to enroll in a statistical study in order to get a reliable result, conditional on controlling sources of bias. This takes us to our main subject.

Thus far we have seen that even before selecting subjects for research can begin, we must satisfy ourselves that the experiment is capable of giving a satisfactory answer to a well-posed question, which will be of some utility or importance.

E. The Population of Possible Research Subjects

The principles for selection of subjects from research can be approached from this direction. It is crucial for the validity and generalizability of the experimental findings that the group of subjects enrolled into the experiment should be composed so as to permit an adequate answer to the experimental question. This will require a statement in the experimental protocol which determines which biological or social characteristics may be relevant to the substance or procedure under test, and why. These are the characteristics which define the population of possible experimental subjects. Other characteristics may be relevant for ethical or social reasons, which permit secondary selection within that experimental population. But people who do not possess the relevant primary characteristics do not belong in the experiment.

It is essential that the primary characteristics are explicit and open to critical scrutiny. For instance, historically, women were (are?) often excluded from many drug trials, and indeed often from trials which would be of benefit almost exclusively to women, on the grounds that they are "pregnant, pregnable, or once pregnable." This was in fact enshrined in many regulatory codes. But it was not clear whether this was for scientific, ethical, or cultural reasons. Women were not approached to be recruited to such trials, although there was nothing in the scientific hypotheses which warranted this. The decision to risk side effects relevant to child-bearing was a secondary tier of selection, properly speaking. To define one's eligible population on social or ethical grounds *before* examining the scientific criteria for determining the study population is to risk greater injustice than one thought to forestall.

It is possible that one might regard gender, or ethnicity or occupation, as a scientific criterion, however. Side effects are a significant part of what one wishes to determine about a treatment or procedure. A series of small trials, each one with relatively exclusive selection criteria related to a specific population, might be regarded as safer, and hence more ethical, because each trial can answer a more precise question, and the influence of more factors can be controlled for.

This is an important argument, but not really relevant here. In the trials where women were excluded for "scientific" reasons, and without consultation, women were often the primary treatment group, not one group among several (two?). And not only were they excluded from a smaller, initial trial on men only, no second trial (on women) occurred. Hence, either the treatment was not licensed for use in women (so that they could not benefit from something which could usually benefit only them), or they were obliged, as a population, to use a treatment untested on them for efficacy or safety. And the only rationale for the test in the first place was to protect women (in fact, anybody) from treatments of unknown safety or efficacy.

The scientific criterion for defining the population of possible research subjects should be interpreted maximally, and in itself will usually involve no explicit selection principle beyond technical criteria such as feasibility of experimentation. It is of course sometimes the case that some "technical" criteria involve suppressed ethical assumptions. In this case the experimental protocol involves an implicit ethical judgment about the appropriateness of experimentation on a certain subpopulation, or a quasi-ethical judgment about the inconvenience (or disutility) of enrolling a certain class of subjects. In such cases, the ethical status of the protocol is subject to scrutiny using the same principles as any proposal involving an explicit selection principle. Ethical good faith requires transparency here, and passing off ethical judgments as scientific may be regarded as suspicious and bad practice.

F. Sample Size and Control Groups

Once the population of subjects for the experimental hypothesis test or inquiry has been determined in principle, the task of determining how many and which subjects to enroll begins. In some experiments, it may be

that only a very few subjects are needed to establish the hypothesis, because all humans are sufficiently similar in the relevant respect that variations between subjects may be ignored. Almost all experimentation on human subjects will involve recognizing that human variation is relevant to the measurement or hypothesis, and so the experiment is statistical in nature. As already remarked, in a statistical experiment a certain mathematical calculation (the "power" calculation or its "Bayesian" analogues) can be used to determine the minimum number of required subjects. An experiment which fails to recruit sufficiently many subjects will produce results which are inconclusive or of uncertain quality. In consequence, an experiment which is unlikely to be able to recruit enough subjects to achieve this minimum size, but which is carried out anyway, is arguably unethical for the same reasons that an inadequately designed experiment is unethical. This objection may be overcome if some reliable method exists of aggregating data from other similar experiments so that a fictional meta-experiment which is of an appropriate size results, but this remains controversial.

The usual method of reducing "sample size" difficulties is to dispense with a control group. Most experimentation using human subjects involves a comparison with a group of subjects who do not receive the substance under test, or who do not undergo the experimental procedure, in order to ensure that any observed effect in the "treatment" group really is due to the new treatment. Usually this means enrolling subjects into the experiment, and assigning some to the treatment group and some to the control (comparison) group by some method (usually random assignment). The alternative is to enroll all subjects into the treatment arm, and to compare each subject with a subject outside the trial (by examination of their medical records, for instance) who is either in a relevantly similar situation now or was in the recent past. This may be done on an individual basis, or it may be done at the level of the group, so that while patient-by-patient similarity may not hold, the two groups share the relevant properties en bloc.

Scientifically, the relevant point is that most statisticians agree that experimentation without a purposely enrolled control group is significantly less reliable than experimentation with such a group. And most agree that unless this group is constructed by assigning subjects at random to the treatment or control groups, the experiment is vulnerable to the effects of unknown, unevenly distributed "confounding" or "nuisance" variables. There are many complex ethical issues related to the ethical merits of these various methods of experimentation which are beyond the scope of this article. Here we should only note that it is important to know how many subjects are required as a minimum, relative to the experimental design (and method of inference) in use.

In some cases no experiment will be possible, simply because the experimental population is smaller than the minimum sample size needed. In this case it may be necessary in *medical* cases to give the treatment to whoever the doctor (in consultation with the patients) deems to require the treatment. Here the patient–subject will be receiving an experimental treatment outside the context of an experiment. This is the situation which obtains in treatments for rare diseases. It will also obtain in the early phase of treatment development, where not enough is known about a treatment for it to be tested experimentally in a rigorous way, but some patients in dire straits may be given the treatment as a last resort, and the consequences observed. Ethically, it is the rarity or extremity of the situation which is relevant, but the issues of risk and benefit and consent are essentially the same as in other circumstances. The distinction should be drawn between these situations and human experimentation proper, because in these situations no true experiment is carried out, although an experimental treatment is used, and data about its effects will be gathered.

Another type of situation where an experiment may prove impossible is the case where a sufficient number of subjects are available, but the rate of recruitment is too low for the experiment to be completed on time. This can be because subjects are unwilling to enroll for ethical or other reasons of their own, or because (in the medical case) physicians are unwilling to put their patients forward as possible subjects. If there is no obvious ethical reason for this, this may represent a case where some inducement might be required. Most experts on research ethics regard inducement as unethical, because it may be considered as either duress or as seducing individuals to act against their own interest or perhaps as destructive of the social virtue of acting altruistically. A distinction may be drawn between overcoming the inconveniences attendant upon participating in an experiment and causing subjects to overlook the additional harms they may undergo as a result of taking part in the experiment. Inducement issues, as with the differential benefits which may accrue to the experimental participant which are unavailable to the nonparticipant, are issues of justice and rights, and I will return to those in the sections on nonscientific issues of selecting subjects.

G. Representativeness

If the necessary minimum sample size is smaller than the size of the experimental subject population, selection on some principle will almost certainly be necessary.

In some cases, where the two sizes are of the same order of magnitude, there may be an argument for including the whole population in the experiment, or rather, offering participation to all members of the population. This would be for reasons of equity. This would be a rare event, however. In most cases, a sample of the same approximate size as the minimum is more appropriate. This will be for reasons of convenience and for other ethical reasons.

A smaller experiment will usually be easier to manage and quicker to complete, and so the results will be easier to determine and turn into policy, from which the whole population will benefit. If one of the treatments is harmful, quick completion (or discontinuation) will mean that the group receiving the harmful treatment will be minimized. It is sometimes possible to terminate an experiment "prematurely," that is, before the minimum size experiment has run its course, in cases where the experimental treatment's harm or benefit becomes statistically "obvious." Usually this is not the case, however. The experiment needs to run its course for the results to have any meaning. In each case, running a controlled experiment where some benefit or harm befalls the subjects because of the experiment due to the novel treatment means using a number of subjects who will, after the fact, be seen to have derived less benefit by reason of their allocation to one or the other group alone. It is not predictable in advance which group this will turn out to be—that is the motivation behind the experiment. In order to minimize this after-the-fact (relative) misfortune, it is usual to run experiments close to the minimum size.

Some subjects will be selected for the offer of participation in the experiment while others will not. In general, more subjects will be offered participation than are statistically necessary, simply because some of those subjects will decline participation. In almost all cases, as we shall see, the subject's consent to participate must be sought, and refusal of this consent makes exclusion of the patient from the study compulsory for the researcher. If a patient, who is judged suitable to be offered participation, accepts the offer, in all but a few exceptional circumstances he or she will be admitted to the study (for instance, in the case of medical treatment, his or her condition might change). This may involve more subjects accepting than predicted, so that the sample size is larger than the minimum. Patients are asked for their consent to receive a treatment under the experimental protocol, and have some right to expect that their consent will admit them to the protocol, unless there is some good reason not to include them. Offering participation and withdrawing the offer at a later stage may be regarded, in many circumstances, as at least rude, and at worst cruel, unless something has changed which alters subjects' suitability for participation or the conditions underlying their own grant of consent. Simply being supernumerary does not satisfy this clause.

Which individuals are to be selected for a given experiment? Scientifically, the criterion is suitability as regards a genuine test of the hypothesis under consideration, or an unbiased measurement of the parameter of interest. In the medical case, patients may be ineligible because their condition is so severe (or so mild) that they would not normally receive either treatment; because there is clear reason to suppose that they would benefit from some specific treatment (and so participation in the experiment would not be in that patient's medical best interests); because they are susceptible to some known side effect of the treatment; or because they are "comorbid" (simultaneously ill with another disease or condition), where this comorbidity either will affect the effectiveness of the treatment under study or will require treatment which will affect the effectiveness of the treatment under study.

The main issue which determines patient suitability is the need to construct a sample which will allow generalization of the results of the experiment upon this sample to the population under study. There is a red herring here: representativeness. It is not necessary to construct a sample where the distribution in the sample of the scientifically important characteristics which define the population reflects their distribution in the population itself. In fact, it is very unlikely that such a construction would be possible, if the intention were to construct this model directly. In most nontherapeutic research it is possible to imitate that distribution by random sampling. This may be necessary, scientifically, if one is attempting to describe the features of the population. In cases where the experiment is intended to be beneficial to some or all subjects, in particular in medicine, selection at random means benefit at random. This is an ethically complex issue, as we shall see later. However, and more crucially, in the medical case, the existence of a stable population of subjects with a certain disease is almost always meaningless, and the experiment takes place under the obligation to treat a patient as and when he or she presents (among other medical obligations and patients' rights). Largely the sample constructs itself, and while the researcher can exclude some patients, it will be difficult to actively include patients who do not present themselves for treatment. In some cases this is possible; in others it is also necessary—for instance, in community-based substance abuse research. The sample which constructs itself in this way has no necessary connection in distribution with the population at large.

If the sample is not connected with the population in distribution, it will almost certainly produce results which are biased, that is, results which are significantly different from those which would be obtained if the experiment involved the whole population (past, present, and future). It is sometimes possible to predict that the bias will be in a certain direction, but inference on correction for bias on this prediction may be unreliable. The bias I have described here is patient self-selection bias; in parallel there is researcher selection bias (where consciously or unconsciously a researcher selects more subjects of one kind than another).

From a simplistic "scientific" point of view, the only factors of interest to enrolling the sample are those which define the population under study, and provided that all subjects are members of the population, use of further criteria of selection are "purely" matters of ethics or social policy. However, the truth is more complicated. It is possible that other factors are relevant to the experiment, whose influence we may or may not know about, and which may affect the results of the experiment. Our method of selection may emphasize the influence of these factors (confounding or nuisance variables), causing our inferences to be in error. In order to control the influence of these variables, it is usual to randomize *within* the sample. That is, subjects are assigned at random to one or the other experimental (or control) group. In this way, it is argued, distribution of the nuisance variables will be the same in each group, and their net influence will be balanced out. There are particular ethical issues to do with randomization in medical trials, which are beyond the scope of this article.

The relevance of these matters to selection issues is that representativeness is not a scientific issue. It may become a scientific issue if the attempt to construct a "representative" selection method causes biased results. A socially representative sample may be constructed, but—as with any active selection principle—it may be a source of bias *unless* steps are taken to construct a control group which is parallel in composition to the treatment group. And given the practicalities of research recruitment, the most effective and fairest way to construct this parallel is randomization.

It is important to underline the distinction between scientific and social criteria of inclusion or exclusion. It may be that it is thought ethically or socially important that some definite group be offered (or refused) the opportunity to participate in some experiment. Does this relate to some feature relevant to the hypothesis under test? In some cases it may: for example, many chemotherapeutic treatments are relevant to the treatment of children's cancers. Ignoring the legal issues, note that in many cases such treatments are not tested upon children, and so may, in general, not be used on children. Yet it is probable that they would benefit from such treatments. Should children be included in the trials of these treatments? The rash answer would be yes.

But, scientifically, the dosage and side effects of the treatments in children would be significantly different. If the tests were to be done on children, they would have to be done in separate pediatric trials. The ethical issues do not relate to the exclusion of children from adult trials, but to the lack of pediatric trials. On the other hand, exclusion of some ethnic minority patients from a trial is unlikely to have any scientific validity, and so may reflect some social injustice (or indeed some other cultural factor).

II. ETHICAL AND SOCIAL ISSUES

In part I, I discussed the ethical necessity for any scientific experiment using human subjects to be scientifically sound in conception. Merely scientifically sound experiments will not generally be ethically legitimate, however. In this part I will discuss ethical issues of design and recruitment to scientific experiments which are relevant to the method of selecting subjects.

A. Protecting Subjects

Historically, the most significant ethical issue in selection of subjects for research has been the protection of subjects. The main ethical standards in human subjects research have been framed and disseminated in the wake of the Nuremberg war crimes trials, and subsequently these have been reinforced by evidence of other "medical" atrocities. The purpose of these standards has been to protect subjects from actual physical and mental cruelty, and from being treated as "mere means" to ends set in the name of Society or Science (or both). The main focus of the Nuremberg inquiry was upon coerced participation by subjects. Later, attention was turned to enrollment of subjects into experiments without their knowledge, in particular, instances of experiments which were performed without any conceivable benefit to the subject, where actual harm would result, and where subjects were offered financial or other inducements to take part in the experiments.

These concerns focus our attention on the issue of enrolling vulnerable individuals into experiments, and onto the issue of risks, harms, benefits, and interests. It has been established that in all but the most extreme or exceptional cases, no one should be enrolled in an experiment without their explicit and voluntary consent. Particular attention has been devoted to those cases where subjects are incompetent to give or incapable of giving explicit and voluntary consent. These include minors (including, on some interpretations, fetuses), the mentally ill and disabled, and the unconscious. In addition to these cases (which for the most part involve special legal considerations as well as ethical ones), attention has been focused on subjects who may be vulnerable to duress or other kinds of pressure to participate. Such duress might on occasion be held to invalidate any consent such subjects may give, because it vitiates the voluntary character of that consent. The original cases where such concerns were raised were prisoners, members of the armed forces, and students and junior staff in research institutions. More recently, concern has been voiced about subjects at a social disadvantage vis-à-vis the researcher or access to health or social services. Throughout the history of these debates attention has returned again and again to the situation of severely ill subjects who may be able to make reflective judgments, but only at the cost of great suffering, or who may not be able to make judgments "in character" because of their suffering.

B. Limits to the Protective Model

It is clear from this focus on the protection of subjects, and on identification of types of vulnerable subjects who particularly need protection, that the emphasis in research ethics lies on the risks of research, rather than on the benefit. This may be a somewhat misleading or distorting emphasis. Arguably, most research exposes subjects to no risk of physical or psychological suffering or injury, even in the short term. Most social research and most psychological research is of this type. A large portion of medical research on minor conditions and on the delivery of healthcare is of this type, too. Levine has distinguished between harm and inconvenience to subject. We may also note that some research is no more or less harmful or inconvenient than "standard" practice, and that if a protective test is to be applied, it should apply not only to the research intervention but also to the standard intervention. This point may be considered irrelevant to experimental sciences where subjects are actively recruited and enrolled, but it is important in research in medicine or management science (say) where research activity is part of or added on to "ordinary" activity, and the subjects are clients or staff of the service. Related to this, in applied sciences where practices are subject to continuous innovation or adaptation to changing circumstances (surgery is a good example) it is hard to determine where research begins and standard practice ends. The distinction may lie in the deliberate intention to alter a practice in a way which, if successful, will lead to a permanent alteration or extension of the practice, so that the purview of the intention is not simply the case in hand, but all future similar cases.

In medical research, many treatments or practices may be both harmful and beneficial. Chemotherapies are usually physically very unpleasant and their effectiveness lies in their being toxic—but more toxic to the cancerous than to the "normal" cells. The aim of a selection principle in this case will be to select patients as subjects for whom the benefits of the innovation will be in proportion to the harms actually to be undergone, or at risk. In addition the principle holds that the researcher (and a respectable body of professional opinion) must believe that the new treatment will turn out to be at least no less effective (that is, beneficial) than any alternative treatment we could offer. In these cases, focus on risk alone is inappropriate.

What is the nature of the protective standard we seek to apply in such cases? In the first case, where the research presents some inconvenience to the subjects, the consent test merely seeks to ensure that subjects are aware of and agree to undergo the inconvenience. Insofar as this is an ethical issue, it is normally the issue of good manners: putting subjects to inconvenience without their consent will make them (and others) less likely to participate in the future, and this is particularly so when they are at first unaware of the experiment or the inconvenience. Where a subject may feel they have been deceived, misled, or spied upon they become at risk of psychological harm (a sense of violation), and misconduct in the practice of the experiment turns the inconvenience into an actual harm. Finally, covert or deceptive research may be regarded as having serious consequences for the nature of the society permitting such research, indicating as it does a disregard for privacy and individual liberty. Most ethics committees are therefore very unwilling to permit such research, however well motivated. In many cases such research is actually illegal.

In the case where the experimental procedure exposes subjects to risks and benefits of the same degree as standard practice (the benefits of the research being to the organization rather than to the subjects directly), we may argue that first, good manners require informing the subjects, and cooperative subjects may be a source of extra information for the study, and second, the consent of subjects may be required for both experimental and nonexperimental procedures where risk is involved.

The role of consent as a protective test is twofold. First, it alerts the subject to the fact of their possible participation in an experiment, which is a situation where some significant risks or benefits or both will be incompletely known. Secondly, it promotes and legitimizes the subjects knowing acceptance of the chances of harm and benefit and any reasonable consequences of the innovation for the subject. The consent test is antipaternalistic in a number of ways. It makes the choice the subject's choice, rather than that of the researcher. In many situations experiments which a sub-

ject might not find objectionable may be ruled out by a too anxious researcher or committee: it may be paternalistic not to offer the subject the choice, in fact. A test based on risk alone may be slightly absurd for the reasons already noted, but it is also seriously expert-centered and paternalistic too. Finally, the consent test, at best, may involve the subject as a partner in the research, rather than as a "subject" in the political sense of the term. Some authors have suggested abandoning the term subject for this very reason, although this may be a case of linguistic usage running ahead of practice.

The consent test does have its pitfalls, however. In Section II.A I mentioned the ways in which consent may not serve the interests of several classes of vulnerable subjects, either by automatically excluding some de facto (the unconscious and the mentally ill or disabled) or de jure (children or prisoners), or by permitting the inclusion of others (the seriously ill or members of vulnerable social groups) against their "best interests" (however they be determined). A further twist is the perception that some professional groups may use the informed consent test as a sort of "buyer beware"—in an exculpatory way. Many patients in medical experiments may regard informed consent cynically (as part of "defensive medicine"), or as the doctor off-loading his responsibility to know, judge, and decide what is best for the patient.

C. Selection and Protection

It is clear that the consent mechanism is a kind of selection mechanism as well as a protection mechanism. Many medical trials have ultimately failed because, while many patients were offered participation, very few accepted the offer. In most cases, the situation is the opposite: the experimental population is large, and enrollment continues until enough suitable subjects have presented and given their consent. The subjects who enroll effectively select themselves from the experimental population. At this point the issue of justice is at least as important as protection. Are certain psychological or cultural types of people more (or less) likely to consent (perhaps because of "intelligence" or "educational level")? Or to present in the first place? Are people of this type unfairly advantaged by their superior access to the benefits of such research? Or unfairly disadvantaged by their exposure to the risks of research? Furthermore, the consent mechanism is one selection mechanism among others.

Subjects in medical trials are "selected" for participation in medical trials of treatment by their misfortune in falling ill with a particular disease, at the given time and in the given place. This is mostly by chance, and may be a blessing or a disadvantage; but in some cases an additional selection factor may combine with this selection by fortune. This is the socioeconomic factor. Being treated at a hospital with a large research load may increase one's chance of participation in a trial, and this may be a good or a bad thing, but one may have a higher prior probability of being treated in such a hospital because of one's inability to pay for care, or alternatively because one lives in the (often) poor neighborhood in which a teaching hospital is found; or perhaps because one has the money and status which make one able to demand "state of the art" treatment. In any of these cases a "market failure" may be occurring which distorts the just selection of subjects for the risks and benefits of research.

The difficulty in negotiating the issue of justice here is heightened by the time asymmetry. After the fact, when the risks and benefits of the experimental procedure are more clear, it is easier to say that some group was harmed by participation or exclusion, because we can identify whether participation entailed access to net harm or net benefit. Doing this before the fact is much harder—in fact, if it were not we would not need to do the experiment in the first place. But it is not always impossible. Any treatment or procedure will involve some knowable consequences, and some unknown consequences. In a trial of radical mastectomy versus lumpectomy, most of the surgical and psychosocial consequences are predictable; what is not predictable is the relative effectiveness of the two procedures. The doctor and the patient must together determine whether the consequences and risks of each treatment are such that the patient and doctor are able to tolerate the predictable outcomes of either treatment to the extent that they are indifferent between them.

In most cases the researcher will screen some subjects out even before informed consent as unsuitable. Recall that we have assumed that the person under consideration as a subject is a member of the experimental population—if they are not then they are automatically ineligible. Of the possible subjects who present themselves, some will be unsuitable for some reason. In a medical trial, their condition may be too severe or to mild, for example. Informally, some subjects may be ruled out as unlikely to comply, as previously discussed. There is a fine line here between pragmatism and paternalism. The suitability of a subject is another, often informal and private, selection criterion.

In many cases, what counts as unsuitability will turn on unsuitability to give consent rather than medical, social, or psychological unsuitability for participation in the study. A patient, otherwise suitable, may be so distressed by his condition that seeking his participation would be cruel. But in this case it might be that giving them the chance to participate is giving them a measure of control over their situation, which they might not otherwise feel they had, and may be therapeutic of itself. In the wider philosophical context, informed consent is usually not taken to stand on its own as a test (as it does in the Nuremberg Code), but as part of "respect for autonomy" (or self-determination). Here we see a case where overzealous respect for the letter of consent, combined with a somewhat paternalistic desire to protect a patient from the stress of choice (understood as harmful), may actually interfere with the patient's autonomy. In most discussions autonomy is taken to be a reflexive character trait which adds to the moral quality of a person's life and develops through use, but is fundamental to human being. I have not used the concept previously in this entry because it is not philosophically neutral, and perhaps not widely accepted, even though it is widely presumed accepted.

1. Women: Protecting a Possible Fetus?

While many subjects are excluded from a study because they are judged unsuitable on scientific or pragmatic grounds, others are excluded for philosophical or legal reasons, and others are included against their interests, through compulsion or coercion. Women are often excluded, not on any grounds of their own personal safety, but on the grounds of possible harm to actual or future children they may bear. Many drugs are indeed very harmful for fetal development; radiotherapy can certainly promote mutation, and so on. The philosophical question here is, whose risk is this? Is it the (potential) mother's, to take as she sees fit? Or is it the (potential) child's, from whom no consent can be derived either actually or by substituted judgment (no such inference would permit a child to undergo a harm without benefit to the child)? In some cases the treatment will be medically necessary for the mother's survival, and in cases where her survival is necessary for the survival of the child, a law of double effect may apply. These are still open questions, and much discussed. Furthermore, the methods of (temporary) resolution in this area are methods of the law and the court.

2. Protecting the Powerless

Other important exclusions include the exclusion of prisoners and military personnel, and of students and junior staff working for the researcher, or in the researcher's sphere of influence. The argument here is not that such subject cannot genuinely volunteer, nor that they cannot benefit from experimentation, but that their consent is either the result of coercion by the authorities or is distorted by their membership of an institution which has a powerful influence over their future, and may be regarded as having interests other than those of the individual subject at heart. These situations are different from that of the patient in the hospital, precisely because the hospital is supposed to care for the (medical) interests of the patients. In some cases prisoners have been prevented from taking part in research which would, on their own judgment, have been to their benefit, by overliteral interpretation of regulations forbidding the enrollment of prisoners for their own protection. Yet it is hard to press the charge of paternalism here, simply because the regulations for prisoners and students were framed to curb actual abuses of power.

The way out of this dilemma may be to develop the notion of group or communal consent, not as sufficient for inferring individual consent, but as sufficient for allowing individuals to be approached. This is a subject's rights analogue of the Institutional Review Board or Local Research Ethics Committee. Institutions of this kind have already been established to protect the rights of the Inuit in Canada, Maori communities in New Zealand, and native Americans in the reservations of the United States.

3. Subjects Incompetent to Consent

The most difficult cases in practical research is research on children, unconscious subjects, and the mentally ill or disabled. In such cases the capacity for exercising autonomy or for giving genuinely informed consent is curtailed or absent, yet in other respects the interests are the subjects' own, and are very real. It is arguable that research on these subjects is unethical, if it is not for the benefit of each individual subject him- or herself, and cannot stand the test of what in other circumstances the subject might want and what society in its best intentions might deem decent and dignified. This is an extreme statement. It would rule out child psychological development research as unethical because it has no benefit for the child. So most authorities allow nontherapeutic research which is noninvasive and nonintrusive, and poses "minimal risk" to the child (or other subject).

D. Rights and Duties

The main protective issue in all of these cases is to protect the individual's rights and interests being overridden in the interests of present and future others or "society." This is held to be both respectful of the subject's own rights and person, and also beneficial for the health of society. The main counterargument to this is a mixed authoritarian and utilitarian one, which according to Daniel Rothman was to be found in societies like the USA in the "total wars" of this century.

According to this argument, in some circumstances (plagues, wars, or natural disasters) the rights of the individual were subservient to those of society, simply because without the continued existence of that society those rights would be destroyed. And in exceptional circumstances, that society has the right to hold certain individual rights in abeyance. This is a very dangerous argument, but one still familiar among so-called "realists." Many of the most serious medically inflicted harms have been imposed on members of the armed forces by their own side for this reason, and for the reason that in battle, troops may be injured, mutilated, and killed, often voluntarily, to save comrades or for military necessity, and participation in medical experiments was simply another form of this duty.

The duty to participate in experimentation has not found many defenders, although some authors have advanced powerful arguments. We have seen that all subjects have the right not to participate in research, and hence that all researchers have the obligation to honor this right, as well as the obligations we derived in the first part of this article concerning adequate study design. We have just discussed an argument which holds both that these rights are sometimes properly overturned in the "national interest," and like all national interest arguments they are as hard to resist in the real world as they are hard to defend on paper. The other implication of this argument is that subjects have an obligation to participate when necessary. It is a mistake to conflate these two versions of the argument. It is an important error to assume that to every right corresponds a duty, simply because of grammar. A duty is owed by me to someone, while a right may be considered to bind everyone. This is one reason why there are fewer rights than one might hope. My duty—if I have one—to participate in research is owed to no state, but to humanity. And the state has no right to confuse itself with humanity.

In other words, if there is a duty to participate in research, then it is a social duty, not a civic one. So it is of the same kind as the duty to become a blood donor or to give to charity. Caplan and Jonas have both observed that this must be the nature of such a duty, and have both argued that it exists. It is a matter of moral honor to observe the duty, because it is unenforceable.

Jonas concentrates on the case of nontherapeutic research, and Caplan on therapeutic research. The controversial part of Jonas's argument is that he claims that the severely, perhaps terminally, ill are particularly suited to being research volunteers, and especially in research related to their illness. They would make, on Jonas's theory, excellent Phase I and Phase II trial participants, where Phase I determines toxicity, and Phase II tries to determine the range of effective and tolerable dosage for a drug. Some studies do indicate that certain altruistic terminally ill subjects agree with Jonas in their own case—it feels like paying back something, or leaving something worthwhile behind. It would be stretching a point to call this recognition of a duty. It would be unusual to recognize altruism as a duty. Jonas's argument rests on the harmfulness of the research. Toxicity research is potentially very harmful indeed; would it not be wrong to attempt it on healthy volunteers or curable cases, even with their consent? It may well be that a "good death" might be available in this way, but many people would regard this sort of stoicism in the face of a death made even more painful, rather than less, as inhuman.

Caplan's argument rests on the principle of reciprocity: we all depend on medical research to some degree, and so refusal to participate where one can is a form of "free riding." It is not clear, however, that this rule applies in every case: can one be choosy about which research one participates in? Of course—one should be. And it is certainly clear that the duty is not transferable: just because I would do something gives me no cause to expect my child, my senile relative, or you to do so.

A final issue in this context is the issue of a right to participate in research. This is in effect a justice issue. If I am aware of research on my condition being done, and I believe that it would benefit me, or indeed that nothing else would do so but this research, have I the right to participate? This is part of a larger issue, the issue of whether one has any rights or entitlements to health care in general or of a specific sort. Most human rights theorists argue that everyone has a right to a minimum standard of care, as to food, but not to anything specific (I have no right to caviar). Similarly, I have no right to any particular form of care that I may demand, although I have a right to expect that my doctor will do what is in his power to preserve my life and health. So the right to participation, if it exists, will be of a limited kind, and will rest on equity issues, limited by scientific and pragmatic matters. To these questions I now turn.

E. Justice and Selection

As I have mentioned throughout this paper, the issue of access to experimentation and fair distribution of risks is a crucial topic in research ethics. It is of little importance in minimal risk, minimal benefit research, save perhaps in the sense that enrolling a wide variety of subjects may have a social solidarity effect: encouraging access to, and participation in, science as a part of culture, and reducing the sense that science is a "white man's" game. Even in this sense, however, there is some worry that subjects are subjects rather than participants, and this is an area where development is needed. But in any research where benefits and harms may accrue to subjects, the issue of fair access to the chances of benefit and fair distribution of risks are of obvious importance.

If we recall the discussion in Section I.A of compensation for the inconvenience of participation, it is possible that such compensation may be regarded as an inducement to participate. As such it may be regarded as a benefit in itself, rather than as a reparation of costs, and as a benefit it may be offset against risk or actual harm. It is possible to regard the supply of participants in a trial as a market, where compensation of subjects is a sort of "price" which is to be set to offset the opportunity cost (the economic value of their inconvenience) of the marginal participant (that is, the last to be recruited for the minimum sample size, taken to be the least willing or most inconvenienced). Understood economically, the earlier subjects (ordered by inconvenience) are indeed receiving a direct benefit—a profit of payment over inconvenience. Are they therefore being induced? In this sense they are. Is this harmful? Not if we can distinguish inconvenience and harm. And can we? Only in some cases. In most cases, neither we nor most participants are that sophisticated. Some economists would regard harm as an economic "good" anyway, and as such tradable. These issues are so complex that most research guidelines ignore them, or specify that financial payments in therapeutic research before the fact (and not as compensation for *unexpected* harms) are unethical, and payments to "healthy volunteers" must be small—a gratuity rather than a payment for a service.

Supposing one can distinguish, as Levine does, between risks and harms on the one hand and inconveniences on the other, it is important that, given the definition of the experimental population, and given the protective standards agreed to in our society, any subject should have fair access to the benefits of the experimental process. The issue of priority setting in research is not unrelated, although it is logically distinct. Arguably, as I stated at the end of the last section, no one has a right to participate in a particular experiment. But it seems clear that if experimentation is going on, no one should be systematically excluded from experimentation which they might benefit from participating in. The arguments of women concerning this are well surveyed by Merton. Such exclusion has two forms: first, through formal or informal exclusion of research directly relevant to them, and second, systematic nonperformance of research relevant to some social group. If some group suspects that work relevant to their health or social well-being is regularly assigned low priority, while some other group's sectional interests are regularly given high priority, then that group may have a right to some remedy. This is a matter of justice about the distribution of social power, and may apply to women, children, homosexuals, and members of ethnic or cultural minorities. A controversial instance would be access of the elderly to experimental health care.

The converse of this is where a group is overrepresented in risky research. This can be through deliberate inclusion of the socially marginal or weak (considered expendable) or as an unplanned consequence of some other social injustice. The very poor, who cannot afford health insurance, may be disproportionately exposed to some kinds of research because of their poverty and because of their poverty-related ill health; in many countries, poverty and ethnic group may be linked, so that the injustice acquires (or is caused by) racial discrimination.

In both questions of justice (considered as fairness), there are two elements. The first is the possibility of social, systematic advantage or disadvantage. Whatever one's views about the ethics of access to standard (or even luxury) health care or social services, it is hard to defend inequality in access to experimental risk or harm, in much the same way as it is hard to defend unequal involuntary exposure to infectious disease as a *policy* rather than as a fact of life. As we noted in the scientific aspects of selection there is no necessity for deliberate social representativeness in sampling. But we also saw that there was no harm to validity in seeking just representation, consistent with selecting from the experimental population, provided suitable randomization measures were taken.

The second element is the element of fairness to individuals considered as such. If we rely—as we must, where we can—on consent as a tool to help the individual protect him- or herself, the more important issue is the issue of ensuring fair access to benefits. For the most part this cannot be detached from group justice, as has been noticed, in another context, in the case of affirmative action. Perhaps these two modes of justice conflict, as they seem to in affirmative action. One solution, as has been noted, is the use of group consent as a prior level of protection. This would need to be proactive, in putting forward group members as a class of participants, and defending the members' right to be approached on equal terms. The function of the group consent would be to protect equality rather than to enforce members' individual rights, which, as we have seen, are probably nonexistent. If we suppose that this protection of equality of access is effective, then the fairness to individuals issue can be approached.

Perhaps the fairest way to distribute chances is by lottery. Other methods include ranking subjects by need (of treatment or of benefit), or inversely by vulnerability (the least vulnerable first). In fact, however, this is an open question in research ethics simply because no one has addressed it since Levine's early paper. The question is perhaps analogous to issues about equality of opportunity (as opposed to equality of access to goods). In practice, two methods are used. The first method is the informal allocation of chances, determined largely by technical considerations of suitability and judgments of compliance, to subjects identified as they present themselves to the researcher. There is no guarantee here of fairness, although there is a link in the medical case to the random element in the epidemiology of the condition. The second method is a modification of the former, in which certain groups are actively recruited, where their low participation has been noted and determined to be linked to a social disadvantage or some specific inconvenience. The latter is the outreach approach.

Bibliography

Caplan, A. L. (1991). Is there a duty to serve as a subject in biomedical research? In *If I were a rich man could I buy a pancreas? And other essays on the ethics of healthcare* (Chap. 6). Bloomington: Indiana Univ. Press.

Dula, A. (1994). African American suspicion of the healthcare system is justified: What do we do about it? *Cambridge Quarterly of Healthcare Ethics,* **3,** 347–357.

Jonas, H. (1969). Philosophical reflections on experimenting with human subjects. *Daedalus,* **98,** 219–247.

Levine, R. J. (1978). Appropriate guidelines for the selection of human subjects for participation in biomedical research. In National Commission for the Protection of Human Subjects of Biomedical and Behavioral Research, *The Belmont Report: Ethical guidelines and principles for the protection of human subjects of research* (Appendix 1, part 4). DHEW Publication (OS) 78-0012. Washington, DC: US Government Printing Office.

Levine, R. J. (1986). *Ethics and regulation of clinical research.* Baltimore/Munich: Urban & Schwarzenberg.

Merton, V. (1993). The exclusion of pregnant, pregnable and once-pregnable people (a.k.a. women) from biomedical research. *American Journal of Law and Medicine,* **XIX,** 369–451.

Senn, S. J. (1995). A personal view of some controversies in allocating treatment to patients in clinical trials. *Statistics in Medicine,* **14,** 2661–2674.

Intrinsic versus Instrumental Value

NICHOLAS JOLL

University of Essex

GLOSSARY

absolute value A value that cannot be overridden by other values.

consequentialism A form of normative moral theory that determines the worth of an action or state of affairs on the basis of its consequences. More particularly, a view of the Right action as being one that maximizes or most "promotes" the Good.

critique of instrumental reason Approach developed under this title by the Critical Theorists of the Frankfurt School and, somewhat differntly, by Martin Heidegger.

deontology A form of normative moral theory that invokes "ends-in-themselves," "absolute rules," "side-constraints," or similar notions. In a more specific formulation, a normative theory that understands Right action as action which "honors" the Good—

and which may thus invoke a notion of "absolute value."

ecophilosophy A name sometimes given to "environmental philosophy," particularly when that philosophy is non-anthropocentric, i.e., does not work with substantively human-centered standards.

final value Value pertaining to that "which is always choosable for its own sake and never because of something else" (Aristotle).

functional food (nutraceutical) Roughly, a food designed or produced to confer specific health or metabolic benefits.

the Good, the Right Terms of normative ethics. The Good concerns what is of moral value or of final value, and the Right how one is to act toward such value.

instrumental value Specifically, a value deriving from a thing's use in relation to something else. More loosely, any type of value opposed to intrinsic value.

intrinsic value Divided herein into three main senses: final value, moral value, and objective value.

metaethics Study of the nature, status, and logic of morality or ethics.

moral value As taken here, an obligatory value.

normative ethics Normative moral theory concerns, at a general level, what one should do or value or how one should live.

objective value A value is objective in a strong sense if it subsists independently of practices of valuation.

speciesism The charge of unjustifiably giving moral preference to one's own species.

xenotransplantation Transplantation of bodily parts or tissues from nonhuman animals to humans.

INTRINSIC VERSUS INSTRUMENTAL VALUE is a phrase rich with ambiguity. Indeed, its manifold meanings gesture toward philosophical territory spanning different philosophical fields, approaches, and traditions. The first necessity for any comprehensive treatment, therefore, is orientation. To anticipate in an approximate manner, central issues of intrinsic value concern: (1) what is morally or truly valuable; (2) what underpins such valuations; and (3) the hypothesis that what we take to be valuable in these ways reveals a shift with profound philosophical implications. Each of these topics, as I will illustrate, finds significant application in cases of recently developed technologies. For with these tools one can ask, roughly, what ethics are at stake with a particular technology, what grounds these ethics, and whether indeed our ethics here have been changing behind our backs in a perhaps uncanny way.

I. WHAT IS AT ISSUE WITH "INTRINSIC VALUE VERSUS INSTRUMENTAL VALUE"?

The first concern of this section is to achieve a working degree of clarity on central terminology. It distinguishes different notions of intrinsic value and the different types of value—which might broadly be termed "instrumental"—to which intrinsic value is opposed. Although I draw upon other philosophers here, there is no standard terminology or typology in this area. This is not surprising given the multitude of phenomena encompassed by "intrinsic versus instrumental value." On the basis of the definitional work, the section concludes by dividing these phenomena into three sets of substantive issues.

A. Instrumental Value

A first sense of "intrinsic value" takes it in an initially straightforward contrast to "instrumental value." That which is valued as a means to some end has an instrumental value, while that end has noninstrumental value. If nails and wood are valued because they can be used to make a table, for instance, the nails have instrumental value, while the table has noninstrumental value. But this picture needs complicating.

Noninstrumental value is often a matter of context or degree. In the situation sketched, the table itself might be a means to furnishing a house. In this case the furnishing has noninstrumental value, and the table only

derivatively, to a lesser degree, or only when considered in relation to the nails and not to the furnishing.

However, something can be valued noninstrumentally *tout court*—in a nonrelative fashion. This is so in two distinct senses. First, something may be instrumentally valuable but also have irreducible noninstrumental value. I might value my cat as (or almost as) a piece of the furnishings, for the pleasure its delicate features afford my senses, and I may also value it for itself, utterly without regard to its being a means. Second, something can be valued noninstrumentally in that it is *never* valued as a means, but always and wholly for itself (which excludes the nails, wood, and table *and* the cat). Such a value I call a "final value."

B. Final Value and Moral Value

The concept of final value captures an important sense in which the phrase "intrinsic value" is used; but this sense requires further specification. Aristotle, from whom I adapt my terminology here, spoke of what "we call final without any qualification" as "that which is always choosable for its own sake and never because of something else" (Aristotle 1976, Book I, Sec. vii). It is that which, as of supreme value, is the ultimate reason for all action. As such, determining what has final value—as I am calling it—was for Aristotle the central object of ethics.

"Intrinsic value," however, also can mean moral value—and this is somewhat different. Note first that Aristotle's treatment of ethics includes subjects that many today would exclude from the ethical domain, such as the virtues or excellences of magnificence and wittiness. Indeed, the very meaning of morality as contemporarily understood differs from the ancient Greek conception of ethics as elaborated by Aristotle.

Very roughly, morality as owing to the Judeo-Christian tradition involves a strong notion of obligation. This can be specified through the cardinal philosophical interpretation that the notion of obligation received with Immanuel Kant. In Kantian spirit, moral value will be taken herein as denoting something it is obligatory to value, or in terms of obligatory ends for our actions. To illustrate, being magnificent or witty, just as setting gastronomic world records or climbing mountains, does not appear to be an obligatory end, but helping one's neighbor might be. (And perhaps, in some continuity with Aristotle, so might be being reasonably agreeable company.)

Now the language of obligatory ends suggests moral value be taken as a subset of final value. In Kantian language, if I *must* value something, this valuation cannot be hypothetical on any end I happen, perhaps choose, to have. The categorical nature of such an end, then, implies that it head our web of valuations. Nonetheless, one can consider the possibility of a nonfinal moral value. Drawing on the terminology developed, could "moral value" not characterize values that are noninstrumental and nonrelative without being "final"? It is conceivable that it be obligatory to value something that is also acceptably valued in other ways (like the cat).

C. Objective Value

Objective value names a yet further sense of intrinsic value—a sense that begins to separate the concept of value from that of valuation. To claim that a value is objective is at least to claim that good reasons can in principle be given for holding, or believing in, that value. Thus, when Aristotle means to convince his audience that happiness is the final value (on which more is in Section II below), and to do so with more than the mere plausibility of the orator, he is relying on this degree of objectivity (cf. Aristotle, Book I, Sec. iii).

Some philosophers attempt to treat the question of the objectivity of values wholly in terms of rationality. Yet it is a stronger notion of objective value—for all its difficulties—that will be the focus herein. This conception is of objective value as value that, somehow, subsists independently of practices of valuation. For instance, the conviction that aiding your neighbors is morally valuable may have a deeper basis than that constituted by there being good reasons for such a stance. For perhaps this value is possessed by that activity in itself, regardless of any human recognition of such value. How such objective value more particularly relates to moral, and final, value will be indicated shortly.

D. The Substantive Issues

It is worth recapitulating the major senses of value just discerned. "Final value" denotes something to be valued solely noninstrumentally, or "for itself" in that sense. "Moral value" is defined as a value that, while not necessarily final, is noninstrumental in that it is obligatory. "Objective value" in a strong sense refers to a rather different sense of "intrinsic value": value subsisting somehow independently of practices of valuation.

As these formulations suggest, this article uses the senses or kinds of intrinsic value to call the shots (as against the broadly "instrumental" values). I will address three types of substantive issues arising from this taxonomy of intrinsic value. These issues do not dovetail with the different concepts of value as neatly as one could wish, but the issues do divide among three philosophical approaches or areas of inquiry.

I begin in Section II with normative ethics. Normative

ethics is a self-description of a branch of analytical philosophy. "Normative" means pertaining to norms, or more generally to "oughtness": at the most general level, normative ethics is about what one ought to do or how one is to live. Here we find a first set of problems turning on what is of intrinsic value, for central questions of normative ethics concern what is to be valued for itself, and these issues can be addressed through the notions of final value and moral value. This thinking is illustrated with reference to a new reproductive technology.

Section III brings in metaethics, in order to examine what is to be valued for itself from a higher altitude. Another field of analytical philosophy, metaethics can broadly and somewhat tersely be defined as investigation of the nature, logic, and status of basic ethical concepts. One such concept is objective value. Different construals of objective value, it will be shown, imply grounds or limits to our candidates for what is of moral value—with potentially radical results. The applications illustrated here are xenotransplantation and genetic modification.

Section IV takes instrumental value and noninstrumental value more generally, and somewhat differently, in terms of "the critique of instrumental reason." This critique urges that, as a matter of fact but one of fundamental philosophical import, less and less is being viewed and valued noninstrumentally—taken, that is, as being of final, moral, or even (for this is somewhat more philosophical) objective value. This is made concrete with the phenomenon of functional food.

II. WHAT IS TO BE VALUED FOR ITSELF? NORMATIVE ETHICS

A. Introduction

The preliminary section noted that a central issue of intrinsic value is "What is to be valued for itself?" and that this is the concern of normative ethics. Now the question of what is to be valued for itself is sometimes expressed as that of "the Good." This notion, together with that of "the Right," allows us to stake out the relevant areas of normative ethics.

Owing to the ancient Greeks, yet still current today, the idea of the Good is broad enough to encompass two different understandings of the very idea of something being "valuable in itself" (cf. Section I.B). Thus the first part of this section considers two accounts of the Good differing not only in substance but also in orientation. The first is Aristotle's, which trades in final value—that which is valuable for itself and only for itself. The second is that of the 18th century founder of utilitarianism,

Jeremy Bentham, who brings in moral value as an obligatory value or end.

The next part of the section considers the Right, a modern counterpart to the Good. The issue of the Right is this: how in general terms should moral value be acted upon? The third part applies the issues to a new reproductive technology.

B. The Good: Aristotle and Bentham

According to Aristotle, that which "we call final without any qualification"—"that which is always choosable for its own sake and never because of something else"—is happiness (Aristotle, Book I). Happiness translates the difficult Greek term *eudaimonia*. What does Aristotle understand by it?

It seems that for Aristotle true happiness, in which humankind partakes of the divine, rests in contemplation. However, such a Good requires the support of "the necessities of life," and a man who contemplates cannot be called happy if much misfortune befalls him. Further, "in so far as he is a human being and a member of the society he chooses [he must choose] to act in accordance with virtue" (Book 10, Sec. viii). The more exactly human Good, then, involves such action—or rather, equally, the appropriate virtuous "activity of the soul." And according to Aristotle there are many virtues.

On this admittedly rather breathless ride over difficult interpretative terrain, then, the true Good or happiness is in short the contemplative life, and, it would seem, the Good or happiness for those who cannot achieve this has as many components as there are virtues. Here, the Good or happiness is the proper exercise of the manifold human excellences including courage, temperance, friendliness, and proper ambition.

I turn now to Bentham, who introduces the fundamentals of his philosophy with the following "metaphor and declamation":

> Nature has placed mankind under the governance of two sovereign masters, *pain and pleasure*. It is for them alone to point out what we ought to do, as well as to determine what we shall do . . .. [T]he standard of right and wrong [is] fastened to their throne. (1962, 1)

Bentham considers that his philosophy thereby treats of the "only right and proper, and universally desirable, end of human action" (p. 1).

Bentham and Aristotle show both agreement and disagreement. To claim as Bentham does that happiness is the "universally desirable" end of human action is to concur that it is the final value. Yet, in adding that happiness determines the standard of right and wrong, that it determines what we *ought* to do, Bentham intro-

duces considerations of distinctively moral value lacking in Aristotle. Most obviously, however, Bentham and Aristotle differ over the nature of the happiness that is to be taken—in one way or another—as valuable in itself.

This juxtaposition of Bentham and Aristotle suggests what is at issue with competing conceptions of the Good. Beginning with a fairly internal perspective on these theories, one can observe first that they can present different accounts of the happiness which ostensibly comprises the Good. Aristotle's conceptions obviously challenge Bentham here. On the former's narrower view, *contemplation* constitutes true happiness, even if such happiness requires a not-too-unpleasant life as a kind of base. And the wider Aristotelian view of happiness, unlike Bentham's, involves a reference to virtue. (Indeed, Aristotle believed in *different kinds* of pleasure pertaining to different virtuous activities.)

A second issue is the inclusiveness of conceptions of the Good. It can be charged against Bentham's idea of happiness, for instance, that there are many things to be valued morally despite their not bringing us happiness in any narrow sense. Bentham infamously contended that "pushpin is as good as poetry" on the ground that the children's game and versifying provided similar amounts of pleasure to the relevant parties. Yet although poetry can console and delight, it seems also to possess moral value, or something like it, in virtue of something else. (Here one might consult Aldous Huxley's *Brave New World* and Ray Bradbury's *Fahrenheit 451,* in which the heroes rebel against the hegemony of a kind of happiness. One can also compare Nietzsche's jibe that utilitarianism was a suitable ethics for a nation of shopkeepers.)

Similar points can be put to Aristotle.

Can the kind of value just alleged for poetry, or which might be alleged for art in general, be encompassed by the Good of contemplation? Or even by the wider Good comprising 'activities of the soul in accordance with virtue?' One can wonder, in fact, whether *any* conception of moral or final value well captures what, *in some moral sense,* we value in this area. (These thoughts point back to the concerns of Section I.B., and forward to Section IV.D.)

Yet, third, can the general moral weight of happiness itself not be disputed? Care is needed here. The value of happiness for Aristotle or Bentham is not just the value of *one's own* happiness. Further, their ideas of happiness differ from one another, as seen—and also from other possible construals of this loaded concept. Aristotelian happiness is contemplation; or, on his other conception, it is inseparable from multiple virtues. Benthamite happiness is meant to accommodate what is yielded from pushpin all the way through to poetry. So any appraisal of the value of happiness will have to go through the preliminary steps, sketched above. That is, one needs to clarify what exactly is meant by "happiness"—or by some other description of the Good—and to consider what, so defined, it might possibly include.

There are many more doctrines of the Good. The ancient Cynics understood it in terms of an ideal of simplicity, the Epicureans as a rather narrow construal of happiness as contentment, and the Stoics—close to Aristotle—as wisdom and the ethical life. Modern candidates for the Good include, at perhaps two extremes, beauty and the goods of friendship (the proposal of the influential English philosopher G. E. Moore), and the range of goods accommodated by contemporary "preference" or "welfare" utilitarianism. This latter position allows the Good to range across whatever people in fact find pleasant, good, or valuable.

C. The Right: Consequentialism versus Deontology

The Right concerns how we should act toward the Good. Two theories of the Right are dominant. (As contemporary theories, these tend to consider the Good in terms of moral value.) The first is consequentialism, and the second deontology. Consequentialism is often understood as a normative moral theory that determines the worth of an action or a state of affairs on the basis of its consequences. Yet Philip Pettit has persuasively argued that the essence of such theories is better captured in terms of the Right, as follows.

Consequentialism teaches that whatever the Good may be, we should act to "promote" or maximize it. Understanding the constitution of the Good and what is conducive to it may be complex, but this principle of action is simple. This is because consequentialism is a "teleological" (goal-driven) theory. The correct way to act in a given situation is to perform those actions which most bring about the realization of the Good in the world. Hence the consequentialist may countenance some particular violation of a value to achieve a state of affairs in which this value is more widely upheld.

A nonconsequentialist or deontological theory understands the Right differently. For deontology, as Pettit puts it, what is valued is to be "honored." That is, one should not act just to maximize the Good, but rather or equally to ensure that such values are not violated. Rights and liberties are very often understood this way. A consequentialist defender of free speech, for example, can oppose a talk promoting a closed society on the grounds that it is instrumentally harmful to free speech—to free speech in general. A deontologist with the same value could not straightforwardly do so.

A deontologist, however, is not necessarily committed to defending all of his or her values come what may. It may be that some values can be trumped by others—a value concerning life being more important than free speech in occasions where they clash, for instance. Hence deontology normally invokes a hierarchy of values. At the top of the hierarchy, however, there is normally a special inflexibility of value. Here one often finds a kind of absolute value: a value that is putatively non-overridable, never (or almost never?) to be violated.

Documenting this point faces a complication. Despite the usage adopted here, deontology often finds the language of values uncongenial. Hence Kant, the most famous deontologist (albeit *avant la lettre*), speaks of a person as "an end for the sake of which all else is means" rather than in terms of an absolute or non-overridable value (Kant 1963, 120–121). More recent deontologies, owing much to Kant, use different terms again. Thus Joseph Boyle argues in terms of "absolute rules," whereas Robert Nozick invokes inviolable "side-constraints" on action.

D. Application: Preimplantation Genetic Diagnosis

Developed as a technique of assisted reproduction in the last few years, preimplantation genetic diagnosis begins from *in vitro* fertilization (or IVF). Several of a woman's eggs are fertilized outside the body by sperm. Molecular genetic techniques are then used to ascertain any genetic defects in the resulting embryos ("genetic defect" is of course open to interpretation, and there is the scientific potential to identify *any* genetic characteristic). Any single embryo deemed acceptable or desirable is then implanted into the womb. The rest are destroyed or used for experimentation if the law permits (the whole procedure is illegal in some countries, and there are often limits on how long an embryo can be preserved for any or even no use).

One can object to this practice by arguing that such use of embryos mistakes their value. Such value is often taken to derive from human life as such, or from natural or divine law. Such law or value is often regarded as absolute in the deontological sense given above. If the embryo is a human life, and if human life is a final or absolute value, the destruction of such life involved in preimplantation genetic diagnosis irrevocably condemns it (clearly the first, much-debated "if" is as important as the second).

A consequentialist outlook is more ambivalent. It might condemn this reproductive technique, or at least some uses of it, on the grounds of the future, potential happiness of persons who could develop from embryos, and/or from the distress the practice induces in some sections of society. (These are typical grounds because most consequentialists are utilitarians about the Good.) Alternatively, a consequentialist might countenance the technique given sufficient benefit to medical science and hence the population, or on the basis of the assumed unhappiness resulting from genetically related conditions (or partly resulting from the conditions—it is often pointed out that society plays its part in such suffering).

These considerations begin to show how modern normative ethics would address this reproductive technology. A more in-depth discussion of the science of preimplantation genetic diagnosis and the range of considerations that can be thought morally relevant has been made by Draper and Chadwick.

III. GROUNDS OR LIMITS FOR WHAT IS VALUED FOR ITSELF—METAETHICS

A. Introduction

While normative ethics debates what is of intrinsic value in the senses of moral or final value and how to act toward it, a further notion of intrinsic value underlies these issues. (For simplicity's sake, treatment henceforward is confined to moral value.) This is objective value—value that holds independently of practices of valuation—and the philosophy of such value is a specialty of metaethics. However, the examination to be pursued here perhaps works more often on the cusp between metaethics and normative ethics.

The theories to be considered here show that what has intrinsic value in the objective sense can force a rethinking of moral value. The first and most wholly metaethical approach I consider is a skepticism about objective values as such, implying a rejection of certain candidates or grounds for moral value. The second owes to the animal rights/welfare/liberation movement, and the third to a radical type of environmental philosophy. By examining the nature of objective value, these latter two positions mean to force reconsideration of the possible location of moral value. The practical implications of the relocation of moral value are considered in terms of xenotransplantation and genetic modification.

B. The Skeptical Challenge

The philosopher J. L. Mackie provides a clear case of a metaethical challenge to the content of normative ethics. "There are no objective values," Mackie bluntly puts it (1977, 15). His understanding of such values matches their construal herein: objective values would be values that held wholly independent of the actions of valuers.

Mackie's central weapon against objective value is the somewhat infamous "argument from queerness": "[I]f there were objective values, then they would be entities or qualities or relations of a very strange sort, utterly different from anything else in the universe" (p. 38). On Mackie's view, a truly objective value would resemble one of Plato's heavenly Forms, and we lack grounds to believe in such strange metaphysical entities.

Mackie also expresses his denial of objective values as the denial of "something in the fabric of the world that validated certain kinds of concern" (p. 22). This suggests the possible implications of his position—which he also called "moral scepticism"—for normative ethics. His position entails that it is philosophically misguided to take certain things as morally valuable or warranting certain valuations. Thus if one considers natural law or the sanctity of human life as at root something more than human creations or concerns, they must go the same way as Platonic forms—on the brute grounds that they do not exist.

Mackie was not a moral skeptic in the sense that he thought his conclusions undermining of morality as such. He meant to show that much of what we morally value could be built upon other, human-centered foundations (this approach is outlined in parts two and three of his book). Following Mackie, the degree of revisionism such a metaethics requires of one's ethics has been at stake in a variety of increasingly sophisticated construals and denials of objective value. The approaches to be considered next represent positions on objective value with a particularly partial bent.

First, however, Mackie should be put in context. Seminal for the debate over the existence of objective values is Hume's case that statements of fact and statements of value are separated by a gulf (see Book III of his *Treatise of Human Nature*). David McNaughton has written on this and its influence on the subsequent metaethical debate in *Moral Vision*, but omits recent attempts at rather different, resolutely "unqueer" or non-Platonic construals of objective value. Alasdair MacIntyre has addressed such in his influential *After Virtue*.

C. The Interests of Animals

I turn now to Peter Singer, one of a number of recent philosophers to rethink intrinsic value with a particular eye to the situation of animals. Singer's thinking turns on a certain construal of interests as that which alone has primary moral value, and this appeal can be understood as being based on the objectivity of such interests.

According to Singer, some entity has interests insofar as it is sentient: insofar, that is, as it can "suffer or experience enjoyment or happiness," taking these terms in their widest possible senses (1993, 58). Now certainly, "the strength of the evidence for a capacity to feel pain diminishes" down the evolutionary scale, as, roughly, does that for the degree of suffering possible (1991, 171). Humans and some of the higher mammals have an increased capacity for suffering and happiness—for interests—because they can anticipate, remember, and more widely consider their lives.

On Singer's thinking, this does not change the fundamental point. All and any interests that are there to be considered are to be taken qua degree of happiness or suffering they represent equally with those of humans. As this implies, Singer is also a consequentialist. As a thorough-going utilitarian, he wants the consideration of interests to be made as quantitatively as possible. The moral course is to maximize the aggregate satisfaction of interests. If then, as Singer thinks is the case, eating and experimenting upon animals causes them more harm than the good it does us, we should not do those things.

Singer does not tend to argue for the interest criterion directly. Rather, he aims to rebut alternative criteria for moral value that try to admit humans but disenfranchise sentient animals. Singer points out that such criteria typically concern rationality or autonomy. For example, it is often held that moral value pertains to rational or autonomous creatures alone. Singer argues that such criteria logically end up excluding not only animals but also the insane, those with Alzheimer's, and the newborn. He adds that the moral status of such people cannot be saved simply by virtue of their being human. This would be "speciesism": moral discrimination without a relevant—one might say "objective"—moral difference.

Singer's interest criterion can be challenged in a variety of ways. Aiming to oppose his denial of any special moral distinction between humans and animals, one could take on the challenge to find a foundational value possessed by the former but not by the latter without disenfranchising any human beings and without the arbitrariness of "speciesism." If one could find such a value, Singer could be most fully opposed by construing it deontologically, for if humans have values that are not to be violated (perhaps, as absolute values, *never* to be violated), then basically human value is to be protected irrespective of the cost to animals.

Alternatively, one might reject the interest criterion and attribute deontological values to both humans *and* animals. This is Tom Regan's stance. He criticizes Singer's interest criterion on the grounds that now or in the future it may be that "speciesist" practices produce the most happiness overall. Regan argues instead that "if humans have rights, so do many animals," and that animals are therefore inviolable individuals as much as humans are (in Zimmerman *et al.* 1998, 51).

D. Ecophilosophy

One of the fiercest battlegrounds for competing notions of objective value lies within the relatively new discipline of environmental philosophy. Rejecting the "anthropocentrism" of many traditional positions on value, the type of environmental philosophy sometimes known as "ecophilosophy" seeks to identify bearers or grounds of objective value beyond the human. In this endeavor there is common ground with Singer and Regan, treated above.

Indeed, a survey of environmental ethics identifies "the sentience approach" as one of three major camps endorsing non-anthropocentric objective value. As do Singer and Regan in their different ways, the environmental sentience approach identifies objective value as inhering in sentient life or experience as such. It is the other camps, more radical in their location of value, that will concern us here.

The other two camps can be identified as the "life" and "holistic integrity" approaches. Many differing versions and refinements of these ecophilosophies have been and are being developed; I will consider some relatively classic examples.

Kenneth Goodpaster concludes an argument for the "life" approach as follows (in Zimmerman *et al.*, 65):

> The truth seems to be that the "interests" that nonsentient beings share with sentient beings (over and against "mere things") are far more plausible as criteria of considerability than the "interests" that sentient beings share (over and against "mindless creatures") . . .. [P]sychological or hedonic capacities seem unnecessarily sophisticated when it comes to the minimal conditions for something's deserving to be valued for its own sake.

"Interests" is written in quotation marks because, on Goodpaster's view, it is the needs, or more loosely defined wants, of all forms of life, including those of trees and grass—to grow—that are the locus of value. So this construal of interests as co-extensive with life goes beyond the idea that objective value inheres solely in entities that *experience*. Given this, it is not only those holding metaethics like Mackie's that see here a transgression of the limits to the sensible imputation of objective value (see, for instance, Tom Regan's "Does Environmental Ethics Rest on a Mistake?" in *The Monist* 1992).

Goodpaster is in fact prepared to go further, locating value beyond what we might term "entities" at all. Thus he entertains the idea that "the biosystem as a whole" could be counted as a living thing, and so have moral standing (in Zimmerman *et al.*, 68). This idea tends toward the "holistic integrity" approach, which owes its first formulation in environmental philosophy to Aldo Leopold. Leopold held that "A thing is right when it tends to preserve the integrity, stability, and beauty of the biotic community. It is wrong when it tends otherwise" (p. 99). This conception of objective value means that the locus of moral value is not so much life as an aggregate ("the biotic"), but certain of its "configurations."

The holistic integrity position faces more objections than the life approach. It requires that a relevant transorganism entity or configuration be suitably isolatable as that thing or configuration; but it has been doubted whether, for instance, ecosystems form discrete enough entities. The idea of holistic integrity may also be premised on the contentious notion of an "organic whole": a whole the value of which is irreducible to the value of its parts (cf. O'Neill's "Varieties of Intrinsic Value" in *The Monist* 1992). Lastly, certain versions of this approach—particularly "deep ecology"—have been associated with psychological and spiritual tenets about human relations to nature that many dispute.

This compressed survey hardly does justice to the range and complexity of ecophilosophical approaches. Zimmerman *et al.*'s entry in the Bibliography provides a grounding in the approaches outlined here, and includes positions I have had to omit. Foremost among these are deep ecology (probably a version of the holistic integrity approach) and "ecofeminism," the distinguishing feature of which is the charge of a specifically male character and bias in Western evaluations of nature.

E. Applications: Xenotransplantation and Genetic Modification

Divorcing objective value from the human clearly has implications for the assessment of many new technologies. First, consider xenotransplantation, which is organ or tissue transplantation from animals to humans. A report by the Nuffield Council on Bioethics recognizes that were the charge of speciesism or something like it to be accepted unreservedly, the majority ethical approach to xenotransplantation would fall, for whatever the specifics of the normative reasoning, conventional metaethics compares the harm to the donors and experimental subjects with the benefit to receivers in a heavily loaded way.

The Nuffield report also recognizes that xenotransplantation will be appraised differently by "differing views about the relationship of human beings to nature in general" (Nuffield 1996, Sec. 4.19). Indeed, given that xenotransplantation can involve genetic engineering (to improve the functionality of animal parts in humans, or to ensure basic compatibility), ecophilosophies can take a particularly strong line here.

Such philosophies can take a stance on this technology even in cases where nothing that most would call

"interests" are immediately, or as yet, at stake (although most would admit that some sort of interests are involved in "harvesting" animal parts). Thus the life approach would be able to pass a verdict on genetically engineered food, possibly on genetically engineered seeds, and maybe even on modified microorganisms, in terms of the good of these organisms themselves.

The holistic approach can go even further, particularly with regard to transgenic organisms (transgenic organisms, roughly, being organisms that have genes from outside their normal genetic "pool" which is largely confined to the organism's own species). Such modification seems a prime candidate for assessment in terms of the integrity and stability of the biosphere or some other transorganism entity.

There is a surprising lack of work on the applications of ecophilosophy to modern genetic technology. A contentious account of the potential of genetic modification to remake life on earth is provided by Jeremy Rifkin's *Biotech Century*.

IV. ARE VALUES BEING INSTRUMENTALIZED? THE CRITIQUE OF INSTRUMENTAL REASON

A. Introduction

The idea that values are being "instrumentalized" will here be developed from two sources. These are the "critique of instrumental reason" developed by the early Frankfurt School of Critical Theory and philosopher Martin Heidegger's strikingly similar notion of "calculative thinking." This section will present these ideas in turn, and then bring out how they relate to intrinsic and instrumental value.

Space does not permit a discussion of the wider conceptual genealogy of instrumental reason. Note however that the seeds of the notion lie primarily in the work of the sociologist Max Weber. Moreover, the critique of instrumental reason is still being pursued, in a somewhat different manner again, by contemporary Critical Theorists. Finally, note that the similarities between Heidegger and the Frankfurt School—between whom in general there was no love lost—are only now beginning to be explored.

B. The Frankfurt School's Critique

The central text of the Frankfurt School for our purposes is *Dialectic of Enlightenment*. This was written during the Second World War by Theodor Adorno and Max Horkheimer (the latter coining the phrase "critique of instrumental reason"). A highly unusual work, *Dialectic* is best approached as presenting a certain history of reason.

The opening consideration of the work is the ideal of rationality championed by Francis Bacon. Bacon urged that human endeavor be harnessed to comprehending and thereby mastering nature. Caveats on religion aside, this end was for Bacon the sole purpose of human reason, holding out the possibility of a society almost free from want, unhappiness, and limitation.

Adorno and Horkheimer take it that this conception of reason comes to self-consciousness in the 18th century Enlightenment. They trace it forward, also, to modern times, and, audaciously, back to Homer's Greece. Clearer than the historical forms of this "dialectic" and its exact causality, however, is the general schema of "instrumental reason." Through its historical excursuses and other analyses, *Dialectic* depicts how a wide range of ends or values, concerning knowledge, morality, health, and culture, become increasingly construed in terms of the rationality expressed by Bacon. Ends or values, that is, are becoming interpreted or bound up with the over-riding aim of increasing our power over the natural world and thereby over the human condition.

Yet Adorno and Horkheimer do not think this the advent of utopia. On the contrary, they thought that by the time of their writing, such "rationalization" had progressed in such a way that such ideals had become almost nothing but mechanisms or masks for power and self-preservation. The "triumph of subjective rationality," they write, "is paid for by the obedient subjection of reason to what is directly given" (1979, 26). This rather gnomic sentence intends the following thesis. When reason is orientated in the way described, the power of humanity to change its world for the better is also in fact impoverished. This is a simplified rendering; there will be occasion below to see further what is meant by this diagnosis of the dark side of Enlightenment.

C. Heidegger on Calculative Thinking

Heidegger's account of the worldview of late industrialist modernity presents a perspective similar to that of *Dialectic of Enlightenment*. This account can be presented under Heidegger's label of "calculative thinking" (the term is from Heidegger's *Principle of Reason*). Limiting the treatment to modernity as such—like the Critical Theorists, Heidegger traces the outlook much further back—one can turn to the essay "The Question Concerning Technology" to ascertain what "calculative thinking" consists of.

Heidegger's essay considers the river Rhine and a hydroelectric power plant upon it. The power plant relates to the river in a special way. As a source of power, the Rhine is "ordered to stand by, to be immediately at hand." It becomes, Heidegger claims, "a standing-reserve." To take the river as a standing-reserve is firstly to relate to it in such a way that one cannot apprehend it otherwise. Second, this situation obtains to such a degree and in such a way that "whatever stands by in the sense of standing-reserve no longer stands over against us an object" (1997, 16–17). The Rhine is no longer a river in the landscape, or not in the same way as it was before.

Heidegger's point is to be understood by way of how, prior to the plant, the Rhine and its land has been integrated into human life and meaning. A farmer might in some way have related to the Rhine as a *sustainer*. And the river, perhaps by way of a bridge across it, might have been understood as *placing and defining* one region in relation to another. So, all too roughly: as embedded in such understandings and the practices that go with them, the Rhine upheld them as the specific understandings and practices they were. As a standing-reserve it cannot do so: the river's transformation is—or is caused by—the collapse of possible meanings. What remains is the river as a source of power.

This state of affairs and the technology which brought it about expresses what Heidegger takes as the modern worldview of "calculative thinking." Indeed, the "Question Concerning Technology" suggests that this "worldview" affects the very way in which the world is "revealed" or "disclosed" to us. As another text puts it, the world is disclosed such that "everything can be overseen and everything remains calculable and manageable in terms of utility."

D. The Instrumentalization of Value

The above exposition suggests that both the *Dialectic* and Heidegger's argument present a core model of the instrumentalization of value—where "value" can be taken in any of its "intrinsic" senses as identified in this article. The hypothesis is that, under the aegis of "instrumental reason" (Critical Theory) or "calculative thinking" (Heidegger), (1) that which was previously valued noninstrumentally (e.g. the Rhine or nature in general; knowledge, morality) is now valued only instrumentally, and (2) the noninstrumental values that remain resemble what were previously merely instrumental (e.g. power, "utility" in general).

This formulation, however, elides the notion of disclosure or experience in favor of that of value—or even to conflate the two. Indeed, Heidegger and Adorno were rather allergic to the concept of value as such. They saw in it already a reification or, as Heidegger puts it, "in a certain manner a loss of being" (1977, 142). All too briefly: to separate a value from the practices that give it sense renders it ripe for dissolution into a means or into another, common denominator value. Thus one might argue that theory should abandon *any* sense of intrinsic value, as one environmental philosopher has urged for his own field (A. Weston in *The Monist* 1992).

For criticism of the instrumentalization thesis and its background, see Part III of Andrew Feenburg's, *Questioning Technology* and Lectures V–VI of Jürgen Habermas's *Philosophical Discourse of Modernity*.

E. Application: Functional Food

The critique of instrumental reason finds application and concretion in the phenomenon of functional food. A "functional food" or "nutraceutical"—taking these as equivalent—is generally defined as a food designed or produced to confer specific health or metabolic benefits. The definition will prove instructively problematic.

The line between food selected in a diet for a specific benefit, for instance, oranges or olive oil, and a more *designed* product aiming at the same end—a nutritionally enhanced orange or low-fat margarine—seems vague. Further, the definition does not determine what is to be excluded for reason of not counting as food. Are vitamin tablets "functional food," medicine, or something else?

Although the definition might be sharpened by stipulation, the difficulties of specification suggest a route to a more essential, potentially critical, understanding of functional food. This route begins by nothing that all food is functional in the sense that it functions to keep the consumer alive and (in the case of most foods) healthy. This is one end, purpose, or meaning of food, and in an obvious sense the fundamental one. But this function normally coexists with many other meanings of food.

Some of these meanings have been summarised by Elizabeth Telfer in *Food for Thought*. Since food has a "literally vital importance," giving and sharing food can be a symbol of trust and interdependency, and mark a "generosity of spending time." It can be part of the marking of an occasion or the sharing of traditions (a food festival, Passover). Choosing, providing, or sharing food can be an expression of explicit values (vegetarianism, boycotts). Thus the food *is* in part the symbol of familiarity, the celebration, the tradition, and the presentation of a value.

Normally then, food and eating "are what they are and often many other things as well" (Telfer 1996, 37). What about functional food? Obviously it can be of "vital importance"—by intention even more so than normal food. This may enhance the degree to which the provision of functional food can represent a friendly (even essential) effort. The situation of functional food in the other contexts is more ambiguous. If, for example,

functional ingredients have obviated or lowered the necessity of preparation, might one spend less time on a meal, and so be less generous? Perhaps, but of course one can make an effort—"make a meal of it"—in other ways. Indeed, if it is the spirit that counts, one can celebrate even with a nutritional pill. And with practices such as Passover, it is presumably the symbolism that counts, not the particular foodstuffs.

Yet all of this might be taken as confirming the emerging point, a point in line with the critique of instrumental reason: The more foods are valued, understood, or disclosed in terms of their "literally vital importance," the more has to be done to sustain their other meanings. Functional or other "technological" food may have a tendency, that is, toward making food a mere somatic resource, and as such of purely instrumental value. One could quarrel with this conclusion on the grounds that there might be, or are already, "designer" or "lifestyle" foods that impart new meanings to food. It is unclear, however, if these are of the same order as the meanings brought out by Telfer. Little work, however, has yet been done in this area.

Bibliography

Adorno, T., and Horkheimer, M. (1944/1979). *Dialectic of Enlightenment* (trans. by J. Cumming). London: Verso.

Aristotle (1976). *Ethics,* rev. ed. (trans. by H. Trednnick). London: Penguin.

Baron, M., Pettit, P., and Slote, M. (1997). *Three Methods of Ethics: A Debate.* Oxford: Blackwell.

Bentham, J. (1780/1962). An introduction to the principles of morals and legislation. In *The Works of Jeremy Bentham,* Vol. 1. New York: Russell and Russell.

Draper, H., and Chadwick, R. (1999). Beware! Preimplantation diagnosis may solve some old problems but it also raises new ones. *Journal of Medical Ethics* 25, 114–120.

Fox, W. (1996). A critical overview of environmental ethics. In *World Futures,* Vol. 46, pp. 1–21. Amsterdam: Overseas Publishers Association.

Heidegger, M. (1977). *The Question Concerning Technology and Other Essays* (trans. and introduction by W. Lovitt). New York: Harper and Row.

Kant, I. (1963). *Lectures on Ethics* (L. W. Beck, Ed.). New York: Harper & Row.

Mackie, J. L. (1977). *Ethics: Inventing Right and Wrong.* London: Penguin.

The Monist (April, 1992). "Special topic: The Intrinsic Value of Nature," 75, no. 2.

Nuffield Council on Bioethics (1996). *Animal-to-Human Transplants. The Ethics of Xenotransplantation.*

Singer, P. (1991). *Animal Liberation,* 2nd ed. London: Harper Collins.

Singer, P. (1993). *Practical Ethics,* 2nd ed. Cambridge: Cambridge Univ. Press.

Telfer, E. (1996). *Food for Thought: Philosophy and Food.* London/ New York: Routledge.

Zimmerman, M. E., Callicott, J. B., Sessions, G., Warren, K. J., and Clark, J., Eds. (1998). *Environmental Philosophy: From Animal Rights to Radical Ecology,* 2nd ed. New Jersey: Prentice Hall.

Life, Concept of

MAURIZIO MORI

Center for Research in Politics and Ethics, Milan

GLOSSARY

efficient causality An Aristotelian concept indicating the relation in which an event is brought about by another one occurring earlier (or in some case at the same time), as is usual in physical sciences.

holistic causality A relationship in which an event is brought about not by an earlier event but by its position in a complex whole. In this sense some events depend on the action of an organism as a whole, and this fact is not reducible to any sort of efficient causality.

iatromechanics A school of biology and medicine which flourished in the 17th and 18th centuries and according to which biological phenomena should be explained on the basis of mechanical principles.

reism (or "thing-ism") A commonsensical view according to which reality is made up of mere "things", i.e., spatiotemporal entities delimited by sharp boundaries, and "things" are derived by fusion with other "things."

teleology A basic property typical of living processes, which show a peculiarly goal-directed tendency, so that it appears as if some former events are "pulled" from events occurring later. Also the philosophical study of manifestations of design or purpose in living phenomena.

LIFE is a very varied phenomenon, so instead of speaking of *the* concept of life, it would be more adequate to speak of various concepts of life. In order to make clear such concepts we need to contrast life and nonliving matter, and capture some basic empirical properties which are typical of life in itself. These are commonly recognized to include spontaneity, self-regulation, reproduction, and reactivity and adaptability (within some limits). Since all these properties are present in protoplasmatic activity, life is defined as protoplasmatic activity. There are different theories trying to account for living phenomena. Vitalism and mechanism are two opposing ones, but both in some sense are inadequate. Organicism seems to be more adequate because it can account for the typical teleology manifested by living processes and for different levels of life depending on different degrees of organizational complexity. From an organicismic conception follow some practical implications, such as the distinction between biological life and biographical life, and furthermore some issues concerning the "beginning of life" are clarified.

I. INTRODUCTORY GENERAL DISTINCTIONS

When a person with ethical interests speaks of "life," very likely such a term is used to mean "human life," i.e., the sort of life that you, the reader, have while reading this entry. In this sense, "life" usually refers mainly to mental or intellectual states that a person has simply by existing, and for this reason when we say, "that person is very lively," we mean he or she is witty and has many ideas; similar expressions are used for groups of people. In common usage, "life" mostly indicates a peculiar property which is typical of human persons, and this use is justified by the fact that in our Western culture there is a strong tendency to believe that humans are at the top of the world and radically different from the rest of nature, so that we distinguish between "us" (human persons) and the remaining part of the world, living and nonliving alike. Even if we share with other living beings many characteristics, we tend to assume that our life is so special and peculiar that we forget our common organic basis, and other living beings appear to us mostly as nonliving entities. We assume that our life transcends that of other living entities, and this makes a crucial difference. In fact, from a normative viewpoint our treatment of living beings is similar to that of inorganic entities, since both are at our disposal and we can do with them what we wish. On the other hand, we think that human persons deserve strong protection and that their lives have very high value.

In the last three decades or so, the common meaning of the term has been blurred under the influence of new practical controversies such as those over abortion and brain death, and more recently *in vitro* fertilization. A "pro-*life* movement" has spread throughout the world, and "life" nowadays assumes in some contexts a more biological connotation. Often discussions are focused on the question of when life begins or when life ends, or similar ones that assume "life" to be equivalent to *human* life, and expressions such as "sanctity of life" or "sacredness of life" have become common.

It is not clear what the precise meaning of the term "life" is in such contexts, but it is clear that these controversies compel us to develop a more rigorous and precise concept of life. In the past this problem was merely theoretical and speculative, but nowadays recent scientific and technological advancements require a clearer view on the subject: science and technology work as a sort of magnifying glass which show that such a problem is more complex than we thought, so that a more adequate concept is needed.

When we attempt to carry out this task, the first problem we have to face is whether (human) "life" is a *descriptive* or an *evaluative* notion, i.e., whether we can get a clear idea of life merely by looking at some phenomena, or whether our perspective of these phenomena depends ultimately on our previous values—or thirdly, whether in this case facts and values are inextri-

cably bound so that it is almost impossible to distinguish between them clearly. Much of contemporary debate on the concept of (human) life concerns this controversy, and it is not certain that it can be definitely solved. But certainly the controversy can be dealt with in a better way once we have a clearer concept of what life is and of its peculiar characteristics. For this reason it is useful to start from scratch, and this procedure seems useful since when we depart from the top of the scale of living beings (i.e., from the life of "persons"), the evaluative controversies are less prominent and therefore it is easier to reach a clearer view of the whole subject.

II. EMPIRICAL CHARACTERISTICS OF LIFE

The general concept of "life" has a crucial role in our thinking since it sets the distinction between the living and the nonliving, which is a crucial category of our worldview. In order to make clear the concept it is useful to start with a simple empirical remark: life *in general* (in a strict sense of the term) does not exist since we always see specific pieces of *living matter*. "Life" is a term denoting an abstract entity which, as such, does not exist because in reality there are only concrete and specific living entities. Therefore we must be aware that the term "life" indicates a characteristic or a set of characteristics which are typical of living bodies or organisms. Life is always *individual* in the sense that it occurs in a strictly delimited spatiotemporal place and with a specific form, deriving from another entity by means of specific generation. Life is always *specific* and *particular* since there is no undifferentiated "general living plasm," but always a specific form of life. It has been estimated that on Earth there are about 3,000,000,000 species, of which only a third are known, showing the great variety of living beings in the world. However, we should not forget that the amount of living material is but a fraction of that of inorganic matter—a thin pellicle about 1/1,000,000,000 of the planet's weight.

Looking at living things, we can detect some general features which certainly are constituents of our common concept of "life." The first visible characteristic is that living matter is in a state of constant change which is directed from within: new matter is assimilated and other matter is excreted in a process called "metabolism," so that a sort of internal equilibrium is maintained. This is what is referred to by the latin expression, *vita in motu* (life consists of movement). Some authors point out that such constant change has a peculiar character since it is spontaneous, and that spontaneity of continuous change is a basic property of life. According to other latin dicta, *vita in motu spontaneo* or *vita in motu ab intrinseco* (life consists of spontaneous movement or of movement from within). In this sense, the entire essence of life consists of being a sort of self-starting process.

There is a second visible aspect of life to be considered: living matter is not only self-starting, but it is also able to maintain an *internal equilibrium* by means of a self-regulating process so that the equilibrium depends on self-regulation produced by the whole entity. It is as if the whole living matter controls its own internal equilibrium, which is a function of the whole. In this case a special sort of causality seems to be at work, since it is the whole that determines the effects of the part and its working. For this reason, sometimes we speak of a holistic causality in which the behavior of a part is controlled by the whole and its nature (and not by antecedent conditions of parts as in the *linear* causality typical of physical sciences).

Holistic causality appears clearly in the normal working of self-regulation of a living organism, but it is even clearer in the process of growth. Consider a simple example such as a watermelon growing on a vine: if such a growth consisted only of a number of cell divisions, then the watermelon would turn out to be more or less spherical. But at a certain point, the watermelon "stops" cell division in some parts in order to reach its correct shape. Moreover, at some other point the cells differentiate in function, so that those that are in the center become seeds and those at the border become the rind: apparently this process of differentiation depends on the position occupied by the cell and not on something internal to the cell. Finally, after a period of constant growth the watermelon concludes its growing and after some time decay takes place. So we have a simple case illustrating self-regulation depending on holistic causality, which seems to be typical of living phenomena.

The third visible aspect is that living entities at some time may give origins to another living being similar to them: we say that living things reproduce, while inorganic matter does not. There are several modes of reproduction, and some seem quite peculiar, but reproduction is another basic characteristic of life. Any living entity comes from another, and grows for a certain period and then dies out, completing its life cycle. Not only is any single individual in constant change, but its group is as well since it is possible that some variation occurs through reproduction, resulting in a change in the life of the group.

The fourth (and last) visible aspect of life is the ability of react and adapt to the environment, at least up to a point. Life presents a peculiar inner capacity to react to external stimuli. This characteristic is quite remarkable in animal life, where such a response is different not only from the predictable way in which certain chemicals react with other chemicals, but also because, especially in higher organisms, such reactions may be learned.

When all of the four preceding characteristics are present, we are sure that there is a living thing. But each property may be present in different degrees, and in some cases one may be absent or manifest only at a certain time. Thus it may become problematic to decide whether a given entity is living or not. Let us consider, for instance, a piece of matter presenting only one, two, or three of the four listed characteristics: is it alive or not? Do we need all four characteristics together or only some of them? And if only some, then how many?

III. PROBLEMS OF A PLAUSIBLE DEFINITION OF "LIFE"

The answers to the previous questions are interesting for a number of reasons, the first of which is their relevance for a possible solution to the biological controversy about whether viruses are living entities. A virus is a collective term denoting entities with quite different natures ranging from giant protein molecules, such as the tobacco-mosaic virus, to sets of molecules approaching a bacterial structure, such as the agent causing spotted fever. However, viruses have the property of multiplying by division: for instance, if a plant is inoculated with a few hundred molecules of crystallized tobacco-mosaic virus, we observe an enormous increase of the virus substance throughout the plant. Since viruses are capable of reproducing (covariant reduplication), we may ask whether they are alive or not.

Whichever the answer, this shows that the line between the living and the nonliving is not always as sharp and clear as sometimes we commonly think: we have to realize that between the most complex phenomena of inorganic matter and the simplest ones of organic matter there is no clear-cut boundary. Does this mean that such a distinction is nonexistent and that it is simply an illusion? Certainly not, at least so far as we think that there is a distinction between black and white, even if in certain occasions they merge into gray and it is difficult to draw a clear distinction and say when black ends and white begins. In one sense, it is impossible to discover such a line, because it is somewhat "arbitrary" depending on a decision: we decide that up to a certain point it is "black," and after that it is "gray" and then "white." Something similar happens in the case of "life": it is clear that a piece of iron is different from a cat, but at a certain point the living merges into the nonliving and here we need a definition which is not "arbitrary" and is supported by good reasons showing its alleged adequateness to our task.

For our biological knowledge, an adequate definition

of life seems to require more than mere covariant re-duplication and to include the ability to carry out primary syntheses typical of protoplasm, which is a feature peculiar to a cell as a whole. In fact, one dictionary of biology says that life is "the sum of the properties of protoplasm, namely metabolism, growth, irritability, movement and reproduction as manifested by a cell or a group of cells or an organism by which is distinguished from inorganic or nonliving matter" (E. B. Steen, 1971. *Dictionary of Biology*. p. 278 New York: Barnes & Noble).

I think that there are good reasons to accept this definition, which is adequate as far as we consider life present on our planet, but do we really want to accept it as *the* definition of "life" as such? Let's grant that we limit the word "life" to the sum of the properties of protoplasm and to protoplasmatic activity: what to say then of the possibility of extraterrestrial life? Should we think that varieties of protoplasm are widespread over the universe or can we think of different forms of life in which some combinations of carbon are unecessary? Moreover, what to say of disembodied life? If life is by definition protoplasmatic activity, should we say that a "disembodied life" (i.e., not connected with protoplasm) is simply a self-contradiction? How can we think of a life that is not connected with protoplasm if we assume that it is a property of protoplasm?

A precise and definite answer to such questions is beyond the scope of this entry, but they are important for anyone thinking on various concepts of "life." However, in the rest of this entry we will accept the biological definition of life as protoplasmatic activity, which seems quite uncontroversial among scholars. However, at this point new problems come in if we want to deepen our concept of life. We may ask, "what makes protoplasmatic activity so special and peculiar?" According to one respected biologist, "protoplasm is a bridge anchored at one end in the simple stuff of chemistry and physics, but at the other reaching far across into the mysterious dominions of the human spirit" (E. W. Sinnott, 1950. *Cell and Psyche* (p. 19). New York: Harpor Torch-books), and we may ask whether it is possible to explain protoplasmatic activity. Or should we take it as something unexplicable? N. Bohr seemed to hold such a view, writing that

> the existence of life must be considered as an elementary fact that cannot be explained, but must be taken as a starting point in biology, in a similar way as the quantum of action, which appears as an irrational element from the point of view of classical mechanical physics, taken together with the existence of elementary particles, forms the foundation of atomic physics. (N. Bohr, 1993. "Light and Life", *Nature*, 131 p. 421)

On the other hand, many biologists think that life can (at least in principle) be explained according to the laws of physical nature, and therefore life is nothing so special, being nothing else but a more complicated form of inorganic matter.

IV. ALTERNATIVE THEORIES OF LIFE: VITALISM VS. MECHANISM

Here we face an important philosophical issue which is crucial for different concepts of life and also for the place that we assign to humankind, since we ourselves are living organisms. Such a controversy, which is often labeled as *mechanism* vs. *vitalism,* was already known in ancient Greek where atomists held a form of mechanism according to which life is a kind of self-moving matter, while Platonists defended a sort of vitalism posing a sharp distinction between matter and life. Aristotle tried, in a sense, to reconcile the opposition, observing that any object is constituted by "matter" and "form," and living entities are alive because they have a special form. Therefore, life is not sharply distinguished from matter, since there is a continuous improvement before that matter is prepared to accept its proper form. This holds for the distinction between inorganic matter and life, as well as for various kinds of life.

The Aristotelian view was forcefully challenged in the 17th century by founders of the new science and modern philosophy. Descartes was one of those critics and formulated the theory of the *bête machine* (beast-machine), according to which animals (as well as any form of life) were interpreted as machines of a very complicated kind. Living organisms are certainly more complex than anything else, but in principle they are comparable with man-made machines whose actions are governed by the laws of physics. This is a basic tenet of mechanism: the same laws of physics govern the inorganic world and living entities.

The application of methods of physical sciences to life led to enormous successes, and Descartes initiated the so-called *iatromechanic* school of biology and medicine, which tried to explain the function of muscles and bones, the movement of blood, and other phenomena on the basis of mechanical principles. Harvey's discovery of the circulation of blood looked as a first strong confirmation of iatromechanical views, marking the beginning of modern physiology. According to Descartes, however, man was the only living entity not submitted to the laws of nature, since he assumed (indeed he claimed to have proved) that men are endowed with a spiritual soul capable of free will, a feature sufficient to distinguish human beings from the rest of nature.

Vitalists retorted that apart from man (a unique and very special case in the whole creation), the activity of living things could not be explained by means of the laws of physics simply because organic structures and functions are purposeful and life is clearly a purposeful activity. In other words, life is radically different from nonliving matter because living processes are goal-directed or teleological (from *telos*, goal). In order to make clear this concept, we may consider the following figure:

The arrow t indicates a sequence of time in which event E″ is to be explained. According to our physical model this can be done only by means of "efficient causality," i.e., showing that there is a temporally antecedent event E′ that is causing E″ ("efficient causality" being a sort of "linear causality"). However, in teleological explanation the situation seems to be reversed, since it is the subsequent event E‴ which appears to be "causing" the antecedent event E″, as if E‴ would be "pulling from the future." In biology such teleological explanations are rather common, as when we explain that the heart pulses in order to circulate the blood: in this case the heart's beating seems to be caused by an event (circulation of the blood) which temporally comes after the fact to be explained. In this sense, living processes seem to show a sort of "intention" similar to our conscious intentions, and this fact was interpreted as a confirmation that life is not a mere product of a blind chance (but had to be produced by a supernatural mind). This view stressing that purposeful activity typical of living beings makes them radically different from nonliving matter seemed to find strong support from the fact that organic compounds, which are characteristic of living entities, in nature are found exclusively in living beings and could be produced only in life processes.

This controversy was ongoing when in 1828 Woehler produced urea in the laboratory, the first organic compound ever synthesized, issuing a stroke to vitalism and an important point in support of mechanism. Since then organic chemistry and biochemistry have become prominent fields in modern science and many organic compounds are nowadays synthesized regularly and rather easily. A further point in favor of mechanism came from Darwin's theory of evolution, according to which the origin of purposiveness in the living world can be explained on the basis of chance variations and natural selection, so that the analogy with conscious intentions becomes senseless, as well as the idea of a purposive supernatural agent. For these reasons, in the second half of the 19th century mechanism was triumphant and gained credibility.

Mechanistic biology was at its highest point at the close of the century when Hans Driesch performed his classical experiment which led him to reject the physico-chemical theory of life (even if he was one of the founders of developmental mechanics, the science devoted to experimental investigation of embryonic development). Driesch divided into two halves a sea urchin's fertilized egg at the beginning of its development, and he realized that each half gave origin to a full organism. This fact was flatly contrary to physical laws, from which one should expect that from half an embryo only half an animal would develop. But these physical laws had been violated not only in the indicated sense, but also in the reverse sense, since under certain conditions two different germs could unify themselves into one organism and produce a unitary giant larva. Therefore, Driesch drew the conclusion that "in the embryo, and similarly in other vital phenomena, a factor is active which is fundamentally different from all physico-chemical forces, and which directs events in anticipation of the goal" (L. von Bertalanffy, 1952. *Problems of life* (p. 6). New York: Harper Torchbooks), and Driesch called this special factor, which "carries the goal within itself" so to produce a typical organism in normal as well as in experimentally disturbed situations, entelechy.

Once again we see that it was teleology that distinguished the living from the nonliving, even if it is important to remember that starting from such a remark vitalists concluded that (being goal directed) life has a special creativity and spontaneity which places it outside the realm of any natural science. For this reason life is a sort of peculiar élan vital (or will to live) which cannot be captured by science but only by a special intuition: science looks for regularity and universal laws while life is creative and always changing, and therefore it is unexplicable in scientific terms.

Here we individuate two very different and opposite concepts of life: for mechanism life is nothing else than more complicated inorganic matter, while for vitalism life is a unique phenomenon which can never be generalized because of its spontaneity. However, it is useful to reflect a bit deeper on the vitalist claim in light of an example. If you tap your cat with a stick, you have a range of possible responses: sometimes the cat runs away, other times it is pleased and plays with the stick, and sometimes it responds aggressively. Many more reactions are possible, but for our task those listed are enough to show that there is a variability of response, and our first problem is to know whether such responses are lawful or not. Vitalists deny that there is any lawfulness and regularity, claiming that living organisms have a typical spontaneity that creates novelties violating any law. One can agree that the cat's reaction is hardly explicable with a physical law, but it is difficult to deny that the cat's behavior is not lawful. We can admit that living organisms have a greater range of variability in their responses that exceed our present ability to grasp them, but it is difficult to claim that they cannot be studied scientifically: if such a scientific study was possible, then it would show new regularities and laws which are different from current physical laws, but which are still interesting and significant. This view holds that we have to realize that biology has its own laws, which for the time being are different from those of other natural sciences (mainly physics and chemistry).

This view is compatible with that version of mechanism which is ready to recognize that up to now biology has not been reduced to physical sciences, even if it is *in principle* reducible to it: a first step toward reaching a better unifying theory is to start with a rigorous inquiry of different specific laws in various fields. If this were the core of the controversy, we could obtain a peaceful agreement, but unfortunately it is a stronger view which is held, pointing out that not only are biological laws empirically different from physical laws, but that they are intrinsically different and irreducible to them. In other words, biology is a science which is different from physics.

V. A THIRD THEORY OF LIFE: ORGANICISM

This view is usually supported by holders of the *organismic conception* who want to suggest a third way between the Scylla of vitalism (which is against any scientific inquiry of life which is per se resistent to science) and the Cariddis of mechanism (which has a unique and too narrow view of science, assuming that physics is the model of scientific thinking). According to organicism we must abandom the analytical and summative conception typical of physics, according to which any phenomenon can be explained by splitting it into its elementary units and then recombining these units into the original unit. This method cannot work in biology because organisms are whole in themselves and cannot be split into parts (because they die). As Bertalanffy says,

> it is impossible to resolve the phenomena of life completely into elementary units; for each individual part and each individual event depends not only on condition within itself, but also to a greater or lesser extent on the conditions within the *whole,* or within superordinate units of which it is a part. (1952, 12)

Moreover, the whole presents properties which are absent from its isolated parts, and this fact shows that what is crucial in life is organization. As Bertalanffy again points out,

> the problem of life is that of *organization.* As long as we single out individual phenomena we do not discover any fundamental difference between the living and the non-living.... There is no "living substance".... Rather life is bound to individualized

and organized systems, the destruction of which puts an end to it. (1952, 12–13)

Having stressed the relevance of organization for life, the organicismic view changes quite significantly our common sense concept of life itself because we now have to realize that "life" is a general term indicating a great variety of living matter which may be very different according to various kinds of organization. Apart from those authors who assume that "organization" is a primitive term incapable of any further analysis (probably collapsing in a version of vitalism), usually organization has to do with *relations* between the constituents of a living entity, and it becomes immediately clear that more complex organization will give rise to higher forms of life. Therefore, organicism holds that there are different *levels* of life, corresponding to different complexities of organization, and that higher levels are irreducible to lower levels, just as living phenomena are irreducible to mere inorganic matter. Moreover, any level presents properties typical of the kind of organization involved, and in higher levels there is an *emergence* of new properties. In a sense, here the concept of "organization" seems to substitute for the older Aristotelian concept of "form": for Aristotle any entity was identified by its form, while now different kinds of life are identified by their proper "organization."

In any case, organicism gives an important contribution to our analysis of the concept of life, since in this view life is irreducible to mere inorganic matter and is an organized process pursuing self-maintenance according to the devices proper to the level of complexity manifested by the system. Therefore, life indicates a quite complex and variegated set of phenomena deserving more specific attention, and at this point we have to specify other properties in order to reach the precision needed by the subject.

VI. BASIC THEORETICAL PROPERTIES OF LIFE

When we consider "life" in a more accurate way (as we need to), we have to realize that living processes show the following basic properties: teleology, organization, and various levels of existence corresponding to degrees of organizational complexity.

While both mechanism and vitalism are monochromatic, or "flat," in the sense that both see life as a single phenomenon with only one dimension, even if in opposite directions (respectively reduced to matter and to spirit), organicism is polychromatic and more varied, allowing a more subtle analysis of concepts of life. Particularly interesting for our purposes is the doctrine of various levels of life, comprising three basic levels: (1) the level of the cell, which is the simple system capable of autonomous life; (2) the level of multicellular organizations forming living individuals; and (3) the level of communities consisting of many multicellular individual organisms.

Let us examine each level separately, since the higher presuppose the lower, but present some properties that are irreducible to the lower. In this sense, we can individuate various concepts of life, so that on the one hand we can easily explain the great multiplicity of the living forms we see over the Earth, and on the other we can hope to better understand the problem I pointed out at the opening, concerning whether the concept of (human) life is descriptive or normative (or inextricably both).

The cellular level is certainly the fundamental unit of life since all living beings are constituted by some combination of cells. From a biological viewpoint, characteristic of this level is the fact that

> the organization of the protoplasm is not static but dynamic. The primary orderliness of processes cannot be attributed to pre-established structural conditions. Rather the process as a whole carries its order in itself, representing a self-regulating steady state. Therefore, the system appears to be widely tolerant of disturbances, so long as fixed structural conditions are scarce. (Bertalanffy, 1952, 34)

From our viewpoint, which is more interested in conceptual analysis, it may be remarked that typical of the cellular level is the fact that life is characterized by spatiotemporal contiguity limited by sharp borders. This is clear when we consider a colony of several single cells or an agglomeration of cells: in these cases the entity looks like an individual because its shape is spatiotemporally limited, but in reality there are several individuals, each performing the whole of its functions. If the larger shape is split, no death occurs but a simple scission of different individuals: a worm is a good example, since we all know that if a worm is split it continues to live.

The situation changes significantly at the next level, because the life of multicellular organisms is characterized not only by spatiotemporal contiguity, but also by a *relation* of strict subordination of the parts to the whole, so that a part cannot live detached from the whole. For this reason, multicellular organisms show a *hierarchical order* according to which the different parts of the individual differentiate themselves in order to perform special functions. There is a strong specialization of cells to form different organs, so that a multicellular organism maintains its general goal (self-preservation) by a cooperation of different subsystems, each having a more limited and specific teleology.

This remark is important because the death of a multicellular organism occurs when such a relation breaks down, so that the whole individual falls apart, even if some part may continue to have residual life:

for a certain time there are metabolic activities in the parts even if the whole teleological process has already vanished. On the other hand, the life of a multicellular organism depends on the presence of this relation of strict subordination of the parts to the whole, where such a subordination can be of different kinds according to the degree of complexity of the organism. In this sense, since Aristotle's time it has been common to distinguish three great genera (or reigns) of being, i.e., the vegetal, the animal, and the human (in the strict sense of being capable of intellectual activity). Needless to say that the dividing line between each class is not sharp and linear, but in general the vegetal level is typical of plant life, which is characterized by relatively little specialization, as shown by the fact that we can take a part of smaller branch of a tree and get from it a new sample of the tree. Vegetal life presents only few functions beside metabolism, namely, growth and reproduction. When we pass to the second level, we find animal life, which is much more complex and specialized, at least in higher species such as mammals, which have a central nervous system and sexual reproduction. In this case there is movement and sentience (i.e., the capacity to feel pain and pleasure), and a capacity for learning special responses to stimuli. At the top of this scale there is intellectual life, which according to the Western tradition is reserved only for human persons who are able not only to move rapidly and to feel pain and pleasure, but also to have symbolic activity, abstract thinking, and a complex language in order to communicate efficiently.

VII. PRACTICAL IMPLICATIONS

Having distinguished three differnt levels of life (and their corresponding concepts) of multicellular organisms, we can approach our original question concerning the nature of the concept of "human life" (whether it is a descriptive or a normative concept). First of all, it is of interest to point out that in Western traditions it is usual to invest great value only in *intellectual life,* and attribute little or no value to other forms of life (vegetal and even animal). This is clear if we consider that according to most legislations living organisms of lower kinds are considered as mere "things," i.e., something at disposal of the human person so that any behavior toward it need not be justified. Moreover, it is important to keep in mind that human life includes all the lower levels of organization, and it is important to state the relations among them. This is a source of problems, mainly because technological advancements can keep alive some lower subsystems of the human multicellular organism while the highest are already destroyed or dead. In this case we may ask whether the relation of strict subordination of the parts to the whole is still present or has already disappeared, and if the answer is in the second sense, there is problem with our right conduct. Similar problems arise at the other end of the spectrum, since *in vitro* fertilization allows us to create new human fertilized cells, so making practically relevant (and not only a mere abstract speculation) the question of whether in the very first cells, the relation of strict subordination is required. Apparently the answer seems to be in the negative, since there is no adequate organization, but certainly there is quite a controversy on the issue.

For these reasons, many contemporary authors draw a sharp distinction between *biological life* and *biographical life,* where the latter is equivalent to intellectual life since it presupposes mental states, memories, etc. On the other hand, mere "biological life" is metabolic activity occurring in a human organism which has lost or not yet acquired higher functions because of the absence of the whole brain or significant parts of it. Once accepting such a distinction, its practical consequences depend on whether or not one accepts the already mentioned assumption typical of Western societies that tends to not attribute value to lower levels of life. Those favoring it are inclined to say that mere biological life in itself has little value, even if this does not mean that such a life is valuable because it matters to other people or because (as in the case of the embryo) it matters to a future person.

Others strongly criticize the distinction between biological and biographical life, holding that human life must be always protected in itself, because it is always a great gift. However, this view either introduces a version of vitalism, by saying that "human life" has a special value only because of its intrinsic constitution (independent of its actual organization), or such a view uses the expression "human life" in a wider sense which includes in its meaning also the transmission of life. These two different aspects are of course strongly connected, since if human life had an intrinsic value, then it is plausible to think that also its transmission should deserve a very special care. Since for centuries reproduction and sexuality were considered a sort of sacred field, this distinction did not emerge; but after the "sexual revolution" of the 1960s, diffusion of contraception, and a sort of "secularization" of sexuality (seen as a normal function of the person), no special value seems to be conferred upon transmission of life by most people, and the idea of an intrinsic value of human life as such appears to be outmoded.

However, this general thesis has been reworded, saying that from conception human life ought to be endowed with peculiar value because either the zygote is already a person or conception is the only "nonarbitrary" point concerning the beginning of "human life." There is not enough space for a detailed examination of these arguments here, but from our viewpoint both these positions seem to presuppose an unstated assumption which can conveniently be called "reism" (or "thing-ism"), i.e., the view that reality is made up of "things" which come into being by fusion with other "things" (spatiotemporal entities delimited by sharp borders). So, for instance, when some flour (a "thing") is mixed with some milk and eggs (another "thing"), we have a new compound (a new "thing") which can assume various shapes without any significant (substantial) change. This frame of reasoning is typical of common sense, and for this reason it is familiar: when such a frame is applied to reproduction we are led to think that fertilization is the crucial event. At conception (or fertilization) two separate "things" (the gametes) fuse themselves together, giving origin to a new "thing" (the zygote), and this event seems to point to a "natural" and "nonarbitrary" watershed within a more complex process.

Assuming the reistic viewpoint we can easily understand the attractiveness of the idea that human life begins at conception, but also we understand its basic fault: reism is a commonsensical way of thinking and cannot be used in an alleged scientific explanation of a sophisticated biological process such as reproduction. In other words, reism is wrong and misleading because it introduces without any ado or qualifications a mere commonsensical assumption within a biological theoretical context. Since biology is nowadays a mature science with its own theory, when we face reproduction and we want to explain it, we have to abandon mere common sense (and its reistic assumptions), operate a sort of new "Copernican revolution," and accept a more adequate biological way of thinking. It then becomes clear that in biology there are no "things," but rather processes which are goal-directed and more or less complex. Therefore the question of when life begins is clearly misleading, and a more appropriate one is, "when does the reproductive process begin?"

Once we assume such a biological way of thinking, it is clear on the one hand that conception is just *one* stage (and not the crucial one) within the more complex process of reproduction, which has its own teleology starting with the formation of the gametes and ending at birth, and on the other hand that throughout reproduction the human living process becomes more and more complex, passing various stages before reaching human intellectual life, so that we have to recognize that such a concept is a descriptive one. At the beginning of this entry, I pointed out that many authors hold that such a concept is an inextricable mixture of descriptions and evaluations, and now I have pointed out the reasons for which I think it is a descriptive concept, even if we can confer evaluative meaning on the whole reproductive process or to some special stage.

A good reason to attribute a special value to the transmission of human life can be found in the fact that reproduction is basic for the continuation of life and maintainance of societies and social life. In this sense, we reach the last level of biological organizations, that of superindividual organizations, i.e., societies. This level is crucial since every organism is begotten by its like, and many organisms beget new organisms themselves. In this sense, each living being is part of a larger stream of life, and the hierarchical structure of life does not end up with the organism itself, but may concern various organisms and a part of society. Looking at societies, we have to understand the specific role of each individual. There are many sorts of societies in which each individual has a specific role, as happens in societies of insects, ants, and bees, where animals of different castes (workers, soldiers, reproductors, etc.) appear to be subordinate "organs" working for the whole society.

In higher organisms, however, individuals assume a stronger position and we might have conflicts between the claims of the group and the claims of single individuals. In human society social life receives special atten-tion, and there are opposite theories explaining the nature of social life. There are two major schools: communitarians seem to think that social life has its own criteria, and that single individuals should see themselves as part of the whole instead of social atoms detached from the vital rest. On the other end, individualists deny most of what communitarians believe, holding that human societies are the result of single individuals, and that individual claims come first. However, the concept of *social life* is quite different compared to the others examined in this entry, and must be postponed to another occasion.

Bibliography

Allen, G. (1978). *Life science in the twentieth century.* Cambridge: Cambridge Univ. Press.

Beckner, M. (1959). *The biological way of thinking.* New York: Columbia, Univ. Press.

Coleman, W. (1977). *Biology in the nineteenth century.* Cambridge: Cambridge Univ. Press.

Dobzhansky, T. (1969). *The biology of ultimate concern.* London: Rapp and Whiting.

Matthen, M. & Linsky, B. (Eds.) (1988). *Philosophy and biology. Canadian Journal of Philosophy,* **14** (Suppl.).

Mori, M. (1996). *Aborto e morale.* Milano: le Saggiatore.

Sober, E. (1993). *Philosophy of biology.* London: Oxford Univ. Press.

Waddington, C. H. (1962). *The nature of life.* New York: Atheneum.

Medical Futility

LORETTA M. KOPELMAN

East Carolina University School of Medicine

GLOSSARY

absolutely futile Something lacking any instrumental value for achieving some goal or purpose, or as a means to some end.

Best-Interests Standard The standard that, all other things being equal, directs us to select from among options what most informed, rational people of good would regard as maximizing net benefits and minimizing net harms.

contested cases of futile or useful treatments These are (a) grounded in medical science, (b) value laden, (c) at or near the threshold of utility, and (d) burdensome.

instrumental value Something useful or important for achieving some goal or purpose, or as a means to some end.

intrinsic value Something esteemed or important in and of itself.

major medical goals Traditional medical purposes include enhancing people's well-being through opportunities of prevention, diagnosis, or treatment of dis-ease, pain, death, or disability, or by restoration of function.

presumably or virtually futile Something almost certainly lacking instrumental value for achieving some goal or purpose, or as a means to some end, because of its extreme implausibility or repeated failure.

qualitatively futile Something that may achieve some goal or purpose, but one that is unimportant, being neither instrumentally nor intrinsically good.

MEDICAL FUTILITY is a topic that involves judgments about the futility or utility of medical care. These assessments influence whether people can obtain interventions allowing them to live, die, or flourish. Consequently, deeply felt views may surface among patient, doctors, nurses, and family members about what treatments are useful or futile. These disputes raise issues about the clinician–patient relationship, resource allocation, informed consent, communication, empathy, relief of suffering, autonomy, compassionate care, and our duties to people with a very poor quality of life.

After examining what is meant by medical futility in the contested cases, we consider four important ways for resolving these disputes. As we shall see, some favor giving authority in the contested cases (a) to doctors (the physician autonomy model); (b) to patients or their representatives (the patient or surrogate autonomy model); or (c) to what most people in a society approve (the social approval model). Still others reject giving final authority to any one group, favoring (d) a consensus forged from many sources including consideration

The Concise Encyclopedia of the Ethics of New Technologies

of clinicians' views, established liberties, social needs, and how most people want to be treated (the overlapping consensus model). In the last section, we consider how our disagreement about what is medically useful or futile may result from incommensurate moral convictions about our duties to incompetent people who have a very poor quality of life. The Best-Interests Standard has been used to resolve disputes over medical futility when the patient is incompetent. Critics charge, however, that the Best-Interests Standard should be abandoned because it: (a) is not objective, (b) may be abused, (c) denies hope, (d) violates the duty to respect the sanctity of life, (e) is unknowable, or (f) is too individualistic. Each of these objections seems problematic, and asking what is best for the incompetent person remains a valuable way to help resolve some disputes about medical futility.

I. THE MEANING OF MEDICAL FUTILITY

If we ask if something is futile, we need to inquire, "Futile for what?" Things can be useful or futile as means to different ends. To be in one's home during the last days of life may be futile from the perspective of prolonging life, yet useful for giving comfort. Thus, depending upon one's goal, something's utility or futility may change. Medicine has various important goals, such as preserving life and relieving suffering. Our disputes may arise from how we rank these values when they conflict. Some hold that prolonging biological life, even permanently unconscious life, is intrinsically valuable and always the most important goal in medicine. Most physicians disagree, maintaining that this stance tends to give too little attention to people's autonomy, suffering, or quality of a life (K. Payne *et al.,* 1996. *Ann. Internal Med.* **125,** 104–110; L. M. Kopelman, A. Kopelman, and T. Irons, 1992. In *Compelled Compassion* (A. L. Caplan, R. H. Blank, and J. C. Merrick, Eds.), pp. 237–266. The Humana Press, Totowa, NJ).

A series of court cases revealed sharp controversies among clinicians, patients, or families about what is medically useful or futile (L. M. Kopelman, 1995. *J. Med. Philos.* **20,** 109–121). These disputed cases include that of Nancy Cruzan, a woman in a coma or persistent vegetative state (PVS) whose family wanted her feeding tube removed; Baby Doe, an infant with Down's Syndrome whose parents did not want surgery to correct anomalies incompatible with life (Kopelman *et al.,* 1992); Baby L, with multiple disabilities including blindness, deafness, quadriplegia, and arrested development at the 3-month-old level whose mother insisted on maximal treatments; Helga Wanglie, a woman in a PVS whose family said she wanted maximal treatment; and

Baby K, an infant with anencephaly (lacking a higher brain) whose mother insisted on maximal treatments; and Baby J, a severely mentally and physically disabled infant whose doctors refused treatments the parents wanted for their infant (F. H. Miller, 1994. *Lancet* **343,** 1584–1585). Examining these contested cases shows that disputes about medical futility are characterized by four features (Kopelman, 1995).

A. Grounded in Medical Science

The first feature these and other contested cases have in common is that they are *grounded in medical science.* These court cases about medically futile treatments arise in the context of standard medical care, so stable scientific information must justify claims about patients' diagnoses, treatments, or prognoses. Reliable information, for example, must support assertions that procedures will fulfill established medical goals such as prolonging life, restoring sentience, or relieving pain. Research and evidence helps distinguish which treatments are useful and which are futile in treating certain conditions. Clinicians should use the best available information to determine when treatments are ideal, standard, innovative, experimental, unverified, or utterly futile. As more information becomes available, our views sometimes change. Earlier disputes about the use of frontal lobotomies to treat severe psychiatric disorders, for example, were resolved with greater information that this procedure caused far more harm than good. Judgments about the utility or futility of medical therapies in the contested cases, then, must be supported by evidence and modified as the relevant evidence changes.

B. Value Laden

Judgments of medical futility, as used in these contested cases, are value-laden judgments incorporating estimates about something's utility relative to some goal, or whether the achievable goal is worth the effort. One person may value life at any price and want maximal therapy, while another believes there are worse things than death. In addition, both professionals and surrogates have many duties to patients, and these express values of respect, fairness, and beneficence (to do good deeds or prevent harm). Yet patients, surrogates, and clinicians may disagree about when therapies are useful or what quality of life should be supported. Disputed cases usually raise troubling questions about how to understand our duties or values, including what constitutes a benefit, and thus whether something is a genuine therapy for a certain condition. These contested cases also arise from concerns about how to fulfill traditional and sometimes conflicting medical goals such as relieving suffering, honoring patients' requests, prolonging life, and being compassionate.

C. At or Near the Threshold

Third, the disputed court cases are *near the borderline between what is considered useless and what is considered beneficial*. Treatments in the contested cases are neither absolutely nor presumably futile (like orthopedic shoes to treat PVS), or those requesting them would not be taken seriously. The dispute is often a matter of whether proposed interventions are qualitatively futile, or whether they achieve an important goal. For example, Baby L's mother understood her daughter would be blind, deaf, quadriplegic, and developmentally arrested at several months of age, but believed that it was paramount to keep her baby alive. In contrast, Baby L's doctors and nurses objected that these painful, life-saving treatments were futile, given her poor prognosis, as well as causing her suffering. Some also mentioned the costs of providing treatments with such dubious, or marginal, benefits.

The problem of distinguishing between the useful and the futile cannot be solved by a different cutoff, since no matter what threshold is selected, there will always be borderline cases. Moreover, the judgment that a treatment is useful or futile is often an amalgam of assessments of different burdens and benefits as measured on different parameters. For example, what is useful to relieve suffering may compromise respiratory function, and what is futile to prolong life for more than a few days may be useful so a family may gather for a death. Consequently, to say that there is no duty to provide futile treatments is an unhelpful guide in the borderline cases because what is contested is whether they are futile.

D. Burdensome

Fourth, these contested cases typically involve *physically, psychologically, or financially burdensome treatments*. If the treatment were merely futile, but not painful or costly, it might simply be provided. Suppose a patient mistakenly believes that he will be cured by getting a daily multivitamin pill. Honoring such simple requests might be very useful, but not because the vitamin itself is considered beneficial by clinicians. Rather, its utility might consist of respecting the patient's desires and lifting his spirits. In contrast, in the contested cases, treatments are generally painful, costly, or rare. Maximal treatment for someone in a persistent vegetative state, such as the previously mentioned cases of Nancy Cruzan and Helga Wanglie, are expensive. In the Cruzan case, the family thought maintaining someone in such a state was wrong, but the doctors insisted upon it. In the Wanglie case, however, the family demanded maximal treatment, and the doctors objected. Some not only opposed the expense of such treatments, but also believed that requiring such support imposed unreasonable suffering, costs, or emotional burden on caregivers or families.

Having clarified what we mean by medical futility, we can see why disputes about medical futility will continue as long as people disagree about patient's needs, resource allocation, how to rank values, or what to do in the borderline cases. Consequently, we turn now to consider alternative proposals for resolving these disputes fairly.

II. PROPOSALS FOR RESOLVING DISPUTES OVER MEDICAL FUTILITY

Ideally, health care choices should represent a socially acceptable consensus among doctors, nurses, and patients or their surrogates about what treatments are reasonable and best suited to the patient. This ideal cannot be fulfilled when, as in the contested cases, people cannot agree about the relevant facts, resource allocation, how to rank values, or what to do in the borderline cases. Several important proposals offer alternative ways of balancing professional norms and social needs with people's rights of self-determination. They give authority in the contested cases (a) to doctors (the physician authority model), (b) to patients or their representatives (the patient or surrogate autonomy model), (c) to what most people in a society approve (the social approval model), or (d) to an overlapping consensus forged from many sources including clinician views, liberties, social needs, and how most people want to be treated (the overlapping consensus model).

A. The Physician Autonomy Model

According to the physician autonomy model, physicians should have unilateral authority to judge what treatments are futile in the contested cases. For example, if a therapy is deemed *qualitatively futile,* or has failed consistently in analogous circumstances, doctors should decide unilaterally not to recommend it as an option and refuse the treatment if requested. There are, however, two very different versions of this view.

A *reductionist version* of the physician autonomy model holds that doctors should have unilateral control of decisions in the contested cases because these disputes over futility are reducible to factual claims in their area of expertise. Decisions about what ought to be done in the contested cases, on the reductionist version of the physician autonomy model, are definable by information about probabilities. If clinicians are experts about when something is quantitatively or absolutely futile, they reason, they should have unilateral authority to resolve disputes in the contested cases.

One difficulty for the reductionist model is that some cases are contested *because* there have been few cases reported from which to draw conclusions. A more fundamental logical problem concerns the difficulty of getting a conclusion about what *ought* to be done from premises stating what *is* the case. The reductionist model suggests that we can draw a moral conclusion about what we ought to do when treatments are minimally beneficial, from probability claims. This is very questionable. The next version of the physician autonomy model avoids the reductionist presuppositions that decisions about what ought to be done in the contested cases can be reduced to factual claims alone.

A professional equipoise version of the physician autonomy model also favors unilateral decision making by physicians but does not claim to eliminate values. Rather it relies upon established values within the medical profession. It acknowledges that some physicians make bad decisions about what is useful or futile, but focuses upon the recommendations of the community of physicians. Clinicians generally agree about what information is relevant and the proper goals of medicine, and this equipoise can offer a stable base for judgments about which treatments are useful or futile. The professional equipoise version of the physician autonomy model avoids the problem we encountered with the reductionist version of this model since it does not attempt to reduce moral decisions to probability claims. It employs values, namely the time-tested traditional goals of medicine to decide how clinicians ought to act. For example, most physicians do not regard maintaining someone in a PVS to be a goal of medicine (Payne et al., 1996). According to the professional equipoise version of the physician autonomy model, doctors should have socially granted authority to refuse unilaterally to do what they consider futile.

This approach to decision-making authority has received social support in countries such as the United Kingdom. For example, *In re J*, the Court of Appeals in London supported the doctors' refusal to treat an infant who was mentally and physically disabled. Lord Balcombe wrote,

> I find it difficult to conceive of a situation where it would be a proper exercise of jurisdiction to make an order positively requiring a doctor to adopt a particular course of treatment unless the doctor himself or herself were asking the court to make such an order. (Miller, 1994, 1585)

The case of Baby K was treated very differently by a Federal Court of Appeals in Virginia. This court rules that doctors had a duty to provide respiratory support to an infant with anencephaly who came to the emergency department despite an explicit Virginia law giving physicians the ability "to refuse to provide treatment that the physician considers medically or ethically inappropriate." The ruling states,

> We recognize the dilemma facing physicians who are requested to provide treatment they consider morally and ethically inappropriate, but we cannot ignore the plain language of the statute [the Emergency Medical Treatment and Active Labor Act] because to do so would transcend our judicial function. (*In re Baby K* 16 F.3d 590 (4th Cir. 1994))

Critics argue that the physician autonomy model, in either form, promotes unjustified paternalism by allowing doctors to control information a reasonable person might want to make treatment decisions (D. Callahan, 1991. *Hastings Center Rep.* **21**, 30–35; R. M. Veatch and C. M. Spicer, 1992. *Am. J. Law Med.* **18**, 15–36). Medical paternalism may threaten the rights of people who, without pertinent information, have no means to exercise self-determination or to protect themselves, or their loved ones, from controversial or unjustified views. The doctrine of informed consent was developed because many people sought more information about their choices and wanted to participate in decisions about their treatments. Defenders of the physician autonomy model point out, however, that society relies upon doctors to make many unilateral decisions, including about which options should be presented to patients. Doctors must use discretion about who gains admission to the hospital, or to treatments, or what treatment options to discuss. Consequently, the charges that the physician autonomy model promotes undue paternalism seems to defenders to be misguided. The doctor's duty is to present the information that a reasonable person would consider material to his or her decision, and if the treatments are genuinely futile, these options should not be presented as options.

This defense by those favoring the physician autonomy model, however, is weakest precisely where there is greatest conflict: in the borderline cases. First, critics argue that the nature and scope of the moral and other values used in these contested cases raise doubts about physicians' claims of expertise to make unilateral decisions (Veatch and Spicer, 1992).

Second, the physician autonomy model presupposes that doctors generally agree about how to resolve contested cases, but this may be untrue. Baruch A. Brody and Amir Halevy examined the statements drafted by a variety of medical professional societies and found that major medical organizations express very different views and values (1995. *J. Med. Philos.* **20**, 123–144). Thus, they questioned whether a sufficient consensus exists about when treatments are medically futile to resolve properly most of the contested cases with unilateral decisions by physicians. Moreover, how would one identify the relevant doctors or community of experts in using the professional equipoise version of the physician authority model? Should the final authority be acclaimed investigators, hardworking clinicians, lay advocates, professional societies, or all of them? As we saw, the professional societies disagree about what constitutes medical futility. Moreover, if the information is so convincing that the treatment is futile, it is likely that administrators will limit physicians discretion; physicians may have very little opportunity to exercise unilateral decision making in such non-borderline cases, their role being supplanted by the administrative decision makers.

Third, agreement among physicians does not guarantee its moral justification. Physicians should agree because their position is worthy; it is not worthy simply because they agree about it. Their agreements might, for example, be wrong because they result from self-interest, bias, or ignorance. Agreement among some groups, then, cannot constitute the final moral appeal about what ought to be done. Despite these criticisms, the professional autonomy model captures the insight that doctors should play a central role in determining which treatments are medically futile.

B. Patient or Surrogate Autonomy Model

Another recommendation is that of the patient or surrogate model which favors giving patients or their representatives authority to decide what ought to be done in the contested cases. Contested cases frequently involve different ways of evaluating and weighing benefits and burdens that are often very personal. Physicians may be medical experts, but they are not experts about people's personal values about what makes life worth living or what constitutes their best interests. It is widely recognized, moreover, that competent patients have moral and legal rights to refuse treatment. When patients are incompetent to express those wishes, their surrogates often have authority to refuse treatments for them.

Patients, or their representatives, however, cannot justifiably interpret the right to refuse treatments as the right to *demand* interventions doctors regard to be futile. Yet, this is what appears to happen in the cases of Baby L and Helga Wanglie. Why do doctors provide futile therapies "on demand"? After all, patients and their representatives cannot usually demand and get such simple things as prompt service or food they like in hospitals; so how can they demand maximal, costly, and scarce treatments and get them? The answer again lies in the nature of the disputes about medical futility. In

the contested cases, what patients or surrogates demand is usually close enough to the borderline of what is considered useful that their claims have force. For example, Baby L was extremely disabled, and her mother insisted upon continuing maximal life-saving treatments. The interventions were beneficial in the sense they were keeping her alive. According to the patient or surrogate autonomy model, in the absence of a social policy about what will or will not be provided to extremely disabled infants, these choices should depend upon the patient's or family's personal values (Callahan, 1991; Veatch and Spicer, 1992).

There are problems with this view. While self-determination is an important value, it does not necessarily result in a morally defensible choice in the disputed cases. Neither scientific nor moral reasoning is just a matter of opinion. Patients and surrogates sometimes express indefensible preferences. If adopted, reliance solely on patient and surrogate direction could (1) make patients suffer needlessly; (2) make professionals act in a way they consider against their conscience and harmful to patients; (3) drive up health care costs; (4) thwart efforts to enact justifiable rationing policies; and (5) hamper triage decisions. Thus, patient and family preference cannot be the final appeal in determining the most morally justifiable position since personal opinion alone does not establish a moral, or scientific, justification.

While it seems problematic to give families, or their surrogates, total authority in these contested cases, this model captures the insight that they should have a central role in shaping these decisions. Patients and surrogates generally bear the consequences of medical decisions, and some options suit certain people better than others. Often the surrogate or family sees itself as defender of the patient in an impersonal institution. Nancy S. Jecker and Lawrence J. Schneiderman (1995) argue that people who want "everything done" rarely desire unlimited or inappropriate technical treatment (1995. *J. Med. Philos.* **20,** 145–163). More likely, they fear abandonment for themselves, or their relatives, or grieve about the anticipated loss of identity through their own death, or that of someone they love. Typically, they want to understand what is happening, and be treated with compassion and respect. We need to minimize conflicts that generate the contested cases and emphasize good communication and compassionate care for patients and families.

C. Social Approval Model

Another proposal of the social approval model is that a social agreement is a morally defensible way to settle what ought to be done in the contested cases. The will of the majority could settle which treatments will be regarded as futile or useful in a certain society or community, thus deciding for themselves how to use their own resources, given their values and priorities. Based upon what most people approve when using public funds for health care within their communities, they generate priorities and recommendations that social agreement should be the basis for resolving disputes. This model, however, would result in the loss of autonomy for both patients and doctors. Either doctors or patients may find something to be useful when the community finds it futile. For example, patients and doctors may view expensive treatments to gain another year of life for a cancer patient to be of great value, while the community using other criteria may decide it is not.

The general problem with the social approval model is that people's agreement may reflect the society's inherent prejudices, mistakes, or unjustifiable biases. A socially approved model could unfairly deny goods, benefits, or services for certain racial or religious groups, or neglect gender-specific illnesses. People might also use democratic means to decide that costly interventions should not be used to treat people with drug addictions, sports-related spinal injuries, or psychotic disorders. Disabled people also experience a great deal of unreasonable prejudice and often need special protections. The *majority*, in short, may overestimate the pain of living with disabilities, use poor information, be unjustifiably biased, or simply gamble that they will never need certain costly services. The goal in moral decision making is to establish and carry out a morally defensible policy or action, not perform the statistically most popular act.

Because social agreements can be wrong, they cannot be the final arbiter of moral disputes. Wide social agreements about policy, however, are important if one hopes to gain public support for them. So while they are not the final appeal, social agreements should play an important role in finding good solutions about how to decide contested cases of medical futility. To resolve these issues, we must also consider people's moral duties and rights, and how to allocate resources fairly.

D. The Overlapping Consensus Model

According to the overlapping consensus model, each of these three previous solutions about how to resolve disagreements over medical futility in the borderline cases contains important insights but fails as a final appeal (Kopelman, 1995). Certainly doctors have been given authority by society to make many decisions about health care, resource allocation, and what constitutes useful and futile treatments. Even if they cannot, in

the end, be unilateral decision makers in the contested cases, or claim immunity from review, their importance in finding solutions is indisputable. Moreover, the roles of the patients, families, or their representatives are also central. Competent patients, or their surrogates, have the right to refuse treatments. While they may not have the moral or legal right to demand costly treatments (or even demand better food or prompt service), they do have some rights to select between reasonable options, even in the borderline cases. Social agreement about policy is also an indispensable ingredient if one seeks public cooperation, even if it must be balanced by basic rights protecting exploited, vulnerable, or other people whom the majority might disvalue. Society must clarify guidelines for the use of public resources, as well as guard the rights and well-being of people. It also has a legitimate role in setting limits to patients' demands as well as physicians' authority.

Moral justification about what ought to be done cannot be guaranteed by giving unilateral authority to doctors, patients or their surrogates, or the majority in a society. None of these groups should be the final appeal about how we ought to guide our action because their views might result from self-interest, bias, or ignorance. Consequently, models for assessing what ought to be done based upon physician autonomy, patient or surrogate autonomy, and social approval fail as final moral appeals about what ought to be done.

Moral decision making represents an ideal which is easier to articulate than to fulfill. It requires gathering the best available and relevant information, as well as taking account of people's salient interests or welfare. This goal also requires giving and defending justifiable reasons for preferences, such as those about our duties, values, or moral principles, and using methodological ideals of clarity, impartiality, and consistency to reach conclusions. Other important, albeit fallible, considerations in making moral decisions include legal, social, and religious traditions, stable views about how to understand and rank important values, and a willingness to be sensitive to the feelings, preferences, perceptions, and rights of others.

While there are many controversies in the sciences, it is generally assumed that they can, in principle, be resolved by better methodologies, studies, data, hypotheses, or theories. In contrast, there may be a diversity of opinion in moral life that cannot be entirely resolved by these means. This does not mean rational discourse is any less important in morality than it is in science, or that science is value-free or theory-neutral. Rather, it means that reasoned differences of opinion in moral life can occur, and they may result in differences in the balance given to conflicting goods, rights, duties, goals, principles, virtues, or values. Insofar as more than one ranking of them can be justified, different views about what counts as useful or futile in these contested cases may be vindicated, or at least understandable. Even when we do not agree in our moral judgments about these contested cases, we can sometimes recognize that alternative views have merit and tolerance of diverse opinions may be appropriate. Thus, the goal in decision making and policy is to have a special kind of agreement. It is one that is morally justifiable. Agreements tend to be morally worthy when born out of a consensus among reasonable and informed people of good will about what ought to be done.

This model presupposes relatively stable agreements of a complex nature. It relies upon general understanding about when to be tolerant of diverse views and procedures for resolving disputes. It presupposes that informed people of good will want to seek morally justified judgments and actions on these matters. Rather than assigning unilateral authority to someone or some group, this view recommends consensus arising from diverse perspectives and informed by procedures and laws to protect people. This view fails if no such areas of general agreement can be reached and coercion rather than cooperation is needed to set limits. For example, a community consisting of religious fundamentalists and secular humanists might have difficulty clarifying the limits of tolerance, or how to proceed when encountering disagreement. Of course, we might equally question if they really form a community when they are unable to reach accommodations to each others important beliefs.

III. UNDERLYING MORAL DEBATES ABOUT PEOPLE'S QUALITY OF LIFE, SUFFERING, AND THE BEST-INTERESTS STANDARD

In the last sections we saw that disputes about the utility or futility of medical treatment cannot be resolved without establishing the goals for judging utility and futility and procedures for conflict resolution. The goals of medicine offer important guidance in assessing what is medically useful or futile, especially if all could agree about their relative importance. These traditional medical aims include enhancing people's well-being and opportunities through prevention, diagnosis, or treatment of disease, pain, death, or disability, or by restoration of function. One uncontroversial ranking of them, however, seems unlikely. At the heart of the futility debate are disputes about what ought to be done when important goals of medicine seem to conflict. For example, what should be done when our duties to relieve suffering and maintain life seem to conflict?

Our disagreements about which treatments are futile sometimes rest upon deeply held views about when a life is worth living and what is best for people. Consequently, they may persist despite good communication and careful policies. As we saw, in the cases of Helga Wanglie and Baby K, the families insisted, while doctors disagreed, that permanently unconscious life was worth supporting maximally. In contrast, in the case of Nancy Cruzan, the doctors held this view, while the family disagreed. Disputes about the intrinsic value of someone's quality of life, then, may underlie subsequent judgments about that person's best interest, and when treatment is futile.

Typically, competent adults can decide what treatments are in their best interest, and what quality of life they wish to support. This is called the self-determination standard. When people are faced with a choice between prolonging life and preventing great suffering, they sometimes believe there are worse things than dying. Most of us would not want to endure a mindless existence of intense and chronic pain with no prospect of improvement. Some people leave advance directives about their desires in such circumstances or designate surrogates to make decisions for them if they become incompetent. Friends and family can help inform these decisions even when no advance directive has been left and no surrogate appointed. They can have a role in determining what they believe the person would have wanted given his or her values, thereby using the substituted judgment standard.

Incompetent people, such as infants and cognitively impaired persons, who have left no directives cannot make decisions for themselves, so it is up to others to decide for them. The contested cases often arise when people are trying to make decisions for incompetent people. A long-standing and well-accepted medical and legal policy is known as the Best-Interests Standard. It relies on picking from among the options most informed and reasonable people of good will would regard as maximizing the person's overall well-being, all other things being equal. It holds that when children, or other people, cannot make choices for themselves, decision makers should try to select the action that is best, or among the best options, for that individual. Initially, surrogates should concentrate upon what is best for the individual, attempting to identify the person's immediate and long-term interests.

The Best-Interests Standard permits complex judgment about what, all other things being equal, is likely to be best for that individual. For example, the benefit of obtaining a long and healthy life would outweigh the burden of enduring intense pain for a short time. The Best-Interests Standard, however, might also be used to justify withholding or withdrawing maximal life-support treatment from an incompetent person when life has no other prospects but severe and chronic pain. The *all other things being equal* clause in the Best-Interests Standard is important because what is absolutely best for someone may not be the right thing to do. It may be best for someone to have a heart transplant, yet wrong because others have a higher claim to this scarce resource.

The Best-Interests Standard is often an important starting point for dispute resolution about medical futility for incompetent people, because most people regard it as an important guide even if they may disagree about how to apply this norm. For example, we agree that suffering is an important consideration in making treatment decisions, because we believe that we should prevent unnecessary pain. We also generally agree that, in deciding how to assess people's best interests when judging what treatments are useful or futile, we should consider how we would wish to be treated in similar circumstances. If we think it would be terrible to be maintained in certain states such as a PVS, we should not inflict this upon others. Arguably, then, we should not have a policy more burdensome for incompetent people than we would want for ourselves.

The Best-Interests Standard has been attacked. If critics succeed in discrediting this norm, we lose an important way to help resolve disputes over medical futility. It is important, then, to consider the weightiness of these objections. Critics charge that the Best-Interests Standard: (a) is not objective, (b) may be abused, (c) denies hope, (d) violates the duty to respect the sanctity of life, (e) is unknowable, or (f) is too individualistic.

A. The Standard Is Not Objective

Some object that an incompetent person's suffering or quality of life should not play a role in establishing what is best for the patient, or in decisions to withhold or withdraw care. These critics worry that families and doctors may misjudge the nature, or length, of someone's suffering, or quality of life, because they are ignorant or fearful. Moreover, decision makers, typically families and doctors, may imagine that they are considering the person's suffering and best interests when they are really worried about their own emotional, social, or financial burdens. Such considerations may not even be conscious, critics hold, but are still dangerous and misguided. These critics typically maintain that dying is a better criteria for withdrawing or withholding maximal treatment because it is an objective, factual decision based on probabilities concerning survival or physiological effects.

In evaluating this objection, we first need to distin-

guish among several meanings of the words "subjective" and "objective." Let us do so with reference to pain. First, when subjective means having some relation to the experiences of subjects, and objective means relating to objects, then pain is subjective. But this does not make it either inappropriate or unreliable as a criterion to use in judgments about someone's quality of life. We try to reduce unnecessary suffering because we believe that people's suffering is important and should be part of our deliberations about what we ought to do. In addition, we would not want to rule out subjective considerations generally. Pleasure and happiness are also subjective in the sense of being experiences of subjects, rather than properties of objects, yet we judge whether treatments should be continued when someone experiences a life with pleasure or happiness.

Thus, the subjectivity of the condition, in the sense of its being an experience of subjects, does not rule it out as an important factor in making treatment decisions for others. Competent adults regard the quality of their lives and the avoidance of suffering as appropriate components in making treatment decisions for themselves. People lacking the capacity to decide for themselves deserve equal consideration or protection.

Second, objective and subjective can be understood differently. What is objective can mean what is intersubjectively confirmable, and what is subjective is not. So understood, some statements about quality of life such as pain are objective. Most people who have been on a ventilator say that it is very unpleasant and being unable to move or speak would be a further burden. The fact that most people would not want for themselves a life of chronic pain and immobility, with no hope of improvement or personal interactions, gives objective confirmation to a judgment that some kinds of lives and experiences would be very painful. If most of us agree that something is painful, we have an objective ground for claiming that something is painful.

Third, objective can also refer to something that is factual, and subjectivity to what is not. Critics use this meaning of objective when arguing that the criterion of biological survival is "more objective" than the Best-Interests Standard in selecting treatments because it is a factual standard. These critics, however, are mistaken about the kind of judgment that they make; they are not offering factual claims, but moral judgments. They are advancing what they believe to be an appropriate goal one ought to use in making decisions about when to withhold or withdraw life-supporting treatments. This is a moral claim about how we ought to act, therefore, not a factual claim about the probability of survival or the physiological effects of procedures. As such, it is not different in kind from other judgments about what ought to be done.

B. The Standard May Be Abused

Critics like former president Ronald Reagan (1986. In *Abortion, Medicine and the Law* (J. D. Butler and D. F. Walbert, Eds.), 3rd ed., pp. 352–358. Facts on File, New York) and his surgeon general, C. Everett Koop (1989. *Hastings Center Rep.* **19**(1), 2–3), believe that the Best-Interest Standard, with its incorporation of quality-of-life considerations, is likely to be abused. For this reason, they sponsored the U.S. federal policy which has come to be called the "Baby Doe Regulations" (Kopelman *et al.,* 1992). Many physicians and others argue these regulations give insufficient recognition to suffering, thereby promoting futile and painful interventions (Kopelman, *et al.,* 1992). Are these judgments, allegedly about when it is in someone's best interest to discontinue or withhold life-sustaining treatments, more likely to be abused than other judgments of comparable complexity?

A. E. Buchanan and D. W. Brock criticize the view that quality-of-life assessments will be routinely abused (1989. *Deciding for Others.* Cambridge Univ. Press, New York). They argue that the courts and others who reject quality-of-life judgments in making decisions for incompetent people have failed to note that there are two ways in which we can understand quality-of-life judgments. Comparative quality-of-life judgments are interpersonal and based on considerations of social worth or the value of a person's life in relation to other people's lives. In contrast, noncomparative quality-of-life judgments try to consider the value of the life to the person, comparing the value of living the individual's life to having no life at all. Both kinds of judments are important, but fundamentally different, and should be made *separately.* We should separately evaluate noncomparative judgments about whether most people would want to live a certain sort of a life. We should also consider separately such matters as comparative quality of life judgments about such matters as if we approve of everyone obtaining a resource in similar circumstances.

Unfortunately, authors often fail to distinguish between these two kinds of quality-of-life judgments when they discuss criteria for withholding or withdrawing care. In failing to separate and justify these radically different considerations, critics' concerns are fueled that the Best-Interests Standard is really a subterfuge for appraisals of the emotional, financial, or social interests of others; it strengthens their convictions that quality-of-life considerations have no place in decisions to withhold or withdraw care from incompetent persons. Critics, then, have some basis for concern about the possible abuse of so-called futility judgments based upon the Best-Interests Standard and quality-of-life considerations if people do not state clearly whose quality of life they have in mind in deciding whether they should withhold or withdraw treatments. Critics are also correct that some people do not justify their claims that others will really suffer chronic pain without compensatory benefits. To determine if noncomparative quality-of-life judgments are especially open to abuse, they should be compared with other kinds of judgments of similar complexity. Critics show why these different kinds of quality-of-life judgments must be kept distinct, and that a good deal of abuse may come from failing to do this. Resource allocation issues are important, but they should not be snuck into noncomparative quality-of-life considerations.

C. The Standard Denies Hope

Critics, such as Ronald Reagan (1986), object to using the Best-Interests Standard and quality-of-life considerations to assess futility or justify withholding or withdrawing treatment because so doing inappropriately denies hope the patient may recover. On this view, unless one can be certain that a biological life cannot be prolonged, there is always a hope of recovery. Physicians have, of course, been mistaken in the past, and some charge that physicians tend to be too gloomy when making predictions. For example, we cannot be *certain* that someone will not improve or some new treatments will not come along to help.

Medicine deals with probabilities, not certainties, so it is true that one cannot be certain about any patient's prognosis. But certainty is an impractical and immoral standard to use. It is impractical because one can never be absolutely certain, short of death, that a patient will not improve; this standard would never let us stop maximal therapy until the person died. Thousands of unconscious or marginally conscious people would have to be supported, perhaps for many years, if we adopted this rule for making decisions. It is an immoral standard if it causes a great deal of unnecessary suffering.

Competent adults would balk at a rule that they were required to suffer with maximal treatment for many months because someone might possibly survive or improve. Adults consider avoiding pain and suffering relevant to their decisions when they have a very small chance for biological survival. Arguably, incompetent people also deserve protection from the kind of burdensome treatments that few adults would want.

If critics respond that they do not mean one must be *absolutely* certain before it is appropriate to make decisions to withhold or withdraw life-support treatments, they have the problem of fairness and deciding where to draw the line. Suppose, for example, that only 1% of infants in some groups would live after spending 20 very painful months in the neonatal intensive care unit. Is it fair to *require* the 99 to suffer for the sake of the 1?

How we assess hope depends upon our goals. Most adults hope to live, but they also hope to live meaningful and conscious lives without severe and intractable pain. When these goals conflict, competent adults sometimes refuse lengthening their lives by aggressive medical treatments. If justifiable, we should treat people lacking decision-making capacity as we would want to be treated, and so give protection and consideration about pain in making choices for them.

D. The Standard Is Opposed to the Duty to Respect the Sanctity of Life

Some critics, such as Koop (1989), object that respect for the sanctity of life demands that biological life must be fully supported unless people are dying and the quality of someone's life is not relevant—only then can treatments be judged futile. Problems of pain and suffering should be addressed, on this view, but not by means of withholding or withdrawing life-support treatments. Some defend this position on religious grounds, arguing that we have a God-given duty to preserve all human life when we can do so, or that God judges us by how we treat the most helpless people under our care (R. Doerflinger, 1998. *Hastings Center Rep.* **19**(1), 16–19). The sanctity-of-life argument is not always grounded in assumptions about God's will or intentions, however. Some defend it on moral grounds alone, arguing that as a community we need to preserve all human life whenever possible in order to teach and show respect for each human life (Koop, 1989; Reagan, 1986; J. Bopp, Jr., 1990. *Hastings Center Rep.* **20**(1), 42–44). "Sanctity of life" is a complex concept whose meaning and use are difficult to clarify, and there are difficulties with appealing to it as a means to fully support people unless they are dying (Clouser, 1973).

First, some use the sanctity-of-life arguments to appeal to a God-given duty to preserve human life when we can do so. But appealing to such an obligation does not tell us how to fulfill that duty. If a life is filled with pain and suffering, and the person can survive only with maximal treatment, is struggling to preserve that life with the full arsenal of medical technology fulfilling or thwarting a God-given duty? Appeals to what people think is a God-given obligation still requires reason and justification because they do not tell us what God actually thinks, but only their views about God's position. People have appealed to God's position to be for and against wars, abortion, euthanasia, and most other controversial stands. If such appeals do not give compelling reasons, they do not help us settle our disagreements about what we ought to do in cases where we disagree.

Such opinions may only harden in people's hearts the belief that they are correct, while failing to give and defend reasons for their correctness.

Second, some assume that we have a duty to save every life we can out of respect for the sanctity of life (Bopp, 1990; Doerflinger, 1989). Yet defenders of the sanctity-of-life position have to give and defend their reasons for believing that maximal care is always the appropriate care for those we are charged to help. Many people sincerely believe that a merciful God would not want someone to be forced to endure medical treatment producing only a life of suffering or minimal consciousness; they do not view death as an evil always to be opposed. Defenders of the sanctity-of-life view thus cannot assume that the appeal to sanctity of life will result in a conclusion supporting their intuitions.

Third, even if we grant that it is of great importance to us as a community to acknowledge the duty to preserve and protect human life from avoidable death, defenders of the sanctity of life need to give and defend their reasons for believing that keeping all people alive is preserving and protecting a human life from an untimely death. Many physicians and others would argue that preserving certain lives is cruel or inappropriate (Payne *et al.*, 1996; Kopelman *et al.*, 1992). Moreover, a community has important responsibilities in addition to those of mutually prolonging each other's biological life. Other important moral goals include preventing unnecessary suffering, promoting empathy, and doing to others what we would have done to us. If we do not acknowledge the suffering that others have to endure, we may lessen our empathy for others and our need to respond to it.

E. Is It Unknowable?

Robert M. Veatch has attacked the Best-Interests Standard, arguing that neither surrogates nor clinicians can be expected to select the option that will be literally best in "promoting the total well-being of the individual" (1994. *Hastings Center Rep.* **25,** 5–12). Certainly none of us know what choices are absolutely in the best interests of ourselves or others, given our biases, ignorance, and the uncertainties of life. It is unclear, however, whether the Best-Interests Standard is used in this way. Even Veatch admits that parents and other surrogates are not held "to a literal best interest standard when they make decisions for their wards" (Veatch, 1994, p. 6), but to a standard of reasonableness. For similar reasons, clinicians cannot be expected to know what is literally the total best for their patients when they make recommendations. If the standard is never meant to require acting in the absolutely best interest of someone, then his criticism does not apply to the standard *as used*. Veatch is correct that clinicians and surrogates may not know the literally best choice for someone, but is that really needed in many cases? A clinician or surrogate may not know what is in someone's total best interest, but still know it is best to tie off someone's spouting artery. Veatch has not shown that our communication, value differences, or uncertainties are so bad that his skepticism is warranted here but not in other matters such as in judging what is best for the environment, public health, schools, etc.

F. Is It Too Individualistic?

Critics argue that the Best-Interests Standard gives us an inappropriate guideline because what is in someone's best interest may be the wrong thing to do (Veatch, 1994). It cannot be right to ignore everyone else's needs and interests in allocating goals, services, or benefits. Many patients may need the same scarce resource, for example, so they all cannot get what is absolutely best. What is marginally beneficial for one individual may not make any sense from the vantage of reasonable allocation decisions. It may be best for the individual to obtain a scarce and expensive resource to extend life for a week; yet it might not be reasonable if, as a result, another whose life which could be extended for many years will be denied access to this treatment. Similarly the family may not be able to set aside every other consideration to do what is best for the sick family member. An elderly, beloved relative might live a few weeks longer if he bankrupted the family, but the family might reasonabley consider their own needs and interests.

Since it would be impossible to employ this individualistic interpretation of the Best-Interests Standard, this cannot reflect how it is used. Such a standard could not guide clinical practice as long as patients have conflicting interests, and clinicians and families have other duties. Yet it is useful to assess, *all things being equal,* what is in each patient's best interest, even when there are often other things to consider in deciding what ought to be done. We cannot ignore that patients' needs may conflict, and some comparative and interpersonal judgments or assessments about discontinuing or withholding marginal benefit treatments may have to be made. This standard is used to set goals and assess means that are reasonable and to serve as a threshold of acceptable behavior by surrogates.

Certain social structures, however, may increasingly pit doctors and families as advocates for their patients' needs and interests against health care systems that do not want to pay for treatments. Governmental agencies, third-party payers, and insurance companies interested in cost savings may set thresholds for what is futile based on what they are willing to fund, and leave doctors uncertain about how to deal with too little money for what they believe to be needed care for their patients.

IV. CONCLUSION

Disputes about medical futility may raise substantive moral questions about the meaning and value of human life. They may also demonstrate the pitfalls of poor communication and failure to take the time to clarify people's concerns, problems, feelings, beliefs, and deeply felt needs. Disagreements about which treatments are futile are likely to continue not only because resources are limited, but because informed people of good will disagree about what constitutes a genuine benefit, what ought to be done in the borderline cases, and what role people's quality of life should play in treatment decisions. When dealing with people lacking decision-making capacity, the Best-Interests Standard has been an important guide for helping people decide when treatments are futile or useful. It helps us understand the threshold of acceptable surrogate choice and to balance people's interest and needs when allocating health care benefits. Judgments about medical futility and utility often represent complex assessments of different burdens and benefits, seen and measured by people with differing interests and convictions about when a life is worth living and how they would want to be treated in similar circumstances. The decisions will continue to be momentous because overtreatment may be burdensome to patients and costly to society, yet undertreatment can compromise the rights and dignity of people seeking help.

Bibliography

Bopp, J., Jr. (1990). Choosing death for Nancy Cruzan. *Hastings Center Rep.* **20**(1), 42–44.

Brody, B. A., and Halevy, A. (1995). Is futility a futile concept? *J. Med. Philos.* **20**, 123–144.

Buchanan, A. E., and Brock, D. W. (1989). "Deciding for Others: The Ethics of Surrogate Decisionmaking." Cambridge Univ. Press, New York.

Callahan, D. (1991). Medical futility, medical necessity: The problem without a name. *Hastings Center Rep.* **21**, 30–35.

Clouser, K. D. (1973). The sanctity of life: An analysis of a concept. *Ann. Internal Med.* **78**, 119–125.

Doerflinger, R. (1989). Assisted suicide: Pro-choice or anti-life? *Hastings Center Rep.* **19**(1), 16–19.

Engelhardt, H. T., Jr. (1996). "The Foundations of Bioethics," 2nd ed. Oxford Univ. Press, New York.

Jecker, N. S., and Schneiderman, L. J. (1995). When families request that "everything possible" be done. *J. Med. Philos.* **20**, 145–163.

Koop, C. E. (1989). The challenge of definition. *Hastings Center Rep.* **19**(1), 2–3.

Kopelman, L. M. (1995). Conceptual and moral disputes about futile and useful treatments. *J. Med. Philos.* **20**, 109–121.

Kopelman, L. M., Kopelman, A., and Irons, T. (1992). Neonatologists, pediatricians and the supreme court criticize the "Baby Doe" regulations. In "Compelled Compassion" (A. L. Caplan, R. H. Blank, and J. C. Merrick, Eds.), pp. 237–266. The Humana Press, Totowa, NJ.

Miller, F. H. (1994). "Infant resuscitation, a US/UK divide. *Lancet* **343**, 1584–1585.

Payne, K., Taylor, R. M., Stocking, C., and Sachs, G. A. (1996). Physicians' attitudes about the care of patients in the persistent vegetative state: National survey. *Ann. Internal Med.* **125**, 104– 110.

Reagan, R. (1986). Abortion and the conscience of the nation. In "Abortion, Medicine and the Law" (J. D. Butler and D. F. Walbert, Eds.), 3rd ed., pp. 352–358.

Schneiderman, L. J., Jecker, N. S., and Jonson, A. R. (1990). Medical Futility: Its meaning and ethical implications. *Ann. Internal Med.* **112**(123), 949–954.

Schneiderman, L. J., and Jecker, N. (1993). Futility in practice. *Issues Law Med.* **9**, 101–102.

Veatch, R. M. (1994). Abandoning informed consent. *Hastings Center Rep.* **25**, 5–12.

Veatch, R. M., and Spicer, C. M. (1992). "Medically futile care: The role of physicians in setting limits. *Am. J. Law Med.* **18**, 15– 36.

Novel Foods

BEN MEPHAM

University of Nottingham

GLOSSARY

functional food Any modified food or food ingredient that may provide a health benefit beyond the nutrients it contains. Some such foods are described as "nutraceuticals," which may be defined as foods providing medical or health benefits, or as "medical foods" or "pharmafoods," which extend the concept more explicitly to the treatment of disease. However, the nomenclature is subject to different interpretations and has no universally acknowledged legal status.

genetically modified food A food or part of a food derived from an organism that has had its genetic material modified in a way that does not occur by natural processes. For example, genes conferring the potential for increased growth or resistance to insects or herbicides can be transferred from an organism of one species into the genome of organisms of a different species, permanently altering the properties of the new (transgenic) organism.

novel food A food is considered to be novel if it (a) has not been used as a food to a significant degree in the European Union (EU) in the past, and (b) is a food or food ingredient which falls into one of the following six categories, namely, it (1) contains or consists of genetically modified organisms (GMOs); (2) is produced from, but does not contain, GMOs; (3) has a new or intentionally modified molecular structure; (4) consists of or is derived from microorganisms, fungi, or algae; (5) consists of or is derived from plants or animals (but excluding those which are obtained by traditional practices and have a history of safe food use); and (6) has come from a novel production process which causes changes affecting the nutritional value, metabolism, or presence of undesirable substances.

substantial equivalence A term introduced in 1993, and endorsed by the UN Food and Agriculture Organization and the World Health Organization, with particular reference to products of modern biotechnology. It codifies the concept that if a food or food ingredient can be demonstrated to be essentially equivalent in composition to an existing food or food ingredient then it can be considered as safe as the conventional equivalent.

NOVEL FOODS is a term used in the European Union (EU) which refers principally, but not exclusively, to foods produced by modern biotechnology, such as those employing genetically modified (GM) microbes, plants, and animals. This article discusses ethical and legal issues associated with their (prospective) use, in both the EU and the United States of America (USA), and describes the relevant government regulatory and advisory committees.

The main focus is on ethical issues raised by the most prominent form of novel foods, GM crops. The analysis, taking account of both utilitarian and deontological criteria, considers the impact of such GM foods on the biotic environment (e.g., in terms of sustainability and biodiversity), producers (both conventional and organic farmers), and consumers (with respect to food safety and choice). Particular attention is paid to the alleged moral imperative for deployment of GM crops in addressing the problem of global hunger. Brief accounts are given of ethical issues specific to animals and functional foods.

I. INTRODUCTION

The term "novel foods" is defined by Regulation (EC) 258/97 of the European Parliament and Council, which took effect on 15 May 1997 and introduced a statutory premarket approval system for such foods throughout the European Union (EU). The other major law concerning novel foods is Council Directive 90/220/EEC, which governs the placing on the market of genetically modified organism (GMO) products that may be described as raw materials. Outside the EU different definitions and different approval systems are employed. For example, in the USA such foods are generally referred to as "foods developed by biotechnology" or, sometimes, "bioengineered foods." This article is written from a United Kingdom (UK) perspective, so that the EU definition will be adhered to; but this does not affect the nature of the ethical questions to be addressed. However, international differences in the emphases accorded to different issues will be illustrated by comparing developments in the UK (and the EU more generally) with those in the USA.

The precise definition of novel foods (see Glossary) is necessarily expressed in legal terminology but, in essence, there are two reasons for considering a food to be "novel." It may (1) have been produced from a novel source which has not been used for food production in the past, or (2) differ in composition from that previously available, e.g., because of the use of a new manufacturing process (such as that involving a new type of heat treatment or novel enzyme). Thus, in sum, novel foods include:

- Foods which consist, totally or in part, of organisms (animals, plants, or microbes) that have been genetically modified (GM), and those which have been derived from such organisms, even when there is no trace of the modified genes in the final food product
- Foods, sometimes described as "functional foods" (see Glossary), which are generally designed to confer some particular advantage, such as a health benefit, to the consumer
- Foods with both the previously mentioned characteristics, i.e., GM functional foods

II. TYPES OF NOVEL FOOD

In this article, the two principal types of novel foods, GM foods and functional foods, will be discussed separately, with most emphasis being placed on GM foods.

A. GM Foods

While several foods originating from GM plants and bacteria are in commercial production, at the time of writing there are no examples of marketed foods derived from GM animals. Nevertheless, currently much research involves GM animals, so that it is important to consider prospective applications, with a view to examining their ethical implications.

The aims of GM foods include:

- Improving yields of crops and animals
- Increasing resistance to pests and diseases, thereby reducing use of agrochemicals and veterinary medicines
- Conferring herbicide resistance, thereby reducing herbicide use
- Improving the nutritional properties of foods
- Improving food qualities (e.g., taste and storage life)

1. GM Microbes

Microorganisms have long been used as producers of foods and food ingredients, but GM microbes have the potential to increase the range of their applications extensively. However, in the UK by 2000 the only licensed food of this type was a "vegetarian" cheese produced using enzymes from GM source organisms, thus obviating the need to employ rennin extracted from calves' stomachs to curdle the milk. In the USA there are several such foods, e.g., those in which enzymes from various GM bacteria are used as food processing aids.

2. GM Crops

The most prominent forms of GM foods are those derived from transgenic crops. By 2000, 14 GM crops had been approved for growing in the EU under the Deliberate Release Directive (90/220/EEC), including various types of transgenic rapeseed, maize, and soya. However, beginning in 1998 the Deliberate Release Directive came under review because of concerns that the existing rules did not address adequately the effects of GMOs on aspects of the environment such as biodiversity. Consequently, the EC Council of Environment Ministers acted to prevent any new GM crop approvals for commercial use until a revised Directive had been introduced. This situation led to a "virtual moratorium" on GM crop approvals. By 2000, only three GM foods had been licensed for use in the UK (Table I).

In contrast, many GM crops have been grown commercially in the USA. Such crops have been genetically engineered to confer resistance to herbicides, insects, or viruses, or to alter the crops' characteristics (such as by delaying ripening or increasing oleic acid content). By 2000, almost 29 million hectares were devoted to GM crops in the USA, which represented 72% of the global total of GM crop plantings. Between 1994 and 2000, over 30 GM foods were approved for marketing in the USA (Table I).

3. GM Animals

The list of GM animals used experimentally with a view to production of food is extensive, e.g., cattle, pigs, sheep, goats, birds (such as chickens and quail), and several species of fish (e.g., salmon, trout, carp). Among the claimed objectives are increased efficiency of animal production (e.g., faster weight gain), improved quality of food products (e.g., leaner meat), disease resistance (conferring both welfare and economic advantages), and

production of novel food substances (e.g., nutraceuticals in milk).

B. Functional Foods

The concept of functional foods originated in Japan. The growing public awareness that many disease conditions, such as cardiovascular disease (CVD), diabetes, and dental caries, are diet-related has coincided with food manufacturers' search for innovative products. For example, the growth in the number of low-fat products is directly related to the perceived link between CVD and high blood cholesterol levels. Indeed, the most common types of functional foods in the UK are yogurts and "spreads" aimed at reducing blood cholesterol levels.

Generally, there is only one active ingredient, such as a bacterial culture or specific fatty acid. Others include those fortified with antioxidants, probiotics (to aid digestion), and a mixture of nutrients designed to improve sporting performance. Types of functional food are illustrated in Table II.

Certain functional foods may be derived from GM crops or animals. For example, GM soybeans, corn, and canola have been developed that produce docosahexanoic acid (DHA), which allegedly improves cardiovascular health and retards senility. Several companies are producing nutraceuticals in GM animals. For example, GM cattle have been developed that secrete lactoferrin in their milk, a protein which has anti-inflammatory and antimicrobial properties and is designed for marketing to immunocompromised patients and premature infants. Another product, bile salt stimulated lipase (BSSL), secreted in the milk of GM sheep, is classed

---

TABLE I GM Foods Licensed for Use in the UK and USA

The only foods derived from GM crops which had been licensed for use in the UK by the year 2000[a]
- GM tomato paste, produced from slower ripening fruit
- GM soybeans; used in a wide range of processed foods
- GM maize; used in a wide range of processed foods

Examples of foods derived from GM crops which had been licensed for use in the USA by the year 2000[b]
- GM tomatoes (with modified ripening and softening properties)
- GM herbicide-tolerant crops (glyphosate-tolerant soybean, cotton, corn, and canola; glufosinate-tolerant canola; bromoxynil-resistant cotton)
- GM pest-resistant crops (virus-resistant squash and papaya; insect-protected corn, potato, and cotton)
- GM vegetable oils with modified fatty acid content

[a] Source: ACNFP (1999).
[b] Source: USFDA (1998).

TABLE II Types of Functional Food

- Enhanced variants of traditional foods (e.g., with processing to improve bioavailability of nutrients)
- Disease-specific foods (sometimes called "nutraceuticals," such as those containing folic acid to prevent neurological disorders in infants)
- Risk-group-specific foods (e.g., gluten-free and nut-free foods for celiacs and allergy sufferers, respectively)
- Aging-retarding foods (e.g., containing docosahexaenoic acid, claimed to retard senility)
- Physical-performance-enhancing foods (e.g., sports diets designed to promote instant energy, optimal hydration, and electrolyte balance)
- Mental-performance-enhancing foods (e.g., those containing choline, claimed to improve short-term memory)
- Mood foods (e.g., those claimed to relieve stress through their choline content)

[a] Information derived from J. T. Winkler in Sadler and Saltmarsh (1998).

as a nutritional supplement for patients suffering from cystic fibrosis and for premature infants.

III. REGULATION OF NOVEL FOODS

Because there are minor differences in the regulation of novel foods between different EU states, arrangements in the UK are chosen to exemplify the situation in the EU and allow subsequent comparison with that in the USA.

A. UK Regulation of Novel Foods

1. Legislation

In the UK the government agency charged with assessing the safety of novel foods is the Food Standards Agency (FSA), set up in 2000, which receives advice primarily from the Advisory Committee for Novel Foods and Processes (ACNFP). The ACNFP operates in accordance with the EC Novel Foods Regulation (258/97), which establishes an EU-wide premarket approval system for all novel foods. Other committees whose advice is sought are the Committee on Toxicity of Chemicals in Food, Consumer Products and the Environment, and the Food Advisory Committee.

According to the outline safety assessment procedures laid down in Regulation 258/97, "wherever possible, the novel food is compared with an existing counterpart, which it may replace in the diet." Much reliance is placed on the concept of substantial equivalence (see Glossary). If a food is not deemed "substantially equivalent" it may need to be subjected to detailed toxicological assessment.

If a GM crop is to be grown in the UK, applications have also to be assessed by the Advisory Committee on Releases to the Environment (ACRE), which reports to the Department for the Environment, Transport, and the Regions. The aim is to prevent or minimize damage to the environment that might arise from the escape of GMOs. The controlling legislation is set out in the Environmental Protection Act (1990), which, together with the GMO (Deliberate Release) Regulations (1992), implements EC Directive 90/220.

A "White Paper on Food Safety" issued by the European Commission (EC) in January 2000 stated that there was a need to clarify procedures for authorizing the marketing of novel foods. In the future, a proposal would be presented to improve the Novel Food Regulation (258/97) in accordance with the revised framework for deliberate releases of GMOs (90/220).

2. Agricultural and Food Policy

In 1999, the UK government initiated a thorough review of developments in biotechnology, consulting 350 organizations with an interest in biotechnology and drawing on results of focus group meetings involving the general public. The results of these investigations suggested that there was little public confidence that ethical and wider social issues were adequately addressed within the existing committee system.

Consequently, the government resolved to establish a new strategic body, the Agriculture and Environment Biotechnology Commission, to advise ministers by addressing "broader issues relating to the acceptability of GM activities and relevant ethical considerations." However, the remit of the Commission specifically excludes "food," a condition apparently explained by the fact the government was simultaneously establishing the FSA (although the terms of reference of the latter do not explicitly encompass ethical concerns). It seems clear that there will be significant interaction between the two new bodies, and that ethical issues will receive greater attention.

B. U.S. Regulation of Foods Produced by Biotechnology

The USA does not have any major statutes specifically addressing biotechnology, so that in general foods derived from GMOs are regulated in the same way as food products developed by traditional plant breeding. This means that legislation relating to agricultural biotechnology falls under the jurisdiction of three regulatory bodies, the U.S. Department of Agriculture (USDA), the Food and Drug Administration (FDA), and the Environmental Protection Agency (EPA). The USDA is responsible for the planting of GM crops, the FDA for food safety (and labeling), and the EPA for establishing tolerance levels for herbicides and insecticides. In 1996, the Animal and Plant Health Inspection Service (APHIS), under the USDA, was charged with regulating the planting, distribution, and harvesting of GM crops.

Although, initially, projected field tests were judged on a case-by-case basis, in 1993 the USDA began relaxing its requirements and new rules now allow companies to field test GM varieties of six common crops (corn, cotton, potatoes, tomatoes, soybeans, and tobacco) without seeking approval from the agency. All that is required is notification 30 days in advance of planting and certification that the tests comply with approved standards. If APHIS determines that a GM plant poses no significant risks to other plants in the environment and is as safe as equivalent traditional varieties, the GM crop may be granted "nonregulated status," which permits its widespread commercialization.

With respect to food safety, the FDA published a policy statement in 1992 explaining how foods are regulated under the Federal Food, Drug, and Cosmetic Act. In some cases, where substantial differences in composition are produced by genetic modification, there is a requirement for premarket approval as food additives, and possibly for labeling to inform consumers of new attributes of the foods. However, only one novel food appears to have been scientifically assessed for safety by the FDA, the so-called "Flavr Savr" (long-life) tomato, which was evaluated at the request of its manufacturers and on the basis of "the safety and nutritional data collected by the company." Following that decision "the FDA has not found it necessary to conduct comprehensive scientific reviews of foods derived from bioengineered plants based on the attributes of these products, but consistent with the 1992 policy expects developers to consult the agency on safety and regulatory questions" (USFDA, 1999).

Thus, in all cases to date, it is considered that these novel foods are substantially similar to foods that have long been consumed, and that premarket approval by the FDA is not required. Such substances are classed as GRAS ("generally recognized as safe by acknowledged experts"). In April 1997, the FDA issued a proposed rule to establish a notification procedure "whereby any person may notify [the] FDA . . . that a particular substance is GRAS," and announced its intention to maintain an inventory of GRAS notices and its responses to them.

C. Functional Foods

For a food manufacturer to see a financial return on investments in designing and developing a functional food, it is essential to be able to make a claim concerning its health benefits. Consequently, legislation on functional foods is intrinsically linked to legislation on health claims. EC legislation (Directives 79/112 and 89/398, and the 1996 Food Labeling Regulations which implement them) prohibits medical claims, i.e., that a food prevents, treats, or cures a disease, so claims need to be made in very general terms that could not be misconstrued. The UK Food Advisory Committee is currently examining the possibility of introducing guidelines on health food claims.

In the USA, in January 2000, the FDA finalized rules for claims on dietary supplements, many of which fall into the category of functional foods. Under the Dietary Supplement Health and Education Act, dietary supplements may bear so-called structure/function claims (i.e., that the product affects the structure or function of the

body) without prior FDA review. However, they may not, without FDA review, claim to "prevent, treat, cure, mitigate or diagnose disease (a disease claim)." Thus, claims such as "maintains a healthy circulatory system" are permissible without review, but those such as "prevents osteoporosis" are not.

In Japan, where with the government's permission manufacturers are allowed to make agreed claims concerning safety and substantiation of health advantages, over 100 such products are licensed.

IV. ETHICAL CONCERNS

Currently, prospective biotechnologies are regulated by reference to assessment procedures that are designed to ensure that they deliver the claimed benefits reliably and without significant risks to people, animals, and the environment. Once these criteria have been satisfied, market forces are generally regarded as the appropriate means for addressing other issues of public concern. While in theory (e.g., in a society where there was a high level of public awareness and trust), the free market might be a satisfactory way of ensuring consumer choice and protection, several aspects of modern-day food production challenge this assumption. For example, many people have concerns about the origins of food, its means of production, its dependence on problematical technologies, and wider impacts on the welfare of consumers, animals, and the environment. These concerns cannot be assessed simply on the basis of economics and chemistry, not least because certain individuals to whom ethical concerns undoubtedly apply (e.g., juveniles; those who, for whatever reason, are unable to appreciate the nature of novel foods; and animals used in the production of novel foods) are unable to express their interests as "stakeholders." A satisfactory form of ethical assessment needs to take account of all these issues.

A. An Ethical Framework

Various approaches have been employed in addressing such issues, and while each has merits and deficiencies, there is insufficient space here to discuss them even superficially. It is common to several approaches that they appeal to a framework (sometimes explicit, sometimes implicit) in which impacts on different "interest groups" (such as farmers, consumers, and farm animals) are examined in the light of acknowledged ethical principles (such as respect for well-being, autonomy, and justice). Typically, such principles derive from consequentialist (e.g., utilitarian) and/or deontological (referring to rights and duties) theories, which many people combine (often unknowingly) in arriving at considered moral judgments. Generally speaking, these are *prima facie* principles, which while acknowledged widely, are not necessarily assigned decisive importance in reaching ethical judgments.

For the purpose of aiding the analysis of the wide range of ethical issues raised by novel foods, in the following discussion the three principles identified above will be applied to the different interest groups ("stakeholders") affected by novel food technologies, thus producing a table, the Ethical Matrix (Fig. 1). Ethical impacts (i.e., (lack of) respect for a defined principle) are assessed for four "interest groups," namely,

- The biota (the flora and fauna of a region, i.e., the "living environment")
- Producers, i.e., farmers
- Consumers
- The treated organisms (in this analysis, animals only)

In brief, the Matrix aims to apply established ethical theory (which may be said to underpin the "common morality") to the interests of the groups affected by any action, in terms that are broadly comprehensible. Thus, the translations of the ethical principles in each "cell" of the Matrix correspond, generally speaking, to familiar concepts. While there is not space here to explain the Matrix in any detail [for a fuller discussion see Mepham (2000) in the Bibliography], it will serve as a frame of reference in the following discussion.

It must be appreciated that the Matrix is not prescriptive, but aims to guide a rational and comprehensive survey of the important issues. It is, of course, not unusual for there to be some conflict in the degrees of respect which different people accord to the different principles. Thus, normative statements tend to be justified by appeal to ethical evaluations that place more or less emphasis on particular principles as they apply to a particular interest group or groups. For example, for some people the safety of food for human consumption is of overriding importance even though concerns for animal welfare or environmental conservation are also acknowledged. Differences of emphasis also arise in respect of the significance of actual or envisaged effects. For example, while particular benefits, costs, or risks might be universally acknowledged, some people might consider them important, while others consider them insignificant. In fact, views on GM foods have tended to become polarized, with proponents seeing great potential benefits and minor or manageable harms, while opponents take the opposite view on both counts.

While it oversimplifies the issues, in the interests of brevity, different views on the impacts of GM foods will often be presented below in terms typically advanced by such "proponents" and "opponents," respectively. It should be recognized that this is a polemical device that is not meant to characterize the view of any particular individual. Moreover, the analyses to be presented serve merely to exemplify the issues: space limitations preclude a more comprehensive account.

Principles featured in the Matrix, such as "respect for biodiversity," are identified in the following discussion by boldface (e.g., **biodiversity**).

B. Intrinsic Concerns

There is, however, some merit in addressing two fundamental questions at the outset, because they have overriding significance for some people. These are the questions of whether we should, as a matter of principle, take it upon ourselves to alter the nature of plants and animals in such a fundamental way (i.e., "Is it wrong to meddle with Nature?") and whether we should engage in a process whereby genetic material from one species is mixed with that of another (most contentiously, when human genes are expressed in food). These are sometimes called "intrinsic concerns."

Fundamentalist religious positions might be thought to have the strongest opinions in this respect, particularly if life in all its forms is seen as God-given. For adherents of such positions, genetic modification represents an immoral attempt to usurp God's purposes. But other religious adherents argue that God has endowed humans with freedom and creativity, and that it is right in such circumstances for us to exercise our talents as "co-creators," particularly when human need appears to demand it. Some religious rituals and traditions do however retain a strong adherence, such as those relating to the prohibition on the use of certain animals, and this is likely to have implications for the ethical (un)acceptability of some GM foods.

Reservations about so fundamentally altering the natural order are not, however, confined to the conventionally religious: others might equally regard respect for the way the biosphere has evolved, and for the dignity of humanity's place in it, as "intrinsic" ethical grounds for rejecting some or all GM technologies. For example, a view having its origins in "virtue theory" appeals to the character ideal of humility, which respects the world as it is and eschews the hubris of scientists who treat other living beings as a mere resource for human exploitation.

Respect for species integrity constitutes another form of intrinsic concern about GMOs. From a rigorously scientific perspective, it is difficult to maintain that species are discrete, objective entities; but it does not necessarily follow that that gives humans *carte blanche* to transgress species barriers on a whim. Perhaps most frequently invoked in relation to genetic manipulation of animals, such views place value on the "telos" of organisms or species (literally their "aim" or "goal"), which is considered to be undermined by procedures which violate natural species barriers and/or the intrinsic nature of individual organisms.

V. GM CROPS: ETHICAL ISSUES

A. The Biota

Environmental impacts of GM crops are fundamental: they might affect farming practices in the short term, the appearance of the landscape in the medium term, and, possibly, the sustainability of the biosphere in the longer term. From a utilitarian perspective, several benefits of GM crops are claimed by proponents to outweigh any conceivable costs or risks. For example, the use of herbicide-resistant crops (HRCs) and insect-resistant crops might improve productivity, both by increasing yields and by reducing herbicide and insecticide usage. Moreover, more efficient weed control might in some cases be achieved by sowing HRCs directly into unplowed land, thereby reducing soil moisture loss and increasing the length of the growing season.

Claimed benefits of GM crops are illustrated by a form of maize, also known as corn (present in food ingredients used for brewing, bakery products, salad dressings, snack foods, and margarines), which contains three transgenes for: (1) resistance to the herbicide glufosinate ammonium; (2) insect resistance; and (3) ampicillin (antibiotic) resistance, i.e., a marker. Farmers growing this GM crop can thus spray it with the herbicide, killing competing weeds but not the crop itself, and because the plants contain the Bt toxin (which is naturally produced by the soil bacterium *Bacillus thuringiensis*), losses from infestation with target insects should also be reduced. (The antibiotic marker gene serves no agronomic purpose, being merely a residuum of the production process of the GM crop.)

Ethical analysis of these claims can be conducted in relation to impacts on **conservation, biodiversity,** and **sustainability** of the biota (Fig. 1). In terms of **conservation,** it is claimed that HRCs will lead to reduced overall use of herbicides, and since their excessive use is environmentally damaging, this should prove beneficial. However, such claims are questioned by a 1999 USDA report which found that in two-thirds of the cases examined, use of GM crops did not lead to decreased herbicide or insecticide use.

Moreover, although, in the short term, there might be reduced need for crop spraying with pesticides, thereby reducing adverse effects on biota, it is possible that subsequently several problems might arise.

- HRCs may be transformed into weeds, as acknowledged by a recent UK Royal Society report. Where native species can cross-fertilize with crops, transfer of genes from GM crops to wild plants is inevitable, and the demonstration that this can occur in a plant that can cross with the rape plant and is a serious weed of 20 crops in over 50 countries illustrates the scale of potential adverse effects. Even where no native species related to a particular crop exist in the country where it is planted, the possibility of unexpected gene transfer occurring elsewhere, e.g., through the accidental transport of seeds, cannot be ignored.

- It is at least possible that herbicide use will be increased (a clear commercial objective) because farmers will be able to spray them more liberally without risk to the crop. Much will depend on economic factors, such as the relative costs of herbicide, seed, labor, and the returns on yield increases. In any event, the herbicides to which resistance is being engineered have several adverse effects. For example, glufosinate, a broad-spectrum weedkiller, is highly soluble and under certain conditions significant runoff or leaching might occur, leading to contamination of ground or surface water. Glyphosate, another herbicide commonly used with GM crops, such as soy, is claimed to endanger many plant species.

Proponents of the use of HRCs often claim that the alternative, use of more and a greater variety of herbicides, has considerably worse environmental impacts. This raises the important question of the kind of agriculture that should be used as a basis of comparison. While some GM crops might be more "environmentally friendly" than conventional intensive systems, opponents argue that they should be compared with even more benign systems such as integrated pest management or organic farming, which might represent viable alternatives. Although the benefits of organic types of farming in relation to soil quality and biodiversity are widely acknowledged, they are often ignored because of the assumed lower yields. But some long-term studies indicate that organic systems can produce equivalent yields to intensive systems, as well as increasing soil fertility.

Loss of **biodiversity** has been identified as the major adverse effect of HRCs. According to a Royal Society report (1998), "the more effective destruction of weeds . . . is likely to reduce the availability of habitats for various insects and invertebrates." The same effect might also result from the incorporation of Bt genes into GM crops in order to kill infecting insects by causing production of an "internal insecticide." Bt toxins, naturally produced insecticides, confer protection from attack by insects such as the European corn borer and the Colorado potato beetle. However, while normal genes code for prototoxins, which are only activated in the larval gut, the GM forms of Bt toxin are constantly active, risking adverse effects on beneficial, nontarget species.

Some recent evidence indicates the potential for harm to nontarget species. For example, larvae of the monarch butterfly reared on milkweed which had been dusted with pollen from GM corn expressing Bt toxin ate less, grew more slowly, and suffered higher mortality than larvae reared on leaves dusted with untransformed corn pollen. Moreover, when aphids fed on GM potatoes expressing snowdrop lectin (another insect toxin) were fed to ladybirds, the females lived only half as long and produced more than twice as many unhatched eggs than ladybirds fed normal aphids. Proponents of this type of GM crop argue that such risks are often exaggerated.

Another critical issue is the possible increase in the evolution of resistant strains of corn borer, since there is no reason to assume that this problem (encountered with conventional insecticides) will not be repeated for GM crops. Since in 1999 Bt crops accounted for about one-third of all GM crops planted worldwide, the probability of Bt resistance arising and causing loss of produce would seem to be significant. Proponents argue that any such problems are manageable, e.g., by resort to alternative pesticides.

Sustainability is a concept which addresses the perceived need for agricultural systems to sustain the Earth's growing population by maintaining the viability of the biosphere. Practices which use renewable or nonrenewable resources at rates which cannot be replaced by renewable resources, or which pollute the environment at rates exceeding the Earth's capacity to degrade, recycle, or absorb them, will prove unsustainable.

Concerns over the impacts of GM technology focus both on the extent to which reliance on intensive, chemically based systems might overexploit nonrenewable resources and on the risks associated with "genetic pollution," whereby—through horizontal gene transfer—the ecological balance might be seriously disturbed. Resistance genes (e.g., herbicides and antibiotics) are known to move readily between organisms so that it is inevitable that some gene transfer will occur from certain crops. Moreover, reliance on GM crops, like intensive agriculture as a whole, leads to increasing dependence on fewer and fewer crops (so-called "genetic erosion"), with the attendant risks of disease epidemics. It is, however, impossible to predict the effects of future developments. Conceivably, use of GM crops could lead to reduced

reliance on chemical inputs but, possibly, their use could increase.

In the UK, the government, in partnership with biotechnology companies, is performing a program of farm-scale trials of GM crops with the aim of assessing their environmental impacts.

B. Producers

Ethical concerns identified in Fig. 1 as respect for **producer income and working conditions, freedom to adopt or not adopt GM technology,** and **fairness in trade and law** are all closely interrelated by economic factors.

1. Impacts on Conventional Farmers

If GM crops prove to be commercially successful, farmers adopting them will benefit financially, because of increased yields and/or a reduced need for spraying with pesticides. Moreover, there might be certain health benefits as a result of reduced dependence on agrochemicals.

However, with increasing use of GM crops it is possible that farmers might subsequently encounter certain problems. For example, HRCs may be transformed into weeds, while HRC "volunteers" may act as reservoirs of pests and diseases, undermining the principles of crop rotation (see Section V.A).

A number of strategies have been proposed to deal with the problem of insect resistance to GM insecticides such as Bt toxin, including the use of "refuges" where non-Bt plants are grown to maintain a pool of sensitive

Respect for:	Well-being (health and welfare)	Autonomy (choice)	Justice (fairness)
The biota (fauna and flora)	Conservation	Biodiversity	Sustainability
Producers	Adequate income and working conditions	Freedom to adopt or not adopt	Fair treatment in trade and law
Consumers	Availability and acceptability of safe food	Respect for consumer choice (e.g., labeling)	Affordability of food
Treated organisms (animals)	Animal welfare	Behavioral freedom	Intrinsic nature

FIGURE 1   The Ethical Matrix.

insects that will prevent a Bt resistance gene from spreading through the entire insect population. This requires that farmers deliberately plant a part of their land with crops they know will either be lost, or will require insecticide treatment—a strategy which some people consider unlikely to be adopted widely.

Conventional farmers who are unwilling to grow GM crops might be financially disadvantaged if their lower yields make them uncompetitive with GM farmers. Hence they might be forced to adopt GM crops to stay in business—an example of the need to join the so-called "technological treadmill" (cf. **freedom not to adopt GM technology**). However, at the time of writing, the strong reaction against GM crops in Europe was dissuading some U.S. farmers who had grown them previously from continuing to do so.

Patenting of GM crops raises several ethical concerns relating to **fairness in trade and law**. Patents, a form of intellectual property, provide legal protection for inventions that are new and useful, giving the owners exclusive rights over exploitation of an invention for up to 20 years. Proponents argue that without patents there would be little incentive for companies to invest time, money, and effort in research, and the progress of biotechnology would be hindered. From this perspective, in terms of both fairness to the inventor and social benefit, patenting is ethically justified.

However, opponents deploy ethical principles to contrary effect, claiming that the granting of patents to life-forms fails to respect the **intrinsic nature** of living beings (particularly in the case of sentient animals). Moreover, patents tend to a result in a concentration of power in the hands of a very few, very powerful, multinational companies. In some cases, products derived from the gene pool of less developed countries may be sold back to them at prices inflated by royalty charges, aggravating existing debt burdens.

2. Impacts on Organic Farmers

One consequence of recent concerns over food safety and integrity in the EU is the greatly increased demand for organic products. With an annual growth rate over 10 years of 25%, it is predicted that 10% of Western European agriculture will be organic by 2005, and 30% by 2010. Yet, in the UK in 2000 only about 2% of land was devoted to organic farming so that most organic food was imported. An important question thus concerns the impact of GM crops on organic agriculture, the official standards for which specifically exclude GM organisms. The potential for adverse impacts is illustrated by the Royal Society report, which noted that "transfer of genes from GM to non-GM crops may also have unwanted effects if the latter are grown organically.

. . . Crops able to outbreed, such as maize . . . will be affected to the greatest extent." A report from the UK John Innes Centre confirmed this risk, stating, "genetic contamination of various kinds is inevitable in field grown crops" so that there is "a need for acceptable levels of contamination of organic crops to be decided, and measures identified to achieve them." This clearly poses a threat to organic farmers whose livelihood (cf. **income**) depends on their being able to assure customers that their products are free of GM material. So, at a time when public demand for organic food in the UK is increasing rapidly, future progress would appear likely to be threatened if use of GM crops became widespread.

Proponents of GM crops suggest that such contamination is not a serious problem, citing the fact that organic standards accept a degree of chemical contamination (e.g., from pesticides). However, opponents claim that the comparison is invalid, since unlike chemical contamination, GM organisms may continue to reproduce indefinitely. Moreover, this is not the only potential problem. The Bt toxin, the gene for which is expressed in several GM crops, is the *only* form of insecticide permissible in organic farming systems. Bt sprays are used quite sparingly, making the evolution and spread of resistance far less likely than if the toxin were persistently present in large numbers of crops. Thus, the risk of development of Bt resistance through use in GM organisms may be another threat to organic farming.

Challenges to organic farming from GM foods are thus not only that they might reduce **consumer choice** (see Section V.C.2) by monopolizing the supermarket shelves but also that, because of the uncontrolled spread of GM genes, the aspirations of those who wish to compete fairly in the marketplace might be frustrated, undermining respect for both **fairness** and **producer freedom not to adopt GM technology**.

C. Consumers

Ethical concerns for consumers are identified (Fig. 1) as the **safety and acceptability of GM food, choice,** and **affordability**. For most people, the **safety** of consuming GM foods is a major concern, and since it would seem to be in nobody's interests for food to threaten consumer health, it might be imagined that every effort would be made to ensure that consumers are not exposed to even remote risks. However, according to a utilitarian analysis, minor risks (or even major risks to very few people) might not significantly diminish the benefits due to GM foods, so it cannot be assumed that according to all ethical viewpoints health risks of GM foods need to be vanishingly small.

A comprehensive utilitarian assessment would entail consideration of a wide range of issues: not only the safety of the foods and any added pleasure, or anxiety, that might be attributable to consuming them, but also questions relating to food prices, the welfare of animals used in testing GM foods, and aesthetic impacts on the appearance of the countryside (i.e., **acceptability**). However, here, the focus is on physical (health and safety) effects of the introduction of GM foods.

1. Safety of GM Foods

Toxins occur naturally in many plants but (obviously) normally at concentrations that present no significant problems to consumers. A GM plant could contain an identical profile of expected toxicants to the natural plant, or conceivably it could contain unexpectedly higher concentrations. The USFDA considers that foods produced by biotechnology are substantially similar in virtually all cases to traditional foods (see Section IIIB), so that no detailed testing or labeling is required.

Essentially the same approach is encompassed by the notion of "substantial equivalence" embodied in the EU directive, although this criterion has been interpreted rather more stringently than in the USA, and labeling is required. However, critics argue that the concept of substantial equivalence is flawed because the degree of acceptable difference has not been defined accurately and because compositional analysis has limitations with respect to the identification of natural toxins and antinutrients. According to this view, since routine compositional analysis of foods cannot reasonably be relied on to detect all possible changes in the levels of toxicants in food, a much more precautionary stance is needed.

Where toxicological testing of GM foods is required, there are also significant difficulties in that the standard procedures used for drug testing (administration at high doses to assess pathological effects) are not applicable to foods. This is because there are clear limitations to how much food an experimental animal can consume, both in terms of the capacity of its digestive system and in terms of palatability. Consequently, testing tends to be conducted on products of the GM gene expressed in bacteria, but this may not closely represent the effects of gene expression in the GM crop.

Some examples illustrate the types of alleged risk.

a. Toxic Products

Clearly, no sane person would deliberately produce a GM food containing a dangerous level of toxins: the concern is, rather, that a novel gene product might accidentally prove to be toxic, perhaps only to a small number of consumers. This also applies to traditional breeding—and some such plants have had to be withdrawn from the market because of unexpectedly high levels of toxins. Even if there were no theoretical reasons why GM crops were any more likely to cause this problem than traditional breeding (though some claim that the random nature of the processes involved in generating GM crops does make it more likely), very little published research has been conducted to address the issue.

b. Allergenicity

Allergic reactions are widely recognized as potential adverse effects of GM foods. GM crops express new proteins, and because proteins can be allergenic, regulatory authorities take measures to screen for allergenicity. For example, since 1992 the USFDA has required premarketing safety testing and labeling of foods containing genes transferred from the 8–10 most commonly allergenic foods. Since USFDA requirements do not apply to foods that are rarely allergenic or to donor organisms of unknown allergenicity, the policy might be said to favor industry over consumer protection. A recent British Medical Association (BMA) report claimed that more research on allergenicity and the possible toxity of GM foodstuffs needs to be undertaken.

c. Antibiotic Marker Genes

Marker genes, which generally take the form of antibiotic resistance genes, are used in the process of developing a GM crop but confer no agronomic advantage. However, the presence of these genes presents a risk that antibiotic resistance will be transferred to humans, and thus might compromise effective treatment of patients. According to the ACNFP, "the risk of transfer is small but finite. Functional gene transfer from bacterial to mammalian cells is certainly possible." Any unnecessary use of antibiotics would thus appear to constitute an unjustifiable risk to human welfare, a view endorsed by both the Royal Society and the BMA. Despite this, the majority of the GM crops in current use contain such genes.

2. Consumer Choice

Respect for **consumer choice** might be said to require that consumers be made aware of any salient differences in physical properties or production methods of a novel food. However, largely because it is thought that labeling might lead to negative discrimination, food producers and retailers have tended to avoid labeling when possible. But in response to consumer concerns, in 1997 the EC banned the importation of unlabeled GM crops from the USA and required labeling of any such crops imported. According to the EU directive, member states have a right to veto the Commission decision to allow importation of GM food only if the product "constitutes a risk to human health or the environment." Three states (Italy, Austria, and Luxembourg) have invoked this provision, claiming the presence of the antibiotic marker gene jeopardizes human health. By contrast, the USA has no labeling requirements, although at the time of writing the FDA appeared to be reconsidering this decision.

In the UK, the Polkinghorne Committee was set up in 1992 to consider future trends in the production of GM organisms and the moral and ethical concerns which might arise from the use of food products derived from them. Several of the report's recommendations have a bearing on labeling of GM foods. The committee "saw no overriding ethical objection which would require an absolute prohibition of the use of organisms containing copy genes of human origin as food," but recommended that ACNFP guidelines should discourage use of "ethically sensitive" genes (e.g., animal genes which might offend vegetarian sensitivities) in food organisms where alternatives are available. The committee also enunciated two new principles: (1) the *"de minimis"* principle (that it is unrealistic to label every last element of a modified food in every product in which it may be incorporated), and (2) derived products of specific nongenetic nature need not be labeled.

In the UK, voluntary labeling to indicate the absence of GM ingredients is permitted and several supermarket retailers have declared that their own-brand foods are "GM-free." But, in the absence of mandatory segregation, labeling does not ensure consumer choice if consumers do not understand the label or if there is no available option.

In January 2000, two amendments to EC regulations were introduced:

- EC 49/2000 (amending Regulation 1139/98 concerning labeling of foods containing GM soy and maize) extended the requirements to companies selling food to catering establishments and introduced a 1.0% *de minimis* threshold for the presence of adventitious GM material in a food product
- EC 50/2000 extended labeling requirements to food additives and flavorings produced from GMOs

The 1.0% threshold was made in recognition of the fact that it is almost impossible to prevent accidental contamination of non-GM soy and maize ingredients and foods at the cultivation, harvest, transport, storage, or processing stages. However, some argue that the threshold is too high, with several supermarket retailers working to maximum levels of 0.1% (only one-tenth of the EC standard).

D. Less Developed Countries

Proponents of GM technology often place great store in the claimed ability of GM crops to ameliorate the world food problem. According to a 1999 report of the UK Nuffield Bioethics Council, "The moral imperative for making GM crops readily and economically available to developing countries who want them is compelling." Nuffield's case appears to rest on the precedent of the Green Revolution, the introduction of high-yielding varieties of wheat and rice, produced by conventional breeding, which occurred in the decades following 1960. However, while it is generally conceded that it did produce more food and enrich some farmers, some analysts consider that, overall, the Green Revolution was an expensive failure, widening the gap between rich and poor, locking farmers into dependence on seed companies, and raising farm costs.

To date, the great majority of GM crops have involved production of HRCs for Western markets, e.g., maize, soy, and canola which are all used in processed foods and in animal feed. Indeed, virtually all the applications of agricultural biotechnology which have emerged have been closely integrated with conventional, capital-intensive agricultural practices employed in North America and Western Europe, a trend which opponents claim will not improve food security in the less developed countries.

Because of reduced government funding, agricultural research on GM crops has now become concentrated in private hands, with five major agrochemical–seed companies controlling most of the agricultural applications worldwide. Unsurprisingly, their commercial objectives appear best served by concentrating on products which satisfy the demands of Western consumers. However, agriculture in developed countries differs markedly from that in the developing world, not least because in comparison with less developed countries the agricultural workforce is such a small percentage of the total population (just 2% in the UK).

It is for such reasons that the Nuffield report recommended that "the UK Government should pre-commit a substantial amount of the UK aid announced in July 1998 to additional spending on the R&D of GM staples grown in developing countries." However, while endorsing the moral obligation for developed countries to greatly increase investment in research and development (R&D) for less developed countries, a report of the UK Food Ethics Council argued that increased funding at the level proposed was unlikely to make any significant impact, because "the concentration of R&D, in the context of patents and emerging IPR (intellectual property rights) regimes, militates against potential benefits going to small scale farmers (and poor countries in general)."

The power exerted by multinational companies is clearly a matter of concern, even for proponents of GM technology. Thus a report of the UK House of Lords

Select Committee in 1998 noted the concern, which they shared with farmers and several witnesses who had provided evidence, that the powers of a few agrochemical–seed companies were already great, and would become greater over the process of producing GM crops.

It might be argued that the most effective way to counter global food shortages and poverty is not to encourage people to grow their own food, but to put their effort into farming for cash crops. This approach, already widely practiced, could favor widespread adoption of GM, because it depends on a limited number of high-yielding crops with defined management regimes. But critics of the approach argue that several problems may arise with this strategy, depending as it does on substantial inputs of expensive fertilizers or pesticides to give marketable yields and its reliance on larger, less labor-intensive farms. The widespread cultivation of cash crops may also lead to a rapid loss of local and indigenous knowledge about food crops, often built up over many generations and transmitted only orally: once lost, this knowledge might be irrecoverable.

Even those GM crops that might be grown in less developed countries seem capable of raising critical problems. Thus, a threat is posed by so-called "terminator technology," which by incorporating two or three novel genes into a plant, causes the second-generation seeds to die in the early stages of germination. By denying farmers the right to reuse seeds saved from previous harvests this could have socially disastrous effects because 1.4 billion people in less developed countries are dependent on crops grown from seed saved by small farmers. Up to 70% of seeds used by such farmers is saved on-farm. The technology has been widely condemned, e.g., by the Indian government, the Consultative Group on International Agricultural Research, and the UK Department for International Development. In 1999 the Monsanto Company announced that it had decided it would no longer pursue research into terminator technology, and a similar assurance was given by AstraZeneca.

VI. GM ANIMAL FOODS: SPECIFIC ETHICAL ISSUES

Although research on GM farm animals began in the early 1980s, few applications have reached the "near market" stage, and those that have are directed to production of pharmaceuticals rather than food. Ethical concerns (Fig. 1) here focus specifically on effects of GM on **animal welfare** (including **behavioral freedom**) and on respect for their **intrinsic nature** (see Section IV.B). Other concerns (e.g., relating to food safety and

effects on producers) correspond to those discussed above for GM crops.

One of the first animal projects involved producing GM pigs expressing human growth hormone genes, which were designed to grow more rapidly than normal pigs and produce leaner meat. However, these so-called "Beltsville pigs" suffered greatly reduced welfare and provided a warning of the dire consequences which could result from this approach.

In 1993, the UK government set up a committee to consider ethical implications of new technologies in the breeding of farm animals, many of which might be employed in the production of novel foods. The committee enunciated three guiding principles which they considered should govern the humane use of farm animals:

- Harms of a certain degree and kind ought under no circumstances to be inflicted on an animal
- Any harm to an animal, even if not absolutely impermissible, nonetheless requires justification and must be outweighed by the good which is realistically sought in so treating it
- Any harm which is justified by the second principle ought, however, to be minimized as far as is reasonably possible
    Ministry of Agriculture, Fisheries, and Food, 1995

Most people would consider premeditated physical or psychological harm inflicted in the interests of "efficient food production" to be ethically unacceptable, but some do regard what they see as minor harms as justified in a utilitarian calculus which benefits humans. However, altering the nature of an animal (e.g., by genetically engineering it to tolerate intensive housing conditions without undergoing stress) presents a quite different type of ethical concern. According to conventional definitions, its welfare might be "improved." Yet, for some people, the infringement of respect for the animal's **intrinsic nature** would make such a modification ethically unacceptable. Such a deontological concern might relate to the respect due to nonhuman animals in their own right and/or to a concept of human dignity which abjures so instrumental a use of fellow sentient creatures.

In purely utilitarian terms, the **animal welfare** "costs" to GM animals range over a "spectrum of harm" from inert proteins expressed in specific organs (e.g., a new type of milk protein, designed to improve the production of cheese), which pose few, if any, welfare problems, to expression of systemically active hormones, with the potential to alter the animal's physiological processes fundamentally. However, in such a calculus of benefits and harms it is important not to ignore the reduced animal welfare entailed in developing a new GM line. GM farm animals are produced by a process of pronuclear microinjection, which is very inefficient, with less than 1% of farm animal embryos generally developing into transgenic animals that could be used as founder animals in a breeding program. Not only are the vast majority of embryos "wasted," but many suffer, prenatally or postnatally, from mutation-induced deformities. Large numbers of animals are also subjected to surgery and hormone treatment as part of the production process.

GM fish are perhaps the animals closest to commercial use, with growth stimulation the principal objective. In some GM salmon, blood concentrations of growth hormone are 40 times the normal level, resulting in 5- to 11-fold increases in weight after one year of growth. Concerns for **animal welfare** arise from the observation that body composition and conformation can be affected, resulting, for example, in an excessive and deleterious deposition of cartilage, a pathological condition analogous to acromegaly in mammals. Other ethical concerns relate to environmental risks (specifically, effects on **biodiversity** and **sustainability**). Unlike farm animals, the possibility of fish escaping into the wider environment is significant, with consequent risks to the natural ecosystems. Research findings to date suggest that GM fish have reduced swimming ability, and hence greater vulnerability to predators and an impaired ability to capture prey. However, due to the unreliability of predictions based on laboratory studies, rigorous physical containment systems have been recommended. Alternatively, fish might be biologically sterilized, preventing breeding.

VII. FUNCTIONAL FOODS: ETHICAL ISSUES

The principal ethical concerns relating to functional foods may be identified as **food safety, choice,** and **affordability** (Fig. 1). While food manufacturers need to be able to inform consumers about research indicating the health benefits of their products, consumers also need to be protected against misleading claims based on inadequate research. Although some diet–health relationships are now well established, many others are unreliable and controversial. Clearly, consumer protection demands that reliable means of authenticating food manufacturers' claims are introduced.

There are also concerns over possible cumulative effects of small changes in the composition of an increasing number of foods. For example, a slight change in the fatty acid content of any individual food might be of little significance, but the consequences of consuming several such food products could be detrimental to health (cf. **food safety**).

Draft guidelines introduced in the UK state that "any health claim should be supported by a dossier of evidence demonstrating the specific physiological effect which is claimed." Such evidence needs to be based on human studies, including appropriate epidemiological evidence. Moreover, the evidence in support of health claims should be available for scientific peer review. However, in contrast to the provisions of the U.S. Nutrition Labeling and Education Act, the UK guidelines do not mention the need to examine the *totality* of evidence available relating to claims. So, in practice, food manufacturers may be selective in presenting evidence substantiating their claims.

The issue of **choice** assumes a particular significance in the case of functional foods. There is a need to ensure that consumers are not misled into believing that a particular food product has unequivocal benefits, especially if they might be secured more effectively, and possibly more cheaply, by other means. For example, a more effective way of avoiding a particular disease might lie in changes in diet, or in lifestyle more generally.

To the extent that functional foods blur the distinction between foods and medicines, it is not difficult to foresee that their increasing popularity could infringe the principle of **affordability**. If functional foods, and consequently improved chances of health, were to become the preserve of the wealthy, those unable to afford them would be just as much the victims of unfairness as if they were unable to afford medical treatment.

Functional foods derived from GM crops and GM animals, respectively, raise correspondingly more complex ethical issues. Prominent examples of such prospective foods are GM rice with enhanced levels of β-carotene (the precursor of vitamin A) and iron (so-called golden rice) and baby foods containing the human milk protein lactoferrin, derived from the milk of GM cattle.

VIII. STEPS TO THE RESOLUTION OF ETHICAL DEBATES

For some people, novel foods present the prospect of increasing wealth and improved standards of living on a global scale. For others, they represent an inappropriate, and possibly dangerous, way to solve problems which demand quite different approaches. Typically, such differences in ethical evaluation are difficult to resolve—and the fact that they are often bound up with cultural, religious, and political factors adds to the intractability of the problem. This has wide-ranging consequences because in a world in which increasing globalization of trade appears inevitable, there are likely to be enormous costs if international agreement cannot be reached on how novel foods should be regulated. After "all the shouting" (or, better, the debate), there is a need for "closure"—for some mutually agreed course of action. Does this analysis suggest ways of achieving closure?

Broadly speaking, ethical judgments on specific issues which have political consequences (e.g., whether to require labeling of a GM food product) depend on two factors, acknowledgment of a general ethical principle (e.g., "people should be free to choose the type of food they eat") and scientific facts (e.g., to pursue this case, whether there is a significant difference between the GM food product and the non-GM product to which it might be claimed to correspond). If there were disagreement as to the validity of the ethical principle (P1), in order to find an acceptable way forward it would be necessary to appeal to some more fundamental ethical principle (P2), from which P1 might be said to have been derived and on which agreement might be found. Rational exploration is needed to determine whether P1 could be derived validly from P2 and also to determine the limitations of P1 (e.g., clearly, infants and the senile are incapable of exercising real choice). The principles which structure the Ethical Matrix are proposed as elements of what has been termed the "common morality" and are designed to facilitate just this form of rational, open, and democratic public debate. However, it needs to be recognized that for many the "common morality" may be an unreflective position, which needs to be tempered by conscientious deliberation.

The second element, the scientific facts, is often more problematical than many scientists are prone to acknowledge. While in some cases the known facts might be accurate and pertinent, in others they might be inaccurate, unreliable, or irrelevant. Clearly, sometimes disputes might be resolved readily by investing resources in acquiring the appropriate missing data, but on other occasions progress might be hampered by a significant degree of uncertainty. For example, in the case of the cattle disease BSE (bovine spongiform encephalopathy), initially, scientists were ignorant of the causative agents (prions). In such cases, there could be said to be a requirement to adopt the "precautionary principle."

In February 2000, the EC adopted a Communication—COM(2000)1—on the use of the precautionary principle (PP), which "covers cases where scientific evidence is insufficient, inconclusive or uncertain and preliminary scientific evaluation indicates that there are reasonable grounds for concern that the potentially dangerous effects on the environment, human, animal or plant health may be inconsistent with the high level of protection chosen by the EU."

The Communication makes a number of notable claims, cited below by paragraph number:

- "The risk assessment on which a measure is based may include non-quantifiable data of a factual or qualitative nature and is not uniquely confined to purely quantitative scientific data" (5).
- "Even if scientific advice is supported only by a minority fraction of the scientific community, due account should be taken of their views, provided the credibility and reputation of this fraction are recognised"(7.2).
- The PP will often need to be invoked in situations where adverse effects will not emerge until long after exposure and/or will affect future generations (7.3.1).
- "Examination of the pros and cons of an action cannot be reduced to an economic cost–benefit analysis. It is wider in scope and includes non-economic considerations. A society may be willing to pay a higher cost to protect an interest, such as the environment or health, to which it attaches priority"(7.3.4).
- As each member of the World Trade Organization (WTO) has the independent right to determine the level of environmental or health protection considered appropriate, a member may apply measures based on the PP which lead to a higher level of protection than that provided for in international standards or recommendations (Annex 2).

Such an approach, addressing the scope and validity of scientific evidence relating to ethical judgments on, *inter alia,* novel foods, represents an important step in promoting rational debate on risks. Discussion of the PP seems certain to provide a critical focus for international negotiations—and an essential avenue to closure.

That the PP is gaining wide international support is illustrated by the agreement by 130 countries to the Biosafety Protocol to the UN Convention on Biodiversity at a meeting in Montreal in January 2000. This agreement allows a country to ban imports of a GM product if it considers that there is insufficient evidence to prove the product is safe. The Protocol will have the same status as the WTO in international law, providing it receives ratification by 50 nations.

IX. CONCLUSIONS

Technological developments relating to novel foods are proceeding at a rapid pace. Such developments pose crucial ethical questions concerning a broad range of issues. Preeminently they impact on the provision (or lack of it) of adequate nutritious food for the hundreds of millions of people in the world who are malnourished. It is estimated that currently 1.3 billion people subsist on less than $1 per day, but even such an alarming degree of human deprivation may be exacerbated by the projected increase in global population to 10 billion in the next 50 years.

Such problems are compounded by others relating to environmental, public health, political, economic, and cultural issues which novel foods have the potential either to aggravate or ameliorate, depending on how they are used, regulated, or abused.

That food is the most basic requirement of life is a truism. However, there is little general appreciation of the insight that rather than technology *requiring* legislation, technology *is* legislation. That realization predicates the need for serious ethical deliberation on new technologies applied to food, the importance of which can hardly be overestimated.

Bibliography

Advisory Committee on Novel Foods and Processes (1999). *Annual Report 1998*. London: Ministry of Agriculture, Fisheries and Food and Department of Health.

British Medical Association (1999). *The Impact of Genetic Modification on Agriculture, Food and Health*. London: British Medical Association.

Food Ethics Council (1999). *Novel Foods: Beyond Nuffield*. Southwell: Food Ethics Council.

Mepham, B. (1996). Agricultural ethics. In *Encyclopedia of Applied Ethics* (R. F. Chadwick, Ed.). San Diego: Academic Press.

Mepham, T. B. (2000). A framework for the ethical evaluation of novel foods: The Ethical Matrix. *Journal of Agricultural and Environmental Ethics* **12,** 165–176.

Mepham, T. B., and Crilly, R. E. (1999). Bioethical issues in the generation and use of transgenic farm animals. *Alternatives to Laboratory Animals* **27,** 1–9.

Ministry of Agriculture, Fisheries and Food (1993). *Report of the Committee on the Ethics of Genetic Modification and Food Use*. London: Her Majesty's Stationery Office.

Ministry of Agriculture, Fisheries, and Food (1995). *Report of the Committee to Consider the Ethical Implications of Emerging Technologies in the Breeding of Farm Animals* (Banner Report). London: Her Majesty's Stationery Office.

Nuffield Council on Bioethics (1999). *Genetically Modified Crops: The Ethical and Social Issues*. London: Nuffield Council on Bioethics.

Royal Society (1998). *Genetically Modified Plants for Food Use*. London: Royal Society.

Sadler, M. J., and Saltmarsh, M. (Eds.) (1998). *Functional Foods: The Consumer, the Products and the Evidence*. London: Royal Society of Chemistry.

Stewart, T. P., and Johanson, D. S. (1999). Policy in flux: The European Union's laws on agricultural biotechnology and their effects on international trade. *Drake Journal of Agricultural Law* **4,** 243–295.

U.S. Food and Drug Administration Center for Food Safety and Applied Nutrition, Office of Premarket Approval, July 1999. *Foods derived from New Plant Varieties Derived through Recombinant DNA Technology: Final Consultations under FDA's 1992 Policy*. Website: http://vm.cfsan.fda.gov/~lrd/biocon.html

Nuclear Power

KRISTIN SHRADER-FRECHETTE
University of South Florida

GLOSSARY

chain reaction A continuous, self-sustaining series of fission reactions in a mass of uranium or plutonium.

fission The only type of nuclear reaction that is controlled by people to produce energy. It occurs when a nucleus of uranium or another heavy element is split into parts.

half-life The time required for half the atoms of a radioactive substance to decay into another substance.

isotopes Different forms of the same element.

nuclear reactor A device for producing nuclear energy by means of controlled chain reactions.

radioactivity The radiation emitted by a radioactive substance such as uranium, radium, thorium, plutonium, iodine-131, strontium-90, cesium-137, or carbon-14. Radiation is ionizing when it is able to remove orbital electrons from other atoms or molecules. The ability of ionizing radiation to change molecular structures is what accounts for its ability to cause cancer, genetic damage, and even death.

NUCLEAR POWER is a means of generating energy, mainly in the form of heat, by using either fission or breeder reactors. Fission reactors create energy by a self-sustaining chain reaction that splits the nuclei of atoms of uranium or plutonium. Breeder reactors convert a uranium isotope to fissionable plutonium and produce more radioactive fuel than they use. Both types of reactors control the heat energy they generate so that it can produce steam and run a turbine. Nuclear weapons such as bombs get their destructive power from uncontrolled fission.

Nuclear generation of electricity has several significant advantages. It avoids use of scarce oil and gas, it requires less fuel than other sources of power, and it releases only minimal chemical or solid pollutants into the air. At present, approximately 30 nations have more than 400 nuclear plants for generating electricity. More than 150 are in European countries, and the United States has the largest number (113 reactors) in the world. The first full-scale, commercial nuclear power plant was opened in England in 1956, and the second plant began operating a year later in the United States. Concerns about safety, however, have stopped the growth of commercial nuclear power in most nations of the world. Reactor accidents, cancer and genetic damage induced by the smallest amounts of radioactive material, and problems with disposal of the high-level radioactive waste (which remains dangerous forever) are the three most important nuclear-related safety issues. Cost overruns and public opposition—driven in part by the 1986 Chernobyl and the 1979 Three Mile Island accidents—also have contributed to the move away from nuclear power. Since 1974, no new commercial reactors (which have not been subsequently cancelled) have been ordered in the United States.

One reason that the health effects of ionizing radiation are so serious is that there is *no threshold* for increased risk as a result of exposure to even small amounts of radiation; all doses of ionizing radiation are harmful. Effects of radiation also are *cumulative*: successive exposures increase one's risk of harm. Exposure to radiation can cause cancer, reproductive failure, birth defects, genetic effects, and death. In 1989 the U.S. National Research Council argued that an acute dose of radiation is three times more likely to cause cancerous tumors and four times more likely to induce leukemia than risks calculated in 1980. Hence, despite the known consequences of radiation, even current standards may be too lenient. Identifying dangerous effects of radiation also is difficult because many of them are latent. For example, at the Oak Ridge National Laboratories, where many persons received high doses of radiation, it took 26 years for the cancer rate to exhibit a statistically significant increase. It took 4 years for thyroid cancers to show statistically significant increases near Chernobyl after the 1986 accident. Much radiation damage also remains undetected because few persons do epidemiological studies to measure its effects on humans, even though no amount of exposure to ionizing radiation is completely safe.

Because of the enormous capital expense of nuclear power and because it supplies about 3% of total global electricity, there is continuing ethical controversy over whether to build any new plants, whether to shut down existing reactors, and how to dispose of radioactive waste so as not to harm members of future generations. Most of these ethical issues focus on problems with equal protection from catastrophic reactor accidents, with due process in the event of nuclear-related harm, and with failure of the public to provide genuine informed consent to the risks associated with nuclear power.

I. HISTORY OF NUCLEAR POWER

In 1895, Roentgen discovered the x-ray, and in 1896, Becquerel and the Curies discovered natural radioactivity. Large volumes of radioactive waste were not created, however, until the atomic bomb program of World War II. During the war, both the United States and the former Soviet Union built uranium enrichment plants and started to develop light-water, nuclear fission reactors (that use enriched uranium fuel) because this was the technology already used in their military efforts. Canada, France, and Great Britain, however, began work on reactors moderated by graphite or heavy water. Using unenriched uranium, these reactors cost more to build than the enriched uranium reactors of the United States, but they were safer by virtue of being better able to withstand a loss of cooling.

The effort to collect the plutonium needed for atomic bombs during World War II was part of a U.S. defense effort, begun in 1942, known as the "Manhattan Engineer District Project." In December of that same year, Enrico Fermi directed a team that produced the world's first nuclear chain reaction. By January 1943, U.S. federal government researchers were overseeing the building of the first atomic bombs at Oak Ridge, Tennessee, and Hanford, Washington. On 16 July 1945, the world's first atomic bomb, using plutonium, was exploded in New Mexico. On 6 August 1945, the United States dropped the first nuclear warhead, employing uranium-235, on Hiroshima. Sixty-five thousand persons perished in the blast. Days later, another atomic bomb was dropped on Nagasaki.

Many of the ethical problems associated with decision making about commercial nuclear power stem from the military origins of the technology. The Manhattan Project, with its military focus, has left a legacy of secrecy, centralization, and technocracy which has dominated nuclear-related decision making for at least three decades. This three-part legacy led to haste, to temporary solutions, and to a lack of public participation in regulating nuclear energy. The legacy also contributed to institutional self-protection and to public policy that was neither properly debated nor scrutinized through scientific peer review. For example, in 1986, in response to public demand, the U.S. Department of Energy released 19,000 pages of formerly classified documents on the operations at the Hanford facility (used to develop the atomic bomb and to store radioactive waste) during the 1940s and 1950s. The documents reveal that U.S. military researchers had conducted radiological experiments on local people, without either their knowledge or their permission. In 1945 alone, the Hanford facility, devoted to wartime production of plutonium, routinely released 340,000 curies of radioactive iodine through its reactor exhaust stacks. (Current regulations permit less than 1 curie of iodine to be released each year at the facility, and even in 1945 this deliberate release far exceeded U.S. government environmental standards.) Numerous other accidents and deliberate releases occurred at Hanford, many of which were designed for the purpose of developing a monitoring methodology for intelligence efforts regarding the Soviet military program. Such experiments were made possible only because of the secrecy that shrouded the nuclear industry in all countries in its early years (U.S. Congress, 1994. *Human Subjects Research: Radiation Experimentation.* U.S. Government Printing Office, Washington, DC).

From 1940 to 1945, the United States spent $2 billion to develop the first atomic bombs used during World War II. Thereafter, the government took 20 years and more than $100 billion in subsidies to develop the first commercial nuclear reactors used to generate electricity. Scientists were optimistic about the Atoms for Peace program; it provided a nonmilitary rationale for continuing the development of nuclear technology and for obtaining weapons-grade plutonium that could be used for military purposes. Nobel-winning scientists, such as Henry Kendall (1991. *Calypso Log* **18,** 9), now claim that the present U.S. government subsidies for commercial reactors are running on the order of $20 billion per year; if these subsidies were removed from nuclear electricity, Kendall and others claim that the costs of fission-generated electricity would double.

In 1951, the Experimental Breeder Reactor at Arco, Idaho, produced the first commercial nuclear electric power. Between 1948 and 1953, the United States built and tested a submarine reactor at Idaho Falls under the leadership of Admiral H. G. Rickover. The submarine *Nautilus* went to sea in 1955, powered by a nuclear reactor using enriched uranium fuel. All of the Polaris missile-carrying submarines of the United States use nuclear reactors. In 1956, the first full-scale nuclear fission power plant began to operate at Calder Hall in northwestern England. Only after the passage of the 1957 U.S. Price–Anderson Act, limiting nuclear liability in the event of an accident, did the U.S. industry agree to begin using atomic energy for commercial generation of electricity. The first U.S. commercial nuclear plant opened in Shippingport, Pennsylvania, in 1957. Also in 1957, the United Nations established the International Atomic Energy Agency to promote the peaceful uses of nuclear power. In 1958, the former Soviet Union opened its first commercial reactor near Chelyabinsk. By 1970, nuclear plants had opened in Canada, France, Great Britain, India, Italy, Japan, the Netherlands, Spain, Switzerland, the United States, and the former Soviet Union. As this brief history of nuclear energy reveals, reactors came into use as part of military efforts to create a bomb. Despite widespread sponsorship and use of this technology, no country evaluated nuclear power, prior to its commercial employment, in order to determine whether or not it was a desirable means of generating electricity. Instead, driven by military applications, nations promoted commercial use of nuclear

technology as a way to support weapons development.

In 1976, 20 years after commercial fission reactors began operating, the *Wall Street Journal* proclaimed them an economic disaster. Nuclear electricity has proved so costly that year 2000 projections for nuclear power plants are now approximately one-eighth of what they were in the mid-seventies. The few U.S. nuclear manufacturers still in business have remained so by selling reactors to other nations, often developing countries seeking fission-generated electricity as a way to obtain nuclear weapons capability, through the plutonium by-product. India exploded its first nuclear bomb, for example, by using plutonium produced by a reactor exported by Canada.

At least since the 1986 reactor core melt in Chernobyl, the future of nuclear power has been in doubt. Since the accident, republics such as Belarus and the Ukraine have been forced to spend 13 to 20% of their annual budget on Chernobyl-related problems. Scientists have confirmed a 100-fold increase in thyroid cancer in Belarus, Russia, and the Ukraine; increases in leukemia in Greece, Sweden, and Finland; and a doubling in germ-line mutations in Belarus children born after Chernobyl. In the face of such massive harms, controversy over continued use of nuclear power has accelerated.

II. TWO ETHICAL FRAMEWORKS FOR EVALUATING NUCLEAR POWER

Both this controversy over nuclear fission and decisions about how to assess societal risks and benefits from commercial reactors are unavoidably and fundamentally ethical. They both address what society "ought" to do in choosing a means of generating electricity, and they presuppose different distributions of the risks, costs, and benefits of generating electricity.

Most people who evaluate ethical issues concerning nuclear power follow variants of two main theoretical approaches, deontological/contractarian ethics or utilitarian ethics. In general, contractarian norms emphasize equity among persons and respecting individual moral rights. Utilitarians, however, typically do not recognize any individual moral rights or duties except to promote the best consequences for the greatest number of people. Often deontologists, such as Immanuel Kant, John Locke, and John Rawls, defend their appeals to duty by means of conceptual or logical analysis. Utilitarians, such as Jeremy Bentham, John Harsanyi, and John Stuart Mill, defend their claims by analyzing the consequences following from particular ethical positions. Contractarians and utilitarians each have important insights. Utilitarians tend to emphasize the economically desirable consequences that may arise from different standards for nuclear power and radiation-related hazards. Contractarians, however, tend to emphasize individual rights to protection from nuclear risks and to argue that any risk imposition is justified only in the context of individual consent, compensation, and equality of risk distribution.

III. NUCLEAR POWER AND THE NATURALISTIC FALLACY

Many scientists and engineers tend to assume that nuclear-related risks are ethically acceptable (or may be involuntarily imposed) if they present the same level of harm as voluntarily chosen risks. As contractarians point out, this assumption errs because ethical acceptability is not only a matter of risk magnitude but also a matter of risk distribution and risk compensation. As utilitarians point out, the ethics of nuclear power and radiation-related risks is also a matter of whether the dangers are associated with important benefits. Hence, when scientists and risk assessors such as Philip Abelson (1994. Reflections on the environment. *Science 263*, 591) erroneously argue that the public ought to be "consistent" in accepting radiation risks that are lower than other risks (such as driving an automobile), they are not really comparing consistent risks. Radiation risks imposed by commercial nuclear power, for example, are typically more catastrophic, less compensated, and less under voluntary individual control than are the risks associated with automobile accidents. Because of these ethical disanalogies, there is no "linear relationship" between risk magnitude and risk aversion. For example, because the radiation risks associated with a nuclear accident often are subject to a liability limit—the U.S. limit is approximately 5% of the total costs of the Chernobyl accident—many persons believe that ethically acceptable radiation risks require a much stricter evaluation than other threats that are fully compensable (K. Shrader-Frechette, 1993. *Burying Uncertainty: Risk and the Case against Geological Disposal of Nuclear Waste,* pp. 9–23, 96–98, 157–158, 205–229. University of California Press, Berkeley).

Similar reasoning applies to the assumption that imposing additional nuclear-related risks of the same magnitude as typical background levels of radiation is always ethically acceptable. Of course, this assumption gains some plausibility because assessors often use "natural standards," in part, to evaluate the acceptability of various pollutants. Natural standards are based on existing or "natural" levels of pollutants. Determining acceptable levels of societally chosen risks on the basis of levels of unavoidable (background) risks, however, is

not ethically defensible because the two classes of risks often are ethically disanalogous. They are associated with different levels of voluntariness, equity of distribution, benefits, compensation, and so on. Moreover, as G. E. Moore showed, to assume that a risk is ethically acceptable just because it is "normal" or "natural" is to commit the naturalistic fallacy (G. E. Moore, 1951. *Principia Ethica,* pp. viii–ix, 23–40, 60–63, 108, 146. Cambridge University Press, Cambridge). A normal rate of automobile accidents is not acceptable, for example, if it is possible and reasonable to lower the rate. Evaluating risks associated with nuclear power thus requires not only knowledge of risk magnitude, but also ethical decisions about issues such as how safe is safe enough, how safe is fair enough, and how safe is voluntary enough. Moreover, all these issues, in part, are matters of citizens' democratic rights to self-determination. They are not matters for only scientific experts to dictate because they are questions of ethics and because they affect human welfare.

Even definitions of "nuclear risk" often reflect the naturalistic fallacy. Some scientists and engineers who study societal risks—such as those associated with nuclear waste facilities—typically define "risk" merely as the probability that some harm, like death, will occur (see National Research Council, 1983. *Risk Assessment in the Federal Government: Managing the Process.* National Academy Press, Washington, DC). However, ethicists, especially contractarians, often argue that risk cannot be reduced merely to a mathematical expression because it also is a function of qualitative components such as citizen consent (K. Shrader-Frechette, 1991. *Risk and Rationality.* University of California Press, Berkeley; D. MacLean, Ed., 1986. *Values at Risk.* Rowman and Allanheld, Totowa, NJ). Thus merely probabilistic definitions of nuclear risk exhibit a "naturalistic fallacy" if they attempt to reduce ethical analyses to purely mathematical and scientific concepts.

IV. NUCLEAR POWER AND UNCERTAINTY

Many estimates and evaluations of nuclear power involve scientific and probabilistic uncertainty. There is, for example, uncertainty regarding actual radiation exposure levels, particular people's sensitivity levels, the likelihood of dangerous consequences in a specific case, the given causal chains of harm, and so on. U.S. Department of Energy (DOE) representatives admit that Chernobyl could cause up to 28,000 fatal cancers in the former Soviet Union, Scandinavia, and Europe over the next 50 years. Other scientists and policy makers claim that the number of fatal cancers caused by Chernobyl will be as high as 475,000. Also, U.S. Nuclear Regulatory

Commission (NRC) data indicate that the United States has a 50-50 chance of having another accident the size of Three Mile Island or larger. For all these reasons, a fundamental ethical issue concerns how risk assessors ought to evaluate situations involving some sort of factual or probabilistic uncertainty. (1) Should one assume that a potential risk imposer is innocent until proved guilty or guilty until proved innocent? Should the burden of proof be on the alleged risk imposer or on the victim? (2) In evaluating uncertain radiation risks is one ethically bound to minimize false positives or false negatives? (3) In evaluating radiation risks, should one assume, in a situation of probabilistic uncertainty, that one is ethically required to follow the utilitarians' rules of maximizing average expected utility or the contractarians' rules of avoiding the worst outcome?

Many assessors evaluating nuclear-related risks in situations of factual or probabilistic uncertainty follow traditional norms of avoiding false positives (type-I error) rather than avoiding false negatives (type-II error). For example, the International Atomic Energy Agency (IAEA) and the International Commission on Radiological Protection (ICRP), which develop principles of radiation protection, require that radiation levels be kept as low as reasonably achievable, even below a particular exposure standard. In so doing, the ICRP and IAEA appear to follow the norm of avoiding false negatives, false assumptions of no harm (ICRP, 1991. *1990 Recommendations of the International Commission on Radiological Protection.* Pergamon, Oxford; IAEA, 1995. *Organization and Operation of a National Infrastructure Governing Radiation Protection and Safety of Radiation Sources.* IAEA, Vienna). However, when U.S. government risk assessors evaluated the radiation exposures of the late Orville Kelly, they avoided false positives and claimed that his death from leukemia was not caused by radiation received when he was Naval Commander of Eniwetok Atoll. Kelly was within 5 miles of ground zero, unprotected, during 23 nuclear weapons tests on the atoll. He died in his thirties, leaving a wife and four children. Indeed, for virtually all of the leukemia and bone cancer victims among 500,000 U.S. servicemen exposed to fallout in nuclear weapons tests in southwestern United States and the Pacific during the 1950s and 1960s, government risk assessors claimed that their high radiation exposures did not cause their cancers (U.S. Congress, 1994). Part of the assessors' rationale was that, in cases of uncertainty, one ought to limit false positives, take a conservative approach, and avoid positing an effect where there may be none. Such a traditional scientific approach places the burden of proof on those—like Orville Kelly—attempting to confirm some harm.

Many deontologists or contractarians, however, claim that traditional scientific norms for dealing with uncertainty are inapplicable to nuclear risk decisions because they affect human welfare. They argue for minimizing type-II error (avoiding false negatives) because doing so gives greater protection to public health and places the burden of proof on risk imposers rather than risk victims. More generally, ethicists have challenged the traditional legal dictum that a potentially dangerous situation (such as exposure to high levels of radiation) is "innocent until proved guilty." They argue that fairness and citizens' rights to equal treatment require risk evaluators to place the burden of proof on the alleged risk imposer. Their reasoning is that causal chains of harm are difficult to prove and that risk victims are less able than risk imposers to bear the medical and economic costs of faulty risk evaluations.

Other ethical controversies concerning nuclear-related uncertainty focus on whether one ought to judge risk acceptability according to Bayesian or maximin rules. Should evaluators follow Bayesian rules and maximize average expected utility (where expected utility is defined as the subjective probability of some outcome times its utility)? Or should evaluators follow maximin rules and minimize the likelihood of the worst outcome (consequences)? Utilitarians like John Harsanyi argue for the Bayesian position on the grounds that "worst cases" of technological risk rarely occur (J. Harsanyi, 1975. Can the Maximin Principle Serve as a Basis for Morality? A Critique of John Rawls' Theory, *American Political Science Review 69,* 594–605). They say that maximin decisions are overly conservative, impede social progress, and overemphasize small probabilities of harm. Egalitarians or deontologists like John Rawls argue for maximin on the grounds that, in cases like radiation risks from commercial nuclear power, the subjective risk probabilities are both uncertain and dwarfed by potentially catastrophic consequences such as nuclear core melts (J. Rawls, 1971. *A Theory of Justice.* Harvard Univ. Press, Cambridge, MA). They also point out the essential asymmetry of zero–infinity risk problems: a small (close to zero) probability of catastrophe does not outweigh infinitely serious risk consequences. Hence, deontological ethical theory—with its emphasis on rights to bodily security and duties to avoid catastrophe—likely explains much aversion to commercial nuclear fission, despite the low probability of catastrophe.

V. EQUITY AND NUCLEAR RISKS

Even if the overall consequences of a nuclear accident are minimal, risk evaluators following deontological ethical theory point out that the hazard may be unacceptable if the risk is inequitably distributed. Psychological studies also reveal that, in their risk evaluations, laypeople are often more averse to a small, inequitably distributed societal risk than to a larger, equitably distributed risk. In general, both deontological and utilitarian ethical theorists argue that people ought not be discriminated against merely because they are in a different location in space and time. For example, Parfit argues that temporal differences are not a morally relevant basis for discounting future costs and thus discriminating against future generations with respect to risks such as radiation. He and others maintain that a risk imposition is less acceptable to the degree that it imposes costs on future persons but awards benefits to present persons. Commercial nuclear fission, for example, benefits mainly present generations, whereas its risks and costs also will be borne by members of future generations who may be affected by long-lived nuclear wastes (D. Parfit, 1983. Energy policy and the further future: the social discount rate. In *Energy and the Future* (D. MacLean and P. Brown, Eds.), pp. 31–37. Rowman and Littlefield, Totowa, NJ).

Every year, each 1000-MW reactor discharges about 25.4 metric tons of high-level waste as spent fuel. For 300 commercial reactors, worldwide, the annual high-level radwaste is 7620 metric tons. Only 10 μg of plutonium is almost certain to induce cancer, and several grams of plutonium, dispersed in a ventilation system, is enough to cause thousands of deaths. Moreover, as even industry experts admit, each of the 7620 metric tons of high-level waste produced annually has the potential to cause hundreds of millions of cancers for at least the first 300 years of storage, and then tens of millions of cancers for the next million years, in the unlikely event of its dispersal. These cancers could be prevented, of course, with isolation of the wastes for a million years. That is why most plans for high-level nuclear waste storage call for defense in depth, for sealing the waste in a ceramic material, and for burying it deep underground in stainless steel or copper canisters. Nevertheless, the U.S. Environmental Protection Agency and the U.S. National Academy of Sciences have warned that we cannot count on institutional safeguards for nuclear waste.

Many economists and utilitarians support the use of nuclear power. They say that policy makers should discount future costs such as deaths caused by radioactive wastes. They also question whether "geographical equity" or "environmental justice" requires risks to be distributed equally across generations, regions, and nations. Proponents of siting nuclear reactors and waste repositories in economically and socially disenfranchised areas argue that such risks have been voluntarily accepted and that they provide employment as well as tax benefits. They say that "a bloody loaf of bread is better than no loaf at all." Opponents of such risk impositions argue that life-threatening hazards rarely bring substantial benefits and that it is unfair for economically, educationally, and socially disenfranchised persons to bear larger burdens of societal and workplace risks. They claim that just as it was wrong, more than a century ago, for U.S. southerners to defend slavery on the grounds that it was necessary to the economy of the South, so also it is wrong for risk evaluators to defend inequitable distributions of radiation and other risks on the grounds that such inequities are necessary to the common good. Deontologists argue that "the end (economic welfare) does not justify any means (inequitable distributions of risk)." Rather, human rights to equal protection are inviolate. They argue that some risks should be kept lower than others and that uniform standards are too lenient in protecting only "average" people rather than those who are especially sensitive, such as children who are victims of the Chernobyl radiation exposures (R. D. Bullard, 1993. *Confronting Environmental Racism.* South End, Boston; K. Shrader-Frechette, 1991, 71–72; 1993).

VI. CONSENT AND NUCLEAR RISKS

Of course, ethics allows inequitable risk impositions if the victims give their consent. The difficulty, however, is that persons most able to give free informed consent to higher societal risks (the wealthy and well educated) usually do not, whereas those least able to give free informed consent are those who usually bear higher societal risks like those from nuclear power. In the United States, for example, the only communities willing to serve as compensated hosts for proposed high-level nuclear waste facilities are Yakima Indian Nation, Nye County (Nevada), and Morgan County (Tennessee), all areas with high unemployment, high poverty, and low levels of education. In such a situation, deontologists argue that it is questionable whether members of the proposed host communities—especially members of future generations—are able to give free informed consent to the radiation risks associated with the facility. Likewise, they argue that because of coercive social conditions, in general poor people ought not be allowed to trade bodily security for higher wages or economic benefits. If the deontologists' arguments are correct, then even small risks that are imposed on people who are unable to give genuine free informed consent are highly questionable. Most modern nations do not recognize the consent—allegedly given by children, the mentally ill, or prisoners—to higher risks of medical experimentation. Likewise deontologist risk evaluators argue that assessors ought not ignore the lack of genuine consent often associated with inequitable impositions of risks such as radiation (see, for example, MacLean, 1986; A. Gewirth, 1982. *Human Rights,* University of Chicago Press, Chicago; Shrader-Frechette, 1993, chap. 8–9).

Some utilitarian ethicists have argued, however, that risk-for-money trade-offs serve the greater good and that no instances of consent are perfect. They also believe that Adam Smith's "compensating wage differential," for example, gives unskilled workers (in riskier occupations) economic opportunities that they otherwise would not have. They claim that contemporary economies require a "politics of sacrifice" in which free informed consent (to all societal and workplace risks) is unrealistic and unattainable (J. Harsanyi, 1975. Can the maximin principle serve as a basis for morality? A critique of John Rawls' theory. *American Political Science Review 69,* 602).

Philosophers who argue for the ethical justifiability of the economics–safety trade-off tend to be utilitarians and to have more lenient conceptions of free informed consent. They also tend to underemphasize sociological differences among those who accept versus those who reject risky jobs and living conditions, and to believe that pursuing neoclassical economics leads overall to social and ethical benefits. Philosophers who argue against the economics–safety trade-off tend to be deontologists or contractarians and to have more stringent conceptions of free informed consent. They also tend to overemphasize the sociological differences among persons allegedly choosing different levels of societal risks and to be more critical of the ethical assumptions underlying neoclassical economics. For example, in the United States most people who allegedly choose to be uranium miners are native Americans with high levels of poverty and unemployment and low levels of education and opportunity. Persons in more sociologically advantaged groups usually do not choose work that exposes them to risks from radiation.

VII. NUCLEAR POWER AND DUE PROCESS

Regardless of whether ethicists take a deontological or utilitarian position on questions of distributive equity, consent, and nuclear-related radiation risks, both groups of theorists generally agree that some form of compensation is required to justify imposing higher societal risks on particular groups, such as uranium miners. Psychological studies likewise reveal that people often are just as concerned about small, uncompensated societal risks as they are about larger risks for which they are certain to be compensated. If these concerns are legitimate, then ethicists ought to assess comparable uncompensated risks more negatively than less compensated risks.

The rationale for such assessments is that lack of full compensation for a societal risk threatens victims' due-process rights and therefore harms them. As a consequence, the acceptability of radiation risks is in part a function of compensation.

At the beginning of the atomic era, industry was reluctant, both on economic and on safety grounds, to use fission to generate electricity. Worried about safety, all major U.S. corporations with nuclear interests refused to produce electricity by means of fission unless some indemnity legislation was passed to protect them in the event of a catastrophic accident. The top lobbyist for the nuclear industry, the president of the Atomic Industrial Forum, has confirmed what numerous government committee reports show. Commercial nuclear fission began mainly because government leaders wanted to justify continuing military expenditures and to obtain weapons-grade plutonium. Moreover, at least in the United States, fission-generated electricity began only because the government provided more than $100 billion in subsidies (for research, development, waste storage, and insurance) to the nuclear industry. Congress also gave the utilities a liability limit (in the Price–Anderson Act) which protects licensees from most of the public losses and claims in the event of a catastrophic nuclear accident. The current U.S. liability limit is approximately $7.2 billion, less than 5% of the cost required to clean up Chernobyl, which was not a worst-case accident. Similar liability limits for commercial nuclear fission exist in other countries, such as Canada, where there is a $75 million government and industry liability limit for nuclear accidents. Since the costs of such accidents can run into the hundreds of billions of dollars, the constitutionality of these liability limits has been challenged in every nation in which they exist (Shrader-Frechette, 1993, 1–27).

At the proposed Yucca Mountain (Nevada) repository for U.S. high-level nuclear waste and spent fuel, site assessors have concluded that the location has met criteria for early site suitability. Nevertheless these evaluators have ignored the fact that the government has not met demands of the proposed host community for full liability for repository accidents. Indeed, the government is willing to compensate the host community for disruptions caused by site studies only if the community withdraws its opposition to the site. (80% of Nevadans oppose the facility. See P. Slovic, J. Flynn, and M. Layman, 1991. *Science* **254,** 1604.) At least part of citizens' opposition to the proposed repository has arisen because government has ignored their rights to full compensation and liability. If citizens' demands for such due-process rights are correct, and most deontologists would claim they are, then Yucca Mountain assessors ought not have ignored these demands. Some utilitarian moral philosophers also reason that if a nuclear waste facility is safe, as the government alleges, then risk imposers have little to lose from allowing full liability for accidents (Shrader-Frechette, 1991, 1993).

VIII. CONCLUSIONS

If the preceding survey of ethical considerations is correct, then several conclusions follow: (1) Evaluations of nuclear-related risk ought to be ethically weighted to reflect the value of deontological factors such as equitable risk distribution, compensation, and consent. Evaluations ought to reflect the fact that such ethical factors are just as important as risk magnitude in determining the acceptability of nuclear power. (2) Although technical assessment is necessary for an ethical evaluation of nuclear power, it is not sufficient. Ethical evaluation also needs to include components such as analysis of democratic preferences, citizen negotiation, and alternative assessments, so that the *procedures*, not merely the *outcomes*, of societal decision making regarding nuclear energy satisfy ethical constraints. (3) Although it is reasonable for good scientists to limit false positives, in evaluating nuclear-related uncertainties societal decision makers also need to limit false negatives and to protect the most vulnerable parties. (4) Although economically realistic risk evaluations, in a society of limited resources, ought not presuppose an infinite value for health and safety, neither ought risk evaluations to presuppose a zero value for even small threats to health and safety, particularly if compensation or consent is inadequate. At least for deontologists, consent and compensation are often necessary to justify risk trade-offs.

Because of the importance of ethical constraints, nuclear power ought not be evaluated by a technological "fix" that addresses only risk magnitude. Likewise, because the public has the right to determine how safe is safe enough, how safe is fair enough, and how safe is voluntary enough, societal risk evaluation cannot be accomplished by a public relations "fix" orchestrated by risk imposers or by the scientific community. Many controversies over evaluation of radiation risks can be solved only by ethical analysis and democratic process. Some of the most important aspects of nuclear power thus are not scientific but ethical.

Bibliography

Dubrova, Y. E., *et al.* (1996). Human minisatellite mutation rate after the Chernobyl accident. *Nature* **380,** 683–686.

Hogan, N. (1994). Shielded from liability. *ABA J.* **80,** 56–60.

International Commission on Radiological Protection (1991). "1990 Recommendations of the International Commission on Radiological Protection." Pergamon, Oxford.

Kendall, H. (1991). Calling nuclear power to account. *Calypso Log* **18,** 8–9.

Marples, D. R. (1996). Chernobyl: The decade of despair. *Bull. At. Scientists* **52,** 20–31.

National Research Council, Committee on the Biological Effects of Ionizing Radiation, Board on Radiation Effects Research Commission on Life Sciences (1990). "Health Effects of Exposure to Low Levels of Ionizing Radiation. BEIR V." National Academy Press, Washington, DC.

Shrader-Frechette, K. (1993). "Burying Uncertainty: Risk and the Case against Geological Disposal of Nuclear Waste." Univ. of California Press, Berkeley.

Shrader-Frechette, K. (1991). "Risk and Rationality: Philosophical Foundations for Populist Reforms." Univ. of California Press, Berkeley.

United Nations Scientific Committee on the Effects of Atomic Radiation (UNSCEAR) (1994). "Sources and Effects of Ionizing Radiation—UNSCEAR 1994 Report to the General Assembly, with Scientific Annexes." United Nations, New York.

U.S. Congress (1994). "Human Subjects Research: Radiation Experimentation." Hearing before the Committee on Labor and Human Resources, United States Senate, 103rd Congr., 1st sess., Jan. 13, 1994. U.S. Government Printing Office, Washington, DC.

Nuclear Testing

STEVEN LEE

Hobart and William Smith Colleges

I. Is the Use of Nuclear Weapons Permissible?
II. Is Nuclear Testing Dangerous?

GLOSSARY

assured destruction A form of nuclear deterrence involving threats to destroy, and the capability of destroying, the opponent's society, if the opponent strikes first.

conditional intention An intention to do something, such as use nuclear weapons, only if some condition obtains, such as an opponent's aggression.

conventional weapons Nonnuclear weapons.

counterforce nuclear deterrence A form of nuclear deterrence strategy in which the threats are made against military rather than civilian targets.

credibility A measure of the effectiveness of deterrence based on the strength of the opponent's belief that the state would carry out its threats.

crisis (in) stability The ability (or inability) of a strategy of nuclear deterrence to ensure the avoidance of war during a political crisis between nuclear opponents.

just-war theory The traditional Western approach to assessing the moral status of military activity, which expresses a special concern for avoiding harm to civilians.

nuclear deterrence A policy of threatening the use of nuclear weapons in order to avoid an opponent's aggression.

nuclear proliferation The spread of nuclear weapons to nonnuclear states.

IS THE TESTING OF NUCLEAR WEAPONS morally permissible? The answer to this is more difficult than the answer to the same question asked about a conventional weapon. The reason is that there are important moral differences between nuclear and conventional weapons. Consider how one would respond to the general question about the moral status of weapons testing. It seems, in general, that weapons testing is permissible if the testing itself is not harmful and if the purpose of the testing is to increase the effectiveness of the weapon for its intended use, given that that use is itself permissible. This could be formulated as *the permissibility principle*: the testing of a weapon is permissible only if (1) its purpose is to make the weapon more effective for the weapon's intended use, which is itself permissible, and (2) the testing itself does not pose a serious risk of harm to others. Is it permissible to test a handgun? In general, the answer is yes, if it is tested at a safe firing range by a person intending to improve in the use of the weapon for legitimate acts of self-defense. But, if it is tested on a busy city street, or by someone intending to use it for an act of terrorism, the testing would not be permissible.

The question about the permissibility of testing conventional weapons is normally fairly easy to answer because it is usually not hard to determine if conditions (1) and (2) apply. The difficulty in applying the permissibility principle to the testing of nuclear weapons is that it is not easy to determine if these conditions apply.

This is the result of the moral differences between nuclear and conventional weapons. In the remainder of
this article, I will consider in turn the difficulties in
applying conditions (1) and (2) to the case of nuclear
weapons testing.

I. IS THE USE OF NUCLEAR WEAPONS PERMISSIBLE?

In regard to condition (1), the main question is
whether the intended use of nuclear weapons is permissible. Many have argued that it is never permissible to
use nuclear weapons, even in self-defense. The reason,
following just-war theory or related lines of argument,
is that the destructive effects of the weapons are so
vast that the use of the weapons would inevitably cause
serious harm to innocent civilians, and that it is never
permissible to cause such harm. Even if, the argument
continues, the use of a single nuclear weapon would not
cause serious harm to civilians, because the weapon was
of a low yield and fired at an isolated area, still, given
the vast number of existing nuclear weapons, the use of
that one weapon would carry a serious risk of escalation,
leading to the death of many civilians. Thus, the use
of even one weapon would be impermissible. If this
argument is sound, then, in the light of the permissibility
principle, the testing of nuclear weapons is impermissible.

But matters are more complex than this. With the
exception of the bombings of Hiroshima and Nagasaki,
the way in which states have used nuclear weapons is
not by firing them on an opponent, but by practicing
nuclear deterrence, using the threat of a retaliatory firing of the weapons to keep an opponent from aggression. What if the purpose of testing the weapons is to
make deterrence more effective? Well, in the light of the
permissibility principle, the question to ask is whether or
not it is permissible to use nuclear weapons to practice
deterrence. Defenders of nuclear deterrence argue that
it is permissible because the intention involved is not
the making of war, but the avoidance of war. But there
is a line of analysis, extending the argument of the previous paragraph, to the effect that the use of nuclear
weapons for deterrence is impermissible. While nuclear
deterrence may involve the intention to avoid war, implicit in the threat to retaliate is the conditional intention
to attack civilians, because nuclear deterrence, as practiced by the major nuclear powers, includes the ultimate
threat to destroy the opponent's society, the threat of
"assured destruction." But, because it is impermissible
to attack civilians, it is impermissible to intend to do
so, and so impermissible to adopt a military policy that
has such an intention, even if the intention is conditional.

If we assume the soundness of this line of argument,
the permissibility principle seems to entail the impermissibility of nuclear testing, even when undertaken for the
sake of more effective deterrence. But there is one line
of defense that proponents of nuclear testing could invoke. Because nuclear deterrence is practiced with two
intentions, one permissible and one not, a defender of
testing might argue that nuclear testing is itself permissible, if it is designed to support the permissible intention
of deterrence, namely, that of avoiding war. Here is an
analogy. The organizers of a prostitution ring may be
said to practice their vocation with two intentions, to
exploit the vulnerable individuals who work for them
and to provide the public with a service. The latter
intention, considered by itself, is permissible, but the
former is not, because of the exploitation involved, and,
for that reason, the practice itself is not permissible.
Now, consider one organizer's decision to require that
condoms be used in the sexual acts with the ring's clients.
This could be said to be morally permissible (perhaps
even morally required), even though the practice as a
whole is impermissible. Why is this? Requiring condoms
supports the permissible intention, in that it makes the
service provided less risky and, in that sense, more effective, without making the nature of the practice in regard
to the impermissible intention any worse. Requiring
condoms does not deepen or extend the exploitation
involved in the practice.

A generalization supported by this analogy is that
when a practice has both a permissible and an impermissible intention, another activity that supports the permissible intention is not itself impermissible in virtue
of the impermissibility of the practice as a whole. This
principle may hold in the case of nuclear testing as well.
Nuclear deterrence is like the prostitution case in that
the practice is impermissible because a permissible end
is pursued through impermissible means. But nuclear
testing designed to make deterrence more effective, like
the organizer's requiring of condoms, supports the permissible intention, the avoidance of war, without making
the practice worse in regard to its impermissible intention, the conditional intention to attack civilians. Thus,
even if nuclear deterrence is an impermissible practice,
nuclear testing designed to make deterrence more effective may not be.

But here another complication arises. The idea of
testing to make deterrence more effective disguises an
important ambiguity. Nuclear testers may seek to make
deterrence more effective in different ways, either (1)
by impressing the potential opponent by making the
capacity to retaliate more evident and the risks of retali-
ation more vivid, (2) by making the weapons more reliable, in order to lessen the risk of accidental detonation
and increase assurances that the threats could be carried

out, or (3) by developing new or more advanced weapons, in order to make possible a different, more effective form of deterrence strategy. All three of these intentions are intentions to make deterrence more effective, but there may be important moral differences among them. In particular, while the first two types of intention seem to raise no special moral problems, the third type of intention bears closer scrutiny.

In regard to the third type of intention, there are two kinds of cases to consider: weapons refinement and nuclear proliferation. First, a state that already has nuclear weapons may test new designs in order to develop weapons that are more advanced, for example, in their explosive yield or in the way in which the energy of the explosion is released, intending to use the improved weapons to adopt a more effective strategy of deterrence. Second, a state without nuclear weapons may seek to acquire them, and to test an initial nuclear device to ensure that it has developed a workable weapon, intending to adopt nuclear deterrence as a more effective form of military deterrence.

Consider the first kind of case, where a state tests a new weapons design in order to apply it to a different form of nuclear deterrence strategy. The example of this from our nuclear history is the development by the major nuclear powers of more advanced weapons that could support a strategy of counterforce nuclear deterrence, where the nuclear threat would be directed more toward military than civilian targets. Proponents of counterforce strategy believe that it would be a more effective form of deterrence, because it would increase the credibility of the nuclear threats. (They also argue that it would be a form of nuclear deterrence that would avoid the conditional intention to attack civilians, because it would allow the weapons to be aimed exclusively at noncivilian targets.) In a similar vein, in regard to the second kind of case, proponents of nuclear proliferation believe that the acquisition of nuclear weapons, and the consequent making of nuclear threats, would allow the state to practice a form of deterrence more effective than conventional military deterrence.

The truth of the beliefs that nuclear weapons refinement or acquisition would lead to a more effective form of deterrence is much debated. The argument that weapons refinement would not lead to a more effective form of deterrence is based on the claim that counterforce deterrence would increase the risk of nuclear war because it would increase crisis instability. In a crisis, counterforce strategy would make each side more likely to believe that the other side was about to strike first, and hence make each side more likely to strike, increasing the risk of war. While counterforce deterrence may make nuclear threats more credible, as its proponents argue, overall it would make war more likely, and hence

deterrence less effective. There is a similar line of argument concerning the acquisition of nuclear weapons. The argument against nuclear proliferation is that the spread of nuclear weapons to states without them would make nuclear war more likely by making the military balance between opponents less stable, because, for one, each side may have a strong incentive to attack in a crisis to attempt to destroy the nascent nuclear capability of its opponent.

II. IS NUCLEAR TESTING DANGEROUS?

But one could argue that even if these criticisms of counterforce deterrence and nuclear proliferation are sound, they do not show that nuclear testing for the sake of weapons refinement or acquisition violates condition (1) of the permissibility principle. For the testing would be carried out with the intention to make nuclear deterrence more effective, even if the intention is based on a false belief. This is where condition (2), the requirement that the testing itself not pose a serious risk of harm, becomes relevant. Consider another analogy with conventional weapons. Imagine that my only intention in having a firearm is to use it in individual self-defense, but that I am testing a more advanced design of my own creation, an automatic weapon, in order to increase the effectiveness of my self-defensive capabilities. Despite the purity of my intentions, one might argue, it would be wrong for me to test the weapon, because automatic weapons pose a serious risk of harm to others. Presumably, this is why citizens in the United States are allowed to own handguns but not machine guns. Testing to create a machine gun is not permissible because they pose a great risk of harm to others.

So, nuclear testing is impermissible, due to condition (2), if the testing leads to a form of deterrence that is less effective, because a less effective form of deterrence creates a greatly increased risk of harm to others, in that it entails an increased likelihood of nuclear war. This is true whatever the intention with which the testing is carried out. The tester may seek to make deterrence more effective by impressing the opponent or by increasing weapons reliability, as well as by seeking to develop a new form of deterrence strategy, but if the result is a less effective form of deterrence, the testing is impermissible in any case. Let us assume for the moment that counterforce deterrence practiced by established nuclear powers and nuclear deterrence practiced by new nuclear proliferators are less effective than the forms of deterrence they replace. Does nuclear testing lead to these less effective forms of deterrence? Clearly it does in the case of new nuclear proliferators: for a state to test its first nuclear weapon is for it to begin to practice nuclear deterrence. While there is not the same tight link in the case of testing by established nuclear powers, our nuclear history suggests that a shift to counterforce strategy as a result of testing is the normal course of events.

But is it correct to assume that counterforce deterrence and nuclear proliferation are dangerous? Are the arguments that these represent less effective forms of deterrence sound? While there is disageement about whether counterforce deterrence is more dangerous, there is fairly strong consensus that nuclear proliferation is dangerous. But this is all the argument needs, for the only way effectively to stop the testing of nuclear weapons by nonnuclear states is to stop all nuclear testing. The reason is that an international ban on testing cannot be enforced against nonnuclear states alone. These states will not be willing to commit themselves not to test, however much they ought to do so, unless the established nuclear powers so commit themselves as well. Thus the testing by established nuclear powers is dangerous, if for no other reason than that it leads nonnuclear states to test as well, and that testing is itself dangerous. Thus there is a strong case, based on condition (2) of the permissibility principle, that nuclear testing is morally impermissible. This, along with other arguments, supports the adoption of a comprehensive test ban treaty, such as was approved by the United Nations in September 1996.

Bibliography

Arnett, E. (1995). *Implementing the comprehensive test ban.* SIPRI Research Report #8 Oxford: Oxford University Press.

Goldblat, J. & Cox, D. (Eds.). (1988). *Nuclear weapons tests: Prohibition or limitation.* Oxford: Oxford University Press.

Lee, S. (1993). *Morality, prudence, and nuclear weapons.* Cambridge: Cambridge University Press.

Shue, H. (Ed.). (1989). *Nuclear deterrence and moral restraint.* Cambridge: Cambridge University Press.

Organ Transplants and Xenotransplantation

RUTH CHADWICK* and UDO SCHÜKLENK†

*Lancaster University and †University of the Witwatersrand

I. Introduction
II. Live Human Donors
III. Cadaver Donors
IV. Organ Transplants and Personal Identity
V. Xenotransplantation
VI. Summary

GLOSSARY

anencephaly Literally meaning "without brain," a congenital absence of all or part of the brain. Babies born with this condition have no hope of normal life or of long-term survival, but may have a functioning brainstem when they are born and are thus not brain dead in accordance with brainstem death criteria.

immunosuppression Reduction of the body's rejection of transplanted organs, for example, by drug therapy.

organ transplantation The act of removing an organ, such as a kidney or heart, from one individual and grafting it into another individual.

xenotransplantation The transplantation of animal tissue between species.

zoonosis Disease communicated to humans from animals.

ORGAN TRANSPLANTATION is the surgical removal of a body organ, such as a kidney, from one individual (the donor) and placement of the organ in another individual for the purpose of improving the health of the recipient.

I. INTRODUCTION

The first successful kidney transplant between human identical twins took place in 1954, followed by the first successful transplant between fraternal twins in 1957. Subsequent development of antirejection drugs led to successful transplants between nonrelatives. The first heart transplant took place in 1967, when Dr. Christiaan Barnard transplanted a heart from a woman into Mr. Washkansky, who died 18 days later.

Since those early experiments not only have well-established heart transplant facilities been put into place (e.g., at Papworth and Harefield in the UK), but there have been developments in a number of areas, including liver transplants, heart-lung transplants, and transplants of tissue into the brain. In 1982 a team of doctors in Salt Lake City implanted an artificial heart into Dr. Barney Clark. The prospect of artificial hearts seemed initially to offer hope of an answer to the perennial problem of shortages; attention has subsequently turned to animals and to cloning technology as possible organ sources for the future.

There have been particular issues surrounding specific organ sources, e.g., fetuses and anencephalic babies. In 1989 in the UK, following considerable discussion of the use of fetal tissue transplants, a Code of Practice on the Use of Fetuses and Fetal Material was published. There have been arguments to the effect that anencephalics should be defined as a distinct category of human being, which it is permissible to use as organ sources.

The practical problems surrounding organ transplants have been of several kinds. In the early days there were psychological problems, associated in particular with heart transplants, because of the symbolism associated with the heart. Rejection occurs in all cases except identical twins, although tissue typing improves chances of success. Immunosuppressive drugs suppress the rejection reaction but also have side effects. Rejection problems in the case of animal to human transplants are considerably greater than those in human to human operations, and research has been and continues to be carried out into ways of overcoming this by genetic modification techniques. It is envisaged that the use of cloning technology would provide a way of bypassing rejection problems.

A major problem has always been the shortage in supply of organs, although from a perspective critical of transplant technology it might be argued that the problem of "shortage" is at least in part a manufactured one, depending on factors such as our inability to accept the finitude of human life and preference for high-tech medicine. Linked with this is another issue that has surrounded organ transplantation, especially heart transplants: the expense. Some cardiologists recommended abandoning the British program in the 1980s, because although it was glamorous high-tech medicine, the opportunity costs were great compared to other forms of treatment. This is an aspect of a more general problem concerning priorities in medicine, and regarding whether doctors should "do all they can" to prolong life.

Given the acceptance of the desirability of organ transplantation, however, it is true that the demand outstrips supply globally, although there is variation between countries. Factors accounting for this variation include legislation prescribing the compulsory wearing of seat belts in cars (e.g., in the UK), which has reduced the organ supply, and cultural differences in attitudes concerning willingness to donate. Donor card systems have not been entirely successful because even when individuals carry a donor card, their wishes are not always respected. The wishes of relatives may be allowed to intervene. Given the shortage in supply, difficult decisions also have to be made about priortization among recipients—not only at an individual level but between different population groups. Issues of equity arise if some ethnic groups are less represented among donors, donees, or both. There are also issues about whether donors should be free to express a preference about the population group from which the donee should be selected.

Where human organ sources are concerned (the word "donor" is not always entirely appropriate), there are two kinds: live sources and cadaver sources. The ethical issues require separate consideration. In the case of some organs, of course, live donation is not a practical possibility. In each case there are potential conflicts of interest to be addressed.

II. LIVE HUMAN DONORS

A patient who needs an organ such as a kidney clearly has an interest in the preservation of his or her life. The potential organ source also has an interest in the preservation of his or her own health and bodily integrity, for donating an organ to someone else is not without risks to the donor. The donor is subjected to a nontherapeutic operation. It may be argued that this problem can be overcome by requirements of free and informed consent, but in intrafamilial cases, in particular, individuals may feel a burden of responsibility and guilt if they refuse.

It might be tempting to think that the interest in preserving life should always take precedence over other interests. After all, life is a necessary condition of having any other interests at all. This is true when we are thinking of the life of any one individual. Things become more complicated when we try to compare the interests of different individuals. Societies which grant the potential donor the chance to consent or refuse to donate do not, as a matter of fact, put the interest of one person in living longer higher than the interests of another in preserving health or bodily integrity.

The reasons for this can be given in the following arguments:

1. The interest in having a say over what happens to our own bodies is central to what gives us a sense of having a life of our own which we can to some extent control. For some people it is even more fundamental than the mere continuance of life, but even if not, it is an interest that we have to accept is a very deep one.

2. The consequences of abandoning the principle that people should be asked to consent to donate would be potentially far-reaching. We have to consider not only the interests of one potential donor and donee, but also the interests of society as a whole. Once we undermine the freedom of people to say what happens to their own bodies we may be undermining respect for human life itself anyway.

There have been cases, e.g., in the USA, of people trying to use legal process to force others to be donors of bone marrow. In a case study discussed in the *Hastings Center Report,* a Mrs. X had actually registered as a potential donor to help a relative. However, Mr. Head, who desperately needed a transplant, found out (through a breach of confidentiality) that she would be a possible suitable match for him, and tried to pressure her into agreeing. This attempt was unsuccessful. Such a precedent could in effect legitimate a new kind of body-snatcher any time organs were needed.

In some legal systems, such as English law, it would be an assault upon the person to perform an operation upon an individual without consent, and there are good reasons for thinking that this belief in freedom to decide is not something which we should give up. To the discussion of this topic there are, however, two qualifications and one criticism. The qualifications concern possible limits to choose what happens to our own bodies.

First, it might be argued that we have a moral obligation, if not a legal one, to help others where we can. Others would say that it is beyond the call of duty to agree to donate but that it provides an opportunity to choose altruism. If freedom of choice is to be a reality, however, it is necessary to be aware of the potential burden of moral pressure.

The second qualification concerns the possibility of a market in organs. In the 1980s there was a number of cases concerning sale of kidneys, for example. Response to this in some jurisdictions was swift. In 1989 the UK government passed legislation to outlaw this. There are arguments to suggest that our ability to control what happens to our own bodies should not extend to selling bits of them. One aspect of this is a view that human beings have an intrinsic worth that puts them beyond price, and that selling human beings or parts of them is incompatible with respect for human dignity. Immanuel Kant, though writing long before organ transplantation became an issue, had arguments for the view that it is wrong to sell parts of oneself—even a tooth. One argument was that this involved a logical contradiction. "Man," he said "cannot dispose of himself because he is not a thing; he is not his own property; to say that he is would be self-contradictory." In selling himself man makes himself a thing and is in effect selling freedom. "We can dispose of things which have no freedom but not of a being which has free will" (I. Kant, 1963. *Lectures on Ethics.* Translated by L. Infield. New York: Harper & Row).

An objection to this view, however, would be that if we are entitled to sell our labor in the market, then where is the dividing line between this in, e.g., physical work, and selling parts of the body? The above line of argument seems ultimately to depend on a view about what is intrinscially degrading to human beings. It also depends on an interpretation of human beings as ends in themselves, an idea which Schopenhauer rejected persuasively as a *contradictio in adjecto* in *On the Basis of Morality.*

If we confine our attention to interests, however, the argument may be that it is not really in the interests of the seller, who may be forced into it because of dire circumstances, and that social change is what is needed to try to prevent such situations from occurring, so that people are not in effect coerced into such transactions. There are also arguments about the wider consequences of a market in organs, analogous to Richard Titmuss's arguments in *The Gift Relationship* regarding a market in blood.

The criticism of giving importance to control over our own bodies might be based on a view that life is more important than anything else. John Harris, in "The Survival Lottery," considered the possibility that whenever two or more persons need organs, one healthy person should be killed to provide them, thus maximizing the number of lives saved. One person would be "sacrificed" (and attitudes in society would need to be changed so that this would be viewed as a noble thing to do) for the good of the greater number. The question is, of course, what implications for society such a strategy would have. There is a variety of negative "side effects" that are not easily quantifiable, such as the anxiety people might feel that they could be sacrificed at any time to save other peoples' lives. These utilitarian cost considerations, as well as the fact that many donees (e.g., in the case of heart transplants) do not survive very long, make Harris's proposal an interesting thought experiment but practically unrealistic.

III. CADAVER DONORS

On the face of it, it might look as if the reasons for giving potential donors the freedom to consent or refuse do not apply when the donor is dead, but this is a view not widely held. Here too there seems to be a potential conflict of interest. Thus there is a commonly held position that it is not acceptable to take a person's organs without first checking whether she carried a donor card, or whether she would have had any objections to donation. Where children are concerned, revelations that hospitals have kept their organs without the parents' knowledge have caused severe distress. What are the arguments for this position?

One reason behind this seems to be that there is an interest in controlling what happens to our bodies, even after our death. There is an opposing view however that the interests a person might have in what happens to her body after death are of less or no relevance when compared with the interests of a person currently alive, but who will not survive much longer without an organ transplant. Indeed, one needs to ask whether it makes much sense to apply the ethical concept of personhood to a dead body. Dead people by definition do not satisfy the typical criteria of moral personhood, and hence their alleged wishes, as put forward possibly by relatives or their advance directives, should not be considered as having the same status as those of living donors. Thus there have been various proposals regarding cadaver donors at various times and in various societies, from requiring that form of consent describable as "opting in," to other ways of handling things, such as "required request," or "opting out."

An "opting out" scheme would indicate that unless a person had positively registered an objection, her organs would automatically be available for use after death. It should be noted that such a scheme still preserves the donor's freedom of choice, but it alters the balance away from the interests of the donor and toward the interests of the donee.

A possible objection to this proposal may be that while it may be irrational to think that we can experience any harm after we are dead, nevertheless while we are alive we do not like to think of certain things happening to our bodies after death. This point is an extension of the point that we have an interest in having a say over what happens to our bodies. After all, they are our bodies. But if this is all it is, it is not clear that such an interest can withstand the competition from those who have an interest in the preservation of their lives. Perhaps we should realize the weakness of the interest in controlling our own dead bodies and move toward an "opting out" scheme.

There are further arguments to consider, however. First, we must take into account the fear of not what will happen to our bodies after death, but of premature death. There is concern not only that mistakes do sometimes occur in determining whether someone is dead, but that the demand for organ transplants has in itself affected the way that death is defined in leading to the establishment of criteria that maximize the chance of retrieving healthy organs suitable for donation. The obvious conflict of interests here is between a person's interest in not being declared dead, or even killed, before her time and a potential donee's urgent need of an organ.

An unsentimental realist might suggest that if the potential donor, who is not quite dead, is so near death that she is never likely to lead a normal life, then the interests of the potential donee should still prevail. But the case is more complicated than this, for all members of society have an interest in people not being killed before their time (with possible limited exceptions such

as voluntary euthanasia). For one thing, faith in the medical profession would be very rapidly undermined if it was thought that doctors viewed patients as potential sets of resources. What we do have to ensure is that the need for organs does not color in an unacceptable way people's judgments in elaborating and applying criteria of death.

Another objection to an "opting out" scheme is that it might reduce the possibility of displaying altruism. On the other hand it might be interpreted as a "push" toward a more altruistic, rather than apathetic, society. It does not deny the possibility of choice but more directly confronts people with it while demanding less than is demanded of a living donor. Both types of conflict of interests considered so far could be much alleviated if the gap between demand and supply was not so wide where organs are concerned.

IV. ORGAN TRANSPLANTS AND PERSONAL IDENTITY

Moral problems surrounding organ transplants are not concerned solely with the interest of one person in staying alive as against the interest of another in having a say in what happens to her own body. Much depends on how we view the human person, and how we regard the relationship of ourselves to our bodies. People have widely varying views about this, of both religious and secular origins. For example, even within the Judeo-Christian tradition, against the Old Testament view of the human being as basically an animated body—a body into which God "breathed life"—we may place some views which have seen the embodied person as an imprisoned soul. Others have thought that the body is needed intact for resurrection, and thus have opposed cremation and regarded dismemberment as a terrible punishment. Such a view could rule out organ transplants altogether. How one views one's relationship to one's body thus has a great impact on how one regards certain forms of medicial treatment. They provide worldviews within which we must consider conflicts of interests.

The modern secular philosophical debate about personal identity tends to concentrate on the merits of the opposing physical and psychological criteria of personal identity. To take the physical citerion, if I think of myself as basically my body, how much of my body can change before I am no longer the same person? There is a standard comparison in the literature here with a physical object such as a ship, of which over the years, various parts wear out and are replaced. Eventually none of the original parts remain. Is the end result the same ship?

If so, what has remained the same? If not, at what point did it cease to be the same ship?

In the case of human beings, the predominant view of the physical criterion of personal identity holds that one remains the same person as long as one has the same brain. This of course would raise problems for the possibility of brain transplants. On the psychological view, it does not matter how many bits of one's body are replaced as long as there is psychological continuity between the person at various stages.

On neither of these views does a kidney transplant raise a deep philosophical problem of personal identity. What may be a problem is how donees feel about their personal identity. So it may be a matter of an interest in feeling secure about what one's identity is, rather than in the problem of personal identity itself. Of course, in the event of the theoretical possibility that at some time in the future we could gradually replace all human parts as they wear out, we should have a problem like that of the ship. In such an event we may have to accept that questions of personal identity are unanswerable.

V. XENOTRANSPLANTATION

Another more recent issue pertains to the potential future possibility of successfully using animal organs for humans in need of replacement organs.

A. Science

Xenotransplantation is the transplant of animal tissue between species. For this article only the issue of transplanting animal organs into humans will be discussed. Xenotransplantation is a means considered to bridge the gap between the organs needed by people and those available to them. As already indicated, the supply of human organs is handicapped by a wealth of ethical, economic, and legal problems, the latter of which apply in most parts of the world. Serious attempts to implant animal organs into humans have been actively researched and tested since the the early 1960s, even though early but equally unsuccessful trials date back to the end of the last century. The ultimate scientific question is the extent to which it will be feasible and safe to use animal organs for use in humans. The problems involved in this pertain to the risk that the human body might reject the animal organ, but also to the risk that infections might be transmitted from the source animal to the tissue recipient, and from the tissue recipient to her immediate contacts and the wider population. Current research focuses on attempting to modify the genetic makeup of pigs by adding human genes, in order to research whether the scientific problems in xenotransplantation can be solved.

B. Ethics

The research into the possibility of xenotransplantation, and xenotransplantation itself, gives rise to a variety of ethical problems.

1. The Use of Sentient Nonhuman Beings as Involuntary Organ "Donors"

Several animal rights activists and philosophers have argued that it is unethical to inflict pain upon sentient nonhuman beings for the sake of helping suffering human beings. Their arguments are either of a utilitarian nature, e.g., P. Singer's *Animal Liberation,* or they are based in one type or another of deontological moral theory, e.g., T. Regan's *Case for Animal Rights.* The utilitarian approach to the use of animals in xenotransplantation argues that the equal consideration of interests requires us to use humans with a similar intellectual capacity to that of animals we might wish to use for xenotransplantation, or to use neither of these. Should we decide that we do not wish to use mentally handicapped people, or newborns, or senile people as organ banks, then it would be incoherent to suggest that we should be ethically entitled to use other higher mammals for this purpose. The reasoning consequentialist ethicists employ is that there is no morally relevant difference between some humans and higher mammals. For instance, they would suggest that we are ethically entitled to use anencephalic babies in which the cerebral hemispheres of the brain are absent in order to provide organs for transplantation purposes. Deontologists, on the other hand, have argued that what qualifies human beings for personhood is also present, to some degree, in other species. Every being that satisfies Tom Regan's "subject of a life" criterion has an inherent value and should not be used merely as a means to certain ends. Regan is clearly attempting to widen the scope of Kantian thinking to include nonhuman animals. He follows Leonard Nelson's line of argument, which proposed to distinguish between subjects of duty, who have a variety of rights and obligations, and subjects of rights, who have certain rights, but no duties because they are incapable of fulfilling these duties. Animals and humans who do not fullfill the criterion of full personhood may well fall into the latter category. Proponents of this view would argue that to treat animals as means to the end of extracting organs from them is ethically unacceptable. In the UK, the Nuffield Council on Bioethics was particularly concerned about the use of primates for transplantation purposes because of the closeness of humans and primates in evolutionary terms. The Council deemed the use of pig organs less problematic simply because society uses these animals already for the purpose of food and clothes production. Even if it is accepted, however, that it is appropriate to use animals as organ sources, there are concerns about animal welfare arising out of methods of production and of keeping the animals that are designed to minimize infection risks.

2. The Interests of Humans as Affected by Xenotransplantation

The history of research on human organ transplantation suggests that early recipients of animal organs have a low chance of survival. On the other hand, there is a large number of patients in desperate need of organs and many of these patients will die before suitable organs become available. A major concern however is that viral infections may jump from animals to humans—this is known as zoonosis—and affect not only the organ recipients but also their contacts. Safety thus arises as an ethical issue as well as a scientific one, and leads to a further ethical issue, namely, surveillance. In the UK, for example, a Steering Group of the United Kingdom Xenotransplantation Interim Regulatory Authority issued in 1999 a consultation report on mechanisms of surveillance involving surveillance of groups such as recipients, their close contacts, and health care workers. The report says that "Xenotransplantation cannot proceed unless there is an agreement to comply with postoperative surveillance." Clearly the surveillance of third parties introduces concerns relating to their consent, in addition to the potential restrictions placed upon the recipient. The Steering Group report suggests that recipients should comply with conditions such as refraining from fathering or bearing a child. While taking the view that a policy of voluntary compliance is the only practical option, they recognize that there may be some emergency which would warrant, for example, detention for testing, and that there are clear human rights implications here. In the longer term and beyond the stage of clinical trials, such a regime would have to be relaxed if xenotransplantation were to become a routine form of therapy. There is therefore a need for regular review.

C. Religion

Some Christian theologians have argued that human beings are fundamentally superior to nonhuman animals. This view is certainly widely shared among most people in Western societies, as a variety of surveys in regard to attitudes toward xenotransplantation indicate. It is based on Genesis 1:27, where human beings are said to "have unique significance and value because only they are made in the image of God." In view of, for instance, the British Christian Medical Fellowship, the painless killing of animals for the purpose of saving human life or for improving its quality would be considered as ethically justifiable. On the other hand some Christian theologians have argued that humans become less and less human the more animal parts we implant into people. They have also pointed out that it would be somewhat frivolous to change God's creation by breeding animals into something more and more human by changing the genetic makeup through the infusion of human genes. This argument is shared to some degree by Islamic theorists.

VI. SUMMARY

A variety of practical problems in regard to live donor organ donations occur which seem to indicate that great care should be taken that no unethical practices arise. In particular, vulnerable populations (i.e., poor living donors selling their organs in order to survive) need to be protected from exploitation. On the other hand, it seems that cultural objections to the use of dead peoples' organs stand in the way of the effective use of such organs for the greater good of a large number of potential donees. It has been suggested that animal organs could help us to avoid the ethical problems raised by the use of human organs. However, to engage in genetic modification of sentient nonhuman beings and expose them, organ recipients, and their contacts to the risk of harm in order to avoid taking organs of nonsentient (i.e., dead) humans may seem to be a policy which is a result of irrational beliefs. Irrespective of these concerns there remains the problem as to how to allocate scarce medical resources.

Bibliography

Advisory Group on the Ethics of Xenotransplantation (1997). *Animal Tissue into Humans.* London: HMSO.

Allan, J. S. (1995). Xenotransplantation at a crossroads: Prevention versus progress. *Nature Medicine* **2**, 18–21.

Chadwick, R. F. (1989). The market for bodily parts: Kant and duties to oneself. *Journal of Applied Philosophy* **6**, 129–139.

Chadwick, R. F. (1993). Corpses, recycling and therapeutic purposes. In *Death Rites: Law and Ethics at the End of Life* (R. Lee and D. Morgan, Eds.). London: Routledge.

Jones, D. G. (1991). Fetal neural transplantation: Placing the ethical debate within the context of society's use of human material. *Bioethics* **5,** 23–43.

Lamb, D. (1990). *Organ Transplants and Ethics.* London: Routledge.

McCarrick, P. M. (1995). Organ transplant allocation (Scope Note 29). *Kennedy Institute of Ethics Journal* **5,** 365–383.

Nuffield Council on Bioethics (1996). *Animal to Human Transplants: The Ethics of Xenotransplantation.* London: Nuffield Council on Bioethics.

Truog, R. D., and Fletcher, J. C. (1990). Brain death and the anencephalic newborn. *Bioethics* **4,** 199–215.

United Kingdom Xenotransplantation Interim Regulatory Authority (1999). Draft report of the Infection Surveillance Steering Group, London (reprint). *Bulletin of Medical Ethics,* Nov.

Playing God

WILLIAM GREY

University of Queensland

I. Secular and Religious Applications
II. The Moral Presupposition
III. Applications in Bioethics
IV. Applications in Genetics
V. Conclusion

GLOSSARY

DNA Deoxyribonucleic acid. The molecular vector of genetic information which determines the attributes and development of organisms.

genetic engineering The deliberate alteration of the structure or hereditary characteristics of organisms, especially by the direct modification of DNA.

germ cells Reproductive cells containing the DNA genetic blueprint which determines the structure, character, and development of an organism. Modifications to an organism's germ cells are passed on to its descendants.

hubris Human overconfidence, arrogance, or overweening pride.

negative genetic engineering Also called therapeutic or corrective engineering. Genetic engineering that aims to rectify some defect, disease, or disability.

positive genetic engineering Also called eugenic or enhancement engineering. Genetic engineering that aims to improve character traits of organisms, for example, increasing disease resistance, robustness, or fertility.

somatic cells The body tissues of an organism apart from the reproductive or germ cells. Somatic changes are not passed on to an organism's descendants.

PLAYING GOD is a phrase frequently used to describe acts or decisons about matters which the speaker believes should either be treated with extreme caution or left well alone. Often, as the phrase suggests, the implied objection to a proposed course of action is based on religious beliefs. Such use of the phrase usually presupposes, or at least alludes to, a divinely ordained order in the physical or moral universe which it would be reckless or impermissible to transgress. There are also secular uses in which the phrase is used metaphorically to indicate that the consequences of an act are exceedingly serious or far-reaching and must therefore be considered with very great care. The phrase may also be used to describe paternalistic or authoritarian decisions, often resented, made by individuals in positions of power.

The most familiar applications of the phrase are in discussions in bioethics, especially in describing decisions about the termination of life. Accusations of "playing God" are frequently encountered, for example, in connection with decisions which have serious and irreversible consequences for individuals, such as decisions about abortion or euthanasia. The expression is also used to describe proposals for genetic manipulation, especially in discussions about the permissibility of modifying human germ cells. What is common to all the various uses of "playing God" is the idea that there is a natural order or structure, perhaps divinely ordained, and that proposals to exceed the limits which this natural order defines should be rejected out of hand—or at least considered very carefully.

When a phrase is used to characterize actions or behavior as morally blameworthy, there is an implicit appeal to an underlying moral principle. It is important to state the principle explicitly, so that its application can be adequately assessed. The major problem with the accusation of "playing God" is the danger that it operates as a rhetorical device to obfuscate rather than to illuminate discussion.

I. SECULAR AND RELIGIOUS APPLICATIONS

The phrase "playing God" is usually encountered in descriptions of acts or decisions which involve arrogating power or control over matters of profound or far-reaching importance, such as decisions about the termination of life. Sometimes the phrase is used rhetorically to indicate that the topic under discussion raises momentous issues which must be addressed with great care. Doctors who have to allocate scarce life-saving medical resources, or decide which fetuses to terminate in unsustainable multiple pregnancies, face invidious decisions which may be characterized as "playing God." Edmund Erde has provided a useful survey of uses of the phrase in biomedical discussion (1989. *Journal of Medical Philos.* **14**, 593–615).

A somewhat different colloquial use of the phrase is to decide the inflexible exercise of authority or "laying down the law," for example, when a supervisor insists on a particular approach to a task, perhaps riding roughshod over the feelings or suggestions of subordinates.

These essentially metaphorical and secular uses of the phrase "playing God" must be distinguished from its use to mark out a zone of choices which supposedly involve overstepping a boundary of legitimate human activity. It is the latter literal and generally condemnatory use of the expression which needs to be examined. This pejorative sense is the one used, for example, in the book title, *Who Should Play God?* (T. Howard and J. Rifkin, 1977. Delacorte, New York).

Often, but not always, the implied objection to a proposed course of action is based on religious belief, and the proposed action is found objectionable because it allegedly involves a morally culpable or *hubristic* transgression into the prerogatives of the deity. As well as being used to indicate that a supposedly divinely ordained limit has been transgressed, the phrase may also be used to indicate a "natural" order. When appeal is made to naturalistic considerations it is suggested that a proposed course of action would upset some supposedly natural cosmic or world order, which, it is supposed,

should be left undisturbed. In these cases the suggestion is typically that we have a power to act which is not matched by the knowledge required to act wisely.

The accusation of playing God, however, is unhelpful and serves to darken rather than to clarify discussion. If it is being used as shorthand to indicate the will of God then an immediate problem is posed by the abundant diversity of opinions about God and the moral order which supposedly receives God's sanction. It is not helpful to be told that an action transgresses the prerogatives of the deity unless we know what these prerogatives are. "Playing God" is equivocal and unhelpful precisely because divine prerogatives are conceived so disparately by different religious authorities. Larue, for example, has recorded an immense diversity of religious opinion both between and within Christian and non-Christian denominations concerning the acceptability of euthanasia or physician-assisted suicide (G. A. Larue, 1996. *Playing God.* Moyer Bell, West Wakefield, RI). There is in general no unified Christian, Jewish, or other denominational position on this issue.

Appeals to divine authority as a basis for moral principles are in any case problematic. Objections to this sort of strategy for establishing secure foundations for morality go back at least as far as Plato's *Euthyphro*.

II. THE MORAL PRESUPPOSITION

Because there is no agreement among the authorities about which acts are divinely sanctioned, there may be argument about whether a particular course of action is serving or usurping the will of God. The confusion generated by the accusation of playing God is well illustrated by the fact that it can be confidently deployed by both sides of a dispute. It can be used, for example, to criticize a decision to withhold life-saving medical treatment as well as to criticize the decision to administer heroic medical treatment. If appeal is made to a divinely sanctioned principle which constitutes the basis of a moral proscription then it is important to state that principle explicitly so that its probity can be properly evaluated.

Serious uses of the phrase "playing God" thus presuppose a moral framework or appeal to a moral principle, and therefore to avoid confusion it is important for the presupposed principle to be stated explicitly so that its application can be properly assessed. The importance of stating the underlying principle can be demonstrated by uncovering the implicit moral principle which lies behind the application of the phrase in its central applications, which concern the domains of bioethics and genetic engineering.

III. APPLICATIONS IN BIOETHICS

The most widespread application of the charge of playing God is in connection with medical technologies, in particular those involving decisions to terminate human life. Most theologies speak of the sanctity of human life. The most significant procedures criticized for not paying proper respect to God's law are euthanasia and abortion, and sometimes also, associated with the latter, prenatal testing. The charge of playing God has also been leveled against the use of new reproductive technologies concerned with establishing pregnancy, such as *in vitro* fertilization (IVF) and gamete intrafallopian transfer (GIFT).

A common basis for objection to applications of medical technology is the belief that allowing something to happen, even though we can control it technically, is to leave it to God's providence, acknowledging that exercising control wisely in some circumstances is beyond our power.

It is difficult, however, to provide a rational justification for this position. After all, much of modern medicine aims precisely to prevent mortality and morbidity which would eventuate if left to providence. Suffering is reduced by the elimination of noxious pathogens such as smallpox, and this is not usually thought of as a violation of God's providential design. If we are prepared to act to prevent disability and disease caused by environmental pathogens, then why balk at measures such as therapeutic abortion which also aim to prevent disability and suffering caused, for example, by severe genetic disorders?

One response is to claim that there is crucial difference between the destruction of pathogens and the destruction of innocent human life. Even if the decision to terminate a human life is motivated by compassion, that does not, on a view widely endorsed by many religious authorities, alter the character of the act as an act of killing. Abortion involves the destruction of a being with an immortal soul, and such an act presumes on the prerogatives of the deity. This objection to abortion may also be extended as an objection to prenatal genetic testing, since a major motivation of prenatal testing is to detect genetic defects with a view to therapeutic abortion if genetic mishap is detected.

Making sense of the relationship between soul and body is notoriously problematic. But even if it is granted that fetuses have immortal souls it is hard to understand how the destruction of a fetus with genes, for example, for Tay-Sachs disease or Lesch-Nyham syndrome is anything but an act of kindness. A physical life will be ended, but a truly immortal soul would continue, perhaps to be reunited with its maker. Indeed precisely this consolation is often provided to parents on the premature death of a child.

It is sometimes suggested that suffering is good for the soul, but this is an implausible way to defend the continuation of a life that will be wretched. It amounts to an objection to virtually the whole institution of modern medical practice, which exists largely to alleviate suffering. Of course this is not a dominant view, and many religious authorities see little or no spiritual merit in suffering, and regard it rather as an affront to human dignity.

It is important to distinguish religious objections to abortion from objections based on the claim that it is absolutely impermissible to terminate innocent human life because that involves the destruction of persons or potential persons. The present concern is only with explicitly religious objections to abortion. Objections based on secular considerations are considered elsewhere.

The other major category of medical practice which attracts the charge of playing God in the termination of life is euthanasia. Sometimes it is alleged that it is contrary to the will of God to fail to treat the terminally ill; but, confusingly, it is also claimed that permitting the terminally ill to refuse heroic medical treatment, that is, to permit what is sometimes called "voluntary passive euthanasia," is precisely to *decline* to play the role of God (Larue, 1996, 10). There is also a more controversial claim, supported by some progressive religious thinkers, that personal autonomy is a God-given right, and that a merciful and compassionate God would have no objection to a terminally ill patient choosing a quick, painless, and dignified death.

Once again confusion surrounds the charge of playing God, and rather than appealing to God's will or God's law it is essential to state explicitly what principle this claim alludes to so that it can be systematically examined. Taking innocent life is always a morally grave matter which needs justification. The justification of end-of-life and ending-life decisions typically involves addressing important distinctions between acts and refrainings, and between killing and letting die, and consideration of the role of individual autonomy. Whether or not moral justification can be provided in particular cases, it certainly cannot plausibly be provided simply by appeal to religious authority.

IV. APPLICATIONS IN GENETICS

Molecular biologists have developed a variety of techniques of genetic engineering to manipulate and control the development, structure, and hereditary characteristics of organisms. In particular manipulation of the genetic material itself (deoxyribonucleic acid or DNA) by gene splicing has led to the development of powerful tools for modifying life-forms and creating new ones. The allegation of playing God has almost become a cliché in the field of genetics and is used, in particular, to describe proposals to modify the human genome.

There are several important distinctions which need to be kept in mind when considering arguments about genetic engineering. First, there is the human–non-human distinction, and the question of whether there are special considerations which constrain the manipulation of human genetic material which do not apply in the case of other species. Secondly, there is a distinction between somatic and germ cell genetic modification: somatic changes affect only the subject organism; modifications to germ cells are passed on to an organism's descendants. Thirdly, there is a distinction between negative (therapeutic or corrective) engineering, which aims at rectifying some disease, defect, or disability, and positive (eugenic or enhancement) engineering, which might aim, for example, to produce a healthier, smarter, more capable, more robust, and longer-lived individual or population.

One way that misgivings about genetic engineering are expressed is in terms of the religious belief that the gene pool is God's sacred creation and should be preserved. Genetic engineering is sometimes characterized as a dangerous Promethean adventure which involves appropriating knowledge which is properly the province of the deity. In our post-Enlightenment state of knowledge this view does not survive inspection. Gene pools are more plausibly seen not as the product of divine providence, but as the piecemeal accretions of billions of years of accident, mishap, and good fortune.

A more secular and more persuasive argument is that biological processes and products have evolved over billion-year geological time scales and have thereby proved their robustness. Natural life-forms come with the quality assurance of exceptionally prolonged testing under the most searching conditions.

There are also self-interested considerations based on the desirability of maintaining biodiversity, which entail that we should take care to ensure that we not reduce biodiversity through genetic tinkering—or any other way.

The argument against the substantial genetic modification of any species which alludes to the alleged wisdom of the evolutionary process is summed up in the phrase, "Nature knows best." This is a modern secular expression of an older religious belief in divine providence. Not everyone, however, is impressed by the result of several billion years of evolution, and it is sometimes suggested that its products might well be improved. Why should we favor the slow, fitful, chancy, piecemeal, small-scale, incremental processes of natural evolutionary change above the rapid and radical changes made possible by genetic technology?

Some critics, such as Howard and Rifkin (1977), suggest that it would be best if genetic engineering were completely prohibited. Howard and Rifkin argue that genetic engineering is inherently dangerous and threatens to transform organisms, including humans, into technologically designed products, and may lead to a new caste system in which social role is linked to genetic makeup.

This alarmist assessment ignores or plays down the fact that genetic engineering of a kind has been practiced by selective breeding in animal husbandry, horticulture, and agriculture for thousands of years. Indeed the principal plant and animal food sources in the human diet are all products of selective breeding.

It is certainly legitimate to be apprehensive about the possible dangers posed by recent genetic technologies, but rather than talk of playing God we need once again to examine what motivates these worries. There may be a serious basis for concern, such as a possible deliberate or accidental release of synthetic pathogens which might devastate a community or ecosystem. Risk should always be assessed prudently, bearing in mind that if there is a potentially hazardous solution to a problem it is always advisable to seek less hazardous alternatives. But the potential benefits of genetic engineering cannot be dismissed, and objections to it need to be selectively directed to particular proposals.

In the case of human genetic modification, negative engineering is generally thought to be ethically unproblematic, while positive engineering in general is not. A basic aim of therapeutic medicine is to eliminate serious disability and *ceterus paribus,* any measures which can help with this project, including genetic technologies, are to be welcomed. A proposal to use gene therapy for the topical treatment of lung tissue of sufferers of cystic fibrosis, for example, seems unexceptionable. But the situation is not always straightforward, and benefits must always be measured against risks.

Michael Ruse accepts negative engineering in principle but argues against producing radical changes through positive genetic engineering, claiming that enhancement engineering could degrade much of what is distinctively human (1984. *Zygon* **19,** 297–316). There is, however, a problem defining the boundary between positive and negative genetic manipulation which features centrally in the discussion by Ruse and others. Negative engineering is corrective, and its aims are spelled out in terms of rectifying dysfunction, disease, or abnormality. While these notions provide a rough guide, they are problematic in particular cases. To take a fairly trifling example, is correcting male pattern baldness a therapeutic treatment or cosmetic enhancement? Ruse may nevertheless be right about the need to restrict or proscribe radical enhancement proposals.

There is no precise, universal, objective criterion which determines what constitutes either a dysfunction or an enhancement. In the case of therapeutic somatic treatment, it may be unexceptionable to allow individuals to make informed autonomous choices about self-regarding changes, though the case of genetic enhancement modification of somatic tissue is less clear. However, if we allow people to change their appearance with breast implants or plastic surgery—and even to mutilate themselves up to a point, for example, with body-piercing and tattoos—perhaps we should allow them to genetically program themselves to grow green hair, or to make even more bizarre choices.

Germ-line modifications are less straightforward because deciding the destiny of others without their consent seems to be precisely one of the objectionable features of an action censured by the charge of playing God. Germ-line engineering, including enhancements, involves making choices for others without their consent, though it is difficult to know exactly how much weight to put on this, since choosing to reproduce at all apparently involves that anyway.

While a strong case can be developed for negative germ-line engineering to eliminate such debilitating genetic disorders as cystic fibrosis, Tay-Sachs disease, and Huntington's chorea, systematic attempts to improve or perfect a species presupposes that we know what constitutes an improvement or perfection. The value of a gene depends on the environmental situation in which it is expressed. Eliminating genetic traits may weaken a species, or make it unable to adapt to changing environmental circumstances which may arise from unforeseen contingencies.

Whereas improving the shelf life of tomatoes does not generate any obviously serious hazards, tinkering with the complex system of the human genome—with an estimated complement of 100,000 genes operating in a sequential and coordinated fashion—may have unforeseen serious or even catastrophic effects, some of which may not be expressed for several generations. This uncertainty argument applies of course to germ-line, but not somatic, genetic engineering.

There are, then, serious considerations which need to be addressed in deciding what forms of genetic engineering to allow; but we need to get behind the rhetoric of the phrase "playing God" and examine the underlying principles to evaluate how seriously these concerns are to be taken.

V. CONCLUSION

The phrase "playing God" is used to characterize actions or behavior that is deemed to be morally blame-worthy, and typically it is intimated, but not explicitly stated, that some immutable moral principle has been violated. It embodies the notion that there are possibilities which should not be realized—choices which should not be made. The accusation also often expresses concern about the *hubris* of tinkering with the sacred.

The power and resonance of the phrase derives ultimately from a deep conviction that there is a providential divine or natural order. This conception of a hierarchically structured benign cosmic order reaches back to Plato, and its classic exposition is *The Great Chain of Being* by Arthur Lovejoy (1936. Harvard Univ. Press, Cambridge, MA). It is deeply entrenched in Western thought and is associated historically with what Lovejoy called the principle of plenitude, according to which every genuine possibility is realized. If inexhaustible divine productivity has left no gaps in nature then we should seek neither to add to nor subtract from the natural order.

This is a potent conception which still has the power to exercise a subtle but substantial subterranean influence on our thinking. The danger presented by this seductive image is that it can come to dominate our thinking and provide a substitute for serious moral thought.

"Playing God" is an expression which is unhelpful as an analytic tool because it suffers from vagueness and multiple ambiguity, and in any case alludes to a dubiously secure foundation for moral principles. Apart from the unexceptionable metaphorical and rhetorical uses of the phrase, noted at the outset, the phrase does more to obfuscate than to clarify.

Bibliography

Erde, E. L. (1989). Studies in the explanation of issues in biomedical ethics. II. On "Playing God." *Journal of Medicine and Philosophy,* **14,** 593–615.

Howard, T., and Rifkin, J. (1977). *Who Should Play God? The Artificial Creation of Life and What It Means for the Future of the Human Race.* Delacorte, New York.

Kitcher, P. (1996). *The Lives to Come: The Genetic Revolution and Human Possibilities.* Simon & Schuster, New York.

Larue, G. A. (1996). *Playing God: Fifty Religions' Views on Your Right to Die.* Moyer Bell, West Wakefield, RI.

O'Donovan, O. (1984). *Begotten or Made?* Oxford Univ. Press, Oxford.

Ruse, M. (1984). Genesis revisited: Can we do better than God? *Zygon,* **19,** 297–316.

Scully, T., and Scully, C. (1987). *Playing God: The New World of Medical Choices.* Simon & Schuster, New York.

Precautionary Principle

JENNETH PARKER
The Hastings Center

GLOSSARY

anthropocentrism The concept that the human species is unique and that environmental issues should thus be examined in the context of human interests and needs.

biocentrism The concept that the human species is one among innumerable species on earth, and that environmental issues should thus be examined in the larger context of all living things rather than within a specifically human context.

biosphere The entire community of living organisms inhabiting the earth, and the physical environment supporting this life.

cocktail effect A term for the principle that certain substances, though not individually deleterious to the environment, may prove harmful in combination.

ecosystem A localized community of diverse organisms and their environment, interacting as one biological unit.

GMO genetically modified organism, an animal or plant whose genetic makeup has been altered through the introduction of foreign genetic material by some laboratory or industrial technique.

indicator species A particular species whose presence (or absence) is regarded as characteristic of a given environment, and whose ability or failure to thrive there is thus thought to be indicative of the overall ecological status of this environment.

In the face of highly uncertain outcomes philosophers should develop alternative systems of ethics which promote environmental quality and are ethically sensitive in the absence of firm ecological knowledge Lemons 1983.

We should recognise our incomplete knowledge of nature and therefore exercise caution and special concern for natural values.

MRS TOBY VIGOD

THE PRECAUTIONARY PRINCIPLE (PP) is wide ranging with many different interpretations and formulations: despite this uncertainty and ambiguity it is incorporated into international law and some domestic laws concerned with the environment. Besides taking a specific form in various laws, the PP in its wider form resembles sustainable development as being an umbrella term covering the growth of a whole discourse and also in that commitments have been made to a principle still in development. The PP also resembles sustainable development in that it is an area of intense contestation in the realm of interpretation in relation to specific decisions and in more theoretical statements attempting to define its scope and commitments. I shall attempt to define the key parameters of the discourse and indicate some of the conflicts, while making my own position clear. I shall argue that, from the perspective of environmental ethics, the principle should be taken in its widest form.

The historical context within which the PP developed was a "permissive" attitude to the environment. Environmental life-support systems were taken for granted as "externalities" and the environment generally was thought to be robust and capable of absorbing the impact of human activities. The growing recognition of a number of dramatic environmental problems began to change these perceptions, and governments who had to license new technologies and substances began to use ecological science to attempt to establish "critical loads" and "safe thresholds." It then became apparent that the task set was impossible since ecological science cannot declare that a new substance or technology is "safe," and equally causal links between actions and effects may either be impossible to establish or else take a very long time to establish, by which time great harm may already have been done.

I. LEGAL FORMULATIONS: STRONG AND WEAK

Overall, in its various legal forms the PP insists that where a substance or a technology is potentially damaging to the environment, regulation should be considered irrespective of "final scientific proof" (B. Dickson, in press. In *Proceedings, International Jacobsen Conference, Feb. 29–Mar. 2, 1996.* Harare: Univ. of Zimbabwe Press).

Within this general legal approach Dickson has identified stronger and weaker versions.

- *Stronger:* For example, the London Declaration of 1987 on the Protection of the North Sea states that substances will be regulated "when there is reason to assume that certain damage or harmful effects on the living resources of the sea are likely to be caused by such substances, even where there is no scientific evidence to prove a causal link between emmissions and effects."
- *Weaker:* For example, the Rio Declaration states that "where there are threats of serious or irreversible damage lack of full scientific certainty shall not be used as a reason for postponing cost-effective measures to prevent environmental degradation."
  This formulation is weaker as it does not unambiguously commit the parties to regulatory action and implicitly suggests that, *although there may be other valid reasons against regulatory action,* scientific uncertainty should not be taken as one such valid reason.

Not only do the various legal forms differ but they are also open to stronger and weaker forms of interpretation; this can be seen even in interpretations of the stronger, less ambiguous form. For example, according to one commentator, Greenpeace has in effect interpreted the principle as meaning that no activity that might impact on the environment should be allowed unless it can be proved to be harmless.

The principle will inevitably be subject to different views of what might constitute environmental harm. For some environmentalists the fact that a substance is not naturally occurring is a *prima facie* reason to suppose that it may have harmful effects on complex natural systems—including the human body: organic farming is, for example, premised upon this view of the precautionary approach. Other interpreters will hold that the "reason to assume ... damage" clause requires some good reason supported by some scientific evidence: what constitutes sufficient scientific evidence to give reason for concern will be contested. The dialogue over these interpretations can be seen played out in the recent controversy over the dumping of oil platforms in the North sea. Greenpeace insists that there is now a presumption against any dumping, while Shell insists that there is no reason to fear any resulting harm.

II. WIDER USES OF THE PP: "WAKING UP IN THE EXPERIMENT"

It is clear that various legal formulations do differ, and yet a range of commentators, lay environmentalists, and NGOs (nongovernmental organizations) insist on referring to *the* precautionary principle. In so doing, I will argue, they are referring to a prototypical ethical principle with important links to other areas of environmental thought.

Some commentators, Dickson in particular, argue for the restriction of the PP to its narrower legal formulations. Fortunately as even applied moral philosophers cannot effectively legislate against the development and use of concepts in the wider world, the PP is developing a life of its own within the range of professional, general environmentalist, and lay discourses that I hope to show will repay attention from environmental ethicists.

For example, in setting out the debate around the release and labeling of genetically modified organisms (GMOs), one lay environmental group, the "Natural Law Party," describes the context of technological intervention in natural systems, including human bodies, as a vast experiment; only unlike more limited medical experimentation, informed consent does not have to be obtained. This change of perspective moves from regarding intervention as unproblematic technological "advance" toward viewing intervention as a potential violation of the natural world and of the rights of humans conceived as part of that world. The PP, while not encapsulating this kind of perspectival shift, is certainly within the same frame of discourse. Concerns about GMOs are formulated within the context of the existing natural system: "The genetic structure of plants has been nourishing mankind for millennia. Tampering with the genetic code of food ... could upset the delicate balance between our physiology and the foods we eat...." (Natural Law Party, 1996). "Dangers of genetically engineered foods." Buckinghamshire, UK: Mentmore).

This perspective of the "vast experiment" indicates a useful possible approach to the PP as a wider application of the degree of precaution that usually prevails in a scientific laboratory—only with the enormously increased complexity obtaining where the whole of the living world constitutes the experimental subject.

In speaking of their concerns about the decision not to label genetically modified foods, the Soil Association claims that this decision shows that "a GMO food is assumed to be harmless unless proved otherwise," viewing this as "a shameless abandonment of the 'precautionary principle.'" Here the principle is viewed as a shifting of the burden of proof: instead of environmentalists having to demonstrate damage after the fact, the PP is viewed as shifting the onus onto the potential polluter to demonstrate that what they propose will not cause damage. In a survey of biologists, lawyers, and administrators it was found that 80% supported this interpretation of the PP. The survey also showed support for varying interpretations and confusion about the status of the principle. I suggest that this does not so much show a lamentable ignorance of the "real" legally limited PP but the emergence of a prototypical ethical principle whose status and assumptions I shall further explore below.

Dickson has claimed that it is in the interests of pragmatism to restrict the PP to the legal forms that have been ratified by states in the real world. In fact the environmental agenda is increasingly set by civil society in its various forms: citizens' organizations, NGOs, and the wider new social movements. For example, the Rio conference itself was initiated by the civil associations of the UN which are to be found around the world. Pragmatism thus dictates that applied ethicists should take note of the debates and perspectives developing in these sectors as these are likely to shape the policies of tomorrow. Equally if current precautionary policies have been "grafted onto a body of policy still based on the presumed assimilative capacity of the environment," as MacGarvin (1994. In T. O'Riordan and J. Cameron, Eds., *Interpreting the precautionary principle*) claims,

"the result will be a contradictory mess." Insofar as applied philosophers seek to clear up contradictory messes, we had better look beyond the current legal actualities.

In summary, therefore, as I have argued, the limited legal form does not express the breadth of the import of the PP in wider environmental discourse, which is the legitimate concern of environmental ethicists. In what follows I will be considering the wider PP—that representing an ongoing debate around the implications of environmental science for the current pattern of our interventions in nature.

III. ENVIRONMENTAL SCIENCE AND THE PP

The PP has been partly founded upon a growing acknowledgment by environmental science of its own limitations. One potent example is that of marine ecology, described with admirable clarity and directness by Malcolm MacGarvin (1994). It has been found to be impossible to establish critical loads of pollutants due to the multiplicity of factors involved in the health of marine ecosystems. The validity of using so-called "indicator species" to judge ecosystem health is now seriously suspect. MacGarvin concludes that "monitoring programmes designed to protect marine habitats do not have a firm scientific foundation." One consequence of this is that there is now a presumption against dumping of *any* pollutants in the ocean—a much stronger precautionary attitude than that based on an earlier (more optimistic) science. The same downshifting of expectations has occurred throughout the whole of environmental science.

A large number of scientific considerations support the PP. Many commentators stress uncertainty, but in order to give an accurate account of the PP it is necessary to place the unknown and the unknowable in the context to what *is* known.

A. Uncertainty

Uncertainty is the product of both the actual current limits of science and the "in principle" limits: the unknown and the unknowable.

Factors currently limiting science (the unknown) include:

- The lack of detailed long-term research in many areas of concern
- The underfunding and marginalization of ecology
- The lack of interdisciplinary research

Factors which limit environmental science in principle (the unknowable) include:

- The intrinsic difficulties of field science in the open system of the environment
- The logistical difficulties of multivariate analysis, e.g., testing for the "cocktail effect"—possibly synergistic effects of chemicals thought to be safe when tested singly
- The impossibility of a "control" environment for comparison as human effects are ubiquitous
- The chaotic nature of many natural processes, e.g., population dynamics and climate
- The intrinsic limits of any intentional process—science designs experiments to look for effects that are expected and information gathering is always limited with respect to a hypothesis (for example, ozone testing equipment in the stratosphere was originally programmed to treat any holes merely as "noise" in the data set)

It should be stressed that this negative assessment of the ability of environmental science to deliver the safety assurances that technological society requires is not just made by a few dissenters: this is the current mainstream opinion. Although it may be understandable that environmental science as a profession has not been proclaiming this from the rooftops, many scientists believe that the situation must be made clear since current unrealistic expectations are threatening to destroy the overall credibility of environmental science.

Dickson emphasizes the dictinction between *risk* and *precaution;* whereas risk applies where there are reliable scientific predictions on the basis of which probabilities can be assigned, precaution applies where uncertainty is prevalent. In view of the previous assessments of environmental science, it seems that situations of risk must be in the minority; this has the interesting corollary that many situations currently referred to in terms of risk (and perhaps a substantial amount of the literature on risk) should be reread in terms of precaution.

B. The Context of Knowledge and Experience

Reasons for supporting precautionary action include the following:

- The overview or model of life as a complex system of mutually interacting subsystems and organisms
- The awareness that ecosystem damage may be irreversible
- The repeated experience of harm being caused in ways that have not been predicted or suspected

- The fact that natural systems do not produce "waste"

The previous limitations of environmental science must be set, against the broader picture that environmental science has been able to assemble. As precaution may be seen as denying benefits, it requires justification. It is the knowledge produced by environmental science that does enable us to rationally state the reasons for concern which are needed to justify precaution.

IV. PRECAUTIONARY IMPLICATIONS OF ENVIRONMENTAL SCIENCE

The implications are many and diverse, and I must refer the reader to the main sources at the end of this article. However, I will here summarize some of the more basic implications to help set the context for discussion. It should be noted that drawing out the precautionary implications of environmental science involves practical judgment guided by an overall aim of preserving life.

- A presumption against any technology at a scale to disrupt natural systems
- A presumption against synthetics (MacGarvin, 1994)
- A presumption in favor of clean production rather than regulation of emmissions
- A general presumption in favor of reducing human impacts on natural systems

V. IS THE PRECAUTIONARY PRINCIPLE A SCIENTIFIC PRINCIPLE?

The exact nature of the relation between environmental science and environmental ethics is controversial and I will only claim here that the PP as a developing ethical principle is *informed* by science. But may it not be that the PP is a scientific principle? I will argue that the PP makes implicit assumptions to do with "taking care" and thus cannot be a purely scientific principle. I will now explore the idea that the PP could be viewed as a form of ecological rationality rather than as a specifically ethical principle.

A. The Precautionary Principle as Ecological Rationality

The concept of "ecological rationality" has been persuasively outlined by Robert V. Bartlett (1986, *Environmental Ethics,* **8**). Lynton K. Caldwell proposes that the test of rationality should be "that which is consistent with continuing health and happiness of man and with the self-renewing tendencies of the planetary life support system" (1971. *Environment.* New York: Doubleday. Quoted in Bartlett, 1986).

It may be that the PP could be seen as a rational principle rather than primarily as ethical: the PP could support an ecological rationality formulated on the recognition of the limits or "bounded" nature of rationality. "Bounded rationality" proposes that "the capacity of the human mind for formulating and solving complex problems is very small compared to the size of the problems whose solution is required for objectively rational behaviour in the real world" (H. A. Simon, 1957. *Models of man.* New York: Wiley. Quoted in Bartlett, 1986). The PP replaces the concept of the *size* of a problem (after all, computers can deal with huge data sets) with that of its unconscionable *complexity.* Thus the PP proposes that a precautionary attitude based on what we do know is rational in the face of unknowns.

Bartlett draws upon Simon's distinction between substantive rationality, "the extent to which appropriate courses of action are chosen," and procedural rationality, "the effectiveness of the procedures used to choose actions." Looked at from this perspective, the wider PP can be seen as proposing procedural ecological rationality, challenging the prevalent procedure of permissiveness toward the environment and implicitly characterizing it as reckless.

Bartlett proposes that ecological rationality will have to be considered along with other forms of rationality (e.g., economic) in any particular decision context. This view would support Dickson's contention that the PP will always be balanced with other considerations in any decision context—that is, it is not absolute. However, Bartlett contends that ecological rationality is in some sense more basic than other forms of rationality, such as economic rationality, in that it provides the ground for other forms to exist.

While the notion of ecological rationality is an interesting conception, it is so precisely because it is a value-enriched conception of rationality where life systems are seen as valuable. The ethical presumptions of ecological rationality are shared by the PP and it is to these that I must now turn to inquire into their status, justification, and relation to environmental ethics.

B. Ethical Assumptions of the PP

Perhaps the most general ethical presumption of the PP is that we ought to exercise rational prudence or take care of things we regard as valuable. This may seem obvious but is not necessarily so: we might take the view that something is valuable precisely in its ephemeral nature and prefer to be fatalistic or whimsical about it; e.g., "Life is just a party and parties weren't meant to last." In this respect arguments in favor of an ethic of care can support the PP (see further on for ethics of environmental care).

A further ethical presumption of the PP (that without which it could not get going) is that living systems are valuable. I do not think that at present there is any clear agreement on exactly on what basis they are valuable (see further on for a reputation of the PP as embodying intrinsic value). This may not matter since, as varieties of moral pluralism such as pragmatism hold, a wide range of values may all legitimately contribute to our ethical decisions without having to be grounded in one overall principle or theory.

We may ask about what environmental science is adding to these general ethical principles. I would say that environmental science has produced knowledge which is helping us to understand some of the ways in which the functioning biosphere is valuable as well as fostering an increased degree of respect for the complexity, longevity, and integrity of the biosphere. I do hold that coming to care for something or someone does involve taking the time to appreciate the relevant qualities. This is not to say that the arts do not also alert us to environmental qualities.

It is against this background of the positive contribution of environmental science that the warnings of the limits of that science have their impact. The PP tells us that caring for our environment is going to involve restraint and details specific presumptions against certain kinds of action. I conclude that the PP represents procedural rules for rational decision making guided by a commitment to environmental care. It is therefore an ethical principle in this sense but is not an ethical *first principle.* I have already claimed that an ethic of care is a necessary presumption for the PP but that this care perspective can accommodate a plurality of views on value. This approach thus avoids the reduction of these values to expressions of a common principle as do monistic and hierarchically ordered systems of ethics. In this way I would propose that an ethic of care has the dual advantage of expressing a broad ethical attitude and accommodating a variety of perspectives. This would seem a very suitable kind of basis on which to found a principle of social decision and conduct which has to be substantive and yet have wide agreement.

Robin Attfield has proposed that there are important relationships between the PP and questions of justice. To value the biosphere is to recognize it as a valuable common inheritance, and this inevitably raises questions of just distribution and "ownership." I do not, however, think that the PP is committed to a particular position on justice. The PP does seem to assume a joint responsibility for environmental care that is not restricted to traditional attributions of moral responsibility. Traditional (Western) models of moral responsibility assume that we are responsible for things that have come about as a result of our actions, or that we have specifically contracted to do (of course there are many forms of mitigation but these need not concern us here). The PP, and, I would argue, the concept of care on which it is based, assumes that our connectedness with living systems involves a necessity to care for them *irrespective* of the relevance or not of traditional moral responsibility. In this way the PP assumes that we all share the problem of excessive human impacts on living systems and the presumption is that we should act to reduce these impacts *whenever we can*—not just when we feel that we are particularly responsible. From my perspective it is care which provides the real support for burden sharing in a time of environmental change, not justice.

C. Support for, and Consistency with, a Biocentric Ethic

In what way can the PP be seen to support a biocentric ethic that views nonhuman species and ecosystems as intrinsically valuable? One might argue that the ethical assumption of care for the environment supports the notion of intrinsic value. This seems an unjustifiable assertion if, as I claim, the PP does not take a particular position as to why the environment should be valued. We could suggest, consistently with the PP, that it would be a good thing if people acted *as if* nonhuman species and ecosystems had intrinsic value—the kind of instrumental use of a biocentric approach or "sanctity" appealed to by a large number of people who apparently believe that this is an argument for intrinsic value.

Equally it would seem possible to have an attitude of "respect" for the qualities of nature without being committed to intrinsic value: one might respect nature in the same way as one respects a potential enemy with whom one wishes to keep on good terms. Insofar as the PP is upheld by science that might be held to limit its possible use to support intrinsic value. If intrinsic value is adopted then it seems that it would be held that the life of least intervention in natural systems and with other species would be morally preferred with or without scientific support.

I do not therefore consider that the PP provides any particular support for intrinsic value as such. However, this is not to say that the PP is not thoroughly *consistent* with a biocentric ethic in that following such an ethic does entail adopting principles of minimum impact or "living lightly."

D. Support for, and Consistency with, an Anthropocentric Environmental Ethic

The PP is clearly consistent with an ethic which takes concern for humans to be central, as we may be morally recommended to take precautionary action merely to protect human interests. However, it could be argued that the science supporting the PP also supports an almost infinite extension of the concept of human interests. On the strong interpretation of the PP the intricacy and interconnectedness of the biosphere together with the intrinsic limits on our knowledge of its workings are such as to justify an *identification* of human interests with the interests of every other aspect of the biosphere. This view is particularly supported by the sometimes spectacular inadequacy of our attempts to pursue human interests by intervening in natural systems; many projects such as dams not only fail to deliver the expected benefits but have negative consequences often unforeseen by planners (but not always unforeseen by local people).

E. Environmental Virtue Ethics

Geoffrey Frasz has claimed that "the thrust of environmental virtue ethics is to foster new habits of thought and action in the moral agent—not just to get the immediate decision made right, but to re-orient all actions henceforth in terms of a holistic, ecologically based way of thinking" (1993. *Environmental Ethics,* **15**). Such an ethic requires much input from the ecologically based sciences in order for agents to be able to think and act within a new environmental paradigm. In this sense a precautionary habit of mind with respect to environmental intervention, based on "respect" for nature, could be seen as an integral part of environmental virtue ethics. Further, proper attention to the PP may begin to clarify the different senses of "virtue" required.

Frasz uses Thomas E. Hall's concept of "humility" as the cardinal environmental virtue. Frasz himself highlights the problems with this formulation when he interprets environmental "humility" as continuous with our human social conceptions of humility, proposing that the appropriately humble person might unfortunately be too reticent to adequately defend nature. Attention to the PP and its scientific underpinnings points to the difference between our ordinary social conceptions of humility (by no means an uncontroversial social "virtue") and the notion that we ought to maintain a properly restricted view of our capacity to knowledgeably intervene in nature as a *species* (or perhaps more accurately as a specific culture). It could be argued that this collective "humility" in the face of nature would be properly expressed in the habitual attitude that the PP could inculcate. The confusion between collective and individual "humility" is a crucial one for Frasz's account; it leads him inevitably to the individualistic conclusion that environmental virtue ethics seeks "workable ways of living with nature that foster changes in personal virtue rather than larger societal changes."

I have argued that the PP is ethically supported by an assumption of care for the environment. An ethics of environmental care can be seen as a form of virtue ethics; for example, Joan C. Tronto discusses the virtues of care. Berenice Fisher and Joan C. Tronto define care as "a species activity that includes everything that we do to maintain, continue and repair our world.... [T]hat world includes our bodies, our selves and our environment, all of which we seek to interweave in a complex, life-sustaining web" (1991).

This suggests that, in parallel with the previous discussion, a care-based environmental ethic may need to beware the transfer of concepts of care drawn from human social contexts to "care for the environment." However, Tronto does not make this error; her concept of "attentiveness" certainly corresponds to the spirit of the PP interpreted in its wider form as a general requirement to exercise caution in intervening in functioning life systems. In that the "care" approach involves a general injunction to look at the widest possible consequences of our actions, seeing them as part of a web of associations and relations, the PP is an harmonious part of an environmental ethics of care.

VI. THE PRECAUTIONARY PRINCIPLE AND THE STATUS QUO

Robin Attfield has made the important point that the PP is not necessarily in favor of the current status quo and may necessitate changing it, giving the example of keeping running "Chernobyl style" reactors. This raises the whole question of the nature of the valued states which we seek to protect, preserve, or reintroduce (ecological restoration). It is indeed true that there are many states and processes which the PP would seek to change. However, this should not lead us to overlook the positive judgment that immensely complex life support systems have been functioning extremely well for millennia and it behooves us to be cautious in bringing about changes. In this sense the PP is for the "status quo" or continuing life.

In this respect it might be argued that humans have been changing ecosystems for a long time, apparently without causing catastrophes (though this is arguable

as some environmental historians seek to show that the collapse of at least some civilizations has been brought about by reckless treatment of natural systems). It is at this point that considerations of qualitative changes or discontinuities are relevant. It can be reasonably argued that there is a qualitative change—in both the nature and the scale of our technologies. I argue that the PP must apply to both: to the nature of our technologies in the ways already reviewed and also to the sheer scale of our interventions in nature. This latter point means that even "traditional" practices may need to be subject to precaution.

Commentators on the PP have sought to include natural and social systems within its remit. I will now argue that this is a mistake which if persisted in gives rise to socially conservative conclusions which are not necessarily implied by the PP. This is important for environmental ethics as it is in part a reforming project—seeking to open up new ethical perspectives on the environment within our culture. The relationship between environmental ethics and environmental social movements has not been openly explored by many writers; there is a lack of attention to the politics of our own discourse. At present there is assumed to be some kind of symbiotic relationship between environmental ethics and related movements attempting to bring about social change. If as environmental ethicists we accept that we should be precautionary in the same way about social change as we are in relation to the environment, this would render our practice highly problematic.

If the PP applies to societies we would then be mandated to respect the complexity and fragility of social systems in the same way as we are told we should respect natural systems. For example, O'Riordan and Cameron claim that "liability and onus of proof shift more and more onto the promoter of social change" (1994. *Interpreting the precautionary principle*); their argument seems to be that this is because scientific uncertainty is "mediated via social contexts." However, it is quite possible to accept the social construction of science without thereby holding that social science can be used in exactly the same way to formulate policy as natural science. This is one of the reasons that it is important to place uncertainty in environmental science against the background of what is known. Notions of a healthy environment are much less problematic than notions of a healthy society. As we have noted, the PP does not just support the status quo; it may support environmental restoration in line with a broad knowledge of environmental health.

To return to the science underpinning the PP, I would argue that there is a *prima facie* case against moving from environmental science to principles that are held to apply to society. But this case can be made much more strongly with the help of a realist philosophy of science. A critical realist perspective (such as that propounded by Bhaskar or Sayer, among others) proposes the objective reality of the structures of the natural world in distinction to the demireality of the structures of the social world while also recognizing the social construction of science. The structures of the social world are viewed as real insofar as they are causal (e.g., the monetary system), but they are subject to human intentions. The structures of the natural world cannot simply be altered by human intention—even genetic engineering and reproductive technologies have to work with natural structures.

The distinction between the realities of the natural world and the human constructs of the social world is at the heart of environmentalism shown, for example, in the New Internationalist dictum, "only when you have killed the last fish and poisoned the last river will you realise that you cannot eat money." This distinction is extremely important in that it implies that a conservative (in the ecological sense) attitude to nature may actually necessitate a radical (in the political sense) attitude toward reforming society. Any conflation of the natural and the social completely undermines the message of environmentalism. To say this is not at all to deny the complexities of social change, but it is to maintain the possibility that human beings can intentionally change society and their relationship to nature—indeed the PP itself is evidence of the belief in this possibility.

The notion that social systems are self-organizing and should be left to do so is most famously associated with Hayek and his concept of "catallaxy." The "new right" has developed this idea in various ways and it does contain some pertinent observations about human societies and the place of regulation. However, it is broadly accepted that Hayek's critics have refuted the extreme claim that human society can be seen as a totally unconscious self-creating phenomenon. In opposition to those who would present the PP as simply the embodiment of a natural conservatism (which might be taken as unproblematically transferable to the social sphere), I have argued that the PP does seek to draw upon our existing knowledge of natural systems to *justify* precaution.

At this point we must return to Attfield's comments about the environmentalist mission to change at least some aspects of the existing situation. Indeed were the PP to be unproblematically extendable to human society there would be a severe dilemma for environmentalists. The fact that some environmentalists believe they do find themselves in this dilemma testifies to the prevalent confusion in this area.

There is an important coda with respect to the previous argument regarding development projects. Development critics, environmentalists, ecofeminists, and indigenous people themselves have sought to present the "other side" to "development." They have stressed the ways in which development projects have often resulted in the destruction of forms of human living which maintained a healthy environment. In the language of ecological rationality, some cultures provide us with "examples of functioning ecological rationality" (Bartlett, 1986).

However, it is precisely because we do have some positive criteria by which to judge ecological rationality that we can argue for the preservation of such forms of living—as well as on the grounds of care, justice, and human rights. With respect to these cultures we might well apply a precautionary principle—but even then it would be different because we would be dealing with human beings who have a perspective which we would have to take into full account. To apply the PP in the same way to such cultures as we might apply it to an ecosystem would be to comply in the oppressive "naturalization" which has operated throughout colonial discourse. In fact political conservatism tends to be equally choosy about which forms of "catallaxy" should be preserved, traditionally holding that subsistence forms of life do not constitute anything worthwhile. An ecological rationality, of which I have argued the PP is a part, leads to a revaluation of such cultures, thus demonstrating the depth of the challenge to at least some traditions of valuation in Western society.

VII. CONCLUSION

The PP rests upon two ethical statements: firstly that we should "take care" of valued things, and secondly that natural living systems (including human bodies) are valuable. The PP interprets these ethical statements in the context of environmental science to deliver a form of procedural rationality—a rule which should help make decisions about our interventions in natural systems. This can also be interpreted as a form of "virtue" or a pattern of caring behavior toward natural systems.

I have interpreted this broad environmental ethics of care as open to a variety of different approaches to environmental value, including, centrally, both biocentric and anthropocentric approaches. This is consistent with the wide approach of the PP which should be retained on pragmatic grounds. The PP is a procedural rule for decisions to do with natural, not social, systems, and hence the association of the PP with conservatism is fallacious. Environmental science has helped us to discover what "taking care" might mean in relation to natural systems and has also enriched our sense of the value of these systems through attention to their specific characteristics.

Bibliography

Attfield, K. (1994). The precautionary principle and moral values. In T. O'Riordan and J. Cameron, Eds., *Interpreting the precautionary principle.* London: Cameron May.

Light, A., & Katz, K. (1996). *Environmental pragmatism.* London: Routledge.

O'Riordan, T., & Jordan, A. (1995). The precautionary principle in contemporary environmental politics. *Environmental Values, 4,* 191–212.

Parker, J. (1995). Enabling morally reflective communities. In Y. Guerrier, N. Alexander, J. Chase, and M. O'Brien, Eds., *Values and the environment: A social science perspective.* New York: Wiley.

Warren, L. M. (1993). The precautionary principle: Use with caution! In K. Milton, Ed., *Environmentalism: The view from anthropology.* London: Routledge.

Reproductive Technologies

LUCY FRITH

The University of Liverpool

GLOSSARY

artificial reproductive technology (ART) A generic term for techniques that artificially assist conception such as IVF and GIFT.

clinical pregnancy The ultrasound evidence of a fetal heartbeat.

donor insemination (DI) The insemination of the woman by donor sperm.

egg retrieval Successful egg collection procedure, used in IVF.

embryo A fertilized egg between two and eight weeks of development.

embryo transfer Transfer of embryos to the uterus from the petri dish (where fertilization has taken place).

gamete The male sperm or the female egg.

gamete intrafallopian transfer (GIFT) Sperm and eggs are mixed together and transferred to one or both of the woman's fallopian tubes, where fertilization takes place.

Human Fertilization and Embryology Authority (HFEA) A body set up in the United Kingdom by the 1990 Human Fertilization and Embryology Act to oversee and regulate new developments in ART.

in vitro fertilization (IVF) Sperm and eggs are collected by egg retrieval and mixed in a petri dish where fertilization takes place, and then the embryo(s) are placed in the uterus.

intracytoplasmic sperm injection (ICSI) A variation of IVF where a single sperm is injected into the inner cellular structure of the egg. This is a technique used for couples where the man has severely impaired or few sperm.

oocyte Unfertilized ovum.

preimplantation genetic diagnosis (PGD) Use of genetic testing on a live embryo to determine the presence or absence of a particular gene or chromosome prior to implantation of the embryo in the uterus.

ARTIFICIAL REPRODUCTIVE TECHNOLOGIES (ARTs) have given urgency to questions that have always perplexed humanity, such as the reasons for wanting children, the implications of infertility, what it means to be a parent, and the strength of the genetic bond between people. In 1978, scientific developments in embryology and embryo transfer culminated in the birth of the world's first baby born as a result of *in vitro* fertilization in Britain. This event prompted extensive ethical debate over the acceptability of ART, with opinions ranging from whole-hearted support to condemnation. The issues raised can be divided into two categories: arguments for and against the use and development of ART, and the ethical dilemmas created by specific aspects of ART. This article will focus on the Anglo-American debate over ART, recognizing that these debates have taken different forms in other countries.

I. THE ARGUMENTS FOR REPRODUCTIVE TECHNOLOGIES

A. Helping the Infertile

Artificial reproductive technologies are often portrayed as a response to the needs of the infertile (approximately 1 in 10 couples, although figures vary). Being unable to have a child can be a significant problem that affects an individual's most important choices about the type of life he or she wishes to live. The frustration of this desire is thought to cause immense suffering and medical science has responded by developing techniques designed to alleviate infertility. This poses the question of why the infertile should be helped. On what basis should the infertile have their desire for a child met? There are three main justifications for the claim that the infertile should be helped.

1. Infertility Is a Disease

The first justification for helping the infertile is that infertility is the kind of condition that merits medical treatment. Robert Winston, a British fertility specialist, has said that "infertility is actually a terrible disease affecting our sexuality and well-being." This position appeals to a particular definition of what it means to be healthy. If health is defined as the optimal physical functioning of an organism then it could be argued that where infertility is caused by some form of physical malfunction, then this should be treated. A wider definition of health could be employed to support the claim that infertility is a disease, such as the World Health Organization's (WHO) definition that health is a complete sense of well-being. It could be argued that infertility treatments are enabling the infertile to function as fully healthy individuals and should be part of health care provision. Conversely, it could be said that infertility cannot be characterized as an disease itself—rather, it is the effect of other conditions, such as fallopian tubal blockage. Infertility treatments do not "treat" infertility, they only ameliorate the effects of other conditions. Nevertheless, many established medical treatments fall into this category—for instance, diabetes is not treated but its unpleasant side effects are managed—so this point in itself does not constitute an argument against the merits of medical treatment for infertility.

2. The Desire for a Child

One of the problems with the WHO's definition of health is that it could include within the health remit anything that contributes to an individual's well-being (and this could mean that we are obliged to provide video recorders, racing bikes, or anything else that enables people to have a complete sense of well-being). Hence, it is necessary to have some mechanism for distinguishing between wants and, more importantly, needs. Leaving aside the problem of what might constitute a distinction between health and social needs (such as welfare benefits), those who are building a case for the importance of helping the infertile need to have an argument to support the claim that having a child is a need rather than a want.

The argument that is often used to support this claim is that humans have a biological imperative to reproduce. Professor Edwards, the doctor involved in the birth of the first *in vitro* fertilization (IVF) baby, considers that the genetic pressures to have children are the very foundation of our nature and that it is such pressures which lie at the heart of most couples' desire for children. By seeing the desire to have children as biologically based, proponents of this view argue that it is a desire that cannot be changed, because it is innate, or at the very least cannot be substantially altered. Thus, in order for people to live fulfilled lives they must have certain basic needs satisfied and having children is one such need.

3. The Right to Reproduce

The final reason for helping the infertile builds on the previous argument and contends that the infertile have the same right to reproduce as the fertile do. This is an area of debate that has been changed radically by the scientific development of ARTs. The right to reproduce, free from interference, is generally viewed as a basic human right. Enshrined in many declarations of human rights is the right of "men and women of full age. . . . to marry and found a family" (UN Declaration of Human Rights). Article 8 of the European Convention on Human Rights states that "everyone has the right to respect for his private and family life." The U.S. Supreme Court has on various occasions supported procreative liberty and has upheld the protection of those within marriage to found a family.

These articles and rulings are usually understood as stipulating what can be called a negative right, that is, a right not to have one's reproductive capacities interfered with against one's will. This negative right is a right to be free from unwarranted interference from the state (or anyone else seeking to impede a couple's reproductive capacities). Such articles were formulated with prohibitive policies in mind, such as enforced sterilizations or the prevention of interracial marriage.

However, in order to exercise their freedom and therefore their right to reproduce, the infertile need some assistance. This assistance should be forthcoming since a couple's interest in reproduction is, arguably, the same no matter how reproduction takes place. "The values and interests underlying a right of coital reproduction are the same, no matter how reproduction occurs. [Such values] strongly suggest a married couple's right to noncoital reproduction as well and, arguably, to have the assistance of donors and surrogates, as needed" (1994. "The Ethical Considerations of the New Reproductive Technologies." Report of the Ethics Committee of the American Fertility Society (ECAFS)). Accordingly, the infertile have to rely on some notion of positive rights that place an obligation on others to provide them with the means they need to reproduce. This raises the question of whether there is a link between the technical possibility of something (in this case a medical technique) and the right to have the resources made available to enable access to that technique.

John Robertson (1994. *Children of Choice.* Princeton University Press) puts forward a right-based argument to support the extended use of ARTs but not, however, any positive measures that would ensure equal access to such technologies. He begins by arguing that the concept of procreative liberty should be given primacy when making policy decisions in this area. Procreative liberty is the freedom to decide whether or not to reproduce. At first sight this appears to be a negative right not to have one's reproductive capacities interfered with. However, Robertson endorses subsidiary enabling rights to procreation; that is, someone has the right to something if it can be regarded as a prerequisite for procreation. For the infertile ART would be a necessary prerequisite to enable them to have children, and hence they have a right to such treatment. This effectively turns a negative right (not to be interfered with) into a positive right (to have something made available to one). This enables the infertile to exercise their reproductive rights and could be said to redress the disparity between the infertile and the fertile.

B. Reproductive Choice

ARTs broaden the range of reproductive options that are available for both the infertile and, in certain circumstances, the fertile. Any extension of choice is frequently portrayed as desirable and this is just as true for repro-

ductive choices. Ronald Dworkin, the philosopher and legal theorist, has developed arguments for what he calls "procreative autonomy." Individuals should be allowed the freedom to extend their reproductive choices unless it can be demonstrated that there are compelling reasons to restrict such choices. It is argued that there are no compelling reasons to restrict the use of ART since the benefits of such procedures far outweight any possible harms. Having children is a desirable end for people to pursue and extending the availability of options to help them fulfill this end is correspondingly a desirable state of affairs. Hence, to extend the reproductive choice of the infertile can be viewed as a positive development. Here a distinction could be made between having children (which would include reproduction and bringing the children up) and the right to reproduce, which could be fulfilled solely by allowing the infertile the means to reproduce, say, by donating their gametes. The infertile would wish to advance the claim that they should be able to have children and this includes, with ART, reproduction. In a letter to *The Lancet,* Professor Edwards states, "It is impossible to put a price on the benefits to society of producing wanted children raised in a caring environment." Hence, ARTs are seen as mechanisms for extending people's reproductive choices, and this is something that is beneficial for both the individual involved and society.

C. Genetic Screening

Another set of arguments in favor of ARTs are that such techniques can be used to screen for genetically inherited diseases and ensure that any child born is free from abnormality. Carl Wood, an Australian fertility specialist, stated that "it may be possible for the test-tube procedure to reduce the incidence of, or eliminate, certain defects from the population." Thus ART could also be of use to those who are fertile but may carry some risk of transmitting a genetic disease. This justification for ART is based on a different set of arguments than those used to justify ART on the grounds of helping the infertile. There is the additional assumption that it is morally important to ensure that only perfectly healthy children or those free of a certain condition are born, on the grounds that we should avoid unnecessary suffering where we can.

Preimplantation genetic diagnosis (PGD) is a technique that enables cells to be removed *in vitro* and tested to detect a genetic disorder. If the embryo is found to be defective then embryo transfer does not take place. This technique will have wider application in the future with the development of the human genome project, and this will make it possible to detect the genes for an increasing number of genetic disorders. ARTs thus enable genetic screening to take place at an earlier stage (before implantation) rather than when the pregnancy is already established, thereby avoiding the physical, emotional, and, some might argue, moral costs of terminating a pregnancy. If genetic screening is seen as desirable then it can be argued that it is preferable to carry it out as early as possible. It is also possible at this stage to determine the sex of the embryo. This can be useful when there is a danger of a sex-linked genetic diseases and only embryos of the relevant gender are discarded rather than foregoing a pregnancy altogether.

D. Conclusion

Those who argue in favor of the use and development of ARTs claim that such techniques bring great benefit to the individuals involved—the infertile. Childlessness is seen as a great handicap and consequently treatment for infertility should be given a high priority. It can also be argued that ARTs are necessary to enable the infertile to exercise their fundamental human right to reproduce. Further, ARTs are heralded as providing benefits for society as a whole because these techniques can help prevent genetically inherited disease and advance scientific knowledge about the human reproductive system that could have, as yet, unforeseen benefits.

II. THE ARGUMENTS AGAINST REPRODUCTIVE TECHNOLOGIES

Some of the arguments against ART will be direct rejoinders to the above case for ART, and some of the arguments will raise different concerns.

A. Criticisms of the Case for ART

1. Infertility Is Not a Disease

It has been argued that infertility is purely a social problem and one that does not need medical intervention. Couples could have counseling to come to terms with this inability, which is an inability to participate in social customs rather than any specific medical problem. Infertility is often seen as a problem for the couple, rather than the individual in the couple. This creates the problem of whether lifestyle choices, such as opting to form a same-sex relationship and hence as a couple being unable to have children, can be seen as a case of genuine infertility when both individuals are perfectly able, physically, to produce children. More generally, infertility is a consequence of certain physical problems, such as blocked fallopian tubes. There is no specific

condition of infertility itself which medicine can be called upon to cure. Hence, the problem of infertility is a response (to a situation), not a condition itself, and such responses could be realigned. It is the effect of these physical problems that are detrimental in a society that purports to value children and the parental role, particularly for women. Thus if medicine seeks a role in the treatment of infertility it could be said that this is medicalizing a problem that is predominantly social in origin. One of the difficulties with this view is that it is hard to distinguish clearly between social and medical problems as the two often have a complex and interactive relationship. The refutation of the claim that infertility is a disease rests heavily on the view one takes as to the importance of having children.

2. The Desire for Children Is Not Genetically Determined

Those who argue that ARTs provide a necessary and vital service for the infertile have often claimed that the desire for a child is biologically conditioned and a central element for humans to flourish. In contrast to this the socially constructed nature of the desire for a child is advanced. Here it is the social pressure to reproduce that creates and determines the desire for a child. This view highlights the existence of pronatalism, an attitude or policy that encourages reproduction and promotes the role of parenthood. Pronatalism particularly affects women who are encouraged to become mothers. In a patriarchal society true femininity is often equated with childbearing and motherhood is thereby regarded as a necessary aspect of womanhood.

Thus, having a child does not have to be a fundamental part of everyone's life. We can make individual choices over whether or not to have children and the importance reproduction will have in our lives. The way in which society pressures women to have children and the focus on genetic relationships can be said to be socially determined ways of constructing our reproductive relationships. It could be argued that we do not have to respond to such pressure to reproduce and it is important to give women the freedom to have other life options that are equally valued and respected.

3. There Is No Positive Right to Reproduce

Robertson's claim that there should be subsidiary enabling rights to help the infertile reproduce—that there should be some form of positive right to reproduce—could clearly be problematic. Laura Purdy (a feminist philosopher) argues that Robertson adopts a position which blurs the distinction between negative and positive rights:

Robertson sees the right not to reproduce and the right to reproduce as two sides of the same coin. From this fact, he seems to infer that the strong right not to reproduce implies an equally strong right to reproduce and also that this strong right to reproduce provides as much support for assisted reproduction as for so-called natural reproduction. (Purdy 1996).

Embedded in this discussion of reproductive rights is the assumption, made by Robertson, that the issues raised by natural reproduction are similar to those raised by artificial reproduction. Christine Overall, a feminist critic of ART, argues that the issues raised by the two different forms of reproduction are fundamentally different and correspondingly the right to use ART needs a different burden of proof, as ARTs might be harmful, from the right to be free from reproductive interference.

It can also be argued that Robertson's focus on procreative liberty could only be used to prevent legislation against ART, thus ensuring the existence of access, rather than upholding any positive rights that could be used to actually enable access and hence make ART available to a greater number of couples.

It is not generally recognized that just because the means are available to achieve some end, people have a right to that means. There exist few positive rights to health care (and it might be objected that ARTs are not strictly health care measures) even of the life-saving nature. Hence, it is by no means clear that there is a good argument to support the claim that there should be a positive right to reproduce (that the infertile should have the means made available for them to reproduce) or what the practical (e.g., financial) implications of such a right would be.

4. Ethical Problems Raised by Genetic Screening

The claim that ART is of benefit to society because it can provide some sort of quality control over the type of children born is by no means accepted by all. PGD can be seen as a form of eugenics that implies that those affected by a particular condition should not have been born or are less valued as individuals. There are further ethical problems with the practical application of PGD, such as misdiagnosis, longer term effects of embryo biopsy on child development, and decisions about which genetic diseases should be identified, how serious these have to be to merit screening, and the extent to which screening should be used.

Both the Human Fertilization and Embryology Authority (HFEA) and the ECAFS have stated that using PGD to determine the sex of an embryo is ethically acceptable as long as it is used to prevent the transmission of a genetic disorder. However, both bodies state that the use of sex selection for any other reasons, such as family balancing or what are broadly termed social

reasons, is unacceptable since nonmedical uses of this technique could result in harm to individuals and would be an unwarranted use of scarce medical resources.

The issue of genetic screening also has implications for recipients and donors of genetic material (eggs and sperm). The HFEA has considered the potential problems raised by the increased use of screening for carrier status of genetic diseases. For instance, "people might not be willing to come forward as donors because they do not wish to be screened for genetic disorders" (HFEA, 1995, *Annual Report*). This could reduce the supply of gametes by curbing donations. This is a concern because gametes, particularly eggs, are already in short supply. There could also be the problem of those who do present themselves to become donors as they may find that they are carriers of some genetic disease creating, possibly, unforeseen devastating implications for them and their families.

B. The Feminist Position

It is important to recognize that feminists do not speak with one voice and not all feminists share the same views of ART. ARTs are important areas for feminist discussion as they directly effect the way some women reproduce and represent another medical involvement in women's lives.

1. Harm to Women

The central tenet of any feminist position is the concern for how ART will affect the individual woman and, at a general level, if these technologies will be harmful or beneficial to women as a group. One school of feminist thought represented by writers such as Gena Corea views ART as intrinsically harmful to women. ARTs are practices constructed by a patriarchal medical and technological establishment to further control and colonize women's bodies. It is emphasized that these technologies reinforce a biologically deterministic view of women that subordinates women's identities to their reproductive role, rather than seeing them as full human persons with a range of interests. It is the social context in which the technologies are developed that makes them inherently harmful to woman. Hence, ARTs are not benign techniques that could possibly be harmful to women—they are deliberately constructed mechanisms of control.

This view has been criticized by supporters of other schools of feminist thought (such as Michelle Stanworth) who see ARTs as benign technologies that could be used inappropriately but are not inherently harmful to women. The crucial issue for this school of thought is that ARTs are adequately controlled, so that regulatory structures ensure that women are protected from abuse and exploitation. This view stresses that ART should operate in an ethical framework—one that recognizes the potential harms to women (and seeks to minimize them) and ensures that women are adequately informed as to the possible risks and side effects of ARTs.

In order for ART to be carried out in an ethically acceptable way, women must be fully informed as to the exact nature of the procedures and possible outcomes. One area that has been heavily criticized is the misrepresentation of the success rates of IVF. It is claimed that women are not given accurate information as to the likelihood of achieving a successful pregnancy and consequently have not given fully informed consent to the procedure.

The overall success rates for IVF treatment are still relatively low (although they are rising). Any couple receiving IVF will have had a history of infertility and have been trying for some time to have a child. Once accepted on the IVF program there is no guarantee that a pregnancy will be achieved. In the United Kingdom, due to the HFEA, national information is available which gives an approximate indication of general success rates. In 1997-1998, out of 34,638 treatment cycles there were 6864 clinical pregnancies and 5687 live births. Per treatment cycle there is an 19.8% clinical pregnancy rate and a 16.4% live birth rate (HFEA, 1999, *Annual Report*).

Due to the low success rates it has been argued, for example, by R. Rowland, that IVF cannot be seen as a therapeutic procedure but rather as an area of research. This raises the question of whether it is ethically justified to offer IVF to women as a therapeutic procedure when they are more accurately taking part in a research project. In order to answer this question it is necessary to come to some decision over what is an acceptable success rate; i.e., when is a success rate so low that the technique should still be considered to be at the research stage? The success rate of IVF could be compared to natural conception rates (approximately 25%) and, arguably, in the light of this the rates are not so low.

It is important that women are given correct information about the success rates and possible risks of the procedure (such as hyperstimulation syndrome and long-term effects of the superovulatory drugs used). Then women will be able to make up their own minds as to whether they wish to undergo a painful and invasive procedure and the level of risk and discomfort they are prepared to accept. If fully informed consent is given then the woman's autonomy has been respected. If IVF is the only way a woman might be able to have a child she may be prepared to undergo the procedure even though the success rate is low, as any percentage of success with IVF is greater than the zero success rate of not having the procedure.

2. Reproductive Choice

While some feminists argue that the existence of ART extends women's procreative choices since women are free to choose another set of options, others argue that women do not freely choose to use these procedures but are responding to pressure that is exerted by society and particularly partners. The overriding pressure, noted earlier, is an effect of the pronatalist context in which ART operates and the corresponding pressures on women to become mothers and provide genetically related offspring for partners.

> The decision to use IVF is carried out within a strong pronatalist context. . . . These kind[s] of ideologies, reinforced by economic structures, pressure women to "choose" motherhood. . . . Though some choices are available within firmly delineated limits, we are encouraged to choose the socially acceptable alternative. (Spallone and Steinberg 1987)

Whether or not one accepts these claims very much depends on which wider theories of individual and societal interaction are held. There are other ways that ARTs can operate as constraining factors on women's procreative choices. It has been claimed that the very existence of ART can constrain and influence choices. In an article describing his own experience of infertility treatments, Paul Lauritzen says, "The problem here might reasonably be described as the tyranny of available technologies. This 'soft' form of coercion arises from the very existence of technologies of control. . . . [O]nce the technology of control exists, it is nearly impossible not to make use of it." Here, what are presented as new options can quickly become seen as the standard of care that women are not really free to refuse. Barbara Rothman has summarized this point: "taking away the sense of inevitability" in relation to infertility and "substituting the 'choice' of giving up does not in any real sense increase such couple's choice and control" (R. Arditti and R. Kline (Eds.). 1984. *Test-Tube Women*. Pandora Press). It is argued that to be childless due to infertility is no longer seen as an acceptable position unless women have tried to conceive by using ART. Even then, if conception does not take place, the feeling of failure and the stress, strain, and costly medical treatment can all take their toll.

C. ART as Unnatural Practices

Another set of arguments against the fundamental principles of ART is the Roman Catholic view. ART is viewed as an unwarranted interference with nature and what is perceived as God's will. ART is a deviation from normal intercourse and, in separating the unitive and procreative aspects of sexual intercourse, devalues the reproductive process. To introduce a third person into this process is seen as defiling the sanctity of marriage and the family. A second area of concern for religious groups is the treatment of the human embryo, and this will be considered in a later section.

D. Conclusion

The arguments against ARTs are various. Some, such as certain feminist and religious views, object to the very principle of ART albeit for very different reasons. Other views are concerned that ART should operate in a safe and well regulated way so that those participating in the treatments will not be unduly harmed. Despite opposition, IVF has now been recognized as a therapeutic technique, and this is exemplified by countries such as Australia having Medicare benefits approved for IVF and gamete intrafallopian transfer (GIFT) by the Federal Budget in 1990. ARTs have been considered and regulated by various legislative bodies (notably the British Parliament in the 1990 Human Fertilization and Embryology Act). Hence, the fundamental principles that lie behind ART have been publically accepted and the debate over whether to have ART, in any form, has largely been superseded by debates concerning how to regulate ART and how to best ensure that progress in ART proceeds ethically and responsibly.

III. PARTICULAR ETHICAL PROBLEMS

Once ARTs become used as therapeutic techniques for alleviating infertility, practical ethical problems arise.

A. The Treatment of Embryos

Embryo research is an integral part of the scientific development of IVF and related techniques, and in enabling embryos to be created outside the body, it has opened up a whole area of debate over the moral status of the embryo. The debates over the moral status of the embryo and the practical problems of research protocols and storage, disposal, and ownership of embryos will be examined here. The focus of this section will be on the embryo up to the eighth week of development and will not be concerned with fetuses at a later stage of development to distinguish this consideration from the abortion debate and concerns over the changing status of the fetus as it develops.

1. The Moral Status of the Embryo

The issue is how embryos should be regarded in the moral sense; that is, some decision needs to be made about what kind of entity they are, so that embryos can be treated in the morally appropriate way. Here the question is one of determining whether the difference between an embryo and an adult is a *morally relevant* difference. At one end of the spectrum, there are those who would claim that as the embryo is a human being, it should be accorded full moral status and therefore be treated with the same regard as any adult. This view is held by the Roman Catholic church, which believes that morally relevant life begins at conception; the embryo has a right to life that must be respected. At the other end of the spectrum, there are those who would contend that only persons should be accorded moral status, and it is this moral status upon which a right to life is based. A person is defined as a self-conscious, thinking, feeling being, and under this definition the embryo does not qualify as a person—it is simply a collection of cells and therefore it has no right to life.

Although there is a lack of consensus over what moral status one should give the embryo, there has been a practical response to this problem by regulatory bodies. In the United Kingdom, *The Warnock Report* (the result of a public committee of inquiry into ART) framed the question in terms of, "how is it right to treat the human embryo?" and concluded that "the embryo of the human species should be afforded some protection in law." This is in effect taking the middle ground—not according the embryo with rights on a par with an adult, but not equating the embryo with a morally insignificant clump of cells.

2. Embryo Research

Embryo research has caused considerable controversy. The issue of the acceptability of research on embryos is closely related to the conclusions reached about the moral status of the embryo. If the embryo is considered to have full moral status, on a par with an adult, then the same considerations that guide research on adults should pertain to the embryo, such as seeking consent from the subject and ensuring that the research only involves reasonable risks. However, the embryo cannot consent and this raises the problem of who should consent on behalf of the embryo. Further, embryo research often involves the destruction of the embryo and it could be argued that this is an unreasonable level of harm for a research subject to undergo.

If the embryo is considered to have no moral status then there will be less concern for the embryo's welfare and it is more likely that research will be permitted.

However, just because the embryo is considered to have limited moral status, in comparison with an adult, it could be argued that this does not imply the embryo is morally insignificant. Embryo research might be permitted if it operates under strict guidelines. An analogy could be drawn here between embryo and animal research. Animal research is permitted but only under carefully supervised conditions and for significant medical benefit.

It is generally thought that embryo research should be subject to time limits (determined as days after fertilization) after which research on the embryo should not be permissable. One of the main practical ethical dilemmas in this area is where to draw such a suitable cutoff point. In the United Kingdom this line has been drawn at 14 days from the day that the gametes were mixed, and therefore an embryo cannot be used after the appearance of the primitive streak. This has been echoed in the USA by the 1994 *Report of the National Institutes of Health Human Embryo Research Panel.*

The appearance of the primitive streak (at 14 days) marks the point at which the embryo becomes morally significant, as it is argued that this is when the embryo has developed irreversible individuality, and reflects the view that the embryo becomes more morally important as it develops. There is a possible problem in determining when, precisely, in its developmental stages an embryo or indeed fetus gains moral status, as biological development is a gradual process and it is impossible to point to one stage where moral status is suddenly conferred. This 14-day cutoff point has attracted criticism on the grounds that it is an arbitrary point of demarcation. Warnock has responded to this by stating that "the point was not however the exact number of days chosen, but the absolute necessity for there being a time limit set on the use of embryos." This illustrates that there is no firm consensus on when the embryo becomes morally significant, but there is a necessity for practical guidelines.

A further dilemma is the source of the embryos for research. Research can utilize embryos created specifically for the purposes of research or spare embryos created during the process of IVF. One objection to creating embryos specifically for research is based on the argument that this reduces the embryo to a commodity, treating it as a mere product and thereby devaluing the procreative process. It could also be argued that if embryos are created specifically for research purposes then increasing numbers will be created to satisfy the demands of researchers. A separation principle could be invoked here (separating the process of creating embryos from their final use) which states that the creation of embryos should not be linked to any future research plans and only spare embryos should be used. A counter

to this would be that if embryo research is acceptable, this is due to claiming that embryos have no moral status and if they have no moral status then the source of the embryos is irrelevant. This tension is summarized by John Harris: "I cannot but think that if it is right to use embryos for research then it is right to produce them for research. And if it is not right to use them for research, then they should not be so used even if they are not deliberately created for the purpose."

3. The Storage of Embryos

In a typical IVF cycle around six embryos will be created, and British law stipulates that a maximum of three embryos should be transferred. There has been concern expressed about the high number of multiple births produced by ART (47% of babies born from IVF come from a multiple pregnancy according to HFEA's 1999 *Annual Report*). Multiple births present a significant risk to both maternal and infant health. The Royal College of Obstetricians and Gynaecologists (RCOG) in the United Kingdom have recommended that, where possible, only two embryos should be replaced, and the HFEA notes that reducing the number of embryos replaced decreases the chances of a multiple birth without reducing pregnancy rates.

The couple undergoing ART have a choice of either having their spare embryos destroyed, used for research, or frozen for future use. If the IVF attempt fails or if future egg retrieval is impossible (due to the woman's age, for example) couples can use these stored embryos for further IVF treatments. There are a number of potential problems with the storage of embryos. How long should embryos be stored? Should there be a time limit or can they be stored indefinitely? Freezing embryos can create problems in the future if there are disputes over ownership of the embryos if the couple split up or if one dies. Who should have the casting decision as to what to do with the embryos? This is a problem of deciding whether the embryo should be treated like an existing child or whether there are different concerns here, as the embryo only has the potential to become a child. One response might be to use a detailed consent form setting up the storage terms to preempt these problems, but no consent form, however well worded, can anticipate every possible future circumstance.

B. The Donation of Gametes

Infertility treatments have used donor gametes to benefit patients whose own gametes are not viable. Factors such as abnormal sperm findings or a woman who has unhealthy occytes but is otherwise able to carry a pregnancy can indicate the need for donated material.

The donation of genetic material has created a large number of ethical dilemmas, such as who is to be considered the parent of the future child and the possible effect on the welfare of the future child.

1. Determining Who Is the Parent and Welfare Concerns of the Child

The use of donor gametes raises the troubling problem of who should be considered to be the parent of the future child. When the sperm donor gives his sperm he also gives up any future parental role and the donor's anonymity is guaranteed. In the United Kingdom it is the husband of the woman undergoing the infertility treatment who is considered to be the father of any resulting child. This provision also extends to couples who are not married. The principle of anonymity is also embraced by the Ethics Committee of the American Fertility Society (1994) as it is seen to be important to encourage men to donate and safeguard them from becoming unwittingly responsible for their genetic offspring. However, some countries, Sweden, Austria, and the Australian State of Victoria, for example, allow offspring access to identifying information about the gamete donor once they have reached a suitable age. It can be argued that children should have the right to know about the circumstances of their conception and the identity of the donor if they so wish. For example, Spain passed a law in 1988 which stipulated that gamete donation should be anonymous and subsequently the law was challenged on the grounds that it violated the basic human right to know of one's origins.

There could arise circumstances where family members or close friends might want to donate gametes to help those they are close to. This could cause potential problems with the future relationship between the donor and the child. Donors may see themselves as the "real" parents, a belief encouraged by regular contact, and may believe they have a legitimate say in how the child is brought up.

The case of egg donation can raise complex problems. Traditionally one could always be certain who one's mother was; however, with egg donation a woman can become pregnant with a child to whom she is not genetically related. It is only through the technique of IVF that the gestational and genetic aspects of motherhood have been able to be separated. So which of these functions is to be held as the true indicator of motherhood? Is one more important than the other?

The ethical problems with all gamete donation are those of potential harm to the donors, the recipients of the gametes, and the future children. Gamete donation is not akin to other forms of donation made in a medical context—it is the giving of genetic material that will be used to create offspring that are genetically related to the donors. The issue here is one of the importance given to the genetic relationship. Donors may always be mindful of the fact they may have children they will never meet—that some part of their genetic inheritance exists that they are not aware of. This might be particularly pertinent for egg donors as women are only able to donate a limited number of eggs. The recipients of the donated material may feel that the child is not really theirs or the future child might not see one or both of his or her parents as a "real" parent.

Conversely, it might be the social relationships between parents and children that are the important defining factor and there would be no particular harm in rearing children to whom you are not genetically related. Adoptions can be successful and many men rear children who are not genetically theirs (even if they are unaware of it). As the Ethics Committee of the American Fertility Society notes, with egg donation the possibility of bonding between the surrogate gestational mother and the fetus in utero is an unresolved issue and this could be an important source of maternal bonding.

In the United Kingdom the mother of the child is deemed to be the pregnant woman and the gestational function determines the legal motherhood, not the genetic relationship. However, there is more ambiguity in the USA, for example, and the Ethics Committee of the American Fertility Society notes that "the legal status of rearing rights and duties in offspring of donor oocytes has not been definitively established."

One solution to the issues raised by sperm donation could be provided by the relatively new technique of intracytoplasmic sperm injection (ICSI). ICSI involves the injection of the sperm cell directly into the egg before transfer into the uterus and is used in couples with male factor infertility, enabling men who previously would not have been able to have their own genetic offspring to do so. The HFEA's 1999 *Annual Report* shows that the use of this technique has increased by 40% since 1997, and donor insemination (DI) has decreased by 11%. A possible problem with ICSI is the effects on the long-term health of the children produced. ICSI enables fertilization to take place artificially and it can be claimed that this circumvents the process of natural selection. However, as yet, the figures from the HFEA do not show any significant difference in outcomes between ICSI and other infertility techniques.

2. Payment of Donors

It is argued by the HFEA that altruistic donation is to be encouraged and that donors of gametes should not receive any payment over and above a minimal compensation for time and inconvenience. It is thought that any payment could have the detrimental effect of encouraging inappropriate motivation (such as merely seeking financial benefit) on the part of the donors. Excessive inducement to donate gametes could possibly exploit the less well off in society, as it is likely that they would be the ones to respond to financial inducement. This concern reflects the general problem of the ethical acceptability of paying people for their body parts (e.g., organs) or body products (e.g., blood). It could be argued that there are additional ethical problems in the case of selling gametes as the vendors are not only selling body products but are also selling their genetic material. Conversely, one could argue, as Harris does in *Wonderwoman and Superman,* that not allowing payment is compounding the problems of those who are less well off by depriving them of a source of income. In the United Kingdom donors are paid £15 and reasonable expenses. However, internationally, with the development of the sale of gametes over the Internet, donors can sell their wares on the open market to the highest bidder.

The issue of payment for donors is important due to the shortage of gametes, particularly oocytes, and ways of encouraging donation are frequently debated. To encourage women to donate eggs they are often given free treatment or free sterilization in exchange for donating their spare eggs. It could be argued that this is ethically on par with providing payment, as such services could operate as an inducement to donate eggs and it is likely that those with limited financial resources would be the ones most susceptible to such inducements and hence more inclined to undergo such procedures. However, the HFEA has stated that they will allow the practice of egg sharing as long as it is correctly monitored.

3. Posthumous Reproduction

In 1996 an English woman, Diane Blood, asked surgeons to remove her husband's sperm while he was in a coma (from which he never recovered and subsequently died). The HFEA refused to allow Mrs. Blood to have infertility treatment with her husband's sperm on the grounds that he had not consented to the removal and subsequent use of his gametes. As the law stands in the United Kingdom, gamete donors and members of the couple undergoing treatment must give explicit consent to any use of their gametes or participation in infertility treatment. After lengthy legal battles Mrs. Blood was allowed to take the sperm to Belgium where she received treatment. There are a number of issues that arise from this and similar cases. First are consent concerns. While the United Kingdom has specific legal regulations concerning this, wider ethical issues are raised. Does the marriage contract imply consent to procreate? What

kind of consent procedures should be adopted by clinics for donors and recipients? Should donors and recipients be able to review or revoke the consent they have given? In an area that is always changing rapidly due to scientific advances, how detailed should the consent be and how can any consent form adequately anticipate all future circumstances? Second, is it ethical to deliberately create orphans? It is argued that while many children have been born after their father has died, to deliberately bring a child into the world knowing that the father is already dead is harmful to the child and society. This question turns on an assessment of the future welfare of the child—a matter of much controversy and political rhetoric, and which will be discussed in the next section.

C. Access to ART

The possibility of infertility treatments has raised the issue of who should have access to such provision. Should ARTs be given to single women, non-heterosexual couples, post-menopausal women, or women who are HIV-positive? This is a concern over their suitability as parents. In the United Kingdom the HFEA does not lay down precise guidelines as to who should or should not receive infertility treatment, but states that treatment centers should have clear written procedures to follow for assessing the welfare of the potential child. Hence, the welfare of the future child is held to be of paramount importance, but there is considerable disagreement over how such welfare is to be protected. However, it could be argued that if the main reason for ART is to alleviate the suffering of the infertile, then the welfare of the child should not be of paramount concern.

1. The Welfare of the Child

To establish which individuals should receive infertility treatment on the basis of who will be a suitable parent is a difficult decision. It could be argued that this is not a medical question and medical practitioners are no more expert in deciding what makes a good parent than anyone else. This kind of decision is particularly vulnerable to the prejudices and preconceived ideas of individual practitioners. The Ethics Committee of the American Fertility Society broadly states that the best interests of the child are served when it is born and reared by a heterosexual couple in a stable marriage. However, they do note that there might be a role for other patterns of parenthood and do not recommend the legal prohibition of ART for nontraditional families. In the USA, as in the United Kingdom, the matter is largely left to the treatment centers and individual practitioners to decide who is suitable to be a parent, and therefore this could give scope for unwarranted discrimination.

2. Particular Cases

In the United Kingdom, with regard to single women and lesbians the HFFA Code of Practice states that attention should be paid to the child's need for a father, and where the child has no legal father, "centres should consider particularly whether there is anyone else . . . willing and able to share responsibility for meeting those needs." This is a relatively progressive provision. Many countries (e.g., France) restrict access to married or stable heterosexual couples.

Another focus of concern has been the issue of whether post-menopausal women should receive infertility treatment. There have been various cases reported in the media, most notably a clinic in Italy that treated a 62-year-old woman, believed to be the oldest ART mother. The question is, is it ethically acceptable to give post-menopausal women infertility treatment?

In answering this question, one possible starting point is to consider the welfare of the child. Is it self-evidently harmful for a child to have parents who are older than the norm? The trend in the developed world seems to be that more women are delaying having children until their late 30s and early 40s. The normal age to have children appears to be slowly rising. Accordingly, it seems to be almost impossible to determine the normal age for childbearing in a society where changing social circumstances and improvements in health can shift that point.

In considering the acceptability of treatment for older women it is a matter of degree. We might think that 60 is too old but 50 is just acceptable. This creates the problem of how we are to justify these two different limits. In response to this, there have been attempts made to determine a suitable cutoff point after which treatment should not be given.

There can be two cases that present themselves to infertility clinics. Those women who require egg donation (for IVF) and those women who can use their own eggs. The women who require egg donation may not necessarily be those in the "older" age bracket, but may have a high risk of transmitting a serious genetic disease or have undergone a premature menopause. However, if a woman is post-menopausal she will require egg donation in order to conceive. It is argued that this could be used as cutoff point for those seeking ARTs. If the woman has reached menopause then her fertile time is over (biologically) and this would provide an objective, testable cutoff point that was not subject to individual practitioners' interpretation and judgment.

However, this seemingly straightforward solution masks two difficult problems. First, some women, albeit a very few, menopause prematurely, sometimes in their 20s. Would we want to say that a woman in this situation is not to receive treatment? The response to this might be that we mean the average age of menopause and not the menopause itself. But this throws the debate back to interpreting norms and deciding who should have treatment purely on the grounds of age. A woman of 40 who has reached menopause may think that it is too early to give up the thought of children, so how are we to distinguish between her and the 20 year-old when deciding who should get treatment if not by age?

Second, current medical research states that it is the age of the oocytes that lead to age-related declines in fecundity, not the age of the uterus. So, women over, say, 40, although they have not yet reached menopause, may wish to have donor eggs provided to minimize the risk of miscarriage. Thus, with the use of donor eggs menopause ceases to be a relevant biological cutoff point. Again, the decision comes back to making a judgment of whether to treat on the basis of an individual's age rather than a biologically determined cutoff point.

These issues frequently come to a head when applied in a practical context, i.e., over how to allocate scarce resources such as donated oocytes. Only giving infertility treatment to those who can provide their own gametes is one way of restricting access and this can be extended to only allowing fertilization between partners (as is the policy in Sweden). Germany only allows treatment if the woman can provide her own gametes.

Any decision on how to promote the future welfare of the child is going to be a difficult and necessarily speculative one. It is often seen as unfair that the infertile have to prove that they will make good parents while the fertile have no such constraints. It is important to base decisions about who should be treated on genuine concerns for the welfare of the child (while recognizing this is very difficult to ascertain), not on prejudice nor on an unwarranted preference for so-called traditional family units.

3. Financial Restrictions on Access to Reproduction Technologies

One of the most of the most hotly debated issues concerning ART is how such treatments should be funded. In the United Kingdom the question concerns the remit of the National Health Service (NHS) and whether infertility treatments should be funded from the collective purse. Certain local health authorities, for example, have decided not to fund infertility treatments or only fund a limited amount of cycles. Often criteria for eligibility for treatment are imposed, such as age limits, only treating married couples, or only treating those who have not had children before. In countries with insurance-based health care systems similar problems arise with debates concerning the extent and type of infertility treatment that a health care plan should fund. Arguments given for restricting funding of infertility treatments range from highlighting the relative expense versus the good outcomes to claiming that infertility is not a medical problem. All these arguments are based on assumptions about the relative importance of infertility and how far societies should go toward helping the infertile.

4. Regulatory Restrictions on Access to Reproductive Technologies

As was demonstrated by the Diane Blood case, there are significant regulatory differences between countries throughout the world and indeed even in Europe. Diane Blood went to Belgium to have a procedure that was illegal in the United Kingdom. Swedish couples travel to Denmark to have DI procedures since it is given anonymously there, whereas in their own country sperm donation is legally nonanonymous and their future offspring would be able to trace the donor once they reached 18. In the United States there is no national and little statewide legislation governing ART, and practitioners follow professional guidelines. Thus, it depends on the country in which one is resident as to what ART you might be offered. Due to such regional variations, a practice of "reproductive tourism" is developing where couples travel to other countries to avail themselves of techniques not available in their own country. This raises questions of whether it is practical or possible to have countrywide legislation, such as the detailed and comprehensive legislation in the United Kingdom, if couples can circumvent the law by moving themselves geographically.

IV. CONCLUSION

Reproductive technologies raise many disturbing and difficult ethical dilemmas for both the practitioners involved with the techniques and the general public. One of the difficulties in both regulating ART and considering the ethical implications is the speed with which new techniques and processes are developed. It is imperative that such developments only proceed after careful consideration as to the short- and long-term implications both for those who use them and for society as a whole.

Bibliography

Alpern, K. D., Ed. (1992). *The Ethics of Reproductive Technology*. Oxford: Oxford Univ. Press.

Andrews, L. (1999). *The Clone Age: Adventures in the New World of Reproductive Technology*. Oxford: Rowmann & Littlefield.

Boling, P., Ed. (1995). *Expecting Trouble: Fetal Abuse & New Reproductive Technologies*. Boulder, CO: Westview Press.

Corea, G. (1988). *The Mother Machine*. London: The Women's Press.

Dolgin, J. (1997). *Defining the Family: Law, Technology and Reproduction in an Uneasy Age*. New York: New York Univ. Press.

Evans, D., Ed. (1996). *Conceiving the Embryo: Ethics, Law & Practice in Human Embryology*. The Hague: Martinas Nijhoff.

Evans, D., Ed. (1996). *Creating the Child: The Ethics, Law & Practice of Assisted Procreation*. The Hague: Martinas Nijhoff.

Godsen, R. (1999). *Designer Babies: The Brave New World of Reproductive Technology*. London: Gollancz.

Harris, J., and Holm, S. (1998). *The Future of Human Reproduction: Ethics, Choice and Regulation*. Oxford: Oxford Univ. Press.

Overall, C. (1993). *Human Reproduction: Principles, Practices, Policies*. Toronto: Oxford Univ. Press.

Purdy, L. (1996). *Reproducing Persons*. Ithaca, NY: Cornell Univ. Press.

Rowland, R. (1993). *Living Laboratories: Women & Reproductive Technologies*. London: Cedar.

Spallone, P., and Steinberg, D. (1987). *Made to Order: The Myth of Reproductive and Genetic Progress*. Oxford: Pergamon.

Stanworth, M. (1987). *Reproductive Technologies*. Oxford: Polity.

Warnock, M. (1985). *A Question of Life: The Warnock Report*. Oxford: Basil Blackwell.

Science and Engineering Ethics, Overview

R. E. SPIER

University of Surrey

I. A Review of the Predisposing Salient Events
II. Toward a Definition of Science and Engineering Ethics
III. A Historical Perspective
IV. Contemporary Issues
V. Conclusion

GLOSSARY

codes Bodies of text that delineate in general terms appropriate behavior.

engineering An activity that requires knowledge, practical skills, and the generation of significant novelty (as in a patent) with an objective of providing benefit for society.

ethical or moral conduct A way of behaving that is acceptable, good, or satisfies one or another ethical objective.

ethics A subject area equivalent to morals dealing with the way humans conduct their lives.

fabrication The construction and reporting of data that were not actually observed.

falsification The modification of observed data.

misappropriation The theft of the intellectual or physical property of another.

misconduct Acts by humans that are held to be unethical or immoral.

misrepresentation The deliberate misleading of another by the presentation of false or incomplete information.

noncompliance The disregarding of laws, codes, regulations, ordinances, and so on.

obstruction The deliberate prevention of the freedom of another to achieve her objectives.

peer review The examination of one's work or proposal by a group of people drawn from a background similar to one's own.

plagiarism The passing off of the work by another as one's own.

science Knowledge that has been tested to determine the level of confidence with which this knowledge may be held.

SCIENCE AND ENGINEERING ETHICS is concerned with the behavior of scientists and engineers with regard to the way they carry out their vocations coupled with the products of those vocations. Whereas it was thought that scientists could generate new knowledge without regard to the ethical implications of their actions, the burgeoning subject areas of biotechnology, computers, and nuclear physics have quickly brought to the attention of the perpetrators of those subjects that their actions have consequences beyond the mere compilation of new knowledge and capabilities. As our societies have increased their wealth, the citizens have determined that their environments and the biota that inhabit those locations can be considered to be integral to the determination of the quality of their lives. These events have occurred during a period when the research and development activities supported by the state have come under increasingly critical scrutiny. Members of the academic research community have been encouraged to compete; they have been urged to associate with industry; they have been rated, graded, reviewed, and evaluated. This has led directly to the action of governmental agencies to prevent misconduct by researchers enjoying state funding for their work. Such actions in turn have led to the emergence of this new area of science and engineering ethics.

I. A REVIEW OF THE PREDISPOSING SALIENT EVENTS

In contemporary Western societies there is a pervading sense that we have lost our way. Gone are the certainties on which constitutions and interpersonal interactions have been based. We are faced with a reexamination of traditional religions, with constitutions preventing beneficial social developments, and with democracies turning into dictatorships. Since the explosion of the atomic bomb at Hiroshima, the social engineering of the Final Solution holocaust, and the catastrophic failure of the nuclear power station at Chernobyl, we have become aware that the practice of science and engineering can lead to situations that threaten the continued existence of humans on planet Earth. These events have led to a resurgence of interest in the way we behave and the factors that control that behavior.

Naturally, this moves us into considerations of ethics. One such set of ethical issues pertains to the suite of activities that goes under the designation of doing "science and engineering." This article focuses on this area and begins by examining how the need to delineate this subject emerged. Once the requirement was set, there follows the necessary establishment of the boundary conditions used to define the field. From this platform we can look both backward at historic events, as well as forward as our contemporary situation melds into the future.

A. The Need to Achieve Accountability

The current political climate in the Western democracies seeks to decrease the pervasiveness of "government" and to return to individuals more control of the resources that they earn or acquire. In cutting public expenditures governments have sought to maintain the level and quality of services that they had supplied at the previous higher expenditure rates. To achieve this end they have become more assiduous in assessing the qualitative nature of publicly purchased services, including those in the education sector that involve science and engineering research in universities and in government research institutes. This development has led to an increase in the requirement for such institutions to demonstrate their responsible deployment of resources allocated to them from the public purse; it has led to the implementation of a greater degree of accountability.

The consequence of these efforts to reduce the public funding of research and development programs, while maintaining the quality and quantity of the output, has been a transformation of the way science and engineering research and development has progressed. By reducing education budgets by 3% per annum the government of the United Kingdom has sought to achieve increases in the efficiency of the production of graduates and research workers. Another way of seeking to promote improved performance with less funding has been the instigation of increased competitiveness both within and between institutions. This has been achieved by the operation of league tables and peer-reviewed assessment of research quality that has, again in the United Kingdom, important consequences on the level of future government funding. A third repercussion has been that universities have been encouraged to enter into active collaborations with the commercial/industrial worlds; this interaction is not without its special ethical problems (R. Spier, 1995).

The intensification of the pressure on university and government research laboratory personnel to perform and raise grant monies from both the public and private sectors on the basis of their research and publication activities has led to the sense that such researchers might act in a manner that was considered inconceivable prior to the application of these contemporary requirements. Issues of fraud, theft, and dissimulation have now become part of the vocabulary of active scientists. As such matters are not dealt with by the normal civil courts, because it is difficult if not impossible to ascertain the monetary value of the damages caused and the people involved do not, in the main, have the necessary monetary ability to mount a lawsuit, it has fallen to the institutes themselves to monitor and control the behavior of their employees. This movement has been augmented by the involvement of public bodies. Since the early 1990s it has become a requirement of the U.S. government agencies of the National Science Foundation (NSF) and National Institutes of Health (NIH) to require grantholders to have taken an approved course in science and engineering ethics prior to having a grant award made to them.

B. Pressing Universities and Industry Together

In efforts to increase the flow of inventions from "basic" scientists to industry and therefore to provide additional justification for the public funding of such investigations, scientists in universities have been cajoled into forming groups, consortia, and consultancies to generate such a flow of knowledge. Few have been successful (Stanford's capitalization on the patents taken out by Cohen and Boyer in 1975 on methods for genetic engineering have been outstandingly productive). However, in the inevitable scrapping around for contracts it is clear that compromises with what would have been considered approved behavior are made. There are numerous examples of such practices, some of which involve conflicts of interest (CoI), while others border on the edge of downright theft or fraud. So, those institutions whose members apply for publicly funded grants have to establish ethics committees and to acquire from putative grantholders any information that would indicate whether or not there would be a CoI were those grantholders to work with public monies in ways that would provide them or their immediate families with "insider benefits."

C. High Profile Cases

In the climate that has come to prevail in the last couple of decades, there have been some notable cases of questionable behavior, often by noted, well-respected and leading scientists. Paul Gallo and his laboratory colleague Mikulas Popovic have only recently had resolved the investigation of their behavior with regard to the discovery of the virus that causes the Acquired Immunodeficiency Disease Syndrome (AIDS) vis à vis the countervailing claims of Montagnier. Recently concluded is the case of Nobel Laureate David Baltimore's association with an unsubstantiated allegation of falsification of data before presentation perpetrated by the research worker, Imanishi-Kari. Other cases involving the painting of black spots on mice and the generating of irreproducible experiments that purport to demonstrate a "phosphate cascade" in the transmission of information between the exterior of the cell to the responding genes, have been well documented (W. Broad and N. Wade (1982). *Betrayers of the Truth*, p. 256. Simon & Schuster, New York). And there are cases where one scientist's work has been plagiarized by another as, for example, in the settled case of Heidi Weissman, whose published work was used, unacknowledged, by her senior.

There are also cases of whistle-blowing in industry. When the space shuttle *Challenger* failed spectacularly (January 28, 1986) on a widely televised launch, tragically killing the seven astronauts (including a teacher), the efforts of Roger Boisjoly to prevent the launch, when made public, raised an outcry as to how his warnings were not heeded (See C. E. Harris Jr., M. S. Pritchard, & M. J. Rabins (1995). *Engineering Ethics*, p. 1. Wadsworth Publishing Company, London. See also S. H. Unger (1994). *Controlling Technology, Ethics and the Responsible Engineer*. Wiley Interscience, New York). The crunch point in this event was when, during a telephone conversation to NASA about the impending launch, the senior vice-president of Morton Thiokol, Gerald Mason, asked the director of engineering, Robert Lund to "... take off your engineering hat and put on your management hat"; this made him change his decision from "do not launch" to "launch." There are other cases where the warnings of staff were taken up by management as in the case of J. Thomas Condie who asserted that his superior John Ninnemann misrepresented data.

All such events became headline news in recent times with a consequence that people began to ask about the training of people who occupy responsible positions in engineering and science. Such questionings were also instrumental in leading to the establishment of science and engineering ethics (SEE) courses in universities, the founding of a journal with the same name, and an increase in interest in this subject area by senior members of industry. This is evidenced by the establishment of the Ethics Officers Association as the appointment of individuals at such a level became a requirement for a company to tender for a large government contract.

D. Environmental Issues

Because fully loaded oil tankers have been breaking up in full view of the television cameras and the resulting pictures of dead and despoiled animals populating our littoral areas and riversides have been brought to viewers instantly, there has arisen a social mood that seeks to eliminate and prevent such happenings. Laws have been enacted. Authorities with powers to investigate and deter potential polluters of the environment have been established, and vigilante groups have been formed to police what they think is the boundary between acceptable (to them) and unacceptable behavior in this regard. At times of social difficulty (such as the depression of the 1930s) or war, environmental issues do not take preference over the need to do everything possible to maintain the existence of the society, even if that means creating a less than desirable environment for the local animals and plants. However, in times of relative plenty, environmental issues do become of increasing social importance and the inculcation of an ethic involving the objective of sustainability becomes necessary.

Industry may be said to be the largest problem in this area. Engineers and scientists working in a "for profit" organization are often presented with CoI issues on the basis of how they call a particular action, which may cause pollution outside the set and tolerated limits. In addition, such individuals may be seen to be working for many masters; their employer, the prevailing society and themselves and their families. Often, while well-meaning individuals can see that environmentally hazardous (or personally jeopardizing) activities may be in train, their notification to their supervising officer can either be ignored, can be denied, can be justified in the interests of the greater good of the company, or can be denigrated as being either irresponsible or naive.

E. Animals Have Become Eligible for Human(e) Treatment

Our use of animals to promote our survival has developed over the years from food source, sentinel early warning system, and toxic material test bed to their use in medical experimentation for new therapeutics and prophylactics, animal competitions for gambling, draft animals for work, and the use of animals to gratify emotional requirements that are not fulfilled by humans. In this process we have learned to rear animals efficiently and intensively. Also, as a corollary, the increased efficiency of our hunting activities has brought many animal species to the point of extinction or to unsustainable recovery from overfishing or hunting; examples abound; whales, boar, buffalo, great apes, manatees, eagles, wolves, and many others are, or have been, on the extinction point. Of course, in highlighting animals we must not overlook the changes to the plant and microbial communities. While these are more contingent to our well-being than the animal cohorts, we tend not to respond to their needs as emphatically, possibly due to our disinterest in anthropomorphizing them.

To investigate the "way animals function" (effected so that we can design improved medicaments for humans (and domestically useful animals)) we now have suites of rules and codes that determine what would be acceptable practices. In sum such regulations seek to decrease the numbers of animals used, to decrease the use of animals whose nearness to humans is evident (even though this means that much research on the action of immunogenic agents in mice has to be completely redone in primates when it comes to justifying trials in humans), and to decrease the pain and suffering to the minimal levels held to be absolutely necessary. All such trials and their protocols have to be agreed in advance, and clear and available records must be kept.

Nonetheless, in spite of the urgings of those who have set up "The Great Ape Project" of a "Homeland" with full sovereign rights for the great apes in some parts of Africa, we may not have to go quite that far. But we cannot deny the need to examine each such issue on its merits and to act accordingly.

F. Engineering with Cells and Genes

In the mid 1970s a suite of abilities was developed that enabled biochemical geneticists deliberately to introduce or delete genetically based characteristics into or from a wide range of micro- and macro-organisms. Such activities may be classified as falling into the purlieu of engineering as defined in Section II.D. The use and widespread distribution of such genetically altered organisms poses a number of ethical issues. There are two outstanding sources of concern. One is that researchers will construct organisms, which, in an uncontrollable manner, would damage our lives or environments. The second worry is that the functions or intentions of some deity will be abrogated and living beings will be created that might be regarded as personifications of the monsters or devils of ancient cultures. And finally there is a sense that genetically engineered organisms are unnatural. In many countries these anxieties have led to the establishment of committees to oversee and vet applications to make genetically engineered organisms. Such committees will examine issues, in addition to the ones mentioned, and from a list such as:

- How much genetic information can or should be passed on to insurance companies when they are asked to quote premiums for life insurance policies?
- Can we replace or modify genes that control human propensity to disease(s)?
  —aging;
  —intelligence (cognitive abilities);
  —height;
  —skin color;
  —sexual potency;
  —mood;
  —aggressiveness, and so on.
- Should such genes be modified on a somatic or gametic cell basis?
- In making animal models of human or animal diseases may we create new genotypes that are designed to experience the pain and suffering we associate with particular disease states, for example, the oncomouse?
- Can we change the balance of the species that occupy particular ecological niches?
- Can we use mood enhancing genes in the absence of the disease condition of depression, and so on?

Another biotechnological area that has generated ethical issues is that of the extra corporeal (*in vitro*) fertilization of human ova and the ability to form clones of animals, including the mammals. This, in the case of humans, has thrown up a set of questions such as:

- the rights of a viable human embryo held in a frozen state in the event that the natural parents disappear or die;
- the use of surrogate mothers to bring to maturity the embryos of others with or without compensation;
- the age of a mother at the time of implantation;
- the use of the sperm or the eggs of named donors who may be living or dead;
- the use of viable embryos for experimental purposes;
- the use of eggs from aborted fetuses or virgins;
- the development and promulgation of cloned humans.

The application of biotechnology can, and probably will, significantly alter the genetic and hence the phenotypic nature of the animals and plants that inhabit planet Earth. Such events will have major implications for our economic state, for the way we organize our societies and for the new value structures we establish. It is important therefore to obtain a view as to the direction such changes may take. This requires that we examine the objectives of our behavior, that is, our ethics, so we can take full advantage of these new and awesome capabilities.

G. Physicochemical/Engineering Issues

The world is in a state of alarm over the prospects of an ubridaled active nuclear confrontation. Visions of a fully robotized battle being fought with satellite information technology coupled with nuclear-tipped rockets and computerized battle logistics, which obviates the footsoldier in the field, are causes for deep concern. In addition to this we have a (relatively) peaceful takeover of the way we conduct our lives by the pervasiveness of information technology systems, as manifested by the Internet and the World Wide Web, as well as the prevalence of closed circuit television cameras surveying our city centers and motor routes to capture on film or video the malefactors in the society. Notwithstanding these serious developments, we have a burgeoning private vehicular transportation system that, at great social expense, moves people to and from their work, shopping venues, and children's educational facilities. A tendency that increases with rising crime rates that, in part, results from engineering decisions made about the way we organize as a society. Machines

that have revolutionized the way the routine tasks of the home are effected have enabled women to become an increasingly important section of the labor force, which has implications on the way children are raised and brought to morality. In short, engineered products have revolutionized the way we live; the ethical implications of these changes have yet to be fully appreciated.

H. Summary

In this introductory section I have outlined some of the contemporary factors and events that have brought ethical issues to the forefront of our attention. Such issues are the need for accountability, especially as this applies to the behavioral aspects of how scientists and engineers perform. The admixture of the universities with industry and the inculcation of a competitive ethos has also conspired to introduce practices that require ethical examination. As we have become richer as societies, we have been able to adopt a more involved and critical appraisal of the way we treat the environment with its human and animal populations. The occurrence of such events has conspired to cause the emergence of an area of active endeavor that is rapidly becoming a new discipline: that of science and engineering ethics.

II. TOWARD A DEFINITION OF SCIENCE AND ENGINEERING ETHICS

Words are only as useful as the meanings that are attributed to them. While it may seem that definitional issues can be dismissed as mere semantics, such concerns are yet the matter for the considerable, weighty, and labored deliberation of much of what is the daily activity of lawyers and jurists. Thus, words and their definitions are important and they do justify critical examination. Furthermore, while there are many books that have dealt with the substance of the nature of science, there is not an equivalent literature dealing with engineering matters. Ethics, of course, has been one of the prime areas of interest of humans as this affects the way they behave with repercussions on the decisions they make and, as a result, on their success or failure in themselves or their communities. In this section I offer the reader definitions of science and engineering that may be slightly at variance with traditional definitions. Yet, while the definition of ethics poses fewer problems, the manifestation of different ethical systems is often the source of socially disruptive conflict that saps the intellectual and social strengths of our modern societies. I will also spell out how I perceive the relationship between knowledge (science) and ethics. This is not a "hands-off" distancing exercise, but rather an intimate, connected, and direct interaction between the worlds of being and the worlds of obligation.

A. A Definition of Science

As a first order definition, science, deriving from the Latin *scientia,* translates directly into knowledge. This then requires us to look into the nature of knowledge as a second order consideration. When we translate the nature of the world outside of our minds (in this I take the mind and its thinking capability to be an activity of the nerve cells of the brain, and the exterior world to include the brain and the rest of our body) into thoughts, ideas, concepts, imaginings, guesses, or other products of an active mind we may be said to be creating or generating new knowledge. From that time, this knowledge has an existence independent of the exterior world and can be used without direct reference to it, as in situations when we reflect, think, or dream. As I hope to show, there are two additional qualifying characterizations of knowledge in the confines of our minds. The first characterizing tag is that of the degree of confidence we have in a new or old mental construct or idea. The second concerns the importance we assign to that knowledge—its value to us as we seek to survive.

It would seem from this definition that all knowledge is science and all science is knowledge. This is almost the case but not wholly. In 1840 Whewell defined science as knowledge acquired by scientists. Such individuals did not generally regard the images, conceptualizations, or constructs that exist as mental phenomena as knowledge. It was only those selfsame images, conceptualization, or constructs that survived a rigorous system of experimentation and testing that could legitimately be lodged in the mind as knowledge: and such knowledge was science. The method by which sense data became knowledge became known as the scientific method and the people who practiced the scientific method were scientists. The scientific method was then described as:

- make repeated observations of the exterior world;
- make a hypothesis as to the relationship between these observations;
- test that relationship by further observations;
- if the additional observations do not require you to change your hypothesis then either continue with the testing or allow your hypothesis to become your knowledge;
- if after exhaustive, rigorous, and stringent testing the hypothesis is not found wanting then it may be considered a theory;
- theories describing particular relationships that have

been extensively tested over long periods by many people may be said to have become laws.

This seemingly seamless rendition of the way observations become knowledge was rattled by K. Popper who, in 1934, asserted that it is impossible to *prove* that a hypothesis is true because there is always the possibility that someday an experiment will be effected that will require the modification of that hypothesis. However, he went on to declare that it is possible to prove that a hypothesis is wrong; for any experiment or test that refutes the hypothesis eliminates it for all time from what can be considered true.

I would add to what Popper has expounded, that it is also *not* possible to *prove* a hypothesis wrong. Even an experiment from which we have to infer that our hypothesis is wrong may itself be a flawed experiment. So if we cannot either prove that the hypothesis is right or that the hypothesis is wrong what can we say about it? What we can attribute to a hypothesis is not a determination of rightness or wrongness but rather *a level of confidence* in the knowledge that the hypothesis presents to the mind. Clearly, hypotheses that withstand the most stringent and exhaustive tests will acquire thereby a high level of confidence while those that fail will be accorded a low or even a vanishingly small level of confidence. As we proceed we continually test all our hypotheses and adjust the associated levels of confidence accordingly. (A similar process may be said to occur when we assign levels of value to each item of knowledge we lodge in our minds (Thomas Bayes 1702–1761)). Clearly, a hypothesis such as "the moon is made of green cheese," would not command a great deal of confidence, particularly because we have had the chance to examine moon rocks returned to earth, but we can argue that such rocks are unrepresentative or are forgeries or trick substitutions for the "real" moon materials, which means that we still have to allow a vanishingly small possibility for the original hypothesis to have some validity. To conclude this exposition; the knowledge or science that we store in our minds cannot be considered to be the truth, or something that is proven nor yet a statement of reality. It is but a guess (a less prosaic word than the equivalent word, hypothesis) at an external reality that we can believe to exist in truth; but we cannot know its nature exactly, or with absolute reliability. So the words truth, proof, fact, certainty, reality, exact, correct, right, and their synonyms cannot be used as qualifiers for the ideas and concepts we acquire, store, and manipulate in our minds.

B. Knowledge and Information

We can now relate knowledge to information. The latter may be said to be preliminary sense data that is on its way to becoming tested so as to obtain the status of knowledge. Are there mental constructs that may contain information that may not be considered to be knowledge? I would propose that information or knowledge that is given to use with an admonition that we must not seek to test this knowledge is therefore not to be considered as part of what we understand as knowledge or science as it has been denied examination by the scientific method as described in Section II.B. There are many important cases where this occurs. For example, the first premises of some religious systems require that the believer accepts, unquestioningly, the existence of a god and/or a holy spirit and/or the giving of the laws governing human conduct by God to Moses. But theology, as a study of gods, can be included in science; for it does not exclude questions. Similarly, the lore about ghosts, sprites, fairies, trolls, and the like who can allegedly effect activities that are not in accord with what we believe to be possible in a system where all effects are caused by the preexisting state of the physical/energetic universe, can be considered science. For such information can, in some circumstances, open itself to testing. (It does not last long in the area of high-level-of-confidence thoughts, except perhaps in the area where the tooth fairy regularly and reliably exchanges money for the milk teeth laid carefully under the pillows of our youngsters).

In concluding this section on the definition of science it is interesting to note that it is possible to discern a number of different types of science. For example, it is clear that those who test hypotheses in laboratories or by experiment in the wider horizons of societies, earthly phenomena, and space do so in a way that is markedly different from individuals who test hypotheses using the published literature deposited in libraries or in the attic of a departed savant. While we may wish to characterize these types of individuals as *laboratory* and *library* scientists, respectively, we can also identify another area where knowledge, guesses, or hypotheses are tested, and that is on the street. Conversations are probably the most common way of "doing science." We test our ideas in speaking with others. I would call this *street* science. Additionally, we test our ideas by relating them to other ideas we already have stored in our minds; surely this may be called *conscious* science. And we even test ideas subconsciously: *subconscious* science. This latter test system can be evidenced by the realization by each driver of a car who gets from A to B without having realized that she or he was both steering the car to keep it in its appropriate position on the road and flexing ankle and calf muscles to depress the accelerator peddle to a degree appropriate to the road conditions at the time, all without recourse to the conscious mind.

C. Constructed Science

The constructivist philosophical movement adds a social dimension to the first-order hypotheses we generate from our excited sense organs. While some have used this movement to denigrate science as "just another construction of reality" it actually extends our concept of what we perceive as it integrates it with aspects of the contemporary society. It should be noted that such social interpretations are themselves derived by a guess/test method and may therefore be considered as part of the knowledge (science) derived by the use of the scientific method. In this sense such considerations might be regarded as a facet of *societal biology,* where the other biological areas are molecular, cellular, and organismal.

In summary, I identify five kinds of science: laboratory, library, street, conscious, and subconscious. The social construction of our knowledge is a special part of this activity where meanings are affected by the associated societal aspects. So, as everybody, including most animals, effects the scientific method, that is "they do science," I have to return to Whewell and define *scientists* as those individuals who, generally in exchange for a stipend, initiate and test guesses in areas that are difficult for most people. These areas may involve either the microscopic or the galactic scales; more detailed or more complex analyses or relating the phenomena of the external world to numerical descriptors.

D. Engineering Requires Genius and Morals

Having practiced as a microbial engineer, it was clear to me that there are four components to the activities that can be construed as engineering:

- engineers use existing and generate new knowledge (when creating and testing new knowledge, they become scientists, and they engage in an activity that is often referred to as engineering science);
- engineers generate products that may either be substantive or intellectual;
- engineers seek to, or intend to, inculcate significant novelty into their products; and
- engineers engage in a social contract that requires them, as their primary objective, to work for the benefit of society (however defined).

While the statements depicted above seem clear and straightforward, in practice they are anything but. For example, the need to define "significant novelty" is a tetchy requirement, as is a clear and agreed view of what is socially beneficial.

Readers may use the definitions of science and scientists as set out in Section II.B. to come to a realization of what an engineer does when addressing a new problem or project. But the issue of the nature of the product of an engineer's endeavors needs a little further elucidation. While it is customary to regard the product of the engineer's work as a bridge, road, ship, car, building, computer, power switch, antibiotic, plastic, and so on, it is not customary to include in such a category a painting, sculpture, musical score, or a propounded philosophy or even a work of literary fiction; yet I would contend that each of the latter items can be considered as the products of an engineering process, if there is a compliance with the conditions as presented above. It is clear that even in each of the latter categories a knowledge base is used or generated; a product is created; the intent of the author, sculptor, painter, philosopher, or composer is to create something that will benefit society in providing an outstandingly novel idea or message. Thus, it is only those individuals who set out with the *intent* of providing social benefit by the generation of something that is significantly novel who are to be included in the category of engineer.

As the word engineer contains the French word *genie,* which means clever or innovative (among other things), it behooves the engineer to express genius. For the purpose of defining engineering, I have taken this to mean the manifestation of *significant novelty.* Clearly, everything everybody does each day is novel. But what distinguishes significant novelty from mere novelty is that the product is, for one, patentable. Second, something that is significantly novel should "surprise one, gifted in the relevant arts and working in the same field." A third test of significant novelty is that it is quite different from anything that has gone before. There may be additional criteria one can adduce to further categorize that which is purported to be significantly novel, but it is not my purpose here to develop this. Rather, having come to this juncture I can proceed to the more difficult issue of social benefit.

One way to determine the nature of social benefit is to discover a social concordance on what it regards as beneficial. This may not turn out to be the case even if agreement has been reached and a referendum or democratic process has been used to make the determination. For example, the subset of people who supported the Third Reich agreed to the elimination of the mentally impaired, gypsies, and Jews, and so they defined social benefit as that which achieved these objectives. In the long term, this determination may not have been beneficial for this society for it failed in war, lost a huge proportion of its intellectual elite, and has voluntarily held back the development of what is one of the hottest areas of pharmaceutical research, that of genetic engineering. So how can social benefit be determined? This requires us to affirm the aims that society seeks to

achieve. These in turn are conditioned by the ethical principle that is adopted as the goal for all behavior; an area that is examined further below.

In concluding this section, I have to refer to another misused word, "technology," which can be translated as the "study of techniques"; a scientific endeavor. I would offer a definition of a technician as a person who effects techniques, albeit doing different or novel things at each turn of the activity, and who makes a product, beneficial to society, without seeking to incorporate anything that is "significantly novel." Such a definition includes most people who would not see themselves as engineers and also some people trained as engineers (with the paper qualifications and the practical experience) who do not seek to or intend to innovate in ways that can, for example, be patented. This means that there are people, often self-designated as inventors, who might actually be eligible for the qualification of engineer, but who have not studied for the normal paper qualifications. Institutions of professional engineers recognize such categories and are prepared, in exceptional cases, to give membership, and hence engineer status, to gifted inventors whose products have been thoroughly tested and have made a significant contribution to the well-being of the community (each year the Engineering Council of the United Kingdom examines for engineer status some 200 to 300 applicants who have not acquired formal qualifications).

E. Introducing Ethics

Ethics is about those aspects of our behavior we categorize as our duties or obligations, or what we consider right or wrong (of actions) or good or bad (of things); it covers much of what we do on a day-to-day basis and it may be equated with its latinized cousin: morals. Clearly our autonomic behavior such as coughing, sneezing, digesting, and breathing do not come under this heading. But those human activities that evoke guidelines for our nonautonomic behavior is the subject of this section. Clearly, in this regard the laws of the land, ordinances of the local community, codes of conduct of the professional institution, and the myriad of rules and regulations that fill the shelves of those who govern our behavior are all part of what we would define as ethics. There is yet another subset of rules that is not written down and whose violation does not immediately result in some compulsory penalty. Such rules might apply when we determine the way we eat, help the less fortunate, obey dress codes, and use language or gestures. In each of these areas, defined or not, we seek to behave so as to maximize some function or principle.

1. Ethical Systems

There are two levels at which the subject of ethics is discussed. The principles, rules, and guidelines that are used to directly control behavior are at the most applied level and are termed *normative ethics*. While the grounding, foundation, or basis for the adoption of such principles, rules, and so on, which is at a more theoretical level, form part of the subject of *metaethics*. In the following treatment, I examine the implications of both transcendental and nontranscendental bases for the determination of ethical principles. For this purpose, I define the transcendental as being a state of affairs that is outside the cause-and-effect system. The latter in turn is defined as those activities that are contingent solely on the interactions of material and energetic entities.

It is in the area of determining which particular principle or principles apply for each individual person where difficulties arise. This encyclopedia is replete with expositions of such principles. For my purpose it is necessary to recognize two sets of such principles. The one is based on what is held to be the wishes or commands of a transcendental entity; a god; a supernatural being; a being that can operate outside the cause-and-effect system. Such a being can appear or disappear at will, it can create material goods on the spot from nothing, it can talk to people while remaining invisible and controlling events and/or providing freedom from the cause-and-effect system, as that being chooses. Individuals who derive their ethical grounding from such transcendental entities may either make dogmatic statements based on the experience of voices, interpretations of that which is regarded as holy writ of deeply felt personal convictions (conscience); such ethics are also regarded as *absolute* ethics. Reason, logic, and common sense does not interfere with those who acquire from transcendental sources convictions of what is right or ethical. However, there is also a subset of people who, while holding to a belief in a transcendental entity, will engage in discourse with their fellow citizens and, providing their basic belief system is not impugned, will come to decisions about behavior that may be consonant with common sense.

The other set of principles is based on the acceptance of the determining nature of the cause-and-effect system. It looks to the observable and tangible events that are incorporated into our knowledge and science. It then moves to determinations of the guidelines for behavior. This scheme does not accept the ruling that is often leveled at such developments, which is that it is not possible to move rationally from statements about the world of being to statements about the world of duty, obligation, or ought. (This is a denial of the validity of the naturalistic fallacy.) It is important to realize that this system of *relative* ethics, where the situation of the world determines our duties and obligations, is itself subject to a number of variants. Many of these variants can be found in this volume (see also Singer, P. (1991). *A Companion to Ethics,* p. 565. Oxford: Blackwell). I summarize below some of the leading contenders for the provision of the principles whereby we decide how to behave:

- Golden Rulers who assert that you should do unto others as you would have others do to you or as Confucious would have it, "do not do to others that which you would not have done to yourself";
- utilitarians who assert that you should do what is most useful, (to you, to the community);
- eudaemonists who assert that you should do what makes you and or the community happiest;
- biological determinists/Darwinians might assert that you should do that which most promotes survival of yourself and/or your community and/or other communities and/or other biotic entities;
- Kantians assert that you should act in accordance with the Categorical Imperative, which requires you to "act only on the maxim through which you can at the same time will that it be a universal law";
- conformity ethics requires you to find out what is acceptable and effect that;
- traditionalists require that you behave in the manner in which your antecedents or progenitors behaved;
- theists hold that you behave in a way defined by the edicts or perceived intentions of a god or deity,
- practical ethicists require you to act naturally;
- contractualists hold that your actions be in accordance with your contract with the society you live in.

It is clear that there is a wide variety of principles from which to choose when making an ethical decision. There are, also, definite differences of emphasis in protecting and/or promoting the interests of the individual versus the interests of the society. While this may not matter when the outcome is the same, when the outcome is different there is a need to resolve the otherwise inevitable conflict. In view of the importance of this issue the section below focuses on how such dissonances may be, in some measure, reconciled.

2. When Ethical Principles Are in Conflict

There is a series of practices that can be put in place to aid the resolution of ethical (and other) issues where the protagonists seem to adopt irreconcilable positions. These reduce to a set of actions that can include one or more of the following:

- define the issue over which there is a dispute so that it is clear that both parties have the same view of the difference between them;
- make sure that what are stated to be the "facts" of the case are indeed the most reliable concepts of the situation at hand;
- by examination of the extreme views of the outcome that might be required by each protagonist it may be possible to move to some compromise position in the middle where neither protagonist obtains all that was originally desired;
- the method of *casuistry* requires that one of the two ends of the possible action spectrum be taken by a solution that is evidently right, while the other end is occupied by a solution that is evident to all to be wrong; we can then move to a solution by interpolating additional cases whose rightness or wrongness are not quite as well defined but on which some agreement can be obtained so as to eventually bracket, contain, and resolve the test case that instigated the examination;
- find a technical solution which solves the ethical dilemma;
- spread the load so that others accept their share of any costs that are incurred as a result of, say, a newly perceived need to improve performance specifications;
- determine and define the dominant ethical issue and act on the basis of that teaching;
- do a calculation of the consequences of the outcomes of the alternative solutions in a common medium (money, lives, dignity) and agree on maximizing the level of this parameter in the solution;
- require each protagonist to attempt to stand outside the system and view it as if they were a member of an arbitration tribunal; compare results and move from this position rather than the *ab initio* situation,
- obtain the agreement of the disputants that the resolution of the difference will provide for both parties benefits that otherwise could not be obtained; once this has been ascertained, the examination of the issues may begin afresh;
- enclose the disputants in a confined space and exert physical pressure on them to resolve their differences (decrease the temperature, quality of the food and beverages, amount of space available, and so on, as per the procedures used for the selection of a new Pope);
- select an abiter (ombudsman) or seek the help of an arbitration service, ethics committee, and so on.

The most difficult disputes to solve are those that involve the lives of humans or animals. For some it is not enough to equate the value of a human life at £750,000 for a road user and £2,000,000 if a railway user, or the assessment of the courts when awarding damages in a civil trial of between £20,000 and £2,500,000 (based on a calculation of lifetime earnings foregone). Others will not, or cannot, put a price on a life saved. All lives are infinitely valuable including those of some (all) animals. Therefore, they would argue it is justified to spend all of our resources to save the life of one cat, canary, or child.

We clearly do not behave in this way. Each structure we build, each car or airplane we construct, or each bridge we design can fail under unforeseen conditions of weather, loading, or component defect. To guard against and to prevent all putative disaster scenarios would require infinite resources, and so compromises are made; risks are taken; and we value our lives in proportion to the level of risk to which we voluntarily expose ourselves. Thus, by an examination of extremes and the ways in which we actually live our lives it should be possible to reach an accommodation with even those absolutists, who assert that the deity made all his creatures of equal value, by asking some frank questions about how they would *actually* behave in particular situations as opposed to how they would *like to* behave were they the possessor of infinite resources.

F. Science and Engineering Ethics

From the definitions I have given above for science and engineering the reader will perceive that I have taken a wide approach to the subject area. I do not confine science to that which is effected in laboratories, nor do I confine engineering to the transformation of materials into products. Rather, I would take the position that teaching is adding value to students, that law makers engineer verbal statements to control our behavior in our best interests, and that jurists engineer verdicts by processes that are themselves the products of other engineering activities. While so engaged scientists and engineers are subjected to situations in which they may not act in accordance with one or another ethical principle. Recently, such aberrations of appropriate behavior have settled into a number of sections that I will elucidate below.

It is useful to further subdivide the area into four categories: (1) process issues in the generation of new knowledge or science; (2) product issues that apply to the use of the so generated knowledge; (3) process issues associated with the practice of engineering; and (4) product issues derived from the consequences of the deployment of material entities resulting from engineering processes.

1. Process Issues in Science

The intent of those engaged in the progression of science is to effect research which will "... extend human knowledge of the physical, biological, or social world beyond what is already known" (1995. *On Being a Scientist*, p. 27. Washington, DC: National Academy Press). In pursuing this goal mistakes are made, due care is not applied, or there are deliberate attempts to obtain personal advancement by the manipulation of observations and people. Some of the commoner forms of misconduct are delineated below. The depiction of these misdeeds should not, however, blind the reader to the hundreds of thousands of scientists who strive diligently to discover new knowledge of the world that they attempt to present to us all in a manner that will command our respect and confidence.

a. Fabrication of Data

Events have come to light wherein an individual has deliberately created data so as to be able to propound with greater conviction a particular hypothesis. Such data were made up and were not derived from empirical observation. Examples include the faking of skin transplants by Summerlin in 1974 and the origin of cancer experiments by Spector in 1980 and 1981.

b. Falsification of Data

Data manipulation or falsification occurs when figures are altered to fit in with a notional guess as to what they ought to have been. Sometimes a predetermined result is aimed at because a senior investigator has achieved such a result previously and a student is under pressure to repeat the supervisor's work in a purportedly analogous situation. (An example is the falsification of results on insulin receptors by Soman in the period from 1978 through 1980). In other cases a degree of acclaim is the objective from the presentation of data that provide evidence for a radically new approach to a subject area.

c. Plagiarism

This involves passing off as one's own the work or ideas of others. Such events are used to acquire prestige or to win a grant application. Examples are the theft of authorship of the Bernoulli equation by the father from the work of the son in 1738, around 60 papers copied by Alsabati between 1977 and 1980, the case of Heidi Weismann, and that of Pamela Berge.

d. Data Selection, Manipulation, and Management

Experimenters generate masses of data or observations. Some of these observations are acquired before the experimenter has learned how to do the experiment in the most effective manner. It is clear that data derived from such "learning" experiments are not required in a final publication although they may be presented in a report to a supervisor or a granting agency. Similarly, some data that are generated in a fully established experimental system may just appear to be so wildly at variance with the flow of the data that the experimenter may choose not to include them in any report. For example, data taken from a culture that harbored a suspected contaminant might be wholly discarded. Or a suspect test tube might have been used in a sensitive enzyme action, with the result that an unknown contaminant was completely inhibited. Or more commonly still, an ingredient of the reaction was inadvertently missed and a second reaction container may have received a double dose. These are common occurrences and experimenters are familiar with handling the suspicious data that occur as a result.

There are other ways of manipulating data that rely more on presentation techniques than on making up new numbers. The choice of a statistical technique can be all important in the determination of the significance of an observation. Or a graphical presentation can be designed to magnify or diminish a particular effect. In such cases fraud is not an issue but the principle of *caveat emptor* might also apply to one's reception of data from the scientific literature.

Data management is a serious factor in the work of scientists. Sometimes the withholding of data from competing scientists may be justified on the basis of its preliminary nature and the situation that, without a number of repeat experiments, such data would be more misleading than helpful. Sometimes a research contract requires the noncommunication of data, and the potential commercial application of some piece of information may consign a piece of research to the part of the library that holds the undisclosable material. But when a scientist knows something with a high degree of confidence that would have materially helped a fellow scientist, albeit a competitor, it could be considered unethical for that individual to retain the information and not divulge it. An even worse situation results when deliberately misleading data are issued. In short, the management of data is often the only way an individual scientist may think it possible to preserve his or her position in a particular subject area. It is short-termism of the highest order. But in today's world with so many young scientists clutching at the straws that are supporting their careers it is to be expected that some practices as outlined above will occur.

e. Conflict of Interest

A conflict of interest applies in both the areas of science and engineering. It happens when an individual is driven by motives other than those that have been overtly declared but which do influence the way a project is effected and in a manner that was not intended and could be antithetical to the requirements of the organization that financed the work originally. An engineer in industry might have a conflict between his or her employers and the benefit of society. Examples are the Challenger incident related above and the conditions outlined in the paper of M. McDonald, "Ethics and Conflicts of Interest" (this appears on the World Wide Web at http://www.ethics.ubc.ca/papers/conflict.htm1).

f. Authorship Issues

The improper assignation of authorship is akin to theft as when intellectual property is improperly allotted, assigned, or abrogated. To be an author one must have contributed to the intellectual content of the work described; it is not appropriate to become an author as a result of the gift of the actual author (courtesy authorship). Examples are the Baltimore case mentioned earlier and the parallel scandal of the Darsee case.

g. Mentoring Issues

It is inappropriate to take advantage of the teacher–student relationship so as to deprive the student of recognition and to abuse the education process in the interest of achieving a research success for the supervisor. An example is the way Lipmann dealt with his research worker in 1960, which lead to the need to withdraw a publication purporting to demonstrate another example of Lipman's original finding.

h. Harassment in the Workplace

In the necessarily close relationship between student and teacher in a research environment it is unethical for a teacher to take advantage of such a situation and, for example, make sexual advances or improper proposals. Examples were presented by Louise Fitzgerald and Myra Strober in a symposium organized by Stephannie J. Bird and Catherine J. Didion for the AAAS in 1994.

i. Discrimination in the Workplace

Discrimination on the basis of race, creed, or color is illegal, but it is clearly not possible to treat everybody in an identical manner. Indeed, we would all suffer were we to try to do so. Nevertheless, it is clear that to deliberately disadvantage an individual on the basis of some nonrelevant criterion should not be permitted. The magnitude of this problem varies in different parts of the world and there are even major differences in the different states in the United States. An example would be the activities of the National Research Council of Canada in allocating jobs and contracts.

j. Peer Review: Misconduct/Theft

Grants are given and papers are accepted for publication on the basis of the reviews of the peers of the proposer or putative author. When the competition for grants and recognition is fierce, the temptation to reject a grant application or steal the idea in such an application is a powerful motivator. It is also possible for a referee of a publication to delay issuing the review until that referee has had a chance to either submit his or her own application or to do extra work to maintain leadership in the field. All such cases are difficult to prove and generally do not get a public airing; yet all those engaged in the grant/publication process have a warchest of stories to justify such suspicions.

k. External Examination

This suffers from many of the potential problems of peer review, but the opportunities for self-benefit are less, and in general there is more than one examiner, which prevents most anomalous excesses.

l. Safety Issues

Scientists working at the laboratory bench find the practices imposed by safety committees, while obligatory, are irksome and are not seen to be preventing probable harm to research workers. Nevertheless, a significant industry has come into being to provide scientists and engineers the specialized equipment to dispense fluids in defined volumes that does not involve mouth pipetting. Centrifuges are fitted with interlocking lids and radioactive materials are controlled with great assiduity. The use of masks, eye protectors, gloves, and containment cabinets are and were de rigueuer before the imposition of the current suite of regulations under the banner of COSHH (Care of Substances Hazardous to Health) regulations in the United Kingdom.

2. Product Issues in Science

a. Information Ethics

Privacy is the right not to have published that which one wishes to keep to one's self or to selected others. Modern computers can process data bases that contain ever increasing amounts of information. It is clearly possible to link computer records involving the health of an individual with court records, with insurance claim records, and with a record of all the items one has bought using the credit card. Who may have access to this information cornucopia? How can it be policed so that individuals who do not have the right to access this data base can be detected and brought to justice? How can we protect individuals from incorrect data being used?

These are examples of the kinds of questions that are being raised with regard to the use of information.

Other issues to be faced in the computer age involve the theft of information using the networks to transport such illicit information. The pirating of computer programs is one such case where an unscrupulous individual might purchase a program, inactivate the protective mechanism, and then make the program available at a price that is below the manufacturer's cost. Other issues involve the electronic scanning of articles in journals and the transmission of the resulting digital information via the Internet or the World Wide Web to whomever is interested in picking it up. An additional area is that of computer viruses. The design and implementation of a computer virus to attack the programs stored in a computer so that the owner of the infected computer must spend considerable amounts of time and money to sort the matter out is not something that is a joke or that can be taken lightly. Rather, we may consider the sources of such viruses as the individuals from whom society (or those infected) might be eligible for due recompense.

An additional issue that has stirred much interest is that of the encryption of messages. On the one hand governments are anxious that messages that promote illegal activities may be transmitted in a way that cannot be intercepted and understood by the authorities. On the other hand commercial and confidential transactions of a legal nature may be coded in a nondecipherable form during the normal course of business dealing. It has been mooted that a special encrypting chip (the clipper chip) made available by the U.S. government might cover the encryption of legal transactions, but users are not necessarily willing for outside observers to become aware of their activities. Indeed, encryption devices that cannot be decoded are available to serve this purpose. The resolution of this dilemma may be found in the way government surveillance of civilian activities is controlled. An open system, available for public scrutiny, under the authority of an individual who is also answerable to the democratically elected assemblies may be a part of the answer; but whichever way we turn the need for cascaded control systems (control systems to control, control systems, to control ...) as occurs within organisms and within collections of organisms is needed.

The ability of computers to control robots to act in precise and defined ways has in some countries redefined the nature of the workplace. Japan and Southeast Asian countries have been actively installing robotic procedures to manufacture goods to a higher technical specification and at less cost than in other countries. This and the implementation of computers to handle the repetitive clerical work in government and industrial bureaucracies has led to a downsizing of such operations and the release of many thousands of "middle managers" and their staff onto the labor market. Job security is threatened across the board and society has yet to develop ways to handle such events with fairness and justice, again highlighting the need for a review of the ethics that we adopt.

b. Knowledge Issues

The traditional credo of the scientist has been that the knowledge generated as a result of the application of the scientific method is value free and is available for others to apply in the practical arena where good or bad effects may be engendered. However, modern science has generated sets of ideas that cause us to reevaluate the basis of the ethical systems which have worked for the last 2 or 3000 or so years. In particular, we now have knowledge about the nature, origin, and evolution of life in which we can place more confidence than the creation myths or stories provided by the ancient religions. Additionally, we can explain by using only our knowledge of material and energetic interactions such phenomena as lightning, thunder, volcanic activity, earthquakes, temporal cycles, infectious disease, and the way our bodies work and develop from a single-celled embryo. This knowledge has weakened our dependence on the books of holy writ that provided the most plausible explanations for these phenomena. Consequentially, it has impugned the link between deities and the provision of the rules for human behavior or conduct. In short, the manifestation of contemporary science has required us to rethink both the basis on which behavior should be founded and also the implications for our conduct on our acceptance of such a grounding. Therefore, we have to take the involvement of science in the realm of ethics to be a subject area for further study and development.

c. Clinical Trials

One knowledge area that is much abused is the area of clinical trials of pharmaceutical, food, and cosmetic products. Whereas the former (pharmaceutical) trials have to withstand the rigorous examinations of a Food and Drug Administration or a Committee of Safety of Medicinal Products, new foods and cosmetics are not as stringently regulated as to their value and safety. The financial implications of the trial results are significant, so product promoters are interested in obtaining as favorable a result as possible to be better able to promote their new product. This raises the profile of the double-blind clinical trial to a high level. Nevertheless, there are many reports of the fraudulent operation of such trials and the "unblinding" of the trial while it is yet in progress.

d. Scientific Predictions

Scientists are often asked questions about the future. Contemporaneously, there are questions as to the safety of British beef, the issue of global warming or cooling and what is responsible for any such change, the prediction of earthquakes, the path of tornadoes, and the eruption of volcanoes. We also have issues in the prevalence of infectious diseases, such as AIDS, tuberculosis, malaria, and others. Much of the information generated and tested by the application of the scientific method bears on such questions, but it is necessary for the lay public and the media to realize that answers to questions about the nature of the future must of necessity bear a probabilistic qualification. Scientists do not claim to make predictions with 100% certainty, although the assertion that the sun will rise tomorrow can be made with a confidence approaching that level. In other cases, such as the timing and location of earthquakes, the degree of certainty is much less, and it varies inversely with the degree of precision of the prediction.

The use of knowledge in making predictions, irrespective of the accuracy of the predictions, can lead to ethical problems. Secrecy is generally thought proper when the affairs of state or the profitability of a company are at risk. It would be less well thought of if antisocial events followed the withholding of information about what could transpire in the future.

3. Process Issues in Engineering

a. Whistle-blowing and Conflicts of Interest

As engineering can involve the generation of new knowledge, many of the process issues as discussed earlier pertain. However, the emphasis in engineering changes because the products of engineering appear in the marketplace. So the issues of whistle-blowing and conflicts of interest exist in a more potent and tangible form because the financial condition of a company and its personnel may depend on closing a particular deal, and the selection of the data and the manner of its presentation can be crucial to achieving such an end. Under such circumstances honesty and the presentation of all the relevant information is the ethic to be followed.

b. Safety

Furthermore, as engineers tend to work at larger and faster scales of operation, issues of safety in the workplace are more pressing than at the relatively benign environment of the laboratory bench. Indeed, the issue of product safety is a matter of concern for all engineers, for it is clear that there will be a legal liability as well as ethical opprobrium were a product resulting from a design that was inherently unsafe to be let into the marketplace.

c. Honesty and Confidentiality

Scientists adopt the view that the communication of research is a priority, whereas engineers working in industry, commerce, or in a political institution may be required to keep much of what they know and do in confidence. The breaching of such a confidence can have financial implications on the share price of a company or the popularity of a political party. Again the divulgence of information outside the preordained channels can constitute an illegal act and one that throws the perpetrator and the company into an ethically reprehensible light.

d. Codes of Practice

Finally, engineers accept that their actions and behavior will be governed by the code of practice set out by their qualifying institution or professional society. Examples of such codes may be obtained from the institutions/societies. The policing of compliance with such a code is an issue that requires more attention. While in the medical, legal, architectural, and media areas there have been actions based on noncompliance with codes, the relationship between the professional, his or her institution, and the legal system is such that the further development of voluntariness in this area is of dubious value. It may be necessary to provide professional institutions with immunity from prosecution for effecting their duties by applying the codes in a legitimate fashion with appropriate sanctions. This would protect societies from litigation resulting from actions taken against a member of the society who felt aggrieved at the treatment received.

4. Product Issues in Engineering

a. Bioethics

The application of ethical principles to humans and animals is part of the general subject of science and engineering ethics as this is just another area where knowledge is generated and used for benefit. This not only applies in the way humans relate to animals and the way humans relate on a one-to-one basis with other humans, but also in cases when humans operate in the societal state. It is clearly a vast subject and is the matter of many learned tomes. (e.g., 1987. *Bioethics*, p. 620. edited by T. A. Hannon, Paulist Press).

Bioethics impinges on the promulgation of science in the use of humans and animals for experimental purposes. Such issues have been dealt with in great detail by bodies set up to monitor, regulate, and control such activities. Codes of behavior have been written and are widely distributed and used as exemplified by the Declaration of Helsinki: World Medical Association, revision of Hong Kong 1989.

Other bioethical issues concerning humans abound. Abortion, contraception, eugenics, euthanasia, homosexuality, infanticide, heroic medicine, prophylaxis versus therapy, informed consent, resource allocation, harassment, and discrimination are all issues that are under constant review in popular and specialist journals. In addition there are the issues referred to earlier dealing with our ability to manipulate genomes and cells where the latter can become human embryos.

b. Ethical Problems with Widgets

While the practice of engineering engenders many ethical issues that are similar to those thrown up in the area of science, there is a subset of problems that are more closely associated with the tangible products of the engineering activity. For example, there is a suite of engineering product issues:

- there are benefits and disbenefits resulting from the exploitation of nuclear energy;
- the mass production of the weapons of war can be an industry devoted to societal defense or an opportunity for profiteering and the instigation of needless conflicts between client states;
- the chemical and oil industries contribute to our well-being in numerous ways but they also provide us with toxic wastes and, in the event of oil tanker catastrophes, with environmental disasters;
- transportation by private or public means as a way to preserve energy, decrease pollution, and generate amenity;
- intensive rearing of animals for food purposes;
- the production of drugs and therapies where prophylactic measures would yield greater social benefit.

c. Ethical Issues in Social Engineering

Ingenuity may also be expressed in the creation and promotion of those cultural organs that control and determine the way we behave as social beings; examples of such activities are:

- the production, execution, and monitoring of laws;
- the criminal justice system;
- the management of an economy so that predefined social or ethical principles can be achieved;
- the operation of an educational system that fits in with the needs of the individual to be provided with an equal opportunity to develop his or her talents to the greatest degree coupled with the need to satisfy the requirement of society for educated individuals;

- the promulgation of eating and exercising habits that drain health care resources to the least amount.

Such issues are coming into focus with increasing clarity and intensity as our modern societies develop without the incursion of a major catastrophe such as a world war, a famine, a cosmic catastrophe (accident), or a universal and devastating plague.

III. A HISTORICAL PERSPECTIVE

A. Data Manipulation

As we do not, nor can we, know with complete certainty the nature of the world outside of ourselves we have to make guesses at what it might be, generally based on some observations or sensate data. Nevertheless, those who have sought to make some sense of what they perceived were on occasion somewhat cavalier with how they handled the raw data. For example, Newton used fudge or adjustment factors to enable his observations to marry to the relationships he devised; Dalton, the formulator of the modern version of the atomic theory, altered his basic data on the weights of the elements that combined with measured weights of other elements, so that he could present his results in the most convincing way. And Millikan, who showed that the charge on the electron was constant and that fractional charges did not exist, selected those experiments that "went right" in order to present his case effectively. (This author has done some similar experiments when investigating the relative charge on hydrated microcarriers and can vouch for the occurrence of the "oddball" observation.) It is also alleged that the data on which Gregor Mendel based the concept of genetics and genes were selected from his observations so that the patterns of heredity were clearly depicted in his reported figures. A similar massaging of the data may also have occurred in the presentation of the statistics regarding the inheritance of mental capabilities in identical and nonidentical twins by C. Burt. Forgery is also not uncommon in the area of paleontology, as evidenced by the Piltdown Skull believed to have been fabricated by C. Dawson and M. A. C. Hinton between 1908 and 1912.

There is little doubt that, in spite of the manipulation of the data by these investigators, the concepts that emanated from their endeavors were worthy and valuable to subsequent workers. It is important to realize that in testing guesses it may be of value to see the data or observations in the light of what they "ought" to be on the basis of some notion or theory of what is out there. While there are pressures to acquire the accolade of being the first individual to demonstrate such a theory or relationship, and that the presentation of data that fits "miraculously" with what one is purporting is a tempting thing to do, it may be more appropriate in the world in which we are increasingly brought to account for our actions to actually do the fudging openly with some comment to the effect that "my methods of making measurements were somewhat unreliable, but within the noisy data I collected, I can discern some relationships that I believe pertain in that reality out there, and these are...."

1. Data/Concept Misappropriation

Not only may data be molded to theories but data or ideas can be "borrowed without acknowledgment" from others. Claudius Ptolemy (2nd century CE) was able to convince the world for almost 1500 years, that this planet was the centre of the universe based on data which, it seems, he purloined from the observations of Hipparchus (who flourished between 146 and 127 B.C.E.). it is suggested that Charles Darwin first saw the ideas implicit in natural selection in the papers of Edward Blyth (of 1835 and 1837), whom he did not acknowledge, whereas he was most careful in giving credit to Malthus for sparking the idea that populations expand in excess of their food supplies, to Wallace for his concomitant realization of the nature of the process of natural selection, and to Spencer for the latter's concept of "survival of the fittest." It has also been suggested recently that, on occasion, Pasteur could be economical with the truth. For when he presented to the public his anthrax vaccine for sheep he told those who assembled to witness the experiment, that he had oxygen-attenuated the material, whereas his notebooks show that he had used the *antiseptic* potassium bichromate (an oxidizing agent). This was because he was reluctant to acknowledge the work of the veterinarian Toussaint who previously showed that organisms killed by the use of the *antiseptic,* phenol or carbolic acid, were effective in making anthrax and other vaccines.

IV. CONTEMPORARY ISSUES

A. Office of Research Integrity (ORI)

The field of science and engineering ethics became recognized as a major area of governmental interest in the United States during the 1980s, culminating in the 1992 formation of the Office of Research Integrity. This body deals with reports of misconduct allegations and oversees the way institutions in receipt of government research grants comply with the recommendations to prevent the occurrence of misconduct and conflicts of interest. Universities are learning to cope with these issues through the formation of the appropriate committees. Bioethics committees abound. Most countries as well as the European Community have bodies that deal with issues resulting from new developments in Medicine and Biotechnology. A recent report of the ORI, entitled "Integrity and Misconduct in Research" (1995. Published by the USDHHS/PHS) defines misconduct as fabrication falsification and plagiarism and includes misappropriation, interference, and misrepresentation as well as obstruction and noncompliance with codes.

B. Ethics Courses

To comply with the need to behave ethically, scientists and engineers in universities and industry have been offered educational courses covering many of the points discussed in this article. It is generally held in this field that the most effective way of purveying this information is by a variegated and structured approach to the subject area. This will include special classes dealing with ethical issues in their historical, operational, and case study aspects. The use of role-playing scenarios engages the involvement of the participants and onlookers, while the formats of seminars, discussions, and debates also serve to instruct. The fundamental problem faced in such interactions is that the proliferation of ethical systems means that some course leaders regard their function as merely showing the students (a) that there are problems and (b) that there are a variety of ways of approaching a solution. The ability to "close-out" a problematic issue is not taken as an objective of the exercise. This causes student dissatisfaction. It therefore becomes of increasing importance to emphasize, portray, and exemplify methods of conflict resolution as I have indicated above. In addition, it is important that ethical issues relevant to subjects that are expounded are brought up during that teaching period. This requires that teachers in all subject areas are familiar with the ethical implications, problems, and pitfalls in their area of specialization, and that they are willing and capable of handling them in the didactic situation. The combination of the provision of the ethical tools coupled with the demonstration of the relevance of such considerations for each area is pivotal in the promotion of ethical thinking and behavior in students, teachers, and practitioners.

A further handle to the didactic situation is slowly coming to the fore. This deals with the efficacy of ethics courses in the subsequent thinking and behaviors of those who have been exposed to such experiences. In the work of Deni Elliot, et al., it is clear that the change in the way some people operate in areas they perceive as vital to their self-interest after having been subjected to a course in ethics is minimal, but measurable.

Although Plato in the Meno comes to the conclusion that virtue cannot be taught, K. D. Pimple and his colleagues at Indiana University have used definitions of the various stages of ethical development to determine the efficacy of ethics teaching practice. Such stages may be depicted as (1) being able to discern the possible actions and their implications when presented with an issue requiring a judgment; (2) the determination of the morally right (fair, just, or good) course of action; (3) such a determination should be above personal values if these militate against the course judged right; and (4) the person should be able to implement the morally correct decision in the face of forces militating against such an implementation.

This may become the paradigm for the future as more effective and relevant ethics courses come on-line and, perhaps what is more important, is that more people in the institution become conscious of ethics and provide examples of ethically appropriate behavior that become accepted as the norm while the ethically suspect behaviors receive general and public disapprobation. It will be of continuing interest to work out ways in which we can measure the efficacy of such courses by the way the participants live their lives subsequent to their exposure to such courses.

C. Whistle-blowing

There are other live issues. Whistle-blowers have been surveyed recently with the result that most of them have recorded that they either lost their jobs, were held back in the promotion stakes, or were held in low esteem and perceived as troublemakers by their colleagues. Of the people who responded to a recent survey, 69% said that they had experienced some negative consequences. It is clear that legislation has to be available to protect the reasonable interests of the whistle-blower. Notwithstanding such protection it should also be clear that deliberate, unjustified, and vindictive victimization of individuals by whistle-blowers should be treated with all due severity (the Hamurabi Code of *ca.* 1700 B.C.E. has it that, if you act as a false witness to a murder then you are subject to the punishment that would have accrued to that murderer).

D. Insurance Issues

As a result of our newly found abilities to effect genetic screening it is possible to determine in advance the propensity of an individual to a range of debilitating disease states. The ethical question at stake is whether such information should be made available to insurance companies who are providing life insurance coverage. This should be compared against an actuarially derived calculation based on the population as a whole rather than the probability of the insured individual making a claim. Such considerations also apply to those who are covered by health management organizations (HMOs). The issue is based on whether society or the individual is responsible for the health and well-being of the individual. One can ask the question as to whether those of us who are born with genes that do not give us a propensity to disease should support financially those who do have such a propensity. Our sense of society and the contract we each have with the body politic would indicate that we shoulder the burden communally; but some individuals would not wish to be part of this: hence the dilemma.

E. Gene Patenting

All the canons of patent law require that there be an inventive step by human intervention in the description of the item or process for which a patent is sought. It is clear that the determination of the sequence of bases in a gene by the application of well-tried-and-tested (often highly automated) procedures cannot be considered to be inventive. Nevertheless, companies, academies, and individuals persist in applying for patents for the sequences of the bases of genes whose function is often unknown. Clearly, such patents cannot be held to be inventions. However, there are situations in which genes are sequenced from particular individuals who may express different genetic properties that could be of benefit to others. Such genes may belong to an individual or a tribe as property and it could be expected that the exploitation of that property requires compensation. Were such gene sequences not patentable then it would be difficult to effect the compensation. This need not be the case, as I may lend anybody some of my property for a fee without the need for patent protection. Where there are modifications to the genes that are not just "cosmetic," then it may be admissible to file for a patent. Or, indeed, if the product of gene expression is modified, formulated, produced, or delivered in a novel way it is also reasonable to expect patent protection of the invention that in this case will be a useful product.

It is possible to patent antibiotics, which are the products of bacterial and fungal organisms, as it requires ingenuity to grow the producing organism in culture in a manner that provides economic yields of the antibiotic. So, it is argued, it should be possible to patent genes that are natural products like antibiotics. Were this to be the case then each material generated by every organism can be patented; this is contrary to the intent of society, which is to strike a contract with an inventor so as to reward personal ingenuity by the granting of a patent monopoly in exchange for making the invention freely available to the public some 10 to 20 years after the granting of the patent. So it behooves the patenter of a gene to show how ingenuity or inventiveness has been expressed, bearing in mind that the sequencing of the gene is not enough.

F. Who Watches the Watchers?
(*Quis custodiet custodies*)

When peers are asked to review manuscripts prior to publication or research grant applications or the quality of contribution of a department or unit to an area of research endeavor then the possibilities of deliberate wrongful judgments and theft of intellectual property are rife. People asked to effect reviews of their contemporaries work are bound by conventions and warnings, yet the public prosecution of any misdeeds is not evident. Nevertheless, there is considerable mistrust of what is held to be the least odious way of apportioning credits. Most people who do not achieve success attribute some of their failure to the misjudgments and intellectual thefts experienced at the hands of the members of reviewing boards. Were it possible to conjure such boards from people who do not have a vested interest in the outcome of the judgments and therefore do not suffer from *a priori* conflict of interest then a considerable service would have been rendered. Such people might be retirees or people from abroad, or groups of individuals, each of whom reviews all the documents before the board and makes a judgment on each document (see also *Science and Engineering Ethics* (1997), Vol. 3, Issue 1, specializing in an examination of the subject of peer review).

G. Cloning and Chimeras

We are fascinated and horrified in turns with our growing capabilities to clone animals and plants. On the one hand we have the possibilities of the high yields of the monoculture, but on the other hand there is the possibility of disease eradicating the whole crop. Other problems pertain to animals. Experiments with ovines and bovines have demonstrated the two possibilities of cloning and chimera formation. From the latter the deified entities that were depicted as part beast, part man come closer to realization. There is little doubt that such developments open up possibilities for humans. While three are proscriptions against the continuance of such work, the myriad of possibilities it presents will probably be made available if, and when, we can be confident that the social and personal control systems we have implemented are worthy and reliable.

V. CONCLUSION

Ethics embraces the law. The law deals with all aspects of theft, fraud, misrepresentation, and injury, and, indeed, were any individual to believe they have cause to be aggrieved, they are free to bring a civil suit to court and claim damages proportional to the injury held to have been suffered. Such cases in the area of science and engineering, as defined above are not common. While civil courts have been involved with issues of plagiarism, for the most part the injuries suffered by scientists (and to a lesser extent by engineers) have just been tolerated. However, in recent times, the pressures on scientists and engineers to perform under a harsher employment regimen than heretofore has meant that officials in governments have been fearful that misconduct would ensue. So attention has been focused on the practices in laboratories with a view to preventing misconduct.

It is clear that there are many "gray" areas where recourse to the courts is futile and where codes can be ignored or "corners cut," yet where conduct can be improved:

- removal, sabotaging (by, for example, making a lense dirty, changing a setting, or jolting a mirror out of alignment), or monopolizing time on a crucial instrument can be done with the intent of depriving a potential competitor of a success;
- holding back seemingly trivial, yet crucial, oral or written information that could help an adversary is a mean but often practiced device;
- providing misleading information (dissimulation) is another stratagem to put a rival off the scent of one's true intentions;
- the use of ideas acquired, but not publicly acknowledged, from conversations, meetings, or reviewing of papers and grant applications is a form of theft that may be effected consciously or subconsciously;
- the abuse of the role of the supervisor in exploiting a research student's educational opportunity;
- the disregard for safety codes in order to work faster can affect not only the perpetrator but also the other members of the laboratory;
- the mistreatment of animals while seemingly complying with codes;
- the disruption of the spirit of collegiality to foster self-interest.

The perpetration of a competitive and financially restrictive environment by governmental agencies has promoted behaviors that are antithetical to the objectives and missions of science and engineering research establishments. It is therefore not surprising that the behavior of individuals will come under scrutiny. The study, development, and promotion of science and engineering ethics, as a subject discipline, will help us become aware of what we are about and hopefully, in spite of the harsh conditions, it will promote behaviors that will enable us to live together with mutual respect and with a view to what it is that we must do for communal as well as personal self-interest.

Bibliography

Cohen, J. (1994). U.S.-French patent dispute heads for showdown. *Science,* **265,** 23–25.

Crossen, C. (1992). *Tainted truth,* p. 272. New York: Simon & Schuster.

Elliott, D., & Stern, J. E. (1996). Evaluating teaching and student's learning of academic research ethics. *Science and Engineering Ethics,* **2,** 345–366.

Gee, H. (1996). Box of bones 'clinches' identity of Piltdown palaeontology hoaxer. *Nature,* **381,** 261–262.

Geison, G. L. (1995). *The private science of Louis Pasteur.* Princeton: Princeton University Press.

Holden, C. (1994). Breaking the glass ceiling for $900,000. *Science,* **263,** 1688.

Jasanoff, S., Markle, G. E., Petersen, J. C., & Pinch, T. (1995). *Handbook of Science and Technology Studies,* p. 820. London: Sage Publications.

Kaiser, J., & Marshall, E. (1996). Imanishi-Kari ruling slams ORI. *Science,* **272,** 1864–1865.

Poole, T., & Thomas, A. De. (1994). Primate vaccine evaluation network recommendations. Guidelines and information for biomedical research involving non-human primates with emphasis on health problems in developing countries. Pub. DGXII/B/4-SDME R2/105.

Marshall, E. (1995). Suit alleges misuse of peer review. *Science,* **270,** 1912–1914.

Rest, J. R., Bebeau, M. J., & Volker, J. (1986). An overview of the psychology of morality. In J. R. Rest (Ed.), *Moral development: Advances in research and theory,* pp. 1–39. Boston: Prager Publishers.

Rhoades, L. J. (1996). Whistleblowing Consequences. *Science,* **271,** 1345.

Spier, R. (1989). Ethical problem? Get a technical fix. *Vaccine,* **7,** 381–382.

Spier, R. (1995). Ethical aspects of the university-industry interface, *Science and Engineering Ethics,* **1,** 151–162.

Spier, R. E. (1995a). Science, engineering and ethics: Running definitions. *Science and Engineering Ethics,* **1,** 5–10.

Spurgeon, D. (1992). Canadian research council found guilty of job bias. *Nature,* **359,** 95.

Uehiro, E. (1974). Practical ethics for You. Tokyo: Rin-yu Publishing Co, Ltd.

Slippery Slope Arguments

WIBREN VAN DER BURG

Tilburg University

GLOSSARY

argument from added authority An argument often (but in my view incorrectly) considered a slippery slope, holding that someone should not be given a certain authority or responsibility because he will probably abuse it.

empirical slippery slope argument A version of the slippery slope argument that argues that doing A will, as the result of social and psychological processes, ultimately cause B.

full slippery slope argument A version of the slippery slope argument that combines various other versions in one complex structure, together with an appeal to a social climate of public opinion.

L_1, or first logical version of the slippery slope argument A version of the slippery slope argument holding either that there is no relevant conceptual difference between A and B, or that the justification for A also applies to B, and therefore acceptance of A will logically imply acceptance of B.

L_2, or second logical version of the slippery slope argument A version of the slippery slope argument holding that there is a difference between A and B, but that there is no such difference between A and M, M and N,..., Y and Z, Z and B, and that, therefore, allowing A will in the end imply the acceptance of B. (M, N, Y, and Z are intermediate steps on the slope.)

slippery slope argument An argument of the following form: if you take a first step A, as a result of a sticky sequence of similar actions by either yourself or by other actors that are relevantly similar to you, action B will necessarily or very likely follow. B is morally not acceptable. Therefore you must not take step A.

sorites (or paradox of the heap) This is an argument holding that if one grain is not a heap and one more grain cannot make the difference between a heap and not a heap, we can never speak of a heap.

SLIPPERY SLOPE ARGUMENTS hold that one should not take some action (which in itself may be innocuous or even laudable) in order to prevent one from being dragged down a slope toward some clearly undesirable situation. Their typical purpose is to prevent changes in the status quo, and, therefore, they are most common in those fields that are characterized by rapid developments. Slippery slope arguments are easily confused with other types of arguments, like arguments that merely point to long-term effects in general or to side effects. Often they are not so much rational arguments, but expressions of feelings of unease about general trends in society. In such cases, we had better address those underlying worries directly rather than discuss them in their disguise as slope arguments.

There are various types of slippery slope arguments, and they should be carefully distinguished because the conditions under which they are convincing arguments differ. There are an empirical version and two logical versions, and there is a full or combined version. A second distinction can be made with regard to the contexts in which the slope is supposed to exist. The mechanisms of social dynamics and the role of logic differ in each of these contexts. The conclusion can be that they are only seldom convincing arguments; their most important role is in institutionalized contexts like law. Nevertheless, they are very popular in practical debates. To understand their popularity, we are to address their rhetorical role. The main reason why they are so hard to attack (and to substantiate) is that they are based on controversial interpretations of reality and of future developments, interpretations that are strongly influenced by underlying attitudes, different backgrounds and emotions.

I. INTRODUCTION

Case 1: "Perhaps, in some extreme cases, voluntary euthanasia may be morally justified. Yet, we should never do it, let alone make it legal, because this would be the first step on the slippery slope toward an inhumane society. Further steps could be the killing of severely handicapped newborns and then the killing of persons with a mental handicap, until we finally kill the useless elderly against their will."

Arguments like this are very common in applied ethics. They have the general following form: If we do (or accept) A, which in itself may not be morally wrong, we will start a process which will lead us to a clearly unacceptable result B. In order to avoid B, we must refrain from A.

Slippery slope arguments are frequently encountered in biomedical ethics. Their typical purpose is to prevent undesirable changes, and, therefore, they are most common in those fields that are characterized by rapid developments, like biomedicine. They can, however, be found in all fields of applied ethics. Consider the following examples:

Case 2: "Once public officials cross the line of accepting seemingly innocent gifts like bottles of wine, there is no stopping and the road to corruption is open."

Case 3: "If we allow the Communists to take over Vietnam, they will successively take over each of the countries of southeast Asia."
Case 4: "If we prohibit a meeting of a Nazi party, we will end up with prohibitions of fully democratic organizations."

More examples can easily be found (for a wealth of case material, see D. Walton, 1992. *Slippery slope arguments.* Oxford: Clarendon Press). The most common name nowadays for this type of argument is the slippery slope argument, but it has many synonyms. Various poetic titles have been used, like "the thin end of the wedge," "letting the camel's nose in the tent," "this could snowball," and "the domino theory."

Slippery slope arguments have dubious standing in philosophy; they have often been treated as mere fallacies. But this characterization does not really do justice to them, even though, as I will argue, they are only seldom convincing arguments. Not only do they often have great rhetoric power, but they usually also have a certain intuitive appeal and an initial plausibility, which means that they cannot simply be dismissed as always fallacious. Arguments of this kind can be brought forward against almost every change in the status quo: there is always a possible risk that this action starts an uncontrollable process leading to undesirable consequences. This makes them a strong rhetoric tool in the hands of conservatives. But this broad scope is also the central problem, because the argument is not discriminating enough. It could forestall almost every action and we clearly cannot avoid all the changes in the world we live in, even if we wanted to.

Therefore, the basic question of evaluation should not be, is the slippery slope argument valid and plausible? General answers to this question are impossible. The question should rather be, under what conditions are which types of slippery slope arguments acceptable arguments?

II. DEFINITION

The basic idea of a slippery slope argument may be easy to grasp, yet it is difficult to construe a precise definition. As a starting point, we might begin with a provisional one:

> A slippery slope argument is an argument of the following form: if you take a first step A, as a result of a sticky sequence of events, step B will necessarily or very likely follow. B is clearly not acceptable. Therefore you must not take step A.

(For A and B we can fill in any type of action or omission. In fact, A and B may refer to an action taken by the actor himself or to an action taken by someone else, which the actor allows, accepts, or prohibits. For reasons of style, I will simply talk of doing, allowing, or accepting A.)

This formulation (which is essentially equal to most definitions found in the older literature) is still much too broad. Some further qualifications should be made, because it covers almost all the arguments that refer to possible negative consequences of a suggested action.

The most common suggestion is to add the requirement that A is in itself morally neutral or even justifiable. This does not seem a useful qualification to me. Often the question is precisely whether A is justifiable, because the proposed principles that seem to justify A would justify B as well and might, therefore, not be sound after all. Moreover, the parties in a practical debate often do not agree on the question of whether A is justifiable in itself, and in such situations the opponent of A might use the slippery slope argument as a second line of defense to convince the proponent that A should not be done after all. Consider Case 1: many opponents of legalizing euthanasia consider even voluntary euthanasia (=A) morally wrong as such. They use the risk of a slippery slope as an additional argument in discussions with those who disagree on that point to convince them that, nevertheless, all forms of euthanasia should be legally prohibited in order to prevent terrible consequences.

We should look elsewhere for useful qualifications of the provisional definition. Studying some concrete examples may show which modifications should be made.

Case 5: "The Supreme Court should not assume authority to evaluate the aspects of public policy involved in this case of affirmative action. Though the exercise of this authority is innocuous (perhaps even beneficial) in this specific case, the Court might later abuse it."
Case 6: "The government should not allow a manufacturer to dump PCB-contaminated waste into this small stream, because the PCBs will run into a downstream river. The PCBs would kill the fish and wildlife in that river and pollute the drinking water for those downstream who use the river for that purpose."
Case 7: "You should not use this pesticide to kill mosquitoes, because it will also kill many useful insects."
Case 8: "A grocery shop should not lower its prices in order to attract more customers, because the bakery around the corner will probably respond with a similar action. The resulting price war may lead to a situation in which both lose out."

Under the previous broad definition, each of these four cases would qualify as a slippery slope argument. There are, however, good reasons to exclude at least the first three and, depending on the perspective, perhaps the fourth as well.

Case 5 exemplifies the type of arguments that Walton (1992) and F. Schauer (1985. *Harvard Law Review*, **99**, 361–383) label the argument from added authority. It certainly can be a valid argument, as it draws our attention to the risk of abuse of power—but it is not a slippery slope argument. There is only one relevant action here: the action by which the Supreme Court implicitly or explicitly assumes authority with respect to a certain type of question. Further actions by the Court are of a completely different type: the exercise of that authority, presumably of an increasingly dubious nature.

If we would call this sequence of events a slippery slope, the category would include the warning for abuse against every action which transfers authority or responsibility to a person or institution. It would include lending a car to a potentially dangerous driver, selling a monumental house to a commercial firm, or even granting parental authority to any parents, simply because we know no parent is perfect. It does not seem useful to include this broad category of arguments from added authority or added responsibility under the heading of slippery slopes. In my opinion, it is essential that the first step and the next steps are somehow of a comparable nature. A first additional requirement for calling an argument a slippery slope argument can be distilled from this: sequential events leading from A to B should be of a relevantly similar type.

Case 6 (like case 5 inspired by Walton (1992), who regards both as slippery slopes) exemplifies what I would label a long causal chain argument. It argues that, through a series of events, action A will necessarily result in B. There is, however, apart from allowing the dump, no further action involved. It is perfectly natural to say that dumping such waste causes the death of fish and wildlife and causes the pullution of drinking water, even if the causal chain is quite long and complex. If we would qualify this type of argument as a slippery slope, we would have to include every argument that points to long-term consequences of actions. We may distill a further requirement from this analysis: a mere sequence of events is not enough, there should be a sequence of actions.

Case 7 points to the side effects of an action. These are, like the long-term effects, clearly relevant for evaluating an action. If the prohibition of abortion were to lead to a rise in the number of deaths among pregnant women as the result of illegal abortion practices, this is certainly a strong argument against it. But it is not a typical slippery slope argument. In practical debates, however, arguments referring to side effects are often intertwined with real slippery slope arguments, and careful analysis is needed to disentangle them, because the method of evaluation of both types is different. In fact, the distinction is implicit in the provisional definition if we realize that A and B should be different actions and not merely different descriptions of the same action.

Case 8, finally, is of a more ambiguous nature. It is what we could call a spiraling-down argument. Action A might trigger a downward spiraling movement through a process of action and reaction. From one point of view, this is not a slippery slope. The reaction is not by the grocery, but by a different actor—the bakery need not react in that way. A criterion for calling something a slippery slope could be that the actions should all be by the same person, group, or institution. Such a criterion would also be relevant in a complete analysis of Case 5, if we reformulate it as, "We should not let the Court assume authority...." Most of the examples mentioned in the discussion of that case, like lending a car to someone, also have to do with the fact that the actor conferring the authority is someone other than the actor exercising the authority.

From a different perspective, however, we might argue that the grocery and the bakery are relevantly similar and belong, in a sense, to the same group of actors, that of bread-selling shops. In this sense, we could say that the grocery does start on a slippery slope, just as an individual judge may take the first step, even though he is not involved in further steps taken by other judges.

This analysis indicates a further requirement. Not only should A start a series of further relevantly similar actions leading to B, but these actions should also be actions taken by the same person, institution, or group, or they should be the actions taken by persons, groups, and institutions that are relevantly similar. What counts as similar both with respect to the actors and to the actions can, as this example illustrates, be a matter of controversy and will sometimes depend on the perspective taken, but we should at least stick to the criterion.

With the help of these four further requirements, we can now formulate a final definition as follows:

> A slippery slope argument is an argument of the following form: if you take a first step A, as a result of a sticky sequence of similar actions by either yourself or by other actors that are relevantly similar to you, action B will necessarily or very likely follow. B is morally not acceptable. Therefore you must not take step A.

III. TYPES OF SLIPPERY SLOPE ARGUMENTS

There are various types of slippery slope arguments. A standard distinction is that between the logical (or

conceptual) and the empirical (or psychological or causal) version. The logical form of the argument holds that we are logically committed to accept B once we have accepted A. We can further subdivide the logical version with the help of the criterion of whether there is a relevant difference between A and B or not. The empirical form tells us that the effect of accepting A will be that, as a result of psychological and social processes, we sooner or later will accept B.

In the literature, we find many further distinctions; some of them are, in fact, based on distinctions in the context of application (see the next section) rather than the form of the argument itself. A framework of three basic types (one empirical and two logical ones) and a combined version, as suggested below, will usually be sufficient for practical analysis.

A. The First Logical Slippery Slope Argument, L_1

The first logical version—I will call it L_1—states either that there is no relevant conceptual difference between A and B, or that the justification for A also applies to B, and, therefore, acceptance of A will logically imply acceptance of B. A and B need not be identical, but the differences are not relevant from a normative point of view. If L_1 is correct, this is a very strong argument. The moral demand of universalizability (which, according to many ethical theories, is central to morality) or the more general demand of consistency requires us to treat A and B in a similar way. If there is no relevant difference between A and B, and if B is clearly unacceptable, we should regard A as unacceptable as well. If, in Case 2, accepting a bottle of wine and accepting a $100,000 gift are not essentially different, as they are both to be seen as forms of corruption, and if accepting the larger bribe is clearly morally wrong, we should also refuse the bottle of wine.

Because the argumentative power of L_1 is primarily based on universalizability, one might refuse to call it a proper slippery slope argument and regard it as merely a slippery slope argument in disguise (a position I once took). Yet, there are good grounds to call it a slippery slope argument. Only after careful analysis, only when the debate is over, one can sometimes conclude that the argument boils down to an appeal to universalizability. At the start of the debate or the analysis, however, it is often difficult to say so, because the question of whether there is any relevant conceptual difference between A and B is yet unclear. In that phase of the discussion, it is often not (yet) possible to distinguish the two logical versions. Perhaps it will even be only after we have fully gone down the slope that we will finally be convinced that, after all, there was no relevant difference between A and B or that there was a distinction which we only noticed when we were beyond it. Though, theoretically, the distinguishing criterion between L_1 and L_2 is simple, in practical debates it is not always so.

B. The Second Logical Slippery Slope Argument, L_2

The second logical version holds that there is a difference between A and B, but that there is no such difference between A and M, M and N,..., Y and Z, or Z and B, and that, therefore, allowing A will in the end imply the acceptance of B. (M, N, Y, and Z are intermediate steps on the slope.) There may seem to be a clear distinction between aborting a 3-month-old fetus and killing a newborn child, but this distinction collapses as soon as we realize there is no such distinction between a 3-month-old fetus and a 3-month-and-one-day-old fetus, and so forth. This version is the practical analogue of the sorites problem in logic: if one hair less cannot make a man bald, how can we ever call a man bald?

The crux in L_2 is that there is a gray zone. We know A is black and B is white, but we cannot tell where A stops and B begins. Some men are clearly bald and some are clearly not, and there is an intermediate category that we might as well call bald as not-bald. In this gray zone, there is no nonarbitrary cutoff point, but the need to set a cutoff point somewhere is not arbitrary. This means that, if we are able, somehow arbitrarily but authoritatively, to set a cutoff point, any point will do. "Driving too fast" is a vague concept, but if we can authoritatively make it more concrete by stating that 30 mph is too fast on this specific road, this may be a reasonable solution. It is reasonable simply because a line has to be drawn somewhere in the gray zone. If it has been arbitrarily set too low, for instance, at 5 mph, it would have been unreasonable, because it would have been in the white zone. This nonarbitrary setting of an arbitrary cutoff point is not always possible, however.

The gray zone in L_2 is usually the result of both semantic indeterminacy and epistemic indeterminacy (R. C. Koons, 1994. *Mind*, **103**, 439–449). It is partly the result of the vagueness of our language. This can sometimes be countered by using more precise language, like in the case of speed limits. But it is also partly the result of a deficiency in our knowledge, both empirical and moral (I assume that we can speak of moral knowledge, if only metaphorically). We simply do not know in advance what the safe dose of a new drug will be for human beings, or what general criteria to set for a bargain in order for it to be considered "unfair." Making language more precise to counter this

epistemic indeterminacy would only be an apparent solution and often be counterproductive.

The L_2 version (and the L_1 version as well) can usually be applied in two directions: as an argument both for and against a certain position. If we start from the intuitive idea that killing a newborn baby is clearly wrong, and then go backward by small steps, we will end up proving that killing an embryo is equally wrong. If we start from the intuitive idea that killing an embryo *in vitro* is not wrong, because an embryo is not yet a human person, we can go forward and defend that killing an older fetus is not the killing of a person and therefore not objectionable either. One line of argument thus leads to a prohibition of abortion at all stages of fetal development, and the other to a defense of legal abortion at all stages.

C. The Empirical Slippery Slope Argument

The empirical version argues that doing A will, as the result of social and psychological processes, ultimately cause B. The causal processes suggested vary from changes in the attitude toward killing held by physicians practicing euthanasia to a general shift in the ethos of a society. One could further subdivide this category by distinguishing the various causal mechanisms, but this does not seem very useful, because usually the various processes are connected.

D. The Full Slippery Slope Argument

The full or combined version combines various versions in one complex structure, together with an appeal to a social climate of public opinion. Walton (1992) demonstrates that in many actual debates slippery slope arguments have this complex nature. Usually the various constitutive elements are not made explicit, so that it remains unclear which versions precisely are combined and how they drive social practice along the various steps of the slippery slope.

The full version is, precisely because of its complexity, hard to evaluate. Especially the vague reference to public opinion makes it a difficult argument both to attack and to defend. The central question, to which we will return in Section V, is whether, and if so how, the combination of the various versions adds to its strength, or whether it is merely an argumentative chain that is as strong as its weakest link. In order to evaluate it, we must carefully disentangle the various subarguments and analyze them separately.

E. The Apocalyptic Slippery Slope Argument

A last type of slippery slope, only mentioned to be discarded again, is the Apocalyptic or Doomsday argument. A horrible situation is sketched that is so highly speculative that the cogency of the argument—insofar as it exists—depends more upon horror than upon its likelihood. Though it is frequent in public debates and has high demagogical power, it has no merits of its own. Insofar as it seems to embody an argument that should be taken seriously, it can better be reformulated as one of the other versions.

IV. CONTEXTS OF APPLICATION

A second important distinction is to be made with regard to the contexts in which the slope is supposed to exist and—in connection with this—the actors that take the first step. Is it the judge who takes the first step on a legal slope when, in an extreme case, he acquits the physician who practiced euthanasia? Is it society at large that, in its social practice, becomes more lenient toward dodging taxes? Or is it perhaps the individual official who accepts small presents from business relations as a small slide in her personal morality?

There are two reasons why this context of application is important. First, the mechanisms of social dynamics and the role of logic differ in each of these contexts. Second, for a slippery slope argument it is essential to discern the distinctive step A that leads us on to the slope; some contexts have more easily identifiable actions of identifiable actors that may count as the first step, like the passing of a bill by the legislature.

Usually, slippery slope arguments are vague about the precise context, or refer to a combination of contexts. We should make this explicit and analyze which contexts could be relevant and how plausible the various versions of slippery slopes are in those specific contexts, given the role of logic and social mechanisms in each of them. Case-by-case decision making in courts is, for instance, highly vulnerable to the L_2 version, because every judicial decision sets a new precedent. This new precedent in turn may be a good reason for taking further decisions that would not have been justified without the precedent. Legislation, on the other hand, is itself not vulnerable to L_2, but it may facilitate a slope of the L_2 version in judicial practice, if the language used in a statute is vague and leaves broad discretion to the judiciary. The opposite possibility is that legislation prevents a slippery slope by setting clear limits and standards, like 50 mph or the strict prohibition of experiments with embryos beyond 14 days after conception.

If various actors and contexts are combined, this will sometimes mean that it is their joined force that irresistibly drags society as a whole down the slope. (In such cases, we should, however, doubt that if all these actors and contexts have the same tendency, we really could

avoid taking the first step at all.) The combination, then, is a negative factor. But the combination may also result in a careful social process, which helps us develop new standards that are more acceptable than the old ones and that constitute a sound guarantee against slides down the slope.

Thus, the interaction between the judiciary and the legislative (in connection with a broader public debate) can, in favorable circumstances, lead to defensible new lines. The judiciary may, through its case-by-case method that can take full account of all the relevant details of concrete situations, fulfill an important role in the careful exploration of new territory, e.g., by dealing with euthanasia cases and gradually developing criteria for cases in which euthanasia can be considered acceptable. This judicial "experimentation" might engender a broad public debate that may sometimes lead to refinement and retraction by the judiciary, and sometimes to further steps. Once this course of judicial experimentation and public debate has led to a broader consensus on some clearer standards, legislation may more strictly formulate these new standards as authoritative. (To make my point somewhat clearer, I would hold that the Dutch developments on euthanasia have largely, though not completely, followed this model.)

The second reason why the distinction between contexts and actors is relevant, is that we must be able to discern step A as a separate action for which we can freely choose. If A is not thus discernible, we are probably either already on the slope (there is no free choice) or A is not so much a separate step as part of a more general process. Then, we had better take a more general level of analysis and discuss that broader process to see whether it can be checked. Orientation of the discussion on A will then probably be a useless effort to fight this process at the wrong place. Only if A is a separate action that might as well not be taken does it make sense to discuss slippery slope arguments as an argument against A.

In some contexts, there are very clearly discernible steps that are a matter of free choice. They are usually actions where only one actor is involved (taking the first cigarette) or where the process of decision making is institutionalized, like in law. It makes perfect sense to say that the legislature made the first step toward the restriction of free speech when it accepted a statute prohibiting hate speech, or that the Supreme Court made the first step toward an inhumane society when it ruled in *Roe v. Wade*. (To avoid misunderstanding, I should add that, though these arguments make sense, they need not be valid or plausible.) With respect to a personal morality, a first step on the slope is usually also easily identifiable, for instance, when a public official accepts the first gift of a business relation. It is much more difficult to discern such a step when we discuss the social practice and positive morality of society at large.

How far should we go in distinguishing various types of contexts? In theory, the number of contexts is endless, because no context is completely identical, but the rough categorization below seems to be adequate for most practical purposes:

1. Personal morality, the morality actually accepted and practiced by an individual
2. Positive social morality, the morality actually shared and practiced by a social group or society
3. Critical morality, the general moral principles or ethical theory used in the criticism of actual social institutions including positive morality and law
4. Adjudication, the case-by-case decision making by both courts and other institutions like mediators
5. Legislation and regulation, the production of general rules by legislators both at the level of parliament and at other levels
6. Other institutionalized practices, like public policy making or managing a commercial company
7. Combinations of the former contexts, including other contexts, like practices based on prudence

V. EVALUATING VALIDITY AND PLAUSIBILITY

After these analytical exercises, we are now equipped to deal with the central question: When are slippery slope arguments good arguments? There are no general answers to this question. The only way to deal with them is careful analysis, to distinguish the versions of the argument involved and the contexts in which they are thought to apply, and then evaluate each of the versions in each of the relevant contexts. And even if after this analysis the conclusion is that the argument is not strictly invalid, it is seldom a fully conclusive argument but only a probabilistic argument, which should be considered more or less plausible and which can be overruled by other arguments. Nevertheless, some more general remarks are possible, and I will deal with them in the order of the four versions of the argument. I will not discuss all contexts but only those where some significant conclusions are possible.

A. The First Logical Slippery Slope Argument, L_1

As was noted in Section III.A, it is often not clear at the outset whether an argument is of the L_1 or L_2 version, because we do not know whether there is a relevant conceptual distinction between A and B or not. In many cases, close analysis will then show that an

argument is a complex argument consisting of various versions. But sometimes, even after such an analysis, it remains unclear because we cannot oversee whether there is a reasonable distinction in a field of new phenomena that we do not yet fully understand. Our normative theories may simply not yet be adequate to deal with certain new phenomena. Should an embryo be considered a person or not? Is an obligatory HIV test morally different from an obligatory genetic test? Perhaps years of further study will result in an acceptable answer but, at the moment the decision has to be made, we just do not have adequate insight. In such cases, it seems to be wise to treat the case as one in which both the L_1 and the L_2 argument might hold.

An interesting problem is posed in the situation where there is a relevant conceptual difference on a line somewhere between A and B, but this difference is not so important that it can bear the whole weight of the presumed distinction between A and B. An example is the line of viability in the continuous development from conceptus to person. Surely, it is relevant and it is a reason for some difference in treatment, like a prohibition of abortion beyond that line. Yet, I would hold that the difference between viable and nonviable is not fundamental enough to constitute the basic line that completely marks the switch from an entity that, either legally or morally, is not worthy of protection to an entity that is. It seems to me that this is a gradual process.

When discussing experiments with embryos, viability is—in my view—not the fundamental line; protection against experiments should start much earlier. (If the reader does not agree with me on this example, he may invent other ones with similar characteristics.) This shows that the same line may in some respects be relevant and reasonable, e.g., concerning the question of whether abortion should be allowed, but not in other respects, e.g., concerning the question of whether experiments with embryos should be allowed. Then the conclusion must be that, only with respect to the abortion problem, we have a clear line and a relevant difference between A and B, so that accepting abortion before viability does not logically commit us to infanticide.

But, in my opinion, we do not (yet) have such a clear line with respect to other issues, like experiments with embryos, so that we cannot exclude that accepting those experiments with embryos would logically commit us to accepting similar experiments with babies. This means that both the L_1 argument and the L_2 argument can be countered when we discuss allowing abortion of 3-month-old embryos, but that it would be possible that they cannot be countered with respect to allowing experiments with 3-month-old embryos. It would well be that the point of no return with respect to embryo experiments is somewhere before the 3 months.

If the analysis shows that we can find no reasonable distinction between A and B, the L_1 argument can be a valid argument against A. (We can, however, also avoid this conclusion by arguing that B was not so wrong after all.) Moreover, in certain contexts this might be a very strong, if not conclusive, argument against A; if B is clearly unacceptable, we should consider A unacceptable as well. This will especially be the case in those contexts where consistency and universalizability are important ideals, like in critical morality, because most ethical theories consider universalizability an essential characteristic.

In the context of law, consistency is also an important requirement, but we should note that it has more force in the context of adjudication than in that of legislation. Legislation (and in some respects public policy making as well) can more easily set arbitrary limits than the judiciary, whose integrity is more strongly connected with consistency. A governmental or legislative decision declaring that only the first 10 applicants will get a grant (because financial means are insufficient to allow more grants) can be justified as a matter of public policy and can be laid down in legislation. But a judicial decision, without such a legal basis, stating that only the first 10 applicants will get asylum would be unacceptable.

The conclusion is that it depends on the context of application what force the L_1 argument will have. In debates on critical morality, it can be a valid and highly relevant argument. In institutionalized contexts, and especially in adjudication, it may have some force as well.

B. The Second Logical Slippery Slope Argument, L_2

The L_2 argument is not valid in the context of critical morality. L_2 holds that there is a difference between A and B but there is no nonarbitrary cutoff point on the continuum between them. As long as we are, in a reflective discussion, able to determine where the gray zone begins and ends (these limits need not be a point, but can also be gray zones themselves), we can make a decision to set an arbitrary line somewhere in that zone—every line will be justified. The fact that we do not know what speed exactly (50 or 55 mph?) should be considered too dangerous is no argument for not even allowing a speed of 45 mph.

The L_2 argument may have some force in positive morality, but then only in combination with the empirical version. As long as we can (like in critical morality) not only draw a line in the gray zone in our theoretical discussions but can also effectively uphold that line in our moral practice, there is no problem. Only if empirical factors result in the fact that we cannot effectively uphold the line, the slope becomes a real danger. But

this means that the primary force in this case is the empirical slippery slope; so it is better to discuss it in Section V.C.

The most interesting context for the L_2 argument is law. It has a completely different role in legislation and adjudication: adjudication is highly vulnerable to this argument, whereas statutes may even form an explicit and safe barrier against it. In adjudication, the risk is real that through a series of small steps by different judges, each of them almost nonobjectional in the light of existing case law, but each adding a new precedent, we will end up with B. On the other hand, legislation can often effectively counteract such slippery slopes by setting clear limits like prohibiting driving at 50 mph rather than driving dangerously.

For example, in my view, nowhere on the continuum between conceptus and newborn is there a nonarbitrary cutoff point for the question of allowing experimentation. If the judiciary (or ethics review boards, which are in this respect comparable to the judiciary) were to develop standards case-by-case for situations in which experimentation is acceptable, the risk of the L_2 argument driving us too far might be real. Once the legislature has set a clear line, however, for example, by enacting a prohibition on experiments beyond 7 or 14 days after conception, even if this line is itself an arbitrary one, this may effectively forestall a slide down the slope.

C. The Empirical Slippery Slope Argument

The empirical slippery slope argument can be valid in almost all contexts. Only in the context of critical morality is its validity doubtful, depending upon the type of ethical theory used. It is hard to imagine how the general principles of utilitarianism or Kantianism could change as a result of an empirical process. But those ethical theories that recognize the importance of moral experience or intuitions as relevant in the formation of theories, like reflective equilibrium theories or neo-Aristotelianism, seem more vulnerable to the empirical slope. If our moral experiences change as a result of social processes, we might come to accept what we now think unacceptable. But for a neointuitionist this is not really an objection, precisely because she will accept that our current intuitions are fallible and, therefore, may be wrong. A hundred years ago, it would probably have been a good slippery slope argument against theories that defended votes for women, that this might lead to a female prime minister or president. But nowadays we do no longer have a strong moral intuition that this would be a bad result. Therefore, we can conclude that the empirical slippery slope argument does not apply to the context of critical morality.

In the other contexts, the empirical version may be valid in theory, but it is usually hard to judge whether it is plausible. A starting point for our analysis may be the discussion of two (partly overlapping) situations in which the argument is sometimes used, but in which it must be considered an invalid argument against the acceptance of A.

The first situation is when the acceptance of A is merely a symptom of a broad social process of which the acceptance of B might be the outcome. A is in fact a result of that process as well, without being itself a causal factor in the process leading to the acceptance of B. Attacking the symptom will not stop the process then, and will probably result in an ineffective symbolic campaign. Therefore, it is not a good argument against accepting A to say that, in the end, the same process will lead to accepting B, because not accepting A will not stop the process.

The second situation in which the empirical slippery slope argument is used is when accepting A, though not merely a symptom but part of the social process itself, is seen as a symbol of the process. Though in itself it may appear to be morally neutral or relatively harmless, it should not be allowed because it is part of that broad process which might ultimately lead to B. I think this is too easy a conclusion, for we could say so only if we are sure that all the constitutive and derivative elements of this social process are wrong in themselves. When we look at social processes in history, however, we find that these are always mixtures of good and bad elements. The French Revolution resulted in much violence but also in great reforms (we need only recall the great Napoleonic codifications).

The growing emphasis on autonomy, symbolized in the increasing acceptance of abortion, might lead to more than only growing tolerance of infanticide upon parental request, as some of the opponents of abortion argue. (For the sake of the argument, I assume with some opponents of abortion that it might do so, but I doubt whether this is an empirically sound assumption.) It might also lead to a strengthening of the norm of informed consent in medical treatment. We should not protest against the acceptance of informed consent only because it is part of the process that might lead to infanticide. The simple fact that acceptance of A is part of a social process leading to B can, as such, never be an argument against accepting A.

By allowing A in those situations, we do not step on the slippery slope; we are already on the slope. We only take a next step, but this step must be evaluated as an act or a process in its own right, for we cannot say a priori in which direction it will go. It may be a neutral step sideward. (For example, the acceptance of informed consent might be seen that way if we use only one criterion of the direction we take: Does it lead to infanticide or not?) Or it may even be a step upward. (If we strengthen the norm of informed consent, it might even form an extra barrier to infanticide.) Therefore, we need something more than the simple fact that the acceptance of A is part of a process toward B to establish a sound slippery slope argument.

Sometimes, however, there is some further evidence. Allowing A is a major factor in the process leading to the acceptance of B, or at least a necessary condition. The acceptance that abortion may sometimes be morally justified is a necessary condition for the acceptance of an abortion program based on eugenic purposes. The line between the status quo and A is a clear and effective one (for example, a general prohibition against killing or abortion), but there are no such lines between A and B. Allowing A will then remove a social barrier without instituting a new barrier. Factor A may not be the only factor, and it may not even be the main factor in the process leading to B. But sometimes it is the only factor we can influence, or it is simply the factor that is most easily influenced.

The distinction Bernard Williams makes between a reasonable and an effective distinction may be helpful here (1985. In M. Lockwood, Ed., *Moral dilemmas in modern medicine* (pp. 126–137). Oxford: Oxford Univ. Press). A reasonable distinction is one for which there is a decent argument, while an effective distinction is one that, as a matter of social or psychological fact, can be effectively defended. A reasonable distinction need not be an effective one, and vice versa.

The argument here is not that there is no reasonable distinction; the argument is that, though there may be a reasonable distinction between A and B, it is not enough. What is missing is an effective barrier against accepting B in the way the existing prohibition serves as an effective barrier. The prohibition against killing is effective against involuntary euthanasia, but once we have accepted voluntary euthanasia, there will be no more barriers. The old standard rule against killing is thus weakened, and the new rule that includes the exception for voluntary euthanasia will not be a defensible new barrier—or so the opponent of legalizing voluntary euthanasia might argue.

This is, in a sense, the empirical transformation of the logical versions. If there is no reasonable distinction, then we have the empirical analogue of L_1. If there is a reasonable distinction that is not effective, it is the empirical analogue of L_2. Once we accept A, we will be driven by long strides or by unnoticeable small steps toward B, without any possibility to stop. The reasons that there is no such effective barrier may differ— maybe there is no consensus about the further distinctions to be drawn, or maybe the concepts used in defending A are so vague and ambiguous that the gray zone can easily be made to encompass B.

This version of the empirical argument cannot be dismissed a priori. It is obvious that it may hold, and it has some intuitive appeal. Whether in a discussion it may be considered a sound argument largely depends upon the facts of the case. Some general criteria will be helpful to judge whether the risk of a slope is really a good argument.

1. One has to make plausible that the expected short-term consequences are clear, negative, and probable, and that these follow from or directly have to do with the proposed act or policy
2. The long-term consequences should result from the short-term consequences and be clear and negative as well, but need not be inevitable
3. It must be plausible that while we can stop now, we will not have that same possibility further down the slope
4. There must be an acceptable alternative action that is less susceptible to the slippery slope

These requirements place a heavy burden of proof on the proponent of the slippery slope. Therefore, though it cannot be ruled out a priori, it will only seldom be a really convincing argument, especially not in those situations where we consider A in itself a morally recommendable action. Only when we do not have very strong opinions about the moral quality of A does the empirical version sometimes have enough plausibility to prevent us from doing it.

D. The Full Slippery Slope Argument

The full version depends for its strength on the constituent elements, the various versions out of which it is built. So we have to evaluate each of them separately first, but that does not mean that we should judge them in isolation. The power of the combined version is that to go from A to B one argument need not go the full way. Arguments of consistency may lead the judges from A to M, social processes may then lead society further from M to P, a move later codified in legislation using vague language and then again the judiciary, interpreting that vague language, may lead us from P to B. Each of the mechanisms may result in some steps. Moreover, sometimes the combination of different types of slopes in different contexts may reinforce one another. The step from A to M taken by the judiciary may be supported by similar arguments in ethical and public debates.

Even if combination and mutual reinforcement is possible, the argument is still as strong as its weakest chain. We have to consider whether each of the steps

is likely in itself. Is it, for instance, really probable that in the codification of the move from M to P, Parliament will introduce vague legislation? It may as well create clear lines, for instance, by setting a definite standard of conditions under which euthanasia will be permitted. If this intervention is probable, a further slide down the slope could be prevented.

If on any of the steps between A and B such a stop is possible and not unlikely, it may well be the most effective use of our energy to try to establish an effective new barrier at that point rather than, probably without success, trying to prevent any moves to A at all. If A is in itself morally unobjectionable or even recommendable (as legalizing certain forms of euthanasia is, in my view), and finds support in both critical morality and social morality, it can be counterproductive to try to stop A. Probably there will then be a continuous effort by groups in our society to get A accepted and the arguments against it are then not very strong. Efforts to establish a new reasonable and effective barrier between A and B will then probably be more fruitful.

This shows how important a careful analysis of the slippery slope argument can be. If we try, impressed by the rhetoric of the argument, to stop a development at the wrong point, we will not only lose the battle on that point, but will perhaps also be unable to stop it at the point where it should and could have been stopped, simply because we have misdirected available energies. This is a good warning against too easy and uncritical use of the argument; it will sometimes only result in a short-term Pyrrhic victory that has disastrous consequences in the long run.

VI. THE RHETORICAL DIMENSION: PRACTICAL DEBATES

The conclusion of Section V is that slippery slope arguments are usually hard to substantiate, though in some contexts they may be valid and plausible. However, they are the most common types of arguments in practical debates. To understand this popularity, we have to address other dimensions of the use of these arguments.

One reason why the slippery slope is so frequent in practical debates is its emotional appeal. This makes it very useful for those who, rather than convince their opponents, simply want to win. "A Socialist government will be the first step towards Communist dictatorship" and similar phrases have sometimes been powerful arguments to win the electorate. Especially in political debates, the effectiveness of slippery slope arguments is often reversely correlated to their plausibility. Because of the emotional appeal, rebuttal is usually difficult—

rational criticism only seldom can correct the emotions the argument has produced.

A second reason is that, even in rational practical debates, there is usually no fully conclusive argument that may decide the case. The decision is taken on the basis of a combination of various arguments, some of them stronger and others weaker. Especially more institutionalized debates, like court proceedings, have the character of a continuous shifting of the burden of proof to the opponent. As Walton (1992) has shown, the slippery slope precisely does that in many cases because when it has some initial plausibility, even far from being conclusive, it can shift the burden of proof to the proponent of a policy. Countertactics in a debate, correspondingly, need not always consist of a full critical analysis along the lines of the last section, but can have a more modest goal: to reshift the burden of proof to the other party. Thus, rather than trying to prove that the slope is highly unlikely, one might stress the positive consequences of A or argue that not doing A would have even worse consequences.

A third reason for their frequent use is that often appeals to slippery slopes express some underlying uneasiness about the rapid transformation of society. The slippery slope is then not really a specific argument against policy A, which is probably only the symptom or symbol of these changes. Yet, it can be a signal that there is something more fundamentally wrong about developments in society and we should try to find ways to address this signal. If a vague public distrust of new technology is the real motive behind the fear for a slippery slope, this distrust should be brought to the open, rather than being "rationalized away" by dismissing the appeal to a slippery slope as an invalid argument.

VII. THE SOCIOLOGICAL AND PSYCHOLOGICAL DIMENSIONS: PERCEIVING REALITY

The major factor that makes slippery slope arguments so problematic still has to be addressed. Slippery slope arguments are based on interpretations of social reality and especially of the likelihood of future developments. These interpretations are inherently controversial, and arguments for one interpretation over another are always inconclusive. Is it likely that legalizing early abortions will lead to the gradual acceptance of later and later abortions, and in the end to the acceptance of infanticide? Or is it rather likely that strict enforcement of the abortion law will lead to substantive suffering and the death of many pregnant women as a result of illegal abortion practices? Both speculations have some initial plausibility, but it is hard to tell which is better.

Moreover, even if the facts are clear, it is not always certain that the presupposed causal relationships are inevitable. An illustration is offered by the "Stepping Stone" theory with respect to drugs.

Case 9: "In most Western countries, a large proportion of those using soft drugs like marijuana will end up as addicts of hard drugs like heroin. We should therefore prevent the use of soft drugs like marijuana, even if in themselves they are much less dangerous than accepted drugs like alcohol or nicotine, to prevent further steps to the more dangerous and addictive drugs."

This is a very nice example of an empirical slippery slope argument, in which the normative conclusion seems to follow almost naturally from the empirical facts. Dutch drug policies since the seventies, however, have been based on the hypothesis that these "drugs careers" were largely the result of the contingent fact that both types of drugs were sold in the same illegal subculture. If it were possible to separate the subculture of soft drugs from that of hard drugs, it might be possible to prevent individual users of soft drugs from switching to hard drugs.

Here again, there is no uncontroversial interpretation of the facts, not even in hindsight. According to most Dutch drug experts (and my impression is that they are correct), this element of Dutch drug policies has worked, resulting in relatively low numbers of hard-drug addicts (though it is difficult to uphold the separation between the two subcultures, especially as the production and distribution of soft drugs remains illegal). Nowadays, a much smaller proportion of those who use soft drugs on a regular basis take the step to hard drugs. But according to many opponents, especially politicians from other countries, the effects of the Dutch drug policy (especially the toleration of the use of soft drugs) are disastrous. They do not believe that this is a way to prevent the slippery slope; they even consider the Dutch tolerance of soft drugs as the first step on a different slippery slope (if only because they do not want to distinguish between more and less harmful drugs).

This controversy shows that even appeals to "objective" facts are not sufficient to decide the question of whether there have been slides down the slope. Evaluative and ideological stances seem to color the observations by both parties in the debate (though, in my opinion, not in equal proportions). When this is so with regard to interpreting reality, it will be even more so with regard to interpreting the future. The basic difference between the optimist and the pessimist regarding the question of whether the glass is half full or half empty is even more strongly reflected with respect to the future danger of a slippery slope. Someone with a pessimistic outlook, who believes "everything is getting worse," will interpret the facts in a negative way and will see every new technique as a further step in the wrong direction. The optimist, on the other hand, will interpret new developments as steps in the right direction; the more negative aspects will be seen as accidental and correctable. These outlooks are also reflected in attitudes toward the question of whether one thinks things can be stopped. Someone who is highly critical toward the existing political and legal order will have less confidence in the possibility of stopping future developments than someone with a strong trust in our democratic institutions.

Thus we have an entanglement of, on the one hand, controversial interpretations of social reality and undecidable predictions regarding the future, and, on the other hand, personal emotional, psychological, and moral attitudes, and fundamental outlooks. The entanglement makes discussions of slippery slopes often futile, because parties do not talk about the same facts and predictions.

VIII. A CASE STUDY: EUTHANASIA IN THE NETHERLANDS

I think the last section can best be illustrated by an extensive discussion of what seems currently the most controversial example of a suggested slippery slope: Dutch euthanasia practice. I have the impression that most physicians, lawyers, and ethicists in the United States believe in something like the following story:

Case 10A: "In 1973, the Dutch took the first step on the slippery slope. They tolerated active voluntary euthanasia on request in a case where death was near and where there was unbearable suffering. Subsequently, however, they abandoned each of these criteria by small steps. Now they are even discussing 'euthanasia' without request in cases of comatose patients, psychiatric patients, and severely handicapped newborns. There seems to be no end to this sequence: we may expect them to go further down the slippery slope yet."

On the other hand, most Dutch physicians, lawyers, and ethicists seem to perceive the Dutch history on euthanasia quite differently, somewhat like the following story:

Case 10B: "In the late sixties, we began to realize that modern medical technology is not always beneficial. Life is not always worth living and sometimes suffering is so unbearable or the quality of life so poor that prolongation of life is itself an evil. Over the last 30 years, Dutch society as a whole has been involved in the process of this general discussion on medicine and health care, including topics like medical decisions concerning the end of life. This broad and intense discussion has been long and difficult, but gradually we have been moving toward some general agreement. The consensus started with the relatively easy cases: euthanasia in cases where there is a clear request and unbearable suffering, and where the end of life is near. We went on to discuss the more difficult cases and we are still struggling with them. Examples of the most challenging cases are psychiatric patients who request euthanasia, comatose patients, and handicapped newborns. Discussion is continuing on these cases."

This perception of the Dutch story is not one of a slippery slope, but that of a long and winding road. For many years, the Dutch have been trying to convince their U.S. colleagues of their—and what seems to me the correct—interpretation of the story, though usually (at least until recently) in vain.

Here we have an interesting problem which seems characteristic of many slippery slope arguments: the same reality is perceived in completely different ways. If opinions differ so strongly about the interpretation of a historical process, the differences will be even larger when discussing future developments. For instance, consider the recent initiatives in various U.S. states which would have allowed certain forms of euthanasia or physician-assisted suicide. If discussions about interpreting the Dutch situation have been in vain, how can we expect agreement on the assessment of the risks involved in following these initiatives? To answer this type of question, we cannot exclude psychological and emotional factors. We need to address these factors directly, because ultimately they seem to determine whether some person or group believes in the slippery slope or not.

In the Dutch euthanasia example, these factors may be quite complex. One explanation is that many Americans simply condemn every form of active euthanasia; every step will then clearly be perceived as a step down the slippery slope. A second explanation is that whether one perceives a development as a slippery slope largely depends on basic attitudes of trust in other persons and in society in general. In the United States, there seems to be much more distrust of physicians, lawyers, politicians, and fellow citizens (like family members) than in the Netherlands. The Dutch practice heavily leans on trusting physicians, because legal control of medical euthanasia practice is extremely difficult. Physicians trust fellow physicians, patients trust physicians, and the legal system entrusts physicians with these decisions. If someone with a basic attitude of distrust looks at this situation, he will see an extreme danger of abuse.

A third explanation is that implicitly one always interprets a development in the light of familiar facts and values. In the Netherlands, there is almost equal access to health care and almost no one will have to pay extremely high hospital bills; euthanasia is usually performed in the context of a long-standing physician–patient relationship, and there has been a long, intense, and broad discussion on euthanasia. These facts are essential to understand why the risk of a slippery slope is perceived as minimal in Dutch society. If one lives in a society where the facts are different, one will more easily perceive the risk of a slippery slope.

IX. CONCLUSION

Cases 10A and 10B illustrate nicely many of the problems surrounding slippery slope arguments. The facts do not await us in objective descriptions, nor are they neatly classified; the future is uncertain; and personal attitudes, backgrounds, and emotions strongly influence our perceptions. Slippery slope arguments are often not so much rational arguments as expressions of an underlying feeling of concern about general trends in society. If so, they have to be taken seriously by trying to reformulate them and bringing the underlying concerns into the open public debate.

In those cases where they are proper slippery slope arguments rather than other arguments in disguise, close analysis of the precise versions involved and the contexts in which they are thought to apply is necessary. Even if they are rarely valid and plausible, there may be situations in which some specific versions are convincing, especially in institutionalized contexts like law.

Acknowledgment

Portions of this article have been reproduced from "The Slippery Slope Arguments," *Ethics,* 102 pp. 42–65 (October). Copyright 1991 by The University of Chicago Press. All rights reserved.

Bibliography

Burgess, J. A. (1993). The great slippery slope argument. *Journal of Medical Ethics,* **19,** 169–174.

Holtug, N. (1993). Human gene therapy: Down the slippery slope? *Bioethics,* **7**(5), 402–419.

Lamb, D. (1988). *Down the slippery slope. Arguing in applied ethics.* New York: Croom Helm.

Van der Burg, W. (1991). The slippery slope argument. *Ethics,* **102,** 42–65.

Whitman, J. P. (1994). The many guises of the slippery slope argument. *Social Theory and Practice,* **20,** 85–97.

Index

ISBN 0-12-166355-8

90038